WORLD ACADEMIC FRONTIERS
世界学术研究前沿丛书

Computer Network Security

计算机网络安全

"世界学术研究前沿丛书"编委会
THE EDITORIAL BOARD OF
WORLD ACADEMIC FRONTIERS

中国出版集团公司
世界图书出版公司
广州·上海·西安·北京

图书在版编目（CIP）数据

计算机网络安全：英文 /"世界学术研究前沿丛书"编委会编. —广州：世界图书出版广东有限公司，2017.8
 ISBN 978-7-5192-2462-2

Ⅰ. ①计… Ⅱ. ①世… Ⅲ. ①计算机网络—网络安全—英文 Ⅳ. ①TP393.08

中国版本图书馆CIP数据核字(2017)第040988号

Computer Network Security © 2016 by Scientific Research Publishing

Published by arrangement with Scientific Research Publishing
Through Wuhan Irvine Culture Company

This Edition © 2017 World Publishing Guangdong Corporation
All Rights Reserved.
本书仅限中国大陆地区发行销售

书　　名：	计算机网络安全 Jisuanji Wangluo Anquan
编　　者：	"世界学术研究前沿丛书"编委会
责任编辑：	康琬娟
出版发行：	世界图书出版广东有限公司
地　　址：	广州市海珠区新港西路大江冲25号
邮　　编：	510300
电　　话：	（020）84460408
网　　址：	http://www.gdst.com.cn/
邮　　箱：	wpc_gdst@163.com
经　　销：	新华书店
印　　刷：	广州市德佳彩色印刷有限公司
开　　本：	787 mm×1092 mm　1/16
印　　张：	55.75
插　　页：	4
字　　数：	1060千
版　　次：	2017年8月第1版　2017年8月第1次印刷
国际书号：	ISBN 978-7-5192-2462-2
定　　价：	598.00元

版权所有　翻印必究
（如有印装错误，请与出版社联系）

Preface

Computer network security consists of the policies adopted to prevent and monitor unauthorized access, misuse, modification, or denial of a computer network and network-accessible resources. Network security involves the authorization of access to data in a network, which is controlled by the network administrator. Users choose or are assigned an ID and password or other authenticating information that allows them access to information and programs within their authority. Network security covers a variety of computer networks, both public and private, that are used in everyday jobs; conducting transactions and communications among businesses, government agencies and individuals. Networks can be private, such as within a company, and others which might be open to public access. Network security is involved in organizations, enterprises, and other types of institutions. It does as its title explains: It secures the network, as well as protecting and overseeing operations being done.[1]

In the present book, thirty literatures about computer network security published on international authoritative journals were selected to introduce the worldwide newest progress, which contains reviews or original researches on network security, cloud computing, wireless sensor network, network confidentiality and so on. We hope this book can demonstrate advances in computer network security as well as give references to the researchers, students and other related people.

编委会：
- ❖ 克里斯·钱宁教授，谢菲尔德大学，英国
- ❖ 李允和教授，高丽大学，韩国
- ❖ 约翰·A·斯普林格教授，普渡大学，美国
- ❖ 卢卡斯·邝会教授，香港大学，中国
- ❖ 顾宗华副教授，中国科技大学，中国

March 9, 2017

---

[1] From Wikipedia: https://en.wikipedia.org/wiki/Network_security.

Selected Authors

Christian Senk, University of Regensburg, Regensburg, Germany.

Kristin Glass, Systems Analysis, New Mexico Institute of Mining and Technology, Socorro, USA.

Alvaro Araujo, Electronic Engineering Department, Universidad Politécnica de Madrid, Avda/Complutense 30, Madrid, Spain.

Roland Schwarzkopf, Department of Mathematics and Computer Science, University of Marburg, Hans-Meerwein-Str. 3, D-35032 Marburg, Germany.

Thomas E. Carroll, Pacific Northwest National Laboratory, P.O. Box 999, Richland, Washington, USA.

Paul Ormerod, Software Development, University College London, London, UK.

David M. Laverty, School of Electronics, Electrical Engineering and Computer Science, Queen's University Belfast, 125 Stransmillis Road, Belfast BT9 5AH, UK.

Jingwei Huang, Information Trust Institute, University of Illinois at Urbana-Champaign 1308 West Main Street, Urbana, Illinois 61801, USA.

Paul Watson, School of Computing Science, Newcastle University, Newcastle-upon-Tyne, UK.

Geumhwan Cho, Department of Computer Science and Engineering, Sungkyunkwan University, Seobu-ro 2066, Suwon, Republic of Korea.

Contents

Chapter 1..1
Virtual Network Security: Threats, Countermeasures, and Challenges
by Leonardo Richter Bays, Rodrigo Ruas Oliveira, Marinho Pilla Barcellos, et al.

Chapter 2..45
Adoption of Security as a Service
by Christian Senk

Chapter 3..77
A Quantitative Analysis of Current Security Concerns and Solutions for Cloud Computing
by Nelson Gonzalez, Charles Miers, Fernando Redígolo, et al.

Chapter 4..113
An Analysis of Security Issues for Cloud Computing
by Keiko Hashizume, David G. Rosado, Eduardo Fernández-Medina, et al.

Chapter 5..143
Estimating the Sentiment of Social Media Content for Security Informatics Applications
by Kristin Glass and Richard Colbaugh

Chapter 6..167
If You Want to Know about a Hunter, Study His Prey: Detection of Network Based Attacks on KVM Based Cloud Environments
by Nikolaos Pitropakis, Dimitra Anastasopoulou, Aggelos Pikrakis, et al.

Chapter 7 189
Intelligent Feature Selection and Classification Techniques for Intrusion Detection in Networks: A Survey

by Sannasi Ganapathy, Kanagasabai Kulothungan, Sannasy Muthurajkumar, et al.

Chapter 8 227
Security in Cognitive Wireless Sensor Networks. Challenges and Open Problems

by Alvaro Araujo, Javier Blesa, Elena Romero, et al.

Chapter 9 247
Anticipating Complex Network Vulnerabilities through Abstraction-Based Analysis

by Richard Colbaugh and Kristin Glass

Chapter 10 273
Increasing Virtual Machine Security in Cloud Environments

by Roland Schwarzkopf, Matthias Schmidt, Christian Strack, et al.

Chapter 11 299
Scaling of Wireless Sensor Network Intrusion Detection Probability: 3D Sensors, 3D Intruders, and 3D Environments

by Omar Said and Alaa Elnashar

Chapter 12 321
Security Informatics Research Challenges for Mitigating Cyber Friendly Fire

by Thomas E. Carroll, Frank L. Greitzer and Adam D. Roberts

Chapter 13 353
Security-by-Experiment: Lessons from Responsible Deployment in Cyberspace

by Wolter Pieters, Dina Hadžiosmanović, Francien Dechesne

Chapter 14 ....**381**

Taking Back Control of Privacy: A Novel Framework for Preserving Cloud-Based Firewall Policy Confidentiality

by Tytus Kurek, Marcin Niemiec and Artur Lason

Chapter 15 ....**417**

Terrorist Networks and the Lethality of Attacks: An Illustrative Agent Based Model on Evolutionary Principles

by Paul Ormerod

Chapter 16 ....**435**

Threat Driven Modeling Framework Using Petri Nets for e-Learning System

by Aditya Khamparia and Babita Pandey

Chapter 17 ....**459**

Secure Data Networks for Electrical Distribution Applications

by David M. Laverty, John B. O'Raw, Kang Li, et al.

Chapter 18 ....**481**

Trust Mechanisms for Cloud Computing

by Jingwei Huang and David M. Nicol

Chapter 19 ....**515**

A Multi-Level Security Model for Partitioning Workflows over Federated Clouds

by Paul Watson

Chapter 20 ....**549**

Enhancing the Security of LTE Networks against Jamming Attacks

by Roger Piqueras Jover, Joshua Lackey and Arvind Raghavan

Chapter 21 ....**581**

Combating Online Fraud Attacks in Mobile-Based Advertising

by Geumhwan Cho, Junsung Cho, Youngbae Song, et al.

Chapter 22..601
Fingerprint-Based Crypto-Biometric System for Network Security

by Subhas Barman, Debasis Samanta and Samiran Chattopadhyay

Chapter 23..637
A Secure User Authentication Protocol for Sensor Network in Data Capturing

by Quan Zhou, Chunming Tang, Xianghan Zhen, et al.

Chapter 24..671
An Economic Perspective of Message-Dropping Attacks in Peer-to-Peer Overlays

by Kevin W. Hamlen and William Hamlen

Chapter 25..705
Fluency of Visualizations: Linking Spatiotemporal Visualizations to Improve Cybersecurity Visual Analytics

by Zhenyu Cheryl Qian and Yingjie Victor Chen

Chapter 26..733
Waterwall: A Cooperative, Distributed Firewall for Wireless Mesh Networks

by Leonardo Maccari and Renato Lo Cigno

Chapter 27..757
WRSR: Wormhole-Resistant Secure Routing for Wireless Mesh Networks

by Rakesh Matam and Somanath Tripathy

Chapter 28..785
An Effective Implementation of Security Based Algorithmic Approach in Mobile Adhoc Networks

by Rajinder Singh, Parvinder Singh and Manoj Duhan

Chapter 29..**807**

How to Make a Linear Network Code (Strongly) Secure

by Kaoru Kurosawa, Hiroyuki Ohta and Kenji Kakuta

Chapter 30..**849**

Privacy and Information Security Risks in a Technology Platform for Home-Based Chronic Disease Rehabilitation and Education

by Eva Henriksen, Tatjana M. Burkow, Elin Johnsen, et al.

Chapter 1

Virtual Network Security: Threats, Countermeasures, and Challenges

Leonardo Richter Bays[1], Rodrigo Ruas Oliveira[1], Marinho Pilla Barcellos[1], Luciano Paschoal Gaspary[1], Edmundo Roberto Mauro Madeira[2]

[1]Institute of Informatics, Federal University of Rio Grande do Sul, Porto Alegre, Brazil
[2]Institute of Computing, University of Campinas, Campinas, Brazil

Abstract: Network virtualization has become increasingly prominent in recent years. It enables the creation of network infrastructures that are specifically tailored to the needs of distinct network applications and supports the instantiation of favorable environments for the development and evaluation of new architectures and protocols. Despite the wide applicability of network virtualization, the shared use of routing devices and communication channels leads to a series of security-related concerns. It is necessary to provide protection to virtual network infrastructures in order to enable their use in real, large scale environments. In this paper, we present an overview of the state of the art concerning virtual network security. We discuss the main challenges related to this kind of environment, some of the major threats, as well as solutions proposed in the literature that aim to deal with different security aspects.

Keywords: Network Virtualization, Security, Threats, Countermeasures

1. Introduction

Virtualization is a well established concept, with applications spanning several areas of computing. This technique enables the creation of multiple virtual platforms over a single physical infrastructure, allowing heterogeneous architectures to run on the same hardware. Additionally, it may be used to optimize the usage of physical resources, as an administrator is able to dynamically instantiate and remove virtual nodes in order to satisfy varying levels of demand.

In recent years, there has been a growing demand for adaptive network services with increasingly distinct requirements. Driven by such demands, and stimulated by the successful employment of virtualization for hosting custom-built servers, researchers have started to explore the use of this technique in network infrastructures. Network virtualization allows the creation of multiple independent virtual network instances on top of a single physical substrate[1]. This is made possible by instantiating one or more virtual routers on physical devices and establishing virtual links between these routers, forming topologies that are not limited by the structure of the physical network.

In addition to the ability to create different topological structures, virtual networks are also not bound by other characteristics of the physical network, such as its protocol stack. Thus, it is possible to instantiate virtual network infrastructures that are specifically tailored to the needs of different network applications[2]. These features also enable the creation of virtual testbeds that are similar to real infrastructures, a valuable asset for evaluating newly developed architectures and protocols without interfering with production traffic. [3]For these reasons, network virtualization has attracted the interest of a number of researchers worldwide, especially in the context of Future Internet research. Network virtualization has been embraced by the Industry as well. Major Industry players—such as Cisco and Juniper—nowadays offer devices that support virtualization, and this new functionality allowed infrastructure providers to offer new services.

In contrast to the benefits brought by network virtualization, the shared use of routing devices and communication channels introduces a series of security-related concerns. Without adequate protection, users from a virtual network might be able to access or even interfere with traffic that belongs to other virtual networks, violating security properties such as confidentiality and integrity[4][5]. Additionally, the infrastructure could be a target for denial of service attacks, causing availability issues for virtual networks instantiated on top of it[6][7]. Therefore, it is of great importance that network virtualization architectures offer protection against these and other types of threats that might compromise security.

Recently, attention has been drawn to security concerns in network infrastructures due to the discovery of pervasive electronic surveillance around the globe. Although all kinds of networks are potentially affected, the shared use of physical resources in virtual network environments exacerbates these concerns. As such, these recent circumstances highlight the need for a comprehensive analysis of current developments in the area of virtual network security.

In this paper, we characterize the current state of the art regarding security in network virtualization. We identify the main threats to network virtualization environments, as well as efforts aiming to secure such environments. For this study, an extensive literature search has been conducted. Major publications from the literature have been studied and grouped according to well known classifications in the area of network security, as well as subcategories proposed by the authors of this paper. This organization allows the analysis and discussion of multiple aspects of virtual network security.

The remainder of this paper is organized as follows. Section 2 presents a brief background on the area of network virtualization as well as a review of related literature. Section 3 introduces the taxonomy used to classify the selected publications. Section 4 exposes the security vulnerabilities and threats found in the literature, while Section 5 presents the security countermeasures provided by solutions found in previous proposals. In Section 6, we discuss the results of this study, and in Section 7 we summarize the main current research challenges in the area of virtual network security. Last, in Section 8 we present our conclusions.

2. Background and Literature Review

In this section, we first provide a brief background on the area of network virtualization, highlighting its most relevant concepts. Next, we present a review of literature closely related to virtual network security.

2.1. Background

Network virtualization consists of sharing resources from physical network devices (routers, switches, etc.) among different virtual networks. It allows the coexistence of multiple, possibly heterogeneous networks, on top of a single physical infrastructure. The basic elements of a network virtualization environment are shown in **Figure 1**. At the physical network level, a number of autonomous systems are represented by interconnected network substrates (e.g., substrates A, B, and C). Physical network devices are represented by nodes supporting virtualization technologies. Virtual network topologies (e.g., virtual networks 1 and 2), in turn, are mapped to a subset of nodes from one or more substrates. These topologies are composed of virtual routers, which use a portion of the resources available in physical ones, and virtual links, which are mapped to physical paths composed of one or more physical links and their respective intermediate routers.

From the point of view of a virtual network, virtual routers and links are seen as dedicated physical devices. However, in practice, they share physical resources with routers and links from other virtual networks. For this reason, the virtualization technology used to create this environment must provide an adequate level of isolation in order to enable the use of network virtualization in real, large scale environments.

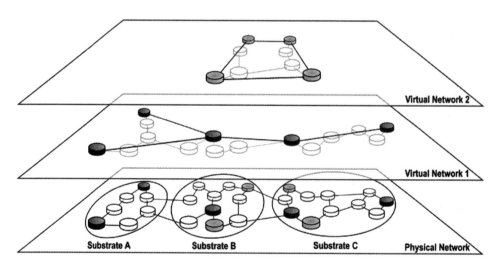

Figure 1. Network virtualizationmodel, denoting a scenario with multiple physical substrates (Substrate A, B, and C) and virtual networks (Virtual Network 1 and 2).

Over the years, different methods for instantiating virtual networks have been used. Typical approaches include VLANs (Virtual Local Area Networks) and VPNs (Virtual Private Networks). Recently, Virtual Machine Monitors and programmable networks have been employed to create virtual routers and links over physical devices and communication channels. These approaches are briefly revisited next.

2.1.1. Protocol-Based Approaches

Protocol-based approaches consist of implementing a network protocol that enables the distinction of virtual networks through techniques such as tagging or tunneling. The only requirement of this kind of approach is that physical devices (or a subset of them) support the selected protocol.

One example of protocol-based network virtualization are VLANs. VLANs consist of logical partitions of a single underlying network. Devices in a VLAN communicate with each other as if they were on the same Local Area Network, regardless of physical location or connectivity. All frames sent through a network are tagged with their corresponding VLAN ID, processed by VLAN-enabled routers and forwarded as necessary[8]. Since isolation is typically based only on packet tagging, this approach is susceptible to eavesdropping attacks.

Another commonly used approach is the creation of Virtual Private Networks. VPNs are typically used to provide a secure communication channel between geographically distributed nodes. Cryptographic tunneling protocols enable data confidentiality and user authentication, providing a higher level of security in comparison with VLANs. VPNs can be provided in the physical, data link, or network layers according to the protocols being employed[9].

2.1.2. Machine Virtualization-Based Approaches

Machine virtualization-based approaches consist of creating virtual networks by means of groups of interconnected virtual machines. Virtual Machine Monitors are used to instantiate virtual routers, and virtual links are established between them, regardless of physical network topology. **Table 1** shows different

Table 1. Virtualization techniques.

Technique	Description	Examples
Full virtualization	The Virtual Machine Monitor emulates a complete machine, based on the underlying hardware architecture. The guest Operating System runs without any modification.	VMware Workstation, VirtualBox
Paravirtualization	The Virtual Machine monitor emulates a machine which is similar to the underlying hardware, with the addition of a hypervisor. The hypervisor allows the guest Operating System to run complex tasks directly on non-virtualized hardware. The guest OS must be modified in order to take advantage of this feature.	VMware ESX, Xen
Container-based virtualization	Instead of running a full Virtual Machine, this technique provides Operating System-level containers, based on separate userspaces. In each container, the hardware, as well as the Operating System and its kernel, are identical to the underlying ones.	OpenVZ, Linux VServer

machine virtualization-based techniques that can be used to create virtual networks, as well as a brief explanation and an example of each.

This alternative is remarkably flexible and relatively cheap, as it allows the use of customized software and does not require the use of specific hardware[1]. However, it is more demanding in terms of resource usage in comparison to previously described protocol-based approaches. Additionally, it may introduce security concerns associated with server virtualization, some of which are mentioned in Sections 4 and 5. A general study on the security issues that arise from the use of machine virtualization was performed by van Cleeff et al.[10].

2.1.3. Programmable Networks

Programmable routers have been used to enable the creation of virtual networks. Although this is not a new concept, research in this area has been recently stimulated by the inception of Software-Defined Networking (SDN). This paradigm consists of decoupling the data plane and the control plane in network devices. More specifically, devices such as routers and links retain only the data plane, and a separated control plane manages such devices based on an overview of the entire network.

OpenFlow[11], one of the most promising techniques for implementing this paradigm, defines a protocol that allows a centralized controller to act as the control plane, managing the behavior of network devices in a dynamic manner. The controller communicates with network devices through a secure connection, creating and managing flow rules. Flow rules instruct network devices on how to properly process and route network traffics with distinct characteristics. Through the establishment of specific flow rules, it is possible to logically partition physical networks and achieve data plane isolation. This isolation enables the creation of virtual networks on top of an SDN environment. OpenFlow gave rise to the Open Networking Foundation, an organization ran by major companies within the area of computer networks that aims to disseminate this type of technology.

2.2. Literature Review

To the best of our knowledge, there have been no previous attempts at characterizing the state of the art regarding security in network virtualization. However, there have been a number of similar studies in other, closely related fields of research. We now proceed to a review of some of the main such studies.

Chowdhury *et al.*[1] provide a general survey in the area of network virtualization. The authors analyze the main projects in this area (both past projects and, at the time of publication, current ones) and discuss a number of key directions for future research. The authors touch upon the issues of security and privacy both while reviewing projects and discussing open challenges; however, as this is not the main focus of this survey, there is no in-depth analysis of security issues found in the literature.

Bari *et al.*[12] present a survey that focuses on data center network virtualization. Similarly to the aforementioned study, the authors survey a number of key projects and discuss potential directions for future work. When analyzing such projects, the authors provide insights on the fault-tolerance capabilities of each one, in addition to a brief discussion on security issues as one of the potential opportunities for future research.

In addition to the general studies on network virtualization presented so far, a number of surveys on cloud computing security have also been carried out. Cloud computing environments tend to make use of both machine and network virtualization, making this a highly relevant related topic for our study. However, while there is some overlap between cloud computing security and virtual network security, we emphasize that cloud computing represents a very specific use case of network virtualization and, therefore, poses a significantly distinct set of security challenges. Zhou *et al.*[13] provide an investigation on security and privacy issues of cloud computing system providers. Additionally, the authors highlight a number of government acts that originally intended to uphold privacy rights but fail to do so in light of advances in technology. Hashizume et al.[14], in turn, focus on security vulnerabilities, threats, and countermeasures found in the literature and the relationships among them.

Last, Scott-Hayward et al.[15] conducted a study on SDN security. As explained in Section 2.1.3, this is one of the technologies on top of which network virtualization environments can be instantiated. The authors first analyze security issues associated with the SDN paradigm and, afterwards, investigate approaches aiming at enhancing SDN security. Last, the authors discuss security challenges associated with the SDN model.

3. Taxonomy

The first step towards a comprehensive analysis of the literature was the selection of a number of publications from quality conferences and journals. To this end, we performed extensive searches in the ACM and IEEE digital libraries using a number of keywords related to network virtualization and security. We then ranked the literature found through this process according to the average ratio of citations per publication of the conferences or journals in which these papers were published. All publications from top tier conferences or journals with a consistent number of citations per publication were considered relevant and, therefore, selected. The remaining papers were analyzed and generally discarded.

Following the aforementioned process, a taxonomy was created in order to aid the organization and discussion of the selected publications. For this purpose, two well known classifications in the area of network security were chosen. Papers are organized according to the *security threats* they aim to mitigate, and afterwards, according to the *security countermeasures* they provide. As different authors have different definitions for each of these concepts, these classifications are briefly explained in the following subsections. The direct connection between them and the area of virtual network security is explained in sections 4 and 5, respectively.

In addition to these broad classifications, subcategories were created in order better organize this body of work. **Figure 2** presents the full hierarchical

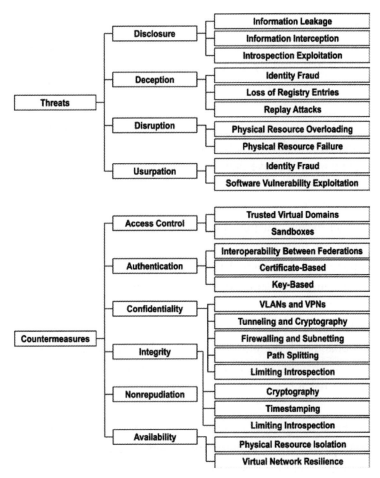

Figure 2. Taxonomy used to classify publications in the area of virtual network security.

organization that will be used in sections 4 and 5. Dark gray boxes represent broad categories used in the literature[16][17], while white boxes denote subdivisions proposed and created by the authors of this paper.

3.1. Security Vulnerabilities and Threats

There are a number of potential malicious actions, or threats, that may violate security constraints of computational systems. Shirey[16] describes and divides the consequences of these threats into four categories, namely *disclosure*, *deception*, *disruption*, and *usurpation*.

Unauthorized *disclosure* is defined as gaining unauthorized access to protected information. Sensitive data may be erroneously exposed to unauthorized entities, or acquired by an attacker that circumvents the system's security provisions.

Deception is characterized by intentionally attempting to mislead other entities. For example, a malicious entity may send false or incorrect information to others, leading them to believe that this information is correct. Fake identities may be used in order to incriminate others or gain illegitimate access.

Disruption means causing failure or degradation of systems, negatively affecting the services they provide. This may be done by directly incapacitating a system component or the channel through which information is delivered, or by inducing the system to deliver corrupted information.

Last, through *usurpation*, an attacker may gain unauthorized control over a system. This unauthorized control may allow the attacker to illegitimately access protected data or services, or tamper with the system itself in order to cause incorrect or malicious behavior.

These threat categories, as well as the previously mentioned subcategories we have created, also cover vulnerabilities and attacks. For ease of comprehension, vulnerabilities and threats are discussed collectively in Section 4. **Table 2** presents the relationships between vulnerabilities and threats in network virtualization environments. This table is organized according to the previously described taxonomy and lists all vulnerabilities found in the literature and the threats associated with each one. Additionally, the terms threat and attack are used interchangeably throughout the paper, as a threat may be understood as a potential attack (while an attack is the proper action that takes advantage of a vulnerability to violate a security policy).

3.2. Security Countermeasures

Due to the existence of the previously described threats, computational systems must provide a series of countermeasures in order to maintain a desirable level of security. Stallings[17] categorizes these essential countermeasures into six

Table 2. Relationships between vulnerabilities and threats in network virtualization environments.

Threat categories		Vulnerabilities	Threats
Disclosure	Information Leakage	Lack of ARP table protection	ARP table poisoning
		Placement of firewall rules inside virtual nodes	Subversion of firewall rules
	Information Interception	Lack of ARP table protection	ARP table poisoning
		Transmission of data in predictable patterns	Traffic Analysis attacks
		Uncontrolled handling of multiple, sequential virtual network requests from a single entity	Inference and disclosure of sensitive topological information
		Unprotected exchange of routing information among virtual routers	Disclosure of sensitive routing information
	Introspection Exploitation	Uncontrolled Introspection	Data theft
Deception	Identity Fraud	Improper handling of identities: - within individual networks;	Injection of malicious messages with forged sources
		- among federated networks;	Privilege escalation
		- during migration procedures.	Abuse of node removal and re-addition in order to obtain new (clean) identities
	Loss of registry entries	Uncontrolled rollback operations	Loss of registry entries
	Replay attacks	Lack of unique message identifiers	Replay attacks
Disruption	Physical Resource Overloading	Uncontrolled resource allocation	Performance degradation
			Abusive resource consumption
		Uncontrolled handling of virtual network requests	Exhaustion of resources in specific parts of the infrastructure
		Lack of proactive or reactive recovery strategies	Denial of Service attacks
	Physical Resource Failure	Lack of proactive or reactive recovery strategies	Failure of virtual routers/networks
		Uncontrolled resource reallocation after failures	Overloading of remaining virtual routers after failures
Usurpation	Identity Fraud	Improper handling of identities and associated privileges	Privilege escalation
	Software Vulnerability Exploitation	Privilege escalation in Virtual Machine Monitors	Unauthorized control of physical routers

subdivisions (referred to by Stallings as "security services"), namely *access control*, *authentication*, *data confidentiality*, *data integrity*, *nonrepudiation*, and *availability*.

Access control allows a system to administer which entities will be able to access its functions, and what permissions each of these entities will have. In order to grant individual access rights and permissions, entities must be properly authenticated in the system.

The purpose of *authentication* is to ensure that entities communicating with each other are, in fact, the entities they claim to be. The receiver of a message must be able to correctly identify its sender, and an entity must not be able to impersonate another.

Providing adequate *data confidentiality* means ensuring that third parties do not have access to confidential information being transmitted between two entities. Additionally, the system should inhibit attackers from deriving information by analyzing traffic flow characteristics.

Data integrity has the purpose of assuring that data stored by entities or transmitted through a network are not corrupted, adulterated or destroyed. Attacks such as duplication, modification, reordering, and replay of messages must be prevented. Furthermore, mechanisms for recovering from data corruption may also be provided.

In communications between peers, *nonrepudiation* provides a way to settle disputes when an entity denies having performed a certain action. The goal of this service is to prevent entities from falsely denying participation in any (possibly malicious) network-related activity.

The last security countermeasure is *availability*. System resources must be available upon request by an authorized entity, and the system must also conform to its performance specifications. In order to maintain availability, countermeasures against attacks such as *denial of service* must be provided.

4. Security Vulnerabilities and Threats

In this section, we present a comprehensive list of vulnerabilities and threats found in network virtualization environments. The interested reader should refer to **Table 2** for a systematic review of such vulnerabilities and threats.

While some of the threats listed in this section are a result of accidental actions, we emphasize that all threats—intentional or accidental—have an effect on security. As an example of an accidental attack, it is common for virtual routers to attempt to use all available resources (as virtualization tends to be transparent and virtual routers are typically not aware that they are not running on dedicated physical hardware). If the network virtualization environment does not adequately limit the resource usage of each virtual router, even this unintentional abuse may cause disruption on other networks hosted on the same substrate or cause the degradation or failure of critical services provided by the virtualization environment.

4.1. Disclosure

In an environment where physical resources are shared between a number of virtual networks, there is a series of behaviors that may result in undesired disclosure of information. Threats related to disclosure of private or sensitive information are explained next.

4.1.1. Information Leakage

Cavalcanti et al.[18] mention the possibility of messages being leaked from one virtual network to another. In this type of attack, an entity may disclose private or sensitive information to members of other virtual networks, who should not have access to such information. The authors state that this may be achieved through ARP table poisoning. For example, a malicious user may spoof the IP address of a node that is able to send messages to the virtual network with which it intends to communicate. Wolinsky et al.[19] describe a similar attack, in which virtual nodes send messages to outside the boundaries of a network virtualization environment. This would make it possible for messages to reach physical nodes that not only do not belong to any virtual network, but are hosted outside of the virtualized network infrastructure. According to the authors, if data isolation is achieved by means of firewall rules, malicious users may be able to subvert such rules by escalating privileges and gaining root access on a virtual node.

4.1.2. Information Interception

Attackers in a virtual network environment may capture messages being exchanged between two entities in order to access their content. This type of attack, often referred to as "eavesdropping" or "sniffing", may lead to theft of confidential information[4][5][20]. Wu et al.[20], specifically, mention ARP table poisoning as a means of achieving this. In contrast to the ARP poisoning attack described by Cavalcanti et al.[18] (explained in Section 4.1.1), in this case the attack would be used in order to mislead physical routers into forwarding packets meant to one entity to another one, allowing a malicious entity to sniff such packets. This is a common threat in any networking environment, but the use of shared physical resources by multiple virtual networks further exacerbates this problem. According to these and other authors, such as Cui et al.[21], networking solutions provided by virtual machine monitors may not properly isolate data belonging to different virtual networks. This means that members of one virtual network may be able to access data being transferred by other virtual networks sharing the same substrate.

Even if data inside network packets is protected (e.g. through the use of cryptography), entities may be able to derive sensitive information by analyzing them. In traffic analysis attacks, described by Huang et al.[22], entities acquire such information by analyzing characteristics of traffic flows between communicating entities in virtual networks. These characteristics include which entities communicate with which other entities, frequency of communication, and packet sizes, among others. For example, an entity that is involved in frequent, short communications with a high number of other entities may be a central point of control in the network. Knowing this, a malicious user could launch an attack directed at that entity, aiming to cause a considerable amount of disruption with limited effort. As previously mentioned, this attack is effective even if traffic is encrypted, making any type of virtual networking environment a potential target.

In addition to the previously detailed forms of information interception, which may also affect traditional network environments, other forms are specific to network virtualization. One such form is the use of multiple virtual network requests to disclose the topology of the physical infrastructure, explored by Pignolet et al.[23]. This constitutes a security threat, as infrastructure providers typically do not wish to disclose this information. The authors demonstrate that by sequentially requesting a number of virtual networks with varying topological characteristics and analyzing the response given by the infrastructure provider (i.e., whether the request can be embedded or not), they are able to gradually obtain information about the physical topology. Moreover, the authors determine the number of requests needed to fully disclose the physical topology on networks with different topological structures (tree, cactus, and arbitrary graphs). Conversely, Fukushima et al.[24] state that the entity controlling a physical network may obtain confidential routing information from virtual networks hosted on top of it. As current routing algorithms require routing information to be sent and received through virtual routers, sensitive information may be disclosed to the underlying network.

4.1.3. Introspection Exploitation

Introspection is a feature present in virtual machine monitors that allows system administrators to verify the current state of virtual machines in real time. It enables external observers to inspect data stored in different parts of the virtual machine (including processor registers, disk, and memory) without interfering with it. While this feature has valuable, legitimate uses (*e.g.*, enabling administrators to verify that a virtual machine is operating correctly), it may be misused or exploited by attackers in order to access (and potentially disclose) sensitive data inside virtual machines[10]. This problem is aggravated by the fact that virtual nodes may be moved or copied between multiple virtual machine monitors, as sensitive data may be compromised through the exploitation of this feature on any virtual machine monitor permanently or temporarily hosting such virtual nodes.

4.2. Deception

We have identified three subcategories of threats that may lead to deception in virtual network environments. These subdivisions—namely identity fraud, loss of registry entries and replay attacks—are explained next.

4.2.1. Identity Fraud

In addition to dealing with unauthorized disclosure, Cabuk *et al.*[5] and Wu *et al.* [20] also describe threats related to deception in virtual network environments. Specifically, virtual entities may inject malicious messages into a virtual network, and deceive others into believing that such messages came from another entity.

Certain characteristics of virtualized network environments increase the difficulty of handling identity fraud. The aggregation of different virtual networks into one compound network, known as federation, is indicated by Chowdhury *et al.*[25] as one of such characteristics. Federation raises issues such as the presence of separate roles and possible incompatibility between security provisions or policies from aggregated networks. Another complicating factor mentioned by the authors is the dynamic addition and removal of entities. An attacker may force a malicious node to be removed and re-added in order to obtain a new identity.

Other characteristics that complicate the handling of identity fraud involve operations such as migration and duplication of virtual nodes, as mentioned by van Cleeff et al.[10]. The study presented by the authors refers to virtualization environments in general. Therefore, in the context of this study, a virtual node may refer to either a virtual router or a virtual workstation. If a virtual node is migrated from one physical point to another, the identity of the machine that contains this virtual node may change. Moreover, virtual nodes may be copied to one or more physical points in order to provide redundancy, which may lead to multiple entities sharing a single identity. Both of these issues may cause inconsistencies in the process of properly identifying the origin of network messages, which may be exploited in identity fraud attacks.

4.2.2. Loss of Registry Entries

Van Cleeff et al.[10] also mention issues related to logging of operations in virtualization environments. If information regarding which entity was responsible for each operation in the network is stored in logs inside virtual machines, entries may be lost during rollback procedures. Likewise, logs of malicious activities performed by attackers may also be lost.

4.2.3. Replay Attacks

Fernandes and Duarte[26] mention replay attacks as another form of deception in virtual networks. In this type of attack, a malicious entity captures legitimate packets being transferred through the network and retransmits them, leading other entities to believe that a message was sent multiple times. The authors explain that virtual routers may launch attacks in which they repeat old control messages with the intention of corrupting the data plane of the attacked domain.

4.3. Disruption

In a network virtualization environment, proper management of resources is crucial to avoid disruption. The main sources of disruption in such environments are related to the abuse of physical resources (either intentional or unintentional) and the failure of physical devices.

4.3.1. Physical Resource Overloading

Physical resource overloading may lead to failure of virtual nodes, or cause the network performance to degrade below its minimum requirements. This degradation may cause congestion and packet loss in virtual networks, as stated by Zhang et al.[27]. In addition to causing disruption in already established networks, overloading may also hinder the deployment of new ones.

Resource requirements themselves can be a point of conflict in virtual network environments. As explained by Marquezan et al.[28], multiple virtual networks may require an excessive amount of resources in the same area of the substrate network. While such prohibitive demands may be unintentional, they may also be due to a coordinated attack. This may not only happen during deployment operations, but also during the lifetime of virtual networks.

It is also possible for one virtual network to disrupt another by using more than its fair share of resources. This concern is explored by a number of authors in their respective publications[26][29]-[31]. Isolation and fair distribution of physical resources among virtual networks are essential to maintain the network virtualization environment operating properly. This includes assuring that the minimum requirements of each network will be fulfilled, as well as prohibiting networks from consuming more resources than they are allowed to.

Overloading may also be caused by attacks aimed at the physical network infrastructure. Attacks may originate from within a virtual network hosted in the same environment, or from outside sources. The most common threats are Denial of Service (DoS) attacks, as presented by Yu et al.[6] and Oliveira et al.[7]. A single physical router or link compromised by a DoS attack may cause disruption on several virtual networks currently using its resources.

4.3.2. Physical Resource Failure

As previously stated, the failure of physical devices is one of the sources of disruption in virtual infrastructures[32]-[34]. Possible causes range from the failure of single devices (a physical router, for example, may become inoperative if one of its components malfunctions) to natural disasters that damage several routers or links in one or more locations[35]. Additionally, further complications may arise as the remainder of the network may be overloaded during attempts to relocate lost virtual resources. In addition to being valuable from the point of view of fault tolerance, countermeasures for mitigating the effect of failures may also be applied in the event of attacks such as DoS, as in both cases there is a need for redirecting network resources away from compromised routers or links.

4.4. Usurpation

In virtual network environments, usurpation attacks may allow an attacker to gain access to privileged information on virtual routers, or to sensitive data stored in them. Such attacks may be a consequence of identity fraud or exploited vulnerabilities, which are explained next.

4.4.1. Identity Fraud

As previously mentioned in Section 4.2, identity fraud attacks can be used to impersonate other entities within a virtual network. By impersonating entities with high levels of privilege in the network, attackers may be able to perform usurpation attacks. As an example, the injection of messages with fake sources mentioned by Cabuk et al.[5] is used for this purpose. By sending a message that appears to have been originated from a privileged entity, attackers may perform actions restricted to such entities, including elevating their own privilege level.

4.4.2. Software Vulnerability Exploitation

Roschke et al.[36] mention that virtual machine monitors are susceptible to the exploit of vulnerabilities in their implementation. According to the authors, by gaining control over a virtual machine monitor, attackers can break out of the virtual machine, obtaining access to the hardware layer. In an environment that uses full virtualization or paravirtualization to instantiate virtual routers, exploiting such vulnerabilities may enable an attacker to have full control over physical routers. By gaining access to physical devices, attackers could easily compromise any virtual networks provided by the infrastructure. As examples of such threats in practice, the Common Vulnerabilities and Exposures system lists a number of vulnerabilities in different versions of VMware products that allow guest Operating System users to potentially execute arbitrary code on the host Operating System[37]-[40].

5. Security Countermeasures

In this section, we explore solutions published in the literature that aim to provide security and protect the environment from the aforementioned security threats.

5.1. Access Control

Access control makes use of authentication and authorization mechanisms in order to verify the identity of network entities and enforce distinct privilege levels for each. This countermeasure is approached in two different manners in the literature, namely Trusted Virtual Domains and sandboxes. While these approaches are closely related to the notion of controlled execution domains, note that access control is performed in order to ensure that entities are granted the appropriate privilege levels.

5.1.1. Trusted Virtual Domains

Cabuk et al.[5] devised a framework to provide secure networking between groups of virtual machines. Their security goals include providing isolation, confidentiality, integrity, and information flow control in these networks. The framework provides the aforementioned security countermeasures through the use of Trusted Virtual Domains (TVDs). Each TVD represents an isolated domain, composed of "virtualization elements" and communication channels between such elements. In Cabuk's proposal, the virtualization elements are virtual workstations. However, the concept of TVDs may be applied to any device supporting virtualization.

Figure 3 depicts a virtual network infrastructure with three TVDs (A, B, and C). Gray routers represent gateways between these domains. While the gateway

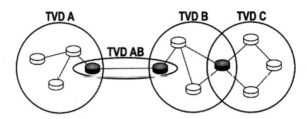

Figure 3. Example of a virtual infrastructure with three Trusted Virtual Domains, as described by Cabuk et al.[5].

between TVDs B and C is simultaneously within both domains, the gateways between A and B are isolated—making use of an auxiliary TVD (AB) in order to communicate.

Access control is performed when virtual machines join a TVD, ensuring that only machines that satisfy a given set of conditions are able to join. This admission control may be applied continuously in case prerequisites to join a TVD are changed. Additionally, TVDs leverage access policies to prevent unauthorized access.

5.1.2. Sandboxes

Wolinsky et al.[19] use virtual machine sandboxes in order to provide security in large scale collaborative environments. Although this work focuses on networked virtual machines hosting virtual workstations, this concept can be extended to virtual networks. Sandboxes are used to limit virtual machine access to physical resources, preventing malicious virtual machines from accessing data within other virtual machines. Moreover, each virtual machine supports IPSec, enabling the creation of secure communication channels, and X.509, providing virtual machine authentication. The authentication process is detailed in Section 5.2.

5.2. Authentication

Authentication aims to ensure that entities in a network environment are who they claim to be. In virtual network environments, providing proper authentication is complicated by factors such as the federation of virtual networks or mobility of virtual routers and links. Approaches that aim to deal with such difficulties are explained next.

5.2.1. Interoperability between Federated Virtual Networks

Although isolation is one of the main security requirements in virtual networking, there are cases in which distinct virtual networks must be able to cooperate. The federation of virtual networks can, for example, enable end-to-end connectivity—through virtual devices of distinct virtual networks—or allow access to distinct services. However, it may not be possible to provide interoperability due to the heterogeneous nature of virtual networks (which may implement different, incompatible protocols). Chowdhury et al.[25] partially tackle this issue with a framework that manages identities in this kind of environment. The main objective of the work is to provide a global identification system. To this end, the authors employ a decentralized approach in which controllers and adapters are placed in each virtual network. Controllers provide functionalities such as address allocation and name resolution, while adapters act as gateways between virtual networks, performing address and protocol translations. The proposed global identification system does not restrict the internal identification mechanisms used locally by virtual networks, allowing each virtual network to keep its own internal naming scheme. Additionally, global identifiers used by this framework are unique, immutable, and not associated with physical location, in order to not hinder the security or mobility of virtual devices.

5.2.2. Certificate-Based

As previously mentioned, the framework presented by Cabuk et al.[5] makes use of Trusted Virtual Domains (TVDs) to provide access control and network isolation. The authentication necessary to support access control is provided by means of digital certificates. These certificates ensure the identity of entities joining the network. Additionally, the system makes use of Virtual Private Networks (VPNs) to authenticate entities in network communications.

Analogously, Wolinsky et al.[19] use IPSec with X.509-based authentication for the purpose of access control in their system. In order to access the system, joining machines must request a certificate to the Certification Authority (CA). The CA responds by sending back a signed certificate to the node. The IP address of the requesting node is embedded into the certificate in order to prevent other nodes from reusing it.

5.2.3. Key-Based

Fernandes and Duarte[26][31] present an architecture that aims to provide efficient routing, proper resource isolation and a secure communication channel between routers and the Virtual Machine Monitor (VMM) in a physical router. In order to ensure efficiency, virtual routers copy routing-related information to the VMM—in this case, the hypervisor. This process is performed by a plane separation module, which separates the data plane (which contains routing rules) and the control plane (responsible for creating routing rules). As a result, packets matching rules in the hypervisor routing table do not need to be redirected to virtual routers, resulting in a significant performance speedup. However, the process of copying routing information needs to be authenticated such that a malicious router is not able to compromise the data plane of another router.

In order to prevent identity fraud, the system requires mutual authentication between virtual routers and the VMM. **Figure 4** depicts a simplified representation of the proposed architecture. The authors consider a Xen (paravirtualization)-based environment, in which virtual routers reside in unprivileged domains (DomUs) while the hypervisor resides within the privileged domain (Dom0). Each virtual router, upon instantiation, connects to the hypervisor following the client–server paradigm and performs an initial exchange of session keys using asymmetrical cryptography. The use of unique keys allows the hypervisor to verify the identity of distinct virtual routers in different unprivileged domains (in this example, DomU1, DomU2, and DomU3) and to isolate traffic between them. After this initial key exchange, the secure communication module is used by other system modules in order to securely exchange messages with the hypervisor.

5.3. Data Confidentiality

As network virtualization promotes the sharing of network devices and links

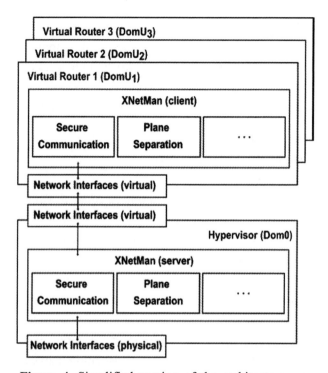

Figure 4. Simplified version of the architecture presented by Fernandes *et al.*[26][31], showing the secure communication modules.

among multiple entities, data confidentiality is a major security-related concern. Next, we explore approaches that leverage different protocols and techniques in order to provide secure communication within virtual networks.

5.3.1. VLANs and VPNs

The security goals approached by Cabuk *et al.*[5] include integrity, data isolation, confidentiality, and information flow control. Other than integrity, the remaining three goals, are directly related, and are tackled by a data confidentiality mechanism. The framework uses TVDs to control data access. However, virtual machines that belong to different TVDs may be hosted in the same physical machine. Therefore, it is necessary to ensure proper isolation, preventing a TVD from accessing data that belongs to another TVD.

The proposed solution for this challenge employs a combination of VLANs and VPNs. VLANs are used to identify packets belonging to different networks, allowing VLAN-enabled devices to route packets to the appropriate network interfaces, thus providing adequate isolation. Untrusted physical channels, however, may require a higher level of security. Therefore, if necessary, VPNs are used to provide data confidentiality by means of end-to-end cryptography.

5.3.2. Tunneling and Cryptography

Wolinsky et al.[19] make use of tunneling in order to isolate network traffic between virtual machines (in this case, virtual workstations). Two tunneling approaches are employed. In the first approach, the host system runs a tunneling software that captures packets incoming from physical interfaces and forwards them to virtual machines. In the second approach, the tunneling software runs inside virtual machines, and traffic is restricted within virtual networks through the use of firewall rules. According to the authors, while the second approach is easier to deploy, malicious users may be able to subvert this firewall, compromising the system. Although the focus of Wolinsky et al. is isolation between virtual workstations, we believe that the techniques used to achieve such isolation could be extended to virtual routers in network virtualization environments.

Fernandes and Duarte[26][31] deal with data confidentiality in communications between a virtual router and the Virtual Machine Monitor (VMM) hosting it. After the authentication process, described in Section 5.2, virtual routers use symmetrical cryptography in order to securely communicate with the VMM.

Huang et al.[22] present a framework that provides secure routing. In the environment presented by the authors, routing information that is propagated through a virtual network is confidential and needs to be kept secret from unauthorized network entities. Routing information is categorized in groups, and group keys are assigned to virtual routers. Therefore, routing information can be encrypted, ensuring that only routers with the correct key are able to decrypt this information. Thus, routing information relative to a given group is protected against unauthorized access from other groups, other virtual networks or the physical network itself.

Similarly to the previously described approach, Fukushima et al.[24] aim to protect sensitive routing information in virtual networks from being disclosed to entities controlling the physical network. To achieve this goal, the authors make use of a strategy based on Secure Multi-party Computation (SMC). SMC allows multiple entities to perform joint computations on sensitive data they hold without disclosing such data. Each entity has access to the result of the global computation, but not to any data held by other entities. This is achieved through the use of one-way functions, which are easy to evaluate but hard to invert. In the context of virtual network routing, SMC allows a virtual router to compute optimal routes without needing to share the information that it holds. As SMC requires full-mesh connectivity between computing nodes, the authors decompose the virtual network into locally connected subsets of routers, called *cliques*. The SMC-based distributed routing algorithm is run locally in each *clique*, and the results of local computations are then shared between *cliques*.

As the employment of cryptographic techniques requires physical devices that are capable of supporting protocols that enable them and generates processing and bandwidth overheads, Bays et al.[4] devise an optimization model and a heuristic algorithm for online, privacy-oriented virtual network embedding. Clients may require end-to-end or point-to-point cryptography for their networks, as well as requiring that none of their resources overlap with other specific virtual networks. Both the optimal and heuristic approaches take into account whether physical routers are capable of supporting cryptographic algorithms in order to ensure the desired level of confidentiality and guarantee the non- overlapping of resources (if requested). Additionally, both methods feature precise modeling of overhead costs of security mechanisms in order to not underestimate the capacity requirements of virtual network requests. This proposal is in line with research performed in the area of virtual network embedding, such as the work of Alkmim et al.[41].

5.3.3. Firewalling and Subnetting

As previously mentioned in Section 5.3.2, Wolinsky et al.[19] make use of firewall rules (in addition to tunneling techniques) in order to prevent communications between different virtual networks. In addition to using firewalls for this purpose, Wu et al.[20] also employ subnetting (*i.e.*, each virtual network is bound to a unique subnet) in order to provide an additional layer of security against unauthorized information disclosure.

5.3.4. Path Splitting

In addition to encryption of routing information, Huang et al.[22] use variable paths in virtual networks to propagate data flows. **Figure 5** illustrates the employment of path splitting in order to hinder an information interception attack. Communication between a virtual router hosted on Physical Router (PR) 1 and another one hosted on PR 7 is split among two different paths—one passing through PR 3 and 6, and the other, through PR 2 and 4 (represented by dashed lines). Even if traffic between these two virtual routers is not encrypted, the threat is partially mitigated as the attacker only has access to part of the information being exchanged (packets passing through the link between PR 3 and 6). Moreover, when used in combination with encryption (as in the work of Huang et al.), this approach helps mitigate traffic analysis attacks. It is worth noting that while in this example the attacker is only eavesdropping on one physical path, in reality, multiple devices may be compromised. In this case, splitting traffic among an increasing number of paths would lead to progressively higher levels of security (or, conversely, to increasingly higher costs for an attacker to capture the full traffic).

5.3.5. Limiting Introspection

Finally, van Cleeff et al.[10] present recommendations for safer use of virtualization. One of these recommendations is to limit, or even disable, the introspection feature, which allows virtual machine monitors to access data inside virtual machines. While useful, this functionality may be exploited by attackers, as previously explained on Subsection 4.1.3.

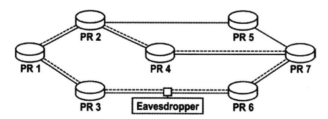

Figure 5. Example of path splitting used to mitigate an information interception attack.

5.4. Data Integrity

Similarly to confidentiality, data integrity is a major concern as a result of shared network devices and communication channels. Next, we describe approaches that aim to establish a desired level of integrity in virtual network environments.

5.4.1. Cryptography

In addition to authentication (*i.e.*, source integrity) and confidentiality, the framework developed by Cabuk *et al.*[5] makes use of VPNs to provide data integrity to virtual networks. The use of cryptographic tunneling protocols prevents malicious entities from manipulating messages going through the network. As previously discussed, the authors use IPSec as the tunneling protocol.

5.4.2. Timestamping

As previously discussed, replay attacks are one of the threats to data integrity that may be present in network virtualization environments. The addition of unique identifiers inside encrypted messages makes it possible to detect duplicated messages, and therefore, replay attacks. For this purpose, the architecture proposed by Fernandes and Duarte[26][31] inserts timestamps inside encrypted messages in order to ensure that messages are non-reproducible.

5.4.3. Limiting Introspection

Besides mitigating information theft, disabling or limiting introspection also prevents data tampering. According to van Cleeff *et al.*[10], this functionality allows the VMM to modify applications running inside it, which may cause inconsistencies. Another recommendation consists of specifically designing applications that facilitate batch processing and checkpointing. According to the authors, this minimizes security issues associated with rollback and restore operations that may otherwise threaten integrity.

5.5. Nonrepudiation

Nonrepudiation provides evidences regarding which (potentially malicious) actions have been performed by which entities. This security countermeasure is highly valuable in the context of network virtualization environments, in which a number of physical devices are shared by different users. Nevertheless, we are not aware of any publication that targets this countermeasure specifically.

5.6. Availability

Last, we present proposals that aim to maintain the availability of network virtualization environments. The key concerns in this area of security are providing proper resource isolation and mitigating attacks that target physical or virtual devices. Approaches aiming to deal with such concerns are explained in the following subsections.

5.6.1. Physical Resource Isolation

One of the main concerns regarding availability is the abuse of physical resources by virtual networks. Virtual networks may attempt to use as much resources as possible in order to maximize their performance. If the environment is not adequately protected, this behavior may lead to the exhaustion of physical resources, compromising the availability of other virtual networks hosted on the same substrate. Therefore, physical resources must be shared in a fair manner, and actions performed by a virtual network must not negatively impact others.

According to Wu et al.[29], the sharing of physical resources by packet processors is usually only performed at a granularity of entire processor cores. The authors claim that finer-grained processor sharing is required in order to provide scalability for network virtualization environments. Thus, the authors propose a system that allows multiple threads to share processor cores concurrently while maintaining isolation and fair resource sharing. However, typical multithreading approaches consider a cooperative environment, which is not the case in network virtualization. The authors devise a fair multi-threading mechanism that allows the assignment of different priorities to each thread. Additionally, this mechanism takes into account the history of how much processing has been performed by each thread. Inactivity times are also considered in order to guarantee that threads will not stay idle for too long. The evaluation performed by the authors shows that the proposed mechanism is able to properly distribute processing resources according to the defined priorities. Furthermore, while it requires more processing power, it is able to provide better resource utilization in comparison to coarse-grained approaches.

Kokku *et al.*[30] propose a network virtualization scheme that provides resource isolation while aiming to maximize substrate utilization. It allows virtual networks to have either resource-based reservations (*i.e.*, reservations calculated as a percentage of available resources in the substrate) or bandwidth-based reservations (*i.e.*, reservations based on the aggregate throughput of the virtual network). Virtual networks are divided in two groups according to the type of reservation required, and treated independently by a scheduler. This scheduler treats flows that belong to different virtual networks with distinct priorities, based on the reservations and average resource usage rate of each network. The authors present an evaluation performed on an implemented prototype, showing that the proposed scheme was capable of ensuring that each virtual network met its reservations.

Fernandes and Duarte[26] present a network monitor that employs plane separation in order to provide resource isolation in network virtualization environments. The system is able to allocate resources based on fixed reservations, as well as to redistribute idle resources between virtual networks that have a higher demand. Additionally, an administrator is able to control the amount of resources to be used by each virtual network, as well as set priorities for using idle resources. The system continuously monitors the consumption of physical resources by each virtual router. If any virtual router exceeds its allowed use of bandwidth, processing power, or memory, it is adequately punished by having packets dropped, or a percentage of its stored routes erased. Harsher punishments are instituted if there are no idle resources available. Conversely, given punishments are gradually reduced if the router stops using more than its allocated resources. This system is capable of adequately preventing physical resources from being overloaded, and packet drops employed by the punishment mechanism do not cause a major impact on network traffic.

In another publication[31], the same authors extend the previously described network monitor. This new system introduces the idea of short term and long term requirements, based on the time frame in which they must be met. Short term requirements may be allocated in an exclusive or non-exclusive manner, while long term requirements are always non-exclusive. In this context, exclusive requirements are always allocated (even if part of the allocated resources is idle), while non-exclusive requirements are only allocated when necessary. The system prioritizes virtual networks that have used the lowest portion of their requirements, and an adaptive control scheme is used in order to improve the probability that long term requirements, if needed, will be met. The presented evaluation shows the improvement of this system over the original[26] in terms of guaranteeing that the demands of each virtual network will be met, as well as reducing resource load on the physical substrate.

5.6.2. Virtual Network Resilience

Even with proper physical resource isolation, maintaining availability remains a challenge in virtualized networks. The virtualization layer must be resilient, maintaining its performance and mitigating attacks in order to sustain its availability. Some of the publications described next approach the issue of virtual network resilience from the point of view of fault tolerance. Nonetheless, we emphasize that the solutions described in these publications may also be used as a response to attacks that cause the failure or degradation of physical devices or links.

The solution presented by Yeow et al.[32] aims to provide network infrastructures that are resilient to physical router failures. This objective is achieved through the use of backups (i.e., redundant routers and links). However, redundant resources remain idle, reducing the utilization of the physical substrate. To minimize this problem, the authors propose a scheme that dynamically creates and manages shared backup resources. This mechanism minimizes the number of necessary backup instances needed to achieve a certain level of reliability. While backup resources are shared, each physical router is restricted to hosting a maximum number of backup instances in order not to sacrifice reliability. The connectivity between each virtual router and its neighbors is preserved in all of its backups, both in terms of number of links and bandwidth reservations.

The illustration on the left side of **Figure 6** shows a simple representation of how backup nodes (represented as circles) may be shared among different virtual networks. For example, the two backup nodes at the right side of this figure are shared between Virtual Network 1 and Virtual Network 3, regardless of whether they belong to one or the other. The right side of **Figure 6**, in turn, depicts in greater detail how backups are allocated to virtual routers. A virtual router $C1$ has virtual routers $B1$ and $B2$ as its backups. Since $C1$ has a virtual link connecting it to another router, $N1$, a virtual link with the same bandwidth reservation (depicted as 1 in the figure) is also established between each backup node and $N1$ in order to preserve the connectivity of the original router.

Meixner et al.[35] devise a probabilistic model for providing virtual networks that are resilient to physical disasters. Disasters are characterized by the occurrence of multiple failures in the physical network, as well as the possibility of correlated cascading failures during attempts to recover network resources. The virtual link mapping strategy guarantees that the failure of a single physical link will not disconnect any virtual network, and aims at minimizing virtual network disconnection in the event of a disaster (*i.e.*, simultaneous failure of multiple links). Additionally, excess processing capacity in the physical network is used to create a backup router for each virtual network, which reduces disconnection in the event of disasters and provides additional processing capacity for the recovery phase. When attempting to recover virtual network resources, the model analyzes all possible virtual router replacements in an effort to replace affected virtual routers in a way that ensures the virtual network will not be disconnected by any post-disaster failures.

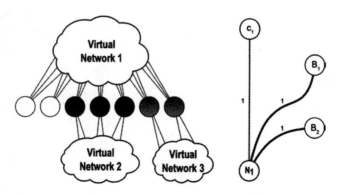

Figure 6. Examples of sharing and mapping of backup instances, used by Yeow et al.[32] to provide resilient virtual networks.

The system presented by Zhang et al.[27] uses redundant virtual networks in order to provide reliable live streaming services. It is able to detect path failures and traffic congestion, dynamically redirecting data flows. Initially, the data flow is distributed equally through available virtual networks. **Figure 7** depicts the distribution of a data flow through virtual networks, using multiple paths between a server and a client. Gradually, the number of packets routed through each virtual network is adapted according to its relative bandwidth capacity. Additionally, an active probing mechanism is used to detect failures in the physical network or routing problems (changes in routing tables, for example, may have a significant impact in live streaming applications). If an issue is detected, the system is able to redirect data flows away from problematic networks and redistribute it among the remaining ones. Experiments performed by the authors demonstrate advantages in using multiple networks instead of a single one, with increasing gains when using up to four virtual networks. Additionally, the authors claim that the bandwidth cost of the probing mechanism is negligible.

Chen et al.[33] propose a virtual network embedding strategy that aims at ensuring survivability. Load balancing is employed in the embedding process in order to balance the bandwidth consumption of substrate links. Moreover, backup links are reserved for each accepted virtual network, but not activated until a failure occurs. Backup links are allocated in physical paths that do not overlap with

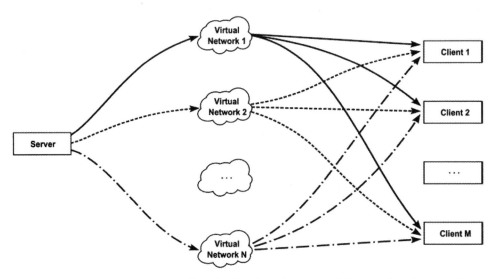

Figure 7. A live streaming data flow is distributed among different virtual networks, a mechanism used by Zhang et al.[27].

the path hosting the original link, guaranteeing that a single physical link failure will not simultaneously affect the original virtual link and one or more of its backups. These backup resources may be shared by multiple virtual networks or reconfigured over time in order to improve efficiency.

Zhang et al.[34] devise a strategy for computing the availability of Virtual Data Centers (VDCs), as well as an algorithm for reliable VDC embedding. In order to determine VDC availability, the authors consider the availability of individual, heterogeneous components, as well as dependencies among them. The embedding mechanism aims at meeting minimum availability criteria while optimizing resource usage. Virtual devices are divided into replication groups (groups in which any virtual device may serve as a backup if another fails). In order to minimize resource consumption, VDCs are embedded on physical devices with the lowest level of availability that still meets the desired level. In a similar way, a minimum number of backups is assigned to each replication group in order to meet availability requirements.

Unlike the previously described approaches in the area of virtual network resilience, Oliveira et al.[7] present a strategy based on "opportunistic resilience", which does not employ backup resources. The bandwidth demand of each virtual link is split over multiple physical paths. As a consequence, physical link failures are less likely to cause a virtual link disconnection (an affected virtual link will remain operational, albeit with less capacity). Additionally, when link failures occur, a reactive strategy is used in order to reallocate the lost capacity over unaffected paths, attempting to fully restore the bandwidth of degraded virtual links.

Distributed Denial of Service (DDoS) attacks are a common threat to the availability of network services. The system proposed by Yu and Zhou[6] aims to detect such attacks on community networks (federated virtual networks that belong to cooperating entities). The devised solution leverages communication between virtual routers that belong to different entities in this collaborative environment to detect possible attacks at an early stage. Virtual routers located on the edges of the community network monitor traffic passing through them and calculate the entropy of its flows. Traffic surges in any of these flows will cause the entropy to drop, indicating a possible attack. If this occurs, other routers are notified and instructed to calculate the entropy rate of this suspected flow. Calculated values are compared, and if they are similar, a DDoS attack is confirmed.

6. Discussion

A number of insights can be obtained from the extensive investigation of the state of the art reported in this paper. First, it is possible to observe that the publications in the area are not equally distributed between the main security categories. **Table 3** and **Table 4** show, respectively, the security threats and security countermeasures approached in these publications. In both tables, publications have been

Table 3. Security threats mentioned in publications in the area of virtual network security.

Publication	Threats			
	DI	DE	DR	US
[4]	×			
[19]	×			
[21]	×			
[22]	×			
[23]	×			
[24]	×			
[10]	×	×		
[20]	×	×		
[5]	×	×		×
[25]		×		
[26]		×	×	
[36]				×
[6]			×	
[7]			×	
[27]			×	
[28]			×	
[29]			×	
[30]			×	
[31]			×	
[32]			×	
[33]			×	
[34]			×	
[35]			×	

From left to right: Disclosure, Deception, Disruption, Usurpation.

Table 4. Security countermeasures provided by publications in the area of virtual network security.

Publication	Countermeasures					
	AC	AU	CO	IN	NR	AV
[4]			×			
[20]			×			
[22]			×			
[24]			×			
[19]	×	×	×			
[5]	×	×	×	×		
[26]		×	×	×		×
[31]		×	×	×		×
[10]			×	×		
[25]		×				
[6]						×
[7]						×
[27]						×
[29]						×
[30]						×
[32]						×
[33]						×
[34]						×
[35]						×

From left to right: Access Control, Authentication, Confidentiality, Integrity, Nonrepudiation, Availability.

grouped together according to the security elements they approach, whenever possible. It is noticeable that disruption and availability—a security threat and a countermeasure that are directly correlated—are approached in the majority of these publications. This is likely due to the high prevalence of attacks aiming at causing disruption. These attacks are relatively simple but can be highly devastating, especially in an environment that makes heavy use of shared resources (as a single physical failure may disrupt several virtual networks). Disclosure and confidentiality follow closely behind, being present in a similar number of publications as disruption/availability. Once again, this is linked to physical resource sharing. Similarly to disruption attacks, such sharing means that a single well-placed sniffer may be able to acquire sensitive information from multiple virtual networks at once. Moreover, there are also privacy concerns between infrastructure providers and virtual network requesters (as the former may have access to data that the latter considers confidential).

Second, only a small number of publications approach more than one threat or countermeasure simultaneously. No single publication has dealt with threats in more than two of the four categories, or presented solutions that provide more than four security countermeasures, out of a total of six. Additionally, one security countermeasure in particular—nonrepudiation—was not approached by any of the publications. The combination of authentication and integrity, which exists in some publications, can be considered as the basis for the provision of nonrepudiation, but this specific countermeasure is not targeted. Nonrepudiation is a highly valuable (albeit challenging) security countermeasure for network virtualization environments, and will be further discussed in Section 7.

Third, we were able to conclude that many of the threats that affect network virtualization environments also affect traditional networks. However, we emphasize that these threats affect traditional and virtual network environments in different ways. In most cases, the effects of these threats are greatly exacerbated by certain characteristics of virtual network environments. Information interception, physical resource overloading, physical resource failure, and software vulnerability exploitation are aggravated by the fact that a number of virtual routers may share a physical router. Therefore, as previously explained, an attack of any of these types targeting a single physical router may affect several virtual networks. Further, it is more difficult to recognize (and, therefore, to perform countermeasures against) identity fraud and replay attacks due to the dynamicity of network virtualization environments, as virtual routers may be freely moved among physical routers and assume different identities. Loss of registry entries and information leakage, as described in the studied literature, are limited to virtual network environments. Moreover, threats related to introspection are also inherent to these types of environments, as this is a (potentially exploitable) feature of virtual machine monitors.

Last, we can observe the employment of different virtualization techniques in some publications. For example, Cabuk et al.[5] implemented a prototype of their framework based on a paravirtualization platform, while Huang et al.[22] consider an underlying network based on programmable routers. Further, Fernandes and Duarte[26][31] build a hybrid solution that combines paravirtualization with plane separation, a core idea of programmable networks. Although the majority of publications do not target specific network virtualization techniques, we emphasize that different types of platforms have their own sets of benefits, as well as security concerns, which need to be taken into account.

7. Challenges

Despite the existence of a sizable body of work in virtual network security, some challenges remain open. In this section, we summarize some of the main research challenges in this area. We emphasize, however, that these challenges should not be considered exhaustive, but rather as a starting point for further discussions in the area.

One clear opportunity for research in virtual network security is the provisioning of nonrepudiation—which, to the best of our knowledge, has not yet been approached. Nonrepudiation requires providing proof of actions performed by entities on a network, which can be used for holding entities accountable for malicious activity. We deem nonrepudiation an essential security countermeasure for virtual networking environments in order to accurately backtrace attacks—not only to ensure that punitive actions will be taken against the attackers but also to properly contain the attacks themselves. In the event of a DDoS attack, for example, this countermeasure could enable administrators to pinpoint the origins of the attack with a high level of precision—which otherwise tends to be a very difficult task. Moreover, nonrepudiation may even prevent attacks, in the sense that malicious users who are aware that such a mechanism is in use may refrain from carrying out attacks in order to avoid exposing themselves. Provisioning nonrepudiation can be challenging for a number of reasons, such as the complexity of securely storing and handling digital certificates—used for proving that an action was, indeed, performed by a given entity—and the negative impact this has on network performance. Moreover, it is necessary to maintain a desired level of privacy for virtual network requesters as well as end users. Nevertheless, we envision that the importance of this countermeasure will grow steadily as network virtualization becomes increasingly prominent in production environments.

In addition to privacy issues related to nonrepudiation, there are also concerns regarding the privacy of general data stored in virtual routers or sent through virtual networks. Although such data may be protected from being intercepted by other entities, infrastructure providers have physical access to all data stored in virtual networks they are hosting. Although this issue has been approached by some authors, their proposed strategies are often based on strong assumptions, such as the ability to choose which physical entity (out of a number of entities controlling the physical substrate) will host each of its routers—a feature that may not commonly be available in practice.

Another opportunity stems from the multiple levels of heterogeneity present in network infrastructures. As previously mentioned, in addition to the use of heterogeneous hardware devices, it is common for network substrates to be composed of a number of physical networks that belong to different entities. As such, there is a need for uniform methods for requesting, negotiating, and enforcing security requirements across devices that may have incompatible interfaces and entities with potentially conflicting policies.

Last, software platforms used to instantiate virtual networks may not always offer adequate protection against security threats. Moreover, although virtualization technologies are gradually evolving and becoming more mature, both hardware and software are susceptible to vulnerabilities that may be exploited by attackers. Consequently, research efforts that build on top of network virtualization need to consider these security issues and, most importantly, overhead costs of additional security mechanisms that may be necessary, in order to ensure that they will be suitable for real world environments.

We emphasize, once again, that this is not an exhaustive list of challenges in the area. The essence of network virtualization is based on layers upon layers with increasing levels of abstraction (e.g., the physical substrate, the virtualization layer, virtual networks, and services running on top of them). Consequently, we envision that a number of other challenges may be present in all of these layers—much like the ones listed in this section.

8. Conclusions

Network virtualization enables the subdivision of a single network infrastructure into multiple virtual architectures. The benefits of this technique apply to a wide range of applications, including the creation of virtual testbeds, community networks, and cloud computing infrastructures. Furthermore, network virtualization has been proposed by researchers as the basis for the creation of a new architecture for the Internet, allowing pluralist network environments that support a number of different network protocols simultaneously.

In spite of the benefits provided by network virtualization, there is a series of security issues that need to be considered. Our study revealed a number of security threats, covering the four categories defined by Shirey[16]. The very act of sharing a physical infrastructure among multiple parties is shown to be the source of several of these threats.

This study shows that there have been several efforts to provide security in virtual networks. However, these efforts were not organized in a comprehensible manner. This study provides a systematic overview of the available research results in the field, categorizing work that represents the state of the art and highlighting different approaches for providing security. Additionally, it also evidences imbalances between different sub-areas of security research in network virtualization, which can be used as guidance for future work in this area. Usurpation and access control, for example, are significantly underrepresented in relation to other security countermeasures, and nonrepudiation is not targeted by any publication. Additionally, while a significant body of work exists in the sub-area of availability, only one publication deals with detection and prevention of attacks. Such gaps may represent valuable opportunities for future work.

To summarize, the categorization of security threats and countermeasures presented in this paper simplifies the analysis of which security aspects have not yet been approached and which types of threats need to be mitigated. Furthermore, it makes it easier to identify a number of existing solutions that aim to provide security in virtual networks.

Endnotes

[1]Machine virtualization is available for personal computers, in commonly used operating systems (e.g., Windows, Linux, and Mac OS X).

Competing Interests

The authors declare that they have no competing interests.

Authors' Contributions

All authors read and approved the final manuscript.

Acknowledgements

The authors would like to thank the anonymous reviewers for their valuable comments and suggestions. This work has been partially supported by FP7/CNPq (Project SecFuNet, FP7-ICT-2011-EU-Brazil), RNP-CTIC (Project ReVir), as well as PRONEM/FAPERGS/CNPq (Project NPRV).

Source: Bays L R, Oliveira R R, Barcellos M P, *et al*. Virtual network security: threats, countermeasures, and challenges [J]. Journal of Internet Services & Applications, 2015, 6(1): 1–19.

References

[1] Chowdhury NMMK, Boutaba R (2010) A survey of network virtualization. Comput Netw 54(5):862–876.

[2] Fernandes N, Moreira MD, Moraes I, Ferraz L, Couto R, Carvalho HT, Campista M, Costa LK, Duarte OB (2011) Virtual networks: isolation, performance, and trends. Ann Telecommun 66(5–6):339–355.

[3] Anderson T, Peterson L, Shenker S, Turner J (2005) Overcoming the internet impasse through virtualization. Computer 38(4):34–41.

[4] Bays LR, Oliveira RR, Buriol LS, Barcellos MP, Gaspary LP (2014) A heuristic-based algorithm for privacy-oriented virtual network embedding. In: IEEE/IFIP Network Operations and Management Symposium (NOMS). IEEE, Krakow, Poland.

[5] Cabuk S, Dalton CI, Ramasamy H, Schunter M (2007) Towards automated provisioning of secure virtualized networks. In: ACM Conference on Computer and Communications Security. New York, USA.

[6] Yu S, Zhou W (2008) Entropy-based collaborative detection of ddos attacks on community networks. In: IEEE International Conference on Pervasive Computing and Communications. IEEE Computer Society, Washington, DC, USA.

[7] Oliveira RR, Marcon DS, Bays LR, Neves MC, Buriol LS, Gaspary LP, Barcellos MP (2013) No more backups: Toward efficient embedding of survivable virtual networks. In: IEEE International Conference on Communications. IEEE, Budapest, Hungary.

[8] LAN/MAN Standards Committee (2006) IEEE Standard for Local and metropolitan area networks—Virtual Bridged Local Area Networks. IEEE Std 802.1Q-2005 (incorporates IEEE Std 802.1Q1998, IEEE Std 802.1u-2001, IEEE Std 802.1v-2001, and IEEE Std 802.1s-2002). http://www.ieee802.org/1/pages/802.1Q-2005.html.

[9] Rosen E, Cisco Systems I, Rekhter Y, Juniper Networks I (2006) RFC 4364: BGP/MPLS IP Virtual Private Networks (VPNs). http://www.ietf.org/rfc/ rfc4364.txt.

[10] van Cleeff A, Pieters W, Wieringa RJ (2009) Security implications of virtualization: A literature study. In: International Conference on Computational Science and Engineering. IEEE Computer Society, Washington, DC, USA.

[11] McKeown N, Anderson T, Balakrishnan H, Parulkar G, Peterson L, Rexford J, Shenker S, Turner J (2008) Openflow: enabling innovation in campus networks. SIGCOMM Comput Commun Rev 38:69–74.

[12] Bari MF, Boutaba R, Esteves R, Granville LZ, Podlesny M, Rabbani MG, Zhang Q, Zhani MF (2013) Data center network virtualization: A survey. Communications Surveys Tutorials, IEEE 15:909–928.

[13] Zhou M, Zhang R, Xie W, Qian W, Zhou A (2010) Security and privacy in cloud computing: A survey. In: Semantics Knowledge and Grid (SKG), 2010 Sixth International Conference On. IEEE, Beijing, China.

[14] Hashizume K, Rosado DG, Fernández-Medina E, Fernandez EB (2013) An analysis of security issues for cloud computing. J Internet Serv Appl 4:1–13.

[15] Scott-Hayward S, O'Callaghan G, Sezer S (2013) Sdn security: A survey. In: Future Networks and Services (SDN4FNS), 2013 IEEE SDN For. IEEE, Trento, Italy.

[16] Shirey R (2000) RFC 2828: Internet Security Glossary. http://www.ietf.org/rfc/ rfc2828.txt.

[17] Stallings W (2006) Cryptography and Network Security: Principles and Practice. Pearson/Prentice Hall, Upper Saddle River, New Jersey, USA.

[18] Cavalcanti E, Assis L, Gaudencio M, Cirne W, Brasileiro F (2006) Sandboxing for a free-to-join grid with support for secure site-wide storage area. In: International Workshop on Virtualization Technology in Distributed Computing. IEEE Computer Society, Washington, USA.

[19] Wolinsky DI, Agrawal A, Boykin PO, Davis JR, Ganguly A, Paramygin V, Sheng YP, Figueiredo RJ (2006) On the design of virtual machine sandboxes for distributed computing in wide-area overlays of virtual workstations. In: International Workshop

on Virtualization Technology in Distributed Computing. IEEE Computer Society, Washington DC, USA.

[20] Wu H, Ding Y, Winer C, Yao L (2010) Network security for virtual machine in cloud computing. In: Computer Sciences and Convergence Information Technology (IC-CIT), 2010 5th International Conference On. IEEE, Seoul, Republic of Korea.

[21] Cui Q, Shi W, Wang Y (2009) Design and implementation of a network supporting environment for virtual experimental platforms. In: WRI International Conference on Communications and Mobile Computing. IEEE Computer Society, Washington DC, USA.

[22] Huang D, Ata S, Medhi D (2010) Establishing secure virtual trust routing and provisioning domains for future internet. In: IEEE Conference on Global Telecommunications, Miami, USA.

[23] Pignolet Y-A, Schmid S, Tredan G (2013) Adversarial vnet embeddings: A threat for isps?. In: IEEE INFOCOM. IEEE, Turin, Italy.

[24] Fukushima M, Sugiyama K, Hasegawa T, Hasegawa T, Nakao A (2013) Minimum disclosure routing for network virtualization and its experimental evaluation. IEEE/ACM Trans Netw PP(99):1839–1851.

[25] Chowdhury NMMK, Zaheer F-E, Boutaba R (2009) imark: an identity management framework for network virtualization environment. In: IFIP/IEEE International Symposium on Integrated Network Management. IEEE Press, Piscataway, USA.

[26] Fernandes NC, Duarte OCMB (2011) Xnetmon: A network monitor for securing virtual networks. In: IEEE International Conference on Communications. IEEE, Kyoto, Japan.

[27] Zhang Y, Gao L, Wang C (2009) Multinet: multiple virtual networks for a reliable live streaming service. In: IEEE Conference on Global Telecommunications. IEEE Press, Piscataway, USA.

[28] Marquezan CC, Granville LZ, Nunzi G, Brunner M (2010) Distributed autonomic resource management for network virtualization. In: IEEE/IFIP Network Operations and Management Symposium, Osaka, Japan.

[29] Wu Q, Shanbhag S, Wolf T (2010) Fair multithreading on packet processors for scalable network virtualization. In: ACM/IEEE Symposium on Architectures for Networking and Communications Systems. ACM, New York, USA.

[30] Kokku R, Mahindra R, Zhang H, Rangarajan S (2010) Nvs: a virtualization substrate for wimax networks. In: International Conference on Mobile Computing and Networking. ACM, New York, USA.

[31] Fernandes NC, Duarte OCMB (2011) Provendo isolamento e qualidade de serviço em redes virtuais. In: Simpósio Brasileiro de Redes de Computadores e Sistemas Distribuídos, Campo Grande, Brazil. (in Portuguese).

[32] Yeow W-L, Westphal C, Kozat UC (2011) Designing and embedding reliable virtual

infrastructures. SIGCOMM Comput Commun Rev 41(2):57–64.

[33] Chen Q, Wan Y, Qiu X, Li W, Xiao A (2014) A survivable virtual network embedding scheme based on load balancing and reconfiguration. In: IEEE Network Operations and Management Symposium. IEEE, Krakow, Poland.

[34] Zhang Q, Zhani MF, Jabri M, Boutaba R (2014) Venice: Reliable virtual data center embedding in clouds. In: IEEE INFOCOM. IEEE, Toronto, Canada.

[35] Meixner CC, Dikbiyik F, Tornatore M, Chuah C, Mukherjee B (2013) Disaster-resilient virtual-network mapping and adaptation in optical networks. In: International Conference on Optical Network Design and Modeling.

[36] Roschke S, Cheng F, Meinel C (2009) Intrusion detection in the cloud. In: IEEE International Conference on Dependable, Autonomic and Secure Computing. IEEE Computer Society, Washington, DC, USA.

[37] Common Vulnerabilities and Exposures (2012) CVE-2012-1516. https://cve.mitre.org/cgi-bin/cvename.cgi?name=CVE-2012-1516.

[38] Common Vulnerabilities and Exposures (2012) CVE-2012-1517. https://cve.mitre.org/cgi-bin/cvename.cgi?name=CVE-2012-1517.

[39] Common Vulnerabilities and Exposures (2012) CVE-2012-2449. https://cve.mitre.org/cgi-bin/cvename.cgi?name=CVE-2012-2449.

[40] Common Vulnerabilities and Exposures (2012) CVE-2012-2450. https://cve.mitre.org/cgi-bin/cvename.cgi?name=CVE-2012-2450.

[41] Alkmim GP, Batista DM, Fonseca NLS (2013) Mapping virtual networks onto substrate networks. J Internet Serv Appl 3(4):1–15.

Chapter 2
Adoption of Security as a Service

Christian Senk

University of Regensburg, Regensburg, Germany

Abstract: *Security as a Service* systems enable new opportunities to compose security infrastructures for information systems. However, to date there are no holistic insights about their adoption and relevant predictors. Based on existing technology acceptance models we developed an extended application-specific research model including formative and reflective measures. The model was estimated applying the *Partial Least Squares* technique to address the prediction-oriented nature of the study. A subsequent online survey revealed that a large number of industries shows significant and steadily growing interest in *Security as a Service*. Adoption drivers were investigated systematically.

Keywords: Cloud Computing, Partial Least Squares, Security as a Service

1. Introduction

Companies face an increasing threat regarding the security and safety of their information systems due to the opening of security domains for web-based access in the course of current technological developments such as *Federated Identity Management*[1] and *Cloud Computing*[2][3]. In this regard, *Cloud Computing* is a model "for enabling convenient, on-demand network access to a shared pool of configurable computing resources [...]"[4]. These resources are referred to .as *Cloud services* and can logically be assigned to the infrastructure, (*Infrastructure as a Service*, IaaS), middleware (*Platform as a Service*, PaaS) or application software layer (*Software as a Service*, SaaS)[5][6]. The *Cloud Computing* model itself not only induces certain security-related risks, it also opens up new opportunities to obtain innovative security solutions in a technically and economically flexible way in order to cope with rising security demands[7]. The outsourcing of security according to SaaS principles is referred to as *Security as a Service* (SECaaS)[3][8]. Such systems are considered to be the next step in the evolution of *Managed Security Services* (MSS) and differ clearly from traditional outsourcing models or on- premises deployments[3][8][9]. According to GARTNER RESE ARCH, the demand for SECaaS will grow significantly and might substantially change existing IT security infrastructure landscapes[10]. However, no deep insights about the current adoption and future developments exist. In this regard, based on an expert-group discussion[a] we defined that the answers to the following research questions (RQ) are important to predict the future of SECaaS:

- **RQ1:** Is there a market for SECaaS enterprise applications in general and for specific application types in particular?

- **RQ2:** Which are the key drivers and inhibitors for the adoption of SECaaS?

- **RQ3:** Which benefits are perceived to be relevant by potential adopters of SECaaS?

- **RQ4:** Which risks are perceived to be relevant by potential adopters of SECaaS?

- **RQ5:** Which organization-specific factors (e.g. company size) affect the acceptance of SECaaS?

The main objective of this paper is to answer these research questions through empirical research in order to gain insights valuable for both potential consumers and providers of SECaaS. The remainder of this paper is structured as follows: Section 2 defines SECaaS and overviews related work regarding the adoption of similar technologies. In Section 3 the research concept is specified and justified. Section 4 gives an overview of the results of the estimation of the research model and the related hypotheses. Afterward, the findings are discussed respecting the specified research questions. Section 5 concludes the paper.

2. Theoretical Background

This chapter provides the theoretical background for the context of the study. This includes the object of adoption (SECaaS), which is defined in Subsection 2.1. Subsequently, an overview of related work regarding the adoption of similar technological innovations is provided in Subsection 2.2 in order to identify adequate research approaches.

2.1. Security as a Service

SECaaS is a service-oriented approach to IT security architecture and thus a consequent evolution of traditional security landscapes[8][9]. It is defined as a model for the delivery of standardized and comprehensive security functionality in accordance with the SaaS model[8][11]. It thus follows the *Cloud Computing* model. Hence, SECaaS systems are delivered in form of Cloud services complying with related principles. This excludes built-in security controls of existing Cloud services[11]. Key attributes of Cloud services contain the following[5][6][12]:

- Application and underlying infrastructure are abstracted and offered through service interfaces;

- Standardized network access by any device;

- Scalability and flexibility of the underlying infrastructure;

- Shared and multi-tenant resources;

- On-demand self-service provisioning and near real-time deployment;

- Flexible and fine grained pricing without up-front commitments.

Based on the market-oriented taxonomy of KARK for outsourced security services[13] and the adaption of SENK AND HOLZ APFEL[14] we classify SECaaS systems as depicted in **Table 1**. This classification scheme was recently validated by a survey of existing SECaaS offerings[14]. According to that survey, the majority of existing SECaaS products cover *Endpoint Security* or *Content Security* applications[14]. The authors further outline existing systems' deficient compliance with Cloud and SaaS design principles. Especially inflexible pricing models often restrict the potential value of existent SECaaS systems[14]. It has to be noted that the granularity of SECaaS offerings can vary from fine-grained basic services addressing highly specific security needs (e.g. biometric user authentication) to coarse-grained solutions covering a broad set of security functionalities.

The delivery of security services according to the SECaaS model differs clearly from traditional MSS provisioning and on-premises deployments (see **Figure 1**). On-premises security systems are deployed, operated, and maintained on the client's side[11]. This requires the allocation of dedicated IT and human resource capacities. Service costs do not scale up or down with the actual degree of capacity utilization. None of the identified Cloud principles apply[14]. Managed Security Services are characterized in that a dedicated security service instance is set up for a client organization by an external service provider. This involves the

Table 1. Classification of SECaaS applications.

Application type	Description
Application security	Secure operation of software applications
	(e.g. application firewalls, code analyzers)
Compliance & IT Security management (ITSM)	Support of the client organization's compliance and IT security management
	(e.g. automatic compliance checks, benchmarking)
Content security	Protection of content data from intended attacks and undesired events
	(e.g. e-mail encryption, filtering of network traffic)
Endpoint security	Protection of servers or client computers in networks
	(e.g. malware protection, host-based intrusion detection)
Identity & access management	Identification of users, provisioning of user identity attributes and assignment of necessary privileges (e.g. single sign-on, multi-factor authentication)
Devices management	Remote management of client-sided security systems
	(e.g. intrusion detection and prevention systems)
Security information & event management (SIEM)	Specific security-related functions for monitoring complex IT systems
	(e.g. archiving and analysis of log-data, forensic analysis)
Vulnerability & threat management (VTM)	Detection of threats apart of eminent internal security incidents
	(e.g. patch management, notifications on current attacks)

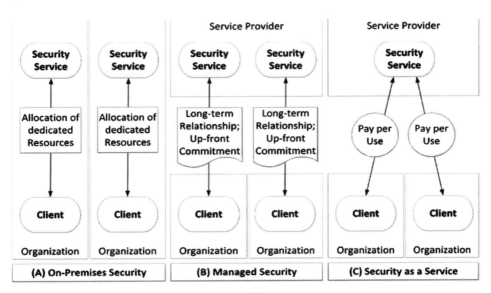

Figure 1. Security service delivery models.

prior negotiation of individual Service Level Agreements (SLA)[11]. In this regard, the provider is responsible for the operation and maintenance of the system[15][17]. Such security services do not provide for native multi-tenancy. Hence, the instant service use is not feasible and economies of scale are not exhausted[18]. Additionally, service usage may involve the deployment of dedicated software and hardware components and due to the initial effort required clients are often bound to providers by fixed-term licenses and up-front commitments[19]. Traditional managed service provisioning thus follows the Application Service Providing (ASP) model[12][14][20]. In contrast, Security as a Service solutions are fully operated and maintained by the service provider with no dedicated client-sided hardware or software necessary[11][14]. Full virtualization of the security service ensures the highest degree of capacity utilization. This makes the service usage highly cost-effective to the customer and enables fine-grained pay-per-use models. A virtualized multi-tenancy architecture not only enables the instant start of service use but also leverages inherent data aggregation benefits for service providers[14]. Moreover, operational and organizational flexibility is improved[11].

2.2. Adoption of Related Technologies

The term *Adoption* can be traced back to ROGERS' (1962) diffusion of innovations theory and is defined as a consumer's positive decision to accept and use an innovation, which ultimately leads to a positive investment decision and actual use[21]. Adopters can be individuals or organizations[22].

There are only a few current insights regarding the adoption of the outsourcing of IT security. GARTNER and FORRESTER RESE ARCH conducted analyses of the MSS market and forecasted a steady and significant growth[7][13]. Moreover, FORRESTER RESE ARCH surveyed IT security decision makers and identified major benefits of MSS[13]: Quality improvements, 24 × 7 support, cost reduction, and decrease of the complexity of security infrastructures. However, the study is not suitable regarding the research questions identified in this paper since the adoption was not investigated holistically and not focused on Cloud systems.

Benlian *et al.* conducted a meta-survey of the adoption of SaaS systems and applied different research theories[23]. They concluded that behavioral theories reveal more consistent results regarding the adoption of SaaS systems than economic or strategic research theories[23]. Behavioral theories include the *Technology Acceptance Model* (TAM)[24], the *Theory of the Diffusion of Innovation*[21], and the *Unified Theory of Acceptance and Use of Technology* (UTAUT)[22]. The results indicate that the adoption of SaaS technologies is mainly influenced by[23]:

- Social influences,

- Attitude toward the technology,

- Uncertainty of adoption,

- Strategic value of respective resources.

However, due to the underlying research design, these results do not provide for causality[23]. BENLIAN ET AL. also concluded that both the adoption and adoption drivers differ across application types, which should be considered in future research[23]. Previous research indicates a higher susceptibility to SaaS adoption for smaller and medium-sized companies[23] and a different perception of risks and potential benefits by large-scale organizations[25], although no correlation was discovered between company size and adoption[23]. Udoh applied a combined model including elements of UTAUT and TAM and observed that the adoption of grid, Cloud and related technologies can be causally explained by four predictors[26]:

- Effort expectations (Perceived ease of use),

- Risk expectations (Trust),

- Performance expectations (Perceived usefulness),

- Individual attitude.

Udoh's model provides a very high level of explanation which indicates a high aptitude for its application in similar technology acceptance studies[26]. Furthermore, its generic constructs can be itemized according to the specifics of subsequent research.

3. Research Design

Based on related studies[23][26], this paper applies the *Structural Equations Modeling* methodology. For the model estimation involved, the *Partial Least Squares* technique is used. The methodology is introduced and justified in Subsection 3.1. Subsequently, in Subsection 3.2, a system of hypotheses—the research model—is developed. In Subsection 3.3, the measurement model is derived from this research model.

3.1. Methodology

Common technology acceptance theories like TAM or UTAUT are based on the development and testing of hypotheses regarding the influences of theoretical constructs on each other[22]. A system of hypotheses can be modeled as a system of equations[27]. A common approach to solving such systems is *Structural Equations Modeling* (SEM)[27]. SEM is defined as "a comprehensive statistical approach to testing hypotheses about relations among observed and latent variables"[27]. Besides the structural model, which primarily prescribes hypothetical relations between latent variables, a measurement model is required to quantize these variables[27].

The measurement model prescribes not directly observed (latent) variables of the structural model by a set of measurable indicators[27]. Measurement models can be reflective or formative. Reflective measurement models assume empirically measurable variables. In this regard, the latent variable causes a set of reflective measurement indicators which correlate highly among each other[28]. In contrast, formative measurement models estimate a latent variable, applying a set of indicators, which are assumed to cause the construct[29]. This facilitates the differentiated analysis of the relevance and strength of certain influences on a theoretical construct[28]. Formative measures are mainly intended to explain the composition of a construct, whereas reflective measures only indicate a construct's outcome [28]. Therefore, on the one hand, formative measures lead to deeper practical insights than reflective ones and are more suitable for practical research applications[28]. On the other hand, such measurement models are restricted regarding the application of quality indicators[28]. To avoid this disadvantage, formative and reflective measures can be combined to form *Multiple Indicators, Multiple Causes* (MIMIC) models[28].

To estimate the comprehensive model either covariance-based approaches (CB-SEM) or the variance-based *Partial Least Squares* (PLS-SEM) technique can be applied[30]. Both approaches provide different benefits and drawbacks that imply their qualification for specific applications in research[28][30]. The PLS-SEM technique is more suitable for the research for this study due to four reasons: 1) the prediction-oriented research goal to explain the adoption of SECaaS (dependent variable) as comprehensively as possible; 2) the formative measurement of perceived overall risks and benefits which is required to get a deep and differentiated understanding of the composition of relevant adoption drivers; 3) the small sample size expected relative to the high complexity of the research model implied by the high number of hypothesized influences; 4) the possibility of applying fewer than four indicators for latent variables which is necessary to keep the study's questionnaire as purposive as possible[28].

The model estimation was performed using the software *SmartPLS* developed by Ringle *et al.*[31]. The tool facilitates the building of both structural and measurement models and was successfully applied in similar studies[32]. Further quality metrics were calculated using the statistics software *SPSS PASW Statistics*[b].

3.2. Research Model

In SEM, hypotheses are relationships between latent variables which are represented by the structural model[27][28]. The system of hypotheses must be theoretically well-grounded[28]. This was assured since its development was based on related literature in the fields of *Cloud Computing*, SaaS and MSS, and continuously validated by an expert group[c] (Below, this expert group is referred to as the *Expert Panel*) using a dedicated online discussion platform (*PBworks*[d]). The labels used for the study's constructs represent the essence of the construct and are assumed to be independent regarding their theorized content. Constructs and hypothesized influences are described and justified below.

3.2.1. Adoption

The endogenous variable Adoption depicts the degree to which a certain entity intends to use SECaaS. This includes both the plan for future deployment[22][26] and the present adoption by an organization[22]. In this regard, the behavioral intention to use a system and actual use can be modeled either separately or using a single construct[33]. We chose the second option to keep the model purposive. The measurement of this variable allows implications about the current state and future development of the SECaaS market. Thus, we utilize Adoption to cope with RQ1. The variable was hypothesized to be driven by the general determinants for grid and Cloud adoption identified[26]. These address RQ2 and represent the first four hypotheses (H) of this study as depicted in **Figure 2**: Adoption is significantly influenced by Perceived Ease of Use (H1), Perceived Usefulness (H2), Trust (H3), and Attitude (H4).

3.2.2. Perceived Ease of Use

This variable is defined as the degree to which the adopter believes that applying SECaaS is effortless[22][26][34][35]. From a client organization's point of

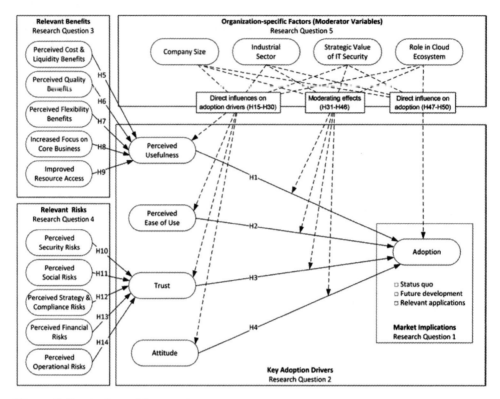

Figure 2. Research model.

view this involves the integration in the IT security infrastructure[11][15] as well as the actual use of the system[36]. Cloud-based security systems promise high ease of use since service interfaces are based on standardized internet technologies and can be accessed ubiquitously via thin clients (e.g. web browsers)[14]. It is questionable whether this fact affects the adoption and whether it is reflected by the perception of the adopters.

3.2.3. Perceived Usefulness

Perceived Usefulness is defined as the degree to which an organizational adopter believes that the application of SECaaS increases the performance of the organization[22][26][34][35]. Performance expectation is a key driver for adoption[22][34]. Based on related literature Benlian *et al.* identified five specific benefit dimensions for SaaS service consumers which are hypothesized for SECaaS according to RQ3[25]:

- Perceived Flexibility Benefits: The SaaS model implies a low organizational dependence of service consumers on service providers. Therefore, switching barriers are low and strategic flexibility regarding IT and IT security architectures is increased[11][25]. Furthermore, service use can be adapted flexibly to actual quantitative and qualitative needs[25].

- Improved Resource Access: Low entry barriers enable easy access to specific resources, skills and technologies of the external service providers[15][25]. Particularly mid-sized or smaller organizations might derive advantages from that when they cannot afford the time and effort involved in roviding sophisticated IT security resources on their own[11].

- Perceived Cost & Liquidity Benefits: Multi-tenancy architectures leverage economies of scale at the service providers' site. At the same time, service consumers' assumed low switching barriers induce a pricing pressure, forcing service providers to share respective savings. This ultimately leads to lower costs of operation and maintenance for service consumers[13][15][25]. Hence, on-demand pricing models enable decreased capital commitment[11][25]. Furthermore, the outsourcing model facilitates the transfer of financial risks of security incidents and thus the reduction of recovery costs[37].

- Perceived Quality Benefits: Security service providers use to be highly specialized, which implies their ability to provide a higher quality of service[11][13][15][16]. In addition, due to low switching barriers, service providers are forced to provide permanent high service quality[25]. Moreover, multi-tenancy architectures enable cross-client data aggregation and the application of business intelligence techniques[14]. Identified patterns can be used to improve quality of service, such as the performance of anti-virus applications, for instance. Lastly, SECaaS services are permanently up-to-date without the necessity of time-delayed updates at the client's site[11].

- Improved Focus on Core Business: The outsourcing of certain systems according to SaaS (or SECaaS) deallocates internal resources[11][25]. These resources can be (re-)allocated to an organization's core business, which might increase overall performance[15][25]—assuming that IT security is not the core competency. Hence, this is also one of the major drivers for IT security outsourcing in general[38].

Many IT outsourcing programs do not yield expected performance outcomes[39][40]. Reasons include exaggerated expectations, poorly developed business cases, deficient change management, non-transparency of vendor performance, and lock-in effects[40]. This so-called "IT outsourcing paradox"[40] might affect the expected usefulness of SECaaS and its influence on the adoption.

3.2.4. Trust

The adoption of grid and Cloud systems is highly influenced by perceived risks[18][23][26][41][42]. This influence is represented by the variable Trust, which is interpreted as a semantic inversion of perceived risk. BENLIAN ET AL. identified SaaS-specific risk dimensions which are hypothesized in analogy to Perceived Usefulness addressing RQ4[25]:

- Perceived Security Risks: The outsourcing of systems according to SaaS implies the loss of control over the processed data and requires the client organization to interface with the external service. This causes risks regarding the enterprise data and affected processes[11][15][25][42]. In this regard, Cloud-specific security risks focus on resource protection, communication and storage security, and authentication and authorization[2].

- Perceived Social Risks: The outsourcing of applications induces social risks including internal resistance or negative influences on the organization's image[25].

- Perceived Strategy & Compliance Risks: The outsourcing of certain systems might involve the loss of critical capabilities[25] and, in turn, the risk of an increased dependency on the service provider[15][37]. Furthermore, the service consumers might lose the control to ensure the compliance with legal and regulative requirements[15].

- Perceived Financial Risks: The deployment of SaaS services might involve unanticipated costs[25]. This includes the organization's own security infrastructure and service customization[11][15].

- Perceived Operational Risks: Since service operation is fully controlled externally there is the risk of the service provider not complying with existing SLAs. This might affect service quality, performance, and availability[15][25][42].

3.2.5. Attitude

The Attitude construct represents an adopter's individual positive or negative behavior toward an innovation and is considered to be independent from the other variables[22][26][43]. It can be prescribed by individual preferences or perceived relative advantage to related technologies[22][26]. Its relevance for SaaS adoption is indicated by previous research[23].

3.2.6. Moderator Variables

The validity of PLS-SEM results can be compromised by heterogeneous and conflicting data[30]. Potential sources for heterogeneity can be modeled and tested by means of moderator analyses[30]. Venkatesh *et al.* propose the use of moderators in addition to key determinants to account for dynamic influences and thus to improve the quality of adoption research models[22]. Moderators are variables that influence the relation between two constructs positively or negatively[22][28]. Moderators at the individual level include demographic characteristics and organizational context (e.g. gender, age)[22]. Since our research focuses on adoption by organizational entities, we hypothesized new moderators to address RQ5. As part of the aforementioned expert workshop in the course of a session of the "IT security solutions" working group of the German Federal Association for Information Technology, Telecommunications and New Media, and based on related literature, we identified four relevant factors: Company size, industrial sector, a company's role in the Cloud ecosystem, and the strategic value of IT security. Moreover, we considered the respondent's job function and the division in which he or she works as potential sources for heterogeneity and modeled respective moderator variables.

3.3. Measures

Based on the constructs of the specified structural model a measurement model was developed. Therefore, an initial literature review was conducted in order to identify and classify the major related indicators which semantically describe the structural model's constructs. These indicators were presented to the *Expert Panel* via the aforementioned collaboration system *PBworks*. The experts actively discussed and supplemented the indicator set which was subsequently revised by the authors of this paper and transformed to the study's online questionnaire. Finally, the *Expert Panel* approved[e] the measurement model (including the online questionnaire).

The measurement model includes both formative and reflective elements to account for the methodological problems mentioned in Subsection 3.1. We developed two-construct MIMIC models separating formative and reflective indicators for the major latent variables Perceived Ease of Use, Perceived Usefulness, Trust, and Adoption. Within the structural equations model the latent variable is represented by a reflective construct; one or more formative constructs model the composition of the variable[44]. For the reflective (refl.) constructs, existing published measures were applied. Formative (form.) constructs and indicators were based on related work and both specified and validated by the *Expert Panel*. The MIMIC models for Perceived Usefulness and Trust each consist of one reflective and several formative elements representing the respective benefit and risk dimensions as depicted in **Figure 2**. In this regard, operational and financial risks were merged to one variable. The remaining two variables were measured linking only one formative element. **Table 2** provides an overview of all indicators of the study's primary variables and the variable Strategic Value of IT Security. The Company Size was determined by means of the company turnover and headcount. The remaining organization-specific factors were each measured by one global indicator with nominal scale.

All indicators were transformed into questionnaire items in German following general construction guidelines[28][46]. As mentioned in the beginning of this section, the supporting expert group validated the wording and soundness of all items as well as the structure of the entire questionnaire from a semantic point of view as suggested by CHURCHILL[47]. SEM requires metrically-scaled data for

Table 2. Metrically measured indicators.

Construct	Indicator(s)	Reference(s)
Adoption (refl.)	Actual use	[22,23,26]
	Use intent short-term (next 3 years)	
	Use intent mid-term (4–7 years)	
	Use intent long-term ($\geq$ 7 years)	
Adoption (form.)	Actual use/intent of Endpoint Security applications	[13,14,23]
	Actual use/intent of content security applications (appl.)	
	Actual use/intent of application security applications	
	Actual use/intent of compliance & IT security management appl.	
	Actual use/intent identity & access management appl.	
	Actual use/intent of managed devices applications	
	Actual use/intent of security information & event management appl.	
	Actual use/intent of threat & vulnerability management appl.	
Perceived ease of use (refl.)	General ease of use	[26,34]
	Ease of learning	
	Ease of target achievement	
Perceived ease of use (form.)	Ease of initial integration/deployment of the service	[13-15,36]
	Usability of the service	
	Ease of customizing the service	
	Comprehensive support by service provider	
Perceived usefulness (refl.)	Increase in performance	[26,34]
	General usefulness	
	Increase in effectiveness	
Perceived cost & liquidity benefits (form.)	Reduction in costs of operation and maintenance	[11,13-15,25,37]
	Variabilization of IT security costs	
	Reduction in recovery costs	
Perceived quality benefits (form.)	Transparency & control of security department	[11,13-16,25]
	Increase in organizational level of security	
	Improvement of legal and regulative compliance	
Perceived flexibility benefits (form.)	Flexibility of IT and security processes	[11,13,14,25]
	Flexibility of business processes	
	Reactivity regarding security-related problems	
Increased focus on core business (form.)	Decrease in employee errors	[11,13-15,25,37,38]
	Time savings in security management	
Improved resource access (form.)	Enablement of access to new technologies	[11,15,25]
	Access to external know-how	
	Independence from dedicated systems	
Trust (refl.)	Overall trust in adoption	[22,23,26]
	Trust in certified service providers	
	Hesitation due to uncertainty	
Perceived security risks (form.)	Vulnerability to unauthorized service access	[2,11,14]
	Deficient data mitigation and security	
	Vulnerability regarding network-based attacks	
	Deficient service continuity	

Continued

Perceived strategy & compliance risks (form.)	Dependence on service providers	[14,15,25,37]
	Inability to comply with obligations to produce supporting documents	
	Non-compliance with data protection regulations	
Perceived social risks (form.)	Internal resistance	[23,25]
	Loss of image	
Perceived financial & operational risks (form.)	Unexpected costs of integration	[11,15,25,42,45]
	Deficient provider's compliance with SLAs	
Attitude (refl.)	General attitude toward cloud technologies	[22,23,26]
	Relative advantage over managed security	
	Relative advantage over on-premises systems	
Strategic value of IT security (refl.)	Criticality of IT security for business	[23]

further analysis[28]. Thus, we applied a systematically constructed seven-point Likert scale, which produces data that can be interpreted metrically for SEM model estimations[28][46].

4. Findings

This section presents the empirical investigation of the research model. In Subsection 4.1, the sample and the process of data collection are described. In Subsection 4.2, implications regarding the market for SECaaS applications are deduced from descriptive data analysis. In Subsection 4.3, the model estimation including quality and hypotheses testing is laid out. Finally, in Subsection 4.4, the results are discussed respecting the research questions of the study.

4.1. Data Collection and Sample

To carry out the data collection, the measurement model was implemented in an online survey tool and validated by a pre-test with 12 voluntary experts of the *Expert Panel*. In the period from 16 February to 15 April 2011 the survey was accessible via a dedicated Internet address. This address was distributed using the network of the German Federal Association for Information Technology, Telecommunications and New Media. The target population for this study was IT and business professionals who would be involved in their organization's decision-making process regarding the investment in SECaaS. The sample included key informants of provider and consumer organizations of IT solutions in Germany, Austria, and Switzerland and is considered to be representative for potential adopters of SECaaS technologies in the German-speaking area. The survey was preceded by a brief registration. The verified e-mail addresses were stored separately from the survey data in order to guarantee anonymity. The survey (see Additional file 1) began with a landing page including a brief definition of SECaaS, the purpose of the study and its target population, the estimated time expenditure, information about incentives, the privacy policy, and contacts. The questionnaire was divided into six sections.

The survey yielded 202 returns. The data was processed and cleaned as suggested by WEIBER AND MÜHLHAUS[28]. Accordingly, incomplete records were excluded. For the remaining records the squared Mahalanobis distances were calculated in order to identify those deviating markedly from the centroid; three outliers were identified and excluded. This left 160 records for further analysis.

The composition of the sample is depicted in **Table 3**, **Table 4**, **Table 5**, **Table 6** and **Table 7** by means of the measured non-metrically scaled organization-specific factors. Of the participating companies, 44.4% originate from the IT sector, which indicates its relatively higher affinity toward SECaaS compared to other industrial sectors (see **Table 3**). Almost half of the sample (48.1%) consists of large-scale organizations. Medium-sized and smaller organizations each represent about a quarter (see **Table 4**). Regarding a company's primary role in the Cloud ecosystem, most respondents (53.8%) evaluate their organization to act as exclusive potential consumer of SECaaS. Further 23.1% see their organization in a hybrid role, potentially doing both consuming and providing SECaaS. Only 23.1% do not consider to consume SECaaS at all (see **Table 5**). **Table 6** shows the composition of the sample grouped by the respondents' job function. The majority (37.5%) has an executive position. Also in this regard, most respondents originate from their organization's IT department. Other divisions are under-represented (see **Table 7**). The variable Strategic Value of IT Security was measured by one reflective item on a seven-point *Likert* scale (Mean: 6.244, standard deviation: 1.080). We assume the sample to be compliant with the study's target population and thus representative for the adoption of SECaaS.

Table 3. Participants by industrial sector.

Class	Percentage	Number
Information technology	44.4%	71
Industry	16.9%	27
Other services	13.1%	21
Public services	10.6%	17
Financial services	8.8%	14
Retail	6.3%	10

Table 4. Participants by company size[f].

Class	Percentage	Number
Small & micro organization	24.4%	39
Medium-sized organization	27.5%	44
Large-scale organization	48.1%	77

Table 5. Participants by role in cloud ecosystem.

Class	Percentage	Number
Cloud service Consumer (exclusively)	53.8%	86
Cloud service Provider (exclusively)	23.1%	37
Cloud service provider and consumer	23.1%	37

Table 6. Participants by job function.

Class	Percentage	Number
Manager	37.5%	60
Employee	21.9%	35
IT security officer	13.8%	22
Other	26.9%	43

Table 7. Participants by division.

Class	Percentage	Number
Steering committee	10.6%	17
Management and support	17.5%	28
Research & development	5.0%	8
Production	2.5%	4
Information technology	50.6%	81
Sales & marketing	12.5%	20
Other	1.3%	2

4.2. Market Implications

The measurement of the study's dependent variable Adoption reveals several implications about the market for SECaaS (RQ1). It was measured by two constructs: a reflective element covering the actual use and use intent of SECaaS in regard to the planning horizon, and a formative one containing different security application types. 18.1% of the respondents indicate that their organization is currently using SECaaS. The data further indicates a steadily rising adoption rate. In the long term, over 30% of the surveyed organizations clearly intend to use SECaaS. Only 16% exclude the possibility of its use entirely. The positive development of the adoption of SECaaS in regard to the organizations' planning horizon is shown in **Table 8**.

Respondents were asked whether their organization uses or intends to use certain SECaaS application types. *Content Security*, *Endpoint Security*, and *Vulnerability & Threat Management* solutions exhibit especially high adoption rates. Whereas the market is already dominated by the first two application types, the diffusion of Cloud-based *Vulnerability & Threat Management* services is relatively low[14]. This indicates a high level of future adoption, especially for this type of application. Hence, the data indicate a very weak current and future interest in *Compliance & IT Security Management* products. Other types of application are middle-ranking. The overall results and detailed findings specific to industrial sector and company size are summarized in **Table 9**.

Table 8. Development of Adoption[f].

Planning horizon	Percentage with strong positive indication for adoption[a]	Standard deviation	Mean
Currently	18.1%	2.800	2.220
Short-term	24.4%	3.444	2.214
Mid-term	26.3%	3.988	1.939
Long-term	31.3%	4.269	1.856

Table 9. Application-specific adoption.

Application type	IT	Public services	Financial services	Industry	Other services	Retail	Large-scale organizations	Mid-sized organizations	Small & micro org.	Overall
Content security	●	●	●	●	◑	●	●	◑	●	●
Endpoint security	●	◑	●	◑	◑	◐	●	◑	◑	◑
Vuln. & threat management	●	◑	◐	◑	◕	○	◑	◑	◐	◑
Application security	◑	◑	◕	◑	◐	◐	◐	◑	◑	◐
Identity & access mgmt.	◑	◑	◕	◕	◐	◕	◑	◕	◐	◐
Security info. & event mgmt.	◑	◐	◐	◕	○	◕	◑	◕	◕	◐
Managed devices	◐	◕	◕	◕	◕	◕	◐	◕	◕	◐
Compliance & ITSM	◐	◐	◕	◕	◕	○	◕	◕	◕	◕
			Industrial sector				Org. size			

Very low (0%, ○), low (25%, ◐), medium (50%, ◑), high (75%, ◕), and very high (100%, ●) adoption rate. The range of mean values (range = [1.857; 4.364]) was linearly grouped into five equal classes.

4.3. Model Estimation

No adequate global indicators for goodness of model fit exist for PLS-SEM[30][48]. Instead, the measurement model should be estimated first, the structural model afterward[30][49].

4.3.1. Evaluation of the Measurement Model

The assessment of reflective measurement models includes their reliability and validity[28][30]. An indicator's reliability can be assumed for a minimal factor loading of 0.7[30][50]. For values between 0.4 and 0.7 indicators should only be eliminated if this increases the construct's *Average Variance Extracted* (AVE) value[30]. Indicators with lower factor loadings indicate deficient reliability and should be removed[30]. According to these requirements, all reflective indicators featured reliability. The reliability of a construct is routinely estimated by means of its *Composite Reliability* (CR) indicating its internal consistency[30]. CR values should be 0.7 or higher[30][49]. This requirement is met for all latent variables. The assessment of the validity of the measurement model includes discriminant validity[g] and convergent validity[h][30]. Discriminant validity was verified by calculating cross loadings[30]. All indicators have a higher correlation with the appertaining latent variable than with others and thus provide for discriminant validity. Furthermore, for each variable the AVE exceeds the minimal value of 0.5, which indicates the convergent validity of the entire measurement model[30][51]. **Table 10** summarizes key metrics of the study's major latent variables including CR and AVE values. For Attitude no AVE value was calculated since no formative measures were applied.

The evaluation of formative indicators aims to support their relevance for the measured construc[30]. This includes the strength and the significance of its influence[28]. The significance can be determined by means of the bootstrap procedure calculating t-values[30]. A formative indicator's influence is assumed to be significant for a t-value = 1.646 (level of significance $\alpha = 10\%$, degrees of freedom df = 1,000). Weights should be 0.1 or higher. Though the results quantitatively indicate the minor relevance of four formative indicators they did not have to be excluded since the *Expert Panel* strongly supported their inclusion from a qualitative point of view[28][30]. **Table 11** summarizes weights and significances of the formative indicators for Perceived Usefulness grouped by the corresponding benefit dimension. Additionally, the significance of the influence of the respective dimension on the core construct Perceived Usefulness (refl.) is shown. In analogy, **Table 12** describes the MIMIC measurement model for Trust. The measurement model does not contain any redundant formative indicators: the *Variance Inflation Factor* (VIF) indicating multi-colinearity was calculated for all indices. All values [range (1.3–4.2)] are smaller than the critical value of 5.0; thus, no indicators had to be reconsidered[30][44].

Table 10. Latent variables.

Variable	Mean	Standard deviation	R^2	AVE	CR
Adoption	3.444	4.733	.710	.729	.915
Perceived usefulness	4.476	2.983	.663	.849	.944
Perceived ease of use	4.151	2.212	.737	.829	.936
Trust	3.910	3.109	.512	.590	.805
Attitude	3.788	3.053	.587	n.a.	.810

Table 11. Formative measurement model for P. usefulness.

Formative construct	Indicator weights	Indicator significances (t-Values)	Significance of construct (t-Value)
Cost & liquidity	.270	2.052	2.319**
Benefits	-.631	-4.576	
Quality	.290	2.420	3.632***
Benefits	-.537	-5.138	
Flexibility	.276	1.957	0.754
Benefits	-.613	-5.717	
Focus on	.492	3.044	0.225
Core business	-.601	-3.690	
Improved	.164	1.830	0.700
Resource access	-.635	-5.055	

Table 12. Formative measurement model for trust.

Formative construct	Indicator weights	Indicator significances (t-Values)	Significance of construct (t-Value)
Security	.182	1.658	4.450***
Risks	-.453	-3.723	
Social	.278	1.660	1.984***
Risks	-.851	-7.291	
Strategy &	.288	1.706	1.742*
compl. risks	-.524	-4.057	
Financial &	.474	3.375	0.841
op. risks	-.755	-6.918	

Since the measurement model meets existing requirements entirely, valid estimations of the study's latent variables can be assumed. This is requisite for the subsequent evaluation of the structural model[28][30].

4.3.2. Evaluation of the Structural Model

The evaluation of the structural model includes the degree of determination of the model's latent variables and the evaluation of the hypothesized relations between them[30]. All independent latent variables meet the required minimal coefficient of determination (R2) value of 0.3 and are thus sufficiently explained[49] (see **Table 10**). Due to the study's predictive research goal the R2 of the dependent variable is of special importance[30]. CHIN suggests a critical value of 0.67 for substantial predictions[49]. This requirement is met for the study's dependent variable Adoption (R2 = 0.71). We additionally proved the model's capacity to predict the dependent variable by means of the Stone-Geisser test (cross-validated redundancy Q2 = 0.489 > 0)[30]. Thus, we consider the adoption of SECaaS to be explained comprehensively by this study's proposed model. To test the significances of the model's hypothesized relations, the bootstrap method (df = 1000) was applied and t-values were calculated[31][52]. In regard to the study's non-directional hypotheses, the influence of one variable on another is considered to be significant when α = 10%[28][30]. Thus, a hypothesis is supported when the corresponding t-value ≥ 1.646 and the respective null hypothesis is falsified [28]. To get a deeper understanding of the relations we tested three levels of significance: α = 10% (*, t-value = 1.646); α = 5% (**, t-value = 1.962); α = 1% (***, t-value = 2.581). Moreover, corresponding path coefficients were calculated indicating both strength and direction of a variable's influence[28]. According to Lohmoeller path coefficients ≥ 0.1 indicate relevance[53].

All in all 10 of the original 50 hypotheses were supported. Of the hypothesized key determinants only Attitude does not have a significant influence on Adoption. This might be caused by the consideration of the individual organization-specific factors; however, we did not find any relations between these variables and Attitude. The investigation of the MIMIC measurement models of Perceived Usefulness and Trust revealed that the respective constructs are determined by very few factors. Only quality, and cost and liquidity benefits matter significantly for Perceived Usefulness. Ex post we identified the significant influence of Trust and Perceived Ease of Use on the variable Perceived Usefulness. Trust is negatively correlated with security risks, social risks, and strategy and compliance risks. Financial and operational risks do not matter. Investigating the organization- specific factors we identified a moderating effect of Role in Cloud Ecosystem on the relationship between Perceived Usefulness and Adoption, and a direct (positive) influence of the Strategic Value of IT Security on Perceived Usefulness. Company Size and Industrial Sector do have neither a direct nor indirect influence on the adoption of SECaaS. However, in the course of a more detailed analysis we found that financial risks matter more for bigger organizations, while social, operational, and strategy and compliance risks are more important to smaller organizations. **Figure 3** depicts the reduced model including supported hypotheses. The path coefficients and significances of the hypothesized key drivers are summarized in **Table 13**.

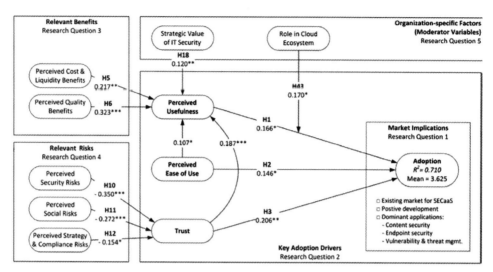

Figure 3. Final model.

Table 13. Estimation of hypothesized key drivers.

Hypo-thesis	Relation	Path coefficient	Significance (t-Value)
H1	Perceived usefulness ⇒ adoption	.168	1.912*
H2	Perceived ease of use ⇒ adoption	.146	1.818*
H3	Trust ⇒ adoption	.207	2.333**
H4	Attitude ⇒ adoption	.121	1.453

4.4. Discussion

Below, the findings are discussed in regard to the research questions considering related findings.

4.4.1. RQ1: Is There a Market for SECaaS Enterprise Applications in General and for Specific Application Types in Particular?

The market for SECaaS applications is still emerging. The study indicates an already significant and steadily growing acceptance by enterprise consumers. The adoption varies across different security service application types, which supports previous findings about SaaS[23]. The market's focus is on applications for *Content Security*, *Endpoint Security*, and *Vulnerability & Threat Management*.

4.4.2. RQ2: Which Are the Key Drivers and Inhibitors for the Adoption of SECaaS?

Key drivers for the adoption of SECaaS are effort expectancies, perceived usefulness, and trust regarding the adoption of respective applications. These results basically confirm UDOH's findings regarding the adoption of grid and Cloud technologies[26]. Only the influence of the adopter's individual attitude toward the technology[23][26] was not supported for this research context. Hence, the influence of perceived risks and thus the uncertainty of SECaaS adoption is more significant than the influence of the other drivers, including perceived usefulness. This supports the findings of BENLIAN ET AL. regarding SaaS adoption[23].

4.4.3. RQ3: Which Benefits Are Perceived to be Relevant by Potential Adopters of SECaaS?

The perceived usefulness of SECaaS is forged by quality as well as cost and liquidity benefits. Quality benefits mainly reflect the expected return in terms of an increased level of security and regulative compliance. Cost and liquidity benefits include the reduction of direct security expenditures and recovery costs. Thus, according to the adopter's perception, SECaaS potentially increases return on security investments[54]. Hence, the expected performance of SECaaS systems is positively correlated with effort expectancies and trust, which supports previous empirical findings[22][55][56].

4.4.4. RQ4: Which Risks Are Perceived to be Relevant by Potential Adopters of SECaaS?

Major barriers to SECaaS adoption are perceived security, social, strategy and compliance risks. Perceived social risks are mainly driven by expected internal resistance. In this context, an inherent problem is the possible fear of the direct loss of competencies in the course of outsourcing certain security systems. The significance of social influences regarding SaaS adoption was already identified by BENLIAN ET AL.[23]. To increase trust and thus future adoption the effectiveness of technical and organizational controls securing Cloud-based security services must be conveyed transparently to potential SECaaS consumers. Specific certification programs for service providers might support this, for example.

4.4.5. RQ5: Which Organization-Specific Factors Affect the Acceptance of SECaaS?

The individual strategic value of IT security for an organization's business directly influences the perceived usefulness of SECaaS. The expected performance is thus higher for organizations with higher demands on IT security from a business point of view. This coherence was already laid out by BENLIAN ET AL. regarding SaaS[23]. Moreover, for the organization's role in the Cloud ecosystem a moderating effect on the relation between the variables Perceived Usefulness and Adoption was identified. This means that perceived benefits matter less for the actual adoption of SECaaS technologies when the organization itself provides Cloud services for external customers, acting in the role of a Value Added Reseller[57]. On the contrary, general organization-specific factors like company size or industrial sector do not have any significant effects on the adoption. This, on the one hand, conflicts with the general rationale that SECaaS is particularly relevant for companies with limited capacities regarding IT security; on the other hand, however, it confirms previous findings regarding the adoption of SaaS[23].

4.4.6. Limitations

The applied methodology (PLS-SEM) is often criticized because calculations tend to be less precise compared to alternative CB-SEM techniques. However, PLS-SEM is more qualified for the application in this study as already laid out in Subsection 3.1. Considering the complexity of this study's research model and the achievable sample size, the application of CB-SEM would not have revealed valid results (compare e.g.[27][28][30]), which supports the authors' research design decision. The sample was selected among organizations with an existing affinity toward IT. Therefore we assume the sample to be representative for potential SECaaS adopters but not for all organizations. The survey explicitly addressed companies in the German-speaking area. Even though it is assumed to provide general insights about the adoption of SECaaS, observations might vary among different markets, for instance due to location-specific data protection regulations. Furthermore, the adoption of SECaaS by private consumers has not been considered and thus remains open for future research.

5. Conclusion

This paper systematically investigates the adoption of SECaaS. An application-specific research model was developed based on existing technology acceptance models. The model was estimated applying the *Partial Least Squares* technique to address the prediction-oriented nature of the study. Based on 160 valid responses from companies in the German-speaking area, we investigated the market potential for SECaaS, key adoption drivers, the relevance of certain risks and benefits, and the influence of organization-specific factors like company size or industrial sector.

The results make valuable contributions for both practice and research. They provide a benchmark for potential adopters of SECaaS. Moreover, the findings support the understanding of the adoption behavior of enterprise consumers. Service providers can use this understanding to direct research, development and marketing programs by considering the significance of perceived security-related risks, for instance. Therefore, this study contributes to driving the future adoption of SECaaS, addressing existing threats to the security of enterprise information systems. Moreover, the developed research model including its measures was validated and can be applied for related future studies.

Future research should reflect the adoption of Cloud- based security services in other markets, survey specific security application types, investigate most relevant application fields and success drivers.

Endnotes

[a]Session of the "IT security solutions" working group of the German Federal Association for Information Technology, Telecommunications and New Media

(BITKOM e.V., see: http://www.bitkom.de last access: 01 August 2012).

[b]http://www.spss.com.hk/statistics/ (last access: 29 September 2012).

[c]16 selected IT and IT security professionals of the German Federal Association for Information Technology, Telecommunications and New Media (BITKOM e.V.).

[d]http://pbworks.com (last access: 01 August 2012).

[e]Therefore, a simple online poll with the options "I do not approve", "I am ok" and "I fully approve" was conducted. Four experts responded "I am ok" and eight fully approved.

[f]According to the European Commission, the number of a company's employees and its turnover (alternatively: balance sheet total) indicate its size. Companies are categorized as follows: "micro", when number of employees < 10 and turnover ≤ €2,000,000; "small", when number of employees < 50 and turnover ≤ €10,000,000; "medium- sized", when number of employees < 250 and turnover ≤ €50,000,000. Larger organizations are labelled as "large- scale". See: http://ec.europa.eu/enterprise/policies/sme/facts-figures-analysis/sme-definition (last access: 01 August 2012).

[g]Discriminant validity is provided for when an indicator's loading with the assigned construct is higher than with remaining constructs. An indicator's loading with non-assigned constructs is referred to as cross loading.

[h]Convergent validity expresses the degree to which a latent variable explains the variance of assigned indicators.

Source: Senk C. Adoption of security as a service [J]. Journal of Internet Services & Applications, 2013, 4(1): 1–16.

References

[1] Hommel W (2007) Architekturund Werkzeugkonzepte für föderiertes Identitäts-

Management. Dr. Hut, München.

[2] Brock M, Goscinski A (2010) Toward a framework for cloud security. In: Hsu CH, Yang L, Park J, Yeo SS (eds) Algorithms and architectures for parallel processing, Lecture Notes in Computer Science, vol. 6082. Springer, Berlin/Heidelberg, pp 254–263.

[3] Rittinghouse J, Ransome J (2010) Cloud computing: implementation, management, and security. CRC, Boca Raton.

[4] Mell P, Grance T (2009) The nist definition of cloud computing. Natl Ins Stand Technol 53 (6): 50. http://csrc.nist.gov/publications/nistpubs/800- 145/SP800-145.pdf.

[5] Furth B (2010) Cloud computing fundamentals. In: Furth B, Escalante A (eds) Handbook of Cloud Computing. Springer, US, Boston, pp 3–20.

[6] Höfer C, Karagiannis G (2011) Cloud computing services: taxonomy and comparison. J Internet Serv Appl 2(2): 1–14.

[7] Gartner Gartner says security delivered as a cloud-based service will more than triple in many segments by 2013 (2008). www.gartner.com/it/page. jsp?id=722307.

[8] Hafner M, Mukhtiar M, Breu R (2009) Seaas—a reference architecture for security services in soa. JUCS 15(15): 2916–2936.

[9] Peterson G (2009) Service-oriented security indications for use. Comput Sci Eng 7: 91–93.

[10] Smith DM (2010) Hype cycle for cloud computing 2010. http://www. gartner.com/DisplayDocument?doc_cd=201557.

[11] Staudenrauss P (2011) Untersuchung und Bewertung von Security-as-a-Service-Diensten. In: Helmbrecht U, Kretzschmar M, Eiseler V (eds) Seminar IT-Sicherheit-Sicherheit und Vertrauen in Cloud Computing. Institut füur Technische Informatik, München, pp 49-74.

[12] Mather T, Kumaraswamy S, Latif S (2009) Cloud security and privacy: an enterprise perspective on risks and compliance. O'Reilly Media Inc, Sebastopol.

[13] Kark K, Penn J, Whiteley R, Coit L (2010) Market overview: Managed security services. http://www.forrester.com/Market+Overview+Managed+Security+Services/fulltext/-/E-RES56068?objectid=RES56068.

[14] Senk C, Holzapfel A (2011) Market overview of security as a service systems. In: Pohlmann N, Reimer H, Schneider W (eds) ISSE 2011 Securing Electronic Business Processes.

[15] Allen J, Gabbard D, May C (2003) Outsourcing Managed Security Services. Carnegie Mellon University Software Engineering Institute. http://books. google.de/books?id=CFGnNwAACAAJ.

[16] Deshpande D (2005) Managed security services: an emerging solution to security In:

Proceedings of the 2nd annual conference on Information security curriculum development, InfoSecCD '05. ACM, New York, pp 107–111.

[17] Keuper F, Wagner B, Wysuwa H (2009) Managed services: IT-Sourcing der nächsten Generation. Gabler. http://books.google.de/books?id= 5J7Qhx?GIYIC.

[18] Martens B, Teuteberg F (2011) Decision-making in cloud computing environments: A cost and risk based approach. Inf Syst Front 14(4): 1–23.

[19] Huber M (2002) IT-security in global corporate networks. Center for Digital Technology and Management, München.

[20] Karyda M, Mitrou E, Quirchmayr G (2006) A framework for outsourcing is/it security services. Inf Manag Comput Secur 14(5): 402–415.

[21] Rogers E (2003) Diffusion of Innovations. 5th Edition. Simon & Schuster, New York.

[22] Venkatesh V, Morris MG, Davis GB, Davis FD (2003) User acceptance of information technology: Toward a unified view. MIS Q 27(3): 425–478.

[23] Benlian A, Hess T, Buxmann P (2009) Drivers of saas-adoption—an empirical study of different application types. Business Inf Syst Eng 1: 357–369.

[24] Davis FD, Bagozzi RP, Warshaw PR (1989) User acceptance of computer technology: A comparison of two theoretical models. Manage Sci 35(8): 982–1003.

[25] Benlian A, Hess T, Buxmann P (2010) Software-as-a-Service: Anbieterstrategien, Kundenbedürfnisse und Wertschöpfungsstrukturen. Gabler, Betriebswirt.-Vlg.

[26] Udoh E (2010). VDM Verlag Dr. Müller e.K., Saarbrücken.

[27] Hoyle R (1995) Structural equation modeling: concepts, issues and applications. Sage Publications, New York.

[28] Weiber R, Mühlhaus D (2010) Strukturgleichungsmodellierung: Eine anwendungsorientierte Einführung in die Kausalanalyse mit Hilfe von AMOS, SmartPLS und SPSS. Springer, Berlin, Heidelberg.

[29] Edwards JR, Bagozzi RP (2000) On the nature and direction of relationships between constructs and measures. Psychol Methods 5(2): 155–174.

[30] Hair J, Ringle CM (2011) PLS-SEM: Indeed a silver bullet. J Mark Theory Pract 19(2): 139–151.

[31] Ringle CM, Wende S, Will A (2005) Smartpls 2.0. http://www.smartpls.de.

[32] Bliemel F (2005) Handbuch PLS-Pfadmodellierung: Methoden, Anwendung, Praxisbeispiele. Schäffer-Poeschel, Stuttgart, Germany.

[33] Hamre LJ (2008) Exploring the use of social capital to support technology adoption and implementation. Ph.D. thesis, University of Bath.

[34] Davis FD (1989) Perceived usefulness, perceived ease of use, and user acceptance of information technology. MIS Q 13(3): 319–340.

[35] Thompson RL, Higgins CA, Howell JM (1991) Personal computing: Toward a conceptual model of utilization. MIS Q 15(1): 125–143.

[36] Cranor L, Garfinkel S (2005) Security and Usability. O'Reilly Media, Inc, Sebastopol.

[37] Böhme R (2010) Security metrics and security investment models In: Echizen I, Kunihiro N, Sasaki R(eds) Advances in Information and Computer Security, *Lecture Notes in Computer Science*, vol. 6434. Springer, Berlin/Heidelberg, pp 10–24.

[38] Schwarze L, Müller, PP (2005) IT-outsourcing-Erfahrungen, status und zukünftige herausforderungen. HMD-Praxis der Wirtschaftsinformatik. http://ephorie.de/pdfs/Schwarze_IT-Outsourcing-Erfahrungen_Status_und_zukuenftige_Herausforderungen.pdf.

[39] Aubert BA, Patry M, Rivard S (1998) Assessing the risk of IT outsourcing In: Proceedings of the Thirty-First Hawaii International Conference on System Sciences, Volume VI. Organizational Systems and Technology. IEEE Computer Society, Washington, U.S.A., pp 685–693.

[40] Rouse AC (2009) Is there an "Information technology outsourcing Paradox"? In: Hirschheim R, Heinzl A, Dibbern J (eds) Information systems outsourcing. Springer, Berlin Heidelberg, pp 129–146.

[41] Duisberg A (2011) Gelöste und ungelöste Rechtsfragen im IT-Outsourcing und Cloud Computing In: Picot A, Götz T, Hertz U (eds) Trust in IT. Springer, Berlin Heidelberg, pp 49–70.

[42] Subashini S, Kavitha V (2011) A survey on security issues in service delivery models of cloud computing. J Netw Comput Appl 34(1): 1–11.

[43] Fishbein M, Ajzen I (1975) Belief, attitude. Addison-Wesley, Reading.

[44] Diamantopoulos A, Winklhofer HM (2001) Index construction with formative indicators: An alternative to scale development. J Mark Res 38(2): 269–277.

[45] Senk C (2010) Securing inter-organizational workflows in highly flexible environments through biometrics In: Proc. of E C I S Pretoria.

[46] Bortz J, Döring N (2006) Forschungsmethoden und Evaluation für Human- und Sozialwissenschaftler Springer-Lehrbuch. Springer.

[47] Churchill GA (1979) A paradigm for developing better measures of marketing constructs. J Mark Res 16(1): 64–73.

[48] Hulland J (1999) Use of partial least squares (pls) in strategic management research: a review of four recent studies. Strateg Manage J 20(2): 195–204.

[49] Chin WW (1998) The partial least squares approach to structural equation modeling. Modern Methods Business Res 295: 336.

[50] Johnson MD, Herrmann A, Huber F (2006) The evolution of loyalty intentions. J

Mark 70(2): 122–132.

[51] Fornell C, Bookstein FL (1982) Two structural equation models: Lisrel and pls applied to consumer exit-voice theory. J Mark Res 19(4):440–452.

[52] Nevitt J, Hancock GR (2001) Performance of bootstrapping approaches to model test statistics and parameter standard error estimation in structural equation modeling. Struct Equation Model Multidisciplinary J 8(3): 353–377.

[53] Lohmoeller JB (1989) Latent variable path modeling with partial least squares. Physica, Heidelberg.

[54] Sonnenreich W, Albanese J, Stout B (2005) Return On Security Investment (ROSI) In: A practical quantitative model Journal of research and practice in information technology. INSTICC Press, Setubal, pp 239–252.

[55] Lee D, Park J, Ahn J (2001) Proceedings of the International Conference of Information Systems 2001 In: On the explanation of factors affecting E-commerce adoption, pp 109–120.

[56] Venkatesh V, Bala H (2008) Technology acceptance model 3 and a research agenda on interventions. Decis Sci 2: 273–315. 39.

[57] Baun C, Kunze M, Nimis J (2010) Cloud computing Web-basierte dynamische IT-services. Springer-Verlag, Berlin and Heidelberg.

Chapter 3
A Quantitative Analysis of Current Security Concerns and Solutions for Cloud Computing

Nelson Gonzalez[1], Charles Miers[1,2], Fernando Redígolo[1],
Marcos Simplício[1], Tereza Carvalho[1], Mats Näslund[3], Makan Pourzandi[4]

[1]Escola Politécnica at the University of São Paulo (EPUSP), São Paulo, Brazil
[2]State University of Santa Catarina, Joinville, Brazil
[3]Ericsson Research, Stockholm, Sweden
[4]Ericsson Research, Ville Mont-Royal, Canada

Abstract: The development of cloud computing services is speeding up the rate in which the organizations outsource their computational services or sell their idle computational resources. Even though migrating to the cloud remains a tempting trend from a financial perspective, there are several other aspects that must be taken into account by companies before they decide to do so. One of the most important aspect refers to security: while some cloud computing security issues are inherited from the solutions adopted to create such services, many new security questions that are particular to these solutions also arise, including those related to how the services are organized and which kind of service/data can be placed in the cloud. Aiming to give a better understanding of this complex scenario, in this article we identify and classify the main security concerns and solutions in cloud computing, and propose a taxonomy of security in cloud computing, giving an overview of the current status of security in this emerging technology.

1. Introduction

Security is considered a key requirement for cloud computing consolidation as a robust and feasible multipurpose solution[1]. This viewpoint is shared by many distinct groups, including academia researchers[2][3], business decision makers[4] and government organizations[5][6]. The many similarities in these perspectives indicate a grave concern on crucial security and legal obstacles for cloud computing, including service availability, data confidentiality, provider lock-in and reputation fate sharing[7]. These concerns have their origin not only on existing problems, directly inherited from the adopted technologies, but are also related to new issues derived from the composition of essential cloud computing features like scalability, resource sharing and virtualization (e.g., data leakage and hypervisor vulnerabilities)[8]. The distinction between these classes is more easily identifiable by analyzing the definition of the essential cloud computing characteristics proposed by the NIST (National Institute of Standards and Technology) in[9], which also introduces the SPI model for services (SaaS, PaaS, and IaaS) and deployment (private, public, community, and hybrid).

Due to the ever growing interest in cloud computing, there is an explicit and constant effort to evaluate the current trends in security for such technology, considering both problems already identified and possible solutions[10]. An authoritative reference in the area is the risk assessment developed by ENISA (European Network and Information Security Agency)[5]. Not only does it list risks and vulnerabilities, but it also offers a survey of related works and research recommendations. A similarly work is the security guidance provided by the Cloud Security Alliance (CSA)[6], which defines security domains congregating specific functional aspects, from governance and compliance to virtualization and identity management. Both documents present a plethora of security concerns, best practices and recommendations regarding all types of services in NIST's SPI model, as well as possible problems related to cloud computing, encompassing from data privacy to infrastructural configuration. Albeit valuable, these studies do not focus on quantifying their observations, something important for developing a comprehensive understanding of the challenges still undermining the potential of cloud computing.

The main goal of this article is to identify, classify, organize and quantify the main security concerns and solutions associated to cloud computing, helping in the task of pinpointing the concerns that remain unanswered. Aiming to organize this information into a useful tool for comparing, relating and classifying already identified concerns and solutions as well as future ones, we also present a taxonomy proposal for cloud computing security. We focus on issues that are specific to cloud computing, without losing sight of important issues that also exist in other distributed systems. This article extends our previous work presented in[11], providing an enhanced review of the cloud computing security taxonomy previously presented, as well as a deeper analysis of the related work by discussing the main security frameworks currently available; in addition, we discuss further the security aspects related to virtualization in cloud computing, a fundamental yet still underserved field of research.

2. Cloud Computing Security

Key references such as CSA's security guidance[6] and top threats analysis[12], ENISA's security assessment[5] and the cloud computing definitions from NIST[9] highlight different security issues related to cloud computing that require further studies for being appropriately handled and, consequently, for enhancing technology acceptance and adoption. Emphasis is given to the distinction between services in the form of software (SaaS), platform (PaaS) and infrastructure (IaaS), which are commonly used as the fundamental basis for cloud service classification. However, no other methods are standardized or even employed to organize cloud computing security aspects apart from cloud deployment models, service types or traditional security models.

Aiming to concentrate and organize information related to cloud security and to facilitate future studies, in this section we identify the main problems in the area and group them into a model composed of seven categories, based on the aforementioned references. Namely, the categories are: network security, interfaces, data security, virtualization, governance, compliance and legal issues. Each category includes several potential security problems, resulting in a classification with subdivisions that highlights the main issues identified in the base references:

1) Network security: Problems associated with network communications and configurations regarding cloud computing infrastructures. The ideal network

security solution is to have cloud services as an extension of customers' existing internal networks[13], adopting the same protection measures and security precautions that are locally implemented and allowing them to extend local strategies to any remote resource or process[14].

a) Transfer security: Distributed architectures, massive resource sharing and virtual machine (VM) instances synchronization imply more data in transit in the cloud, thus requiring VPN mechanisms for protecting the system against sniffing, spoofing, man-in-the-middle and side-channel attacks.

b) Firewalling: Firewalls protect the provider's internal cloud infrastructure against insiders and outsiders[15]. They also enable VM isolation, fine-grained filtering for addresses and ports, prevention of Denial-of-Service (DoS) and detection of external security assessment procedures. Efforts for developing consistent firewall and similar security measures specific for cloud environments[16][17] reveal the urge for adapting existing solutions for this new computing paradigm.

c) Security configuration: Configuration of protocols, systems and technologies to provide the required levels of security and privacy without compromising performance or efficiency[18].

2) Interfaces: Concentrates all issues related to user, administrative and programming interfaces for using and controlling clouds.

a) API: Programming interfaces (essential to IaaS and PaaS) for accessing virtualized resources and systems must be protected in order to prevent malicious use[19]–[23].

b) Administrative interface: Enables remote control of resources in an IaaS (VM management), development for PaaS (coding, deploying, testing) and application tools for SaaS (user access control, configurations).

c) User interface: End-user interface for exploring provided resources and tools (the service itself), implying the need of adopting measures for securing the environment[24]–[27].

d) Authentication: Mechanisms required to enable access to the cloud[28].

Most services rely on regular accounts[20][29][30] consequently being susceptible to a plethora of attacks[31]–[35] whose consequences are boosted by multi-tenancy and resource sharing.

3) Data security: Protection of data in terms of confidentiality, availability and integrity (which can be applied not only to cloud environments, but any solution requiring basic security levels)[36].

a) Cryptography: Most employed practice to secure sensitive data[37], thoroughly equired by industry, state and federal regulations[38].

b) Redundancy: Essential to avoid data loss. Most business models rely on information technology for its core functionalities and processes[39][40] and, thus, mission-critical data integrity and availability must be ensured.

c) Disposal: Elementary data disposal techniques are insufficient and commonly referred as deletion[41]. In the cloud, the complete destruction of data, including log references and hidden backup registries, is an important requirement[42].

4) Virtualization: Isolation between VMs, hypervisor vulnerabilities and other problems associated to the use of virtualization technologies[43].

a) Isolation: Although logically isolated, all VMs share the same hardware and consequently the same resources, allowing malicious entities to exploit data leaks and cross-VM attacks[44]. The concept of isolation can also be applied to more fine-grained assets, such as computational resources, storage and memory.

b) Hypervisor vulnerabilities: The hypervisor is the main software component of virtualization. Even though there are known security vulnerabilities for hypervisors, solutions are still scarce and often proprietary, demanding further studies to harden these security aspects.

c) Data leakage: Exploit hypervisor vulnerabilities and lack of isolation controls in order to leak data from virtualized infrastructures, obtaining sensitive customer data and affecting confidentiality and integrity.

d) VM identification: Lack of controls for identifying virtual machines that are being used for executing a specific process or for storing files.

e) Cross-VM attacks: Includes attempts to estimate provider traffic rates in order to steal cryptographic keys and increase chances of VM placement attacks. One example consists in overlapping memory and storage regions initially dedicated to a single virtual machine, which also enables other isolation-related attacks.

5) Governance: Issues related to (losing) administrative and security controls in cloud computing solutions[45][46].

a) Data control: Moving data to the cloud means losing control over redundancy, location, file systems and other relevant configurations.

b) Security control: Loss of governance over security mechanisms and policies, as terms of use prohibit customer-side vulnerability assessment and penetration tests while insufficient Service Level Agreements (SLA) lead to security gaps.

c) Lock-in: User potential dependency on a particular service provider due to lack of well-established standards (protocols and data formats), consequently becoming particularly vulnerable to migrations and service termination.

6) Compliance: Includes requirements related to service availability and audit capabilities[47][48].

a) Service Level Agreements (SLA): Mechanisms to ensure the required service availability and the basic security procedures to be adopted[49].

b) Loss of service: Service outages are not exclusive to cloud environments but are more serious in this context due to the interconnections between services (e.g., a SaaS using virtualized infrastructures provided by an IaaS), as shown in many examples[50]–[52]. This leads to the need of strong disaster recovery policies and provider recommendations to implement customer-side redundancy if applicable.

c) Audit: Allows security and availability assessments to be performed by customers, providers and third-party participants. Transparent and efficient methodologies are necessary for continuously analyzing service conditions[53] and are usually required by contracts or legal regulations. There are solutions being developed to address this problem by offering a transparent API for automated auditing and other useful functionalities[54].

d) Service conformity: Related to how contractual obligations and overall service requirements are respected and offered based on the SLAs predefined and basic service and customer needs.

7) Legal issues: Aspects related to judicial requirements and law, such as multiple data locations and privilege management.

a) Data location: Customer data held in multiple jurisdictions depending on geographic location[55] are affected, directly or indirectly, by subpoena law-enforcement measures.

b) E-discovery: As a result of a law-enforcement measures, hardware might be confiscated for investigations related to a particular customer, affecting all customers whose data were stored in the same hardware[56]-[58]. Data disclosure is critical in this case.

c) Provider privilege: Malicious activities of provider insiders are potential threats to confidentiality, availability and integrity of customers' data and processes' information[59][60].

d) Legislation: Juridical concerns related to new concepts introduced by cloud computing[61].

3. Cloud Computing Security Taxonomy

The analysis of security concerns in the context of cloud computing solutions shows that each issue brings different impacts on distinct assets. Aiming to create a security model both for studying security aspects in this context and for supporting decision making, in this section we consider the risks and vulnerabilities previously presented and arrange them in hierarchical categories, thus creating a cloud security taxonomy. The main structure of the proposed taxonomy, along with its first classification levels, are depicted in **Figure 1**.

The three first groups correspond to fundamental (and often related) security principles[7] (Chapters 3–8).

The *architecture* dimension is subdivided into network security, interfaces and virtualization issues, comprising both user and administrative interfaces to access the cloud. It also comprises security during transferences of data and virtual machines, as well as other virtualization related issues, such as isolation and cross-VM attacks. This organization is depicted in **Figure 2**. The architecture group allows a clearer division of responsibilities between providers and customers, and also an analysis of their security roles depending on the type of service offered (Software, Platform or Infrastructure). This suggests that the security mechanisms used must be clearly stated before the service is contracted, defining which role is responsible for providing firewalling capabilities, access control features and technology-specific requirements (such as those related to virtualization).

The *compliance* dimension introduces responsibilities toward services and providers. The former includes SLA concerns, loss of service based on outages and chain failures, and auditing capabilities as well as transparency and security assessments. The latter refers to loss of control over data and security policies and configurations, and also lock-in issues resulting from lack of standards, migrations and service terminations. The complete scenario is presented in **Figure 3**.

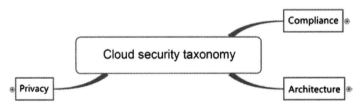

Figure 1. Cloud computing security taxonomy. Top level overview of the security taxonomy proposed, highlighting the three main categories: security related to privacy, architecture and compliance.

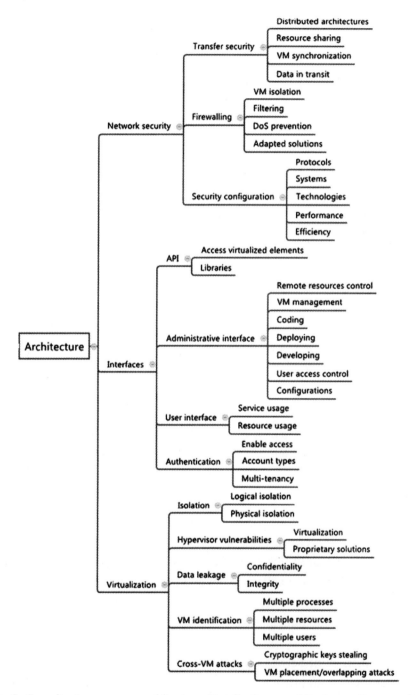

Figure 2. Security taxonomy—architecture. Details from architecture category, which is divided in network, host, application, data (security and storage), security management, and identity and access controls—all these elements are directly connected to the infrastructure and architecture adopted to implement or use a cloud solution.

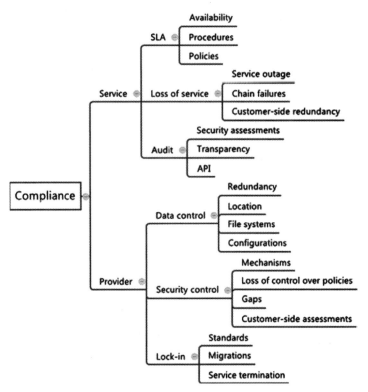

Figure 3. Security taxonomy—compliance. Details from compliance category, divided in lifecycle controls and governance, risk and other compliance related issues (such as continuous improvement policies).

The *privacy* dimension includes data security itself (from sensitive data, regulations and data loss to disposal and redundancy) and legal issues (related to multiple jurisdictions derived from different locations where data and services are hosted). The expansion of this group is represented in **Figure 4**. We note that the concerns in this dimension cover the complete information lifecycle (*i.e.*, *generation*, *use*, *transfer*, *transformation*, *storage*, *archiving*, and *destruction*) inside the provider perimeter and in its immediate boundaries (or interfaces) to the users.

A common point between all groups is the intrinsic connection to data and service lifecycles. Both privacy and compliance must be ensured through all states of data, including application information or customer assets, while security in this case is more oriented towards how the underlying elements (e.g., infrastructural hardware and software) are protected.

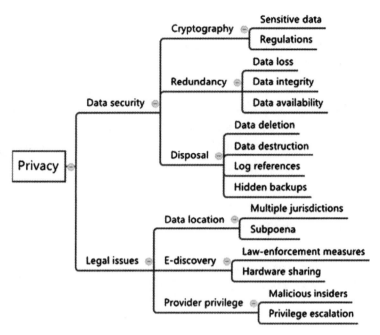

Figure 4. Security taxonomy—privacy. Details from privacy category, initially divided in concerns and principles. Concerns are related to the complete data lifecycle, from generation, use and transfer to transformation, storage, archival and destruction. Principles are guidelines related to privacy in the cloud.

4. Current Status of Cloud Security

A clear perspective of the main security problems regarding cloud computing and on how they can be organized to ease decision making is the primary step for having a comprehensive overview of the current status of cloud security. In this section, we analyze industry and academia viewpoints focusing on strategic study areas that need to be further developed. This study is based on more than two hundred different references including white papers, technical reports, scientific papers and other relevant publications. They were analyzed in terms of security problems and solutions by evaluating the number of citations for each case. We used a quantitative approach to identify the amount of references related to each category of concerns or solutions. Our goal is not to determine if the presented solutions completely solve an identified concern, since most of the referenced authors agree that this is an involved task. Nonetheless, we identify the number of references dealing with each concern, providing some useful insight on which are the concerns that have received more attention from the research community and which have not been so extensively analyzed. Some observations about the analysis method:

1) The references consulted came from different research segments, including academia, organizations, and companies. Due to the article's length limitations, we did not include all the consulted references in the References section. In the following we present some of the main sources of consultation:

a) Academia: conference papers and journals published by IEEE, ACM, Springer, Webscience, and Scipress.

b) Organizations: reports, white papers, and interviews from SANS Institute, CSA, NIST, ENISA, Gartner Group, KVM.org, OpenGrid, OpenStack, and OpenNebula.

c) Companies: white papers, manuals, interviews, and web content from ERICSSON, IBM, XEROX, Cisco, VMWare, XEN, CITRIX, EMC, Microsoft, and Salesforce.

2) Each reference was analyzed aiming to identify all the mentioned concerns covered and solutions provided. Therefore, one reference can produce more than one entry on each specified category.

3) Some security perspectives were not covered in this paper, as each security/concern category can be sub-divided in finer-grained aspects such as: authentication, integrity, network communications, etc.

We present the security concerns and solutions using pie charts in order to show the representativeness of each category/group in the total amount of references identified. The comparison between areas is presented using radar graphs to identify how many solutions address each concern category/group.

4.1. Security Concerns

The results obtained for the number of citations on security issues is shown in **Figure 5**. The three major problems identified in these references are legal issues, compliance and loss of control over data. These legal- and governance-related concerns are followed by the first technical issue, isolation, with 7% of citations. The least cited problems are related to security configuration concerns, loss of service (albeit this is also related to compliance, which is a major problem), firewalling and interfaces.

Grouping the concerns using the categories presented in section "Cloud computing security" leads to the construction of **Figure 6**. This figure shows that legal and governance issues represent a clear majority with 73% of concern citations, showing a deep consideration of legal issues such as data location and e-discovery, or governance ones like loss of control over security and data. The technical issue more intensively evaluated (12%) is virtualization, followed by data security, interfaces and network security.

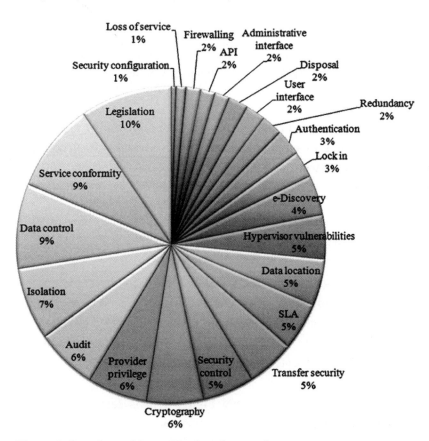

Figure 5. Security problems. Pie chart for security concerns.

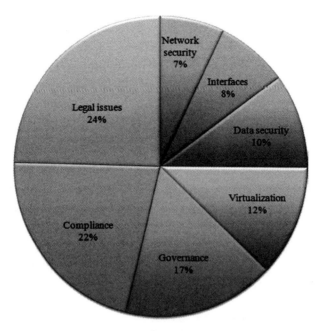

Figure 6. Security problems with grouped categories. Pie chart for security concerns with grouped categories (seven altogether: legal issues, compliance, governance, virtualization, data security, interfaces and network security).

Virtualization is one of the main novelties employed by cloud computing in terms of technologies employed, considering virtual infrastructures, scalability and resource sharing, and its related problems represent the first major technical concern.

4.2. Security Solutions

When analyzing citations for solutions, we used the same approach described in the beginning of this section. The results are presented in **Figure 7**, which shows the percentage of solutions in each category defined in section "Cloud computing security", and also in **Figure 8**, which highlights the contribution of each individual sub-category.

When we compare **Figure 6** and **Figure 7**, it is easy to observe that the number of citations covering security problems related to legal issues, compliance and governance is high (respectively 24%, 22%, and 17%); however, the same

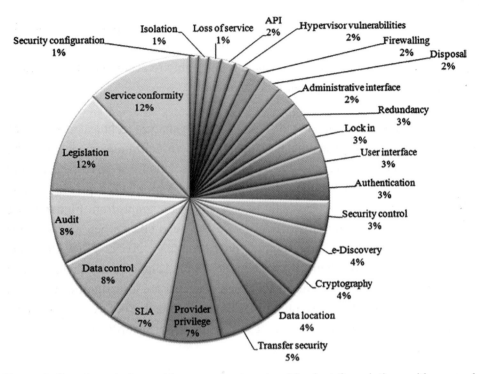

Figure 7. Security solutions with grouped categories. Pie chart for solutions with grouped categories, showing a clear lack for virtualization security mechanisms in comparison to its importance in terms of concerns citations.

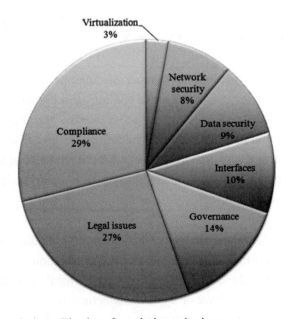

Figure 8. Security solutions. Pie chart for solutions citations.

also happens when we consider the number of references proposing solutions for those issues (which represent respectively 29%, 27%, and 14% of the total number of citations). In other words, these concerns are highly relevant but a large number solutions are already available for tackling them.

The situation is completely different when we analyze technical aspects such as virtualization, isolation and data leakage. Indeed, virtualization amounts for 12% of problem references and only 3% for solutions. Isolation is a perfect example of such discrepancy as the number of citations for such problems represents 7% in **Figure 5**, while solutions correspond to only 1% of the graph from **Figure 8**. We note that, for this specific issue, special care has been taken when assessing the most popular virtual machine solution providers (e.g., XEN, VMWARE, and KVM) aiming to verify their concerns and available solutions. A conclusion that can be drawn from this situation is that such concerns are also significant but yet little is available in terms of solutions. This indicates the need of evaluating potential areas still to be developed in order to provide better security conditions when migrating data and processes in the cloud.

4.3. Comparison

The differences between problem and solution citations presented in the previous sections can be observed in **Figure 9**.

Axis values correspond to the number of citations found among the references studied. Blue areas represent concern citations and lighter red indicates solutions, while darker red shows where those areas overlap. In other words, light red areas are problems with more citations for solutions than problems – they might be meaningful problems, but there are many solutions already addressing them – while blue areas represent potential subjects that have received little attention so far, indicating the need for further studies.

Figure 9 clearly shows the lack of development regarding data control mechanisms, hypervisor vulnerabilities assessment and isolation solutions for virtualized environments. On the other hand, areas such as legal concerns, SLAs, compliance and audit policies have a quite satisfactory coverage. The results for grouped categories (presented in section 4) are depicted in **Figure 10**.

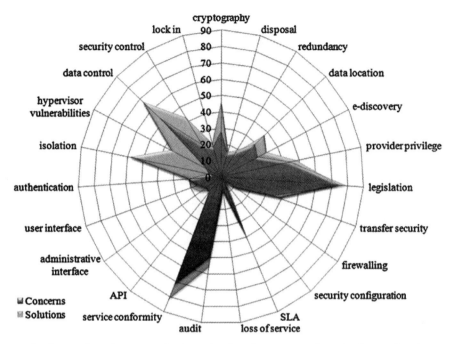

Figure 9. Comparison between citations. Radar chart comparing citations related to concerns and solutions, showing the disparities for each security category adopted.

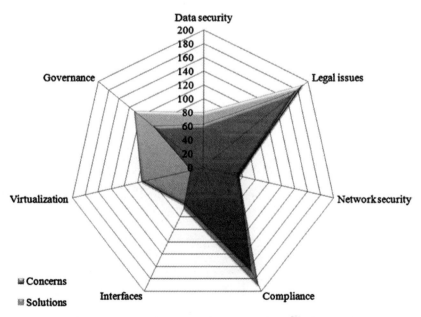

Figure 10. Comparison between citations with grouped categories. Radar chart grouping the categories, showing the difference between citations about concerns and solutions regarding each category.

Figure 10 shows that virtualization problems represent an area that requires studies for addressing issues such as isolation, data leakage and cross-VM attacks; on the other hand, areas such as compliance and network security encompass concerns for which there are already a considerable number of solutions or that are not considered highly relevant.

Finally, considering virtualization as key element for future studies, **Figure 11** presents a comparison focusing on five virtualization-related problems: isolation (of computational resources, such as memory and storage capabilities), hypervisor vulnerabilities, data leakage, cross-VM attacks and VM identification. The contrast related to isolation and cross-VM attacks is more evident than for the other issues. However, the number of solution citations for all issues is notably low if compared to any other security concern, reaffirming the need for further researches in those areas.

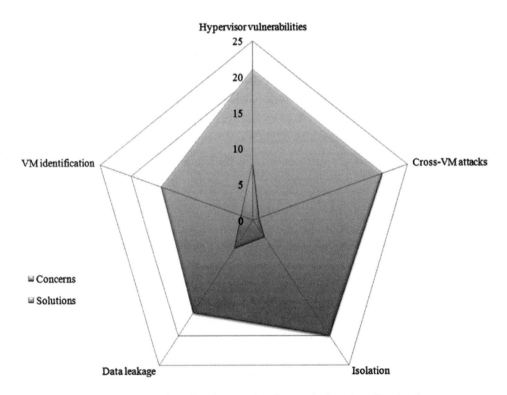

Figure 11. Comparison for virtualization. Radar chart only for virtualization issues.

5. Related Work

An abundant number of related works and publications exist in the literature, emphasizing the importance and demand of security solutions for cloud computing. However, we did not identify any full taxonomy that addresses directly the security aspects related to cloud computing. We only identified some simplified models that were developed to cover specific security aspects such as authentication. We were able to recognize two main types of works: 1) security frameworks, which aim to aggregate information about security and also to offer sets of best practices and guidelines when using cloud solutions, and 2) publications that identify future trends and propose solutions or areas of interest for research. Each category and corresponding references are further analyzed in the following subsections.

5.1. Security Frameworks

Security frameworks concentrate information on security and privacy aiming to provide a compilation of risks, vulnerabilities and best practices to avoid or mitigate them. There are several entities that are constantly publishing material related to cloud computing security, including ENISA, CSA, NIST, CPNI (Centre for the Protection of National Infrastructure from UK government) and ISACA (the Information Systems Audit and Control Association). In this paper we focus on the first three entities, which by themselves provide a quite comprehensive overview of issues and solutions and, thus, allowing a broad understanding of the current status of cloud security.

5.1.1. ENISA

ENISA is an agency responsible for achieving high and effective level of network and information security within the European Union[62]. In the context of cloud computing, they published an extensive study covering benefits and risks related to its use[5]. In this study, the security risks are divided in four categories:

- Policy and organizational: issues related to governance, compliance and reputation;

- Technical: issues derived from technologies used to implement cloud services and infrastructures, such as isolation, data leakage and interception, denial of service attacks, encryption and disposal;

- Legal: risks regarding jurisdictions, subpoena and e-discovery;

- Not cloud specific: other risks that are not unique to cloud environments, such as network management, privilege escalation and logging;

As a top recommendation for security in cloud computing, ENISA suggests that providers must ensure some security practices to customers and also a clear contract to avoid legal problems. Key points to be developed include breach reporting, better logging mechanisms and engineering of large scale computer systems, which encompass the isolation of virtual machines, resources and information. Their analysis is based not only on what is currently observed, but also on what can be improved through the adoption of existing best practices or by means of solutions that are already used in non-cloud environments. This article aims at taking one step further by transforming these observations into numbers – a quantitative approach.

5.1.2. CSA

CSA is an organization led by a coalition of industry practitioners, corporations, associations and other stake-holders[63], such as Dell, HP and eBay. One of its main goals is to promote the adoption of best practices for providing security within cloud computing environments.

Three CSA documents are analyzed in this paper—the security guidance[6], the top threats in cloud computing[12] and the Trusted Cloud Initiative (TCI) architecture[64]—as they comprise most of the concepts and guidelines researched and published by CSA.

The latest CSA security guidance (version 3.0[65]) denotes multi-tenancy as the essential cloud characteristic while virtualization can be avoided when implementing cloud infrastructures—multi-tenancy only implies the use of shared resources by multiple consumers, possibly from different organizations or with different objectives. They discuss that, even if virtualization-related issues can be circumvented, segmentation and isolated policies for addressing proper management and privacy are still required. The document also establishes thirteen security domains:

1) Governance and risk management: ability to measure the risk introduced by adopting cloud computing solutions, such as legal issues, protection of sensitive data and their relation to international boundaries;

2) Legal issues: disclosure laws, shared infrastructures and interference between different users;

3) Compliance and audit: the relationship between cloud computing and internal security policies;

4) Information management and data security: identification and control of stored data, loss of physical control of data and related policies to minimize risks and possible damages;

5) Portability and interoperability: ability to change providers, services or bringing back data to local premises without major impacts;

6) Traditional security, business continuity and disaster recovery: the influence of cloud solutions on traditional processes applied for addressing security needs;

7) Data center operations: analyzing architecture and operations from data centers and identifying essential characteristics for ensuring stability;

8) Incident response, notification and remediation: policies for handling incidents;

9) Application security: aims to identify the possible security issues raised from migrating a specific solution to the cloud and which platform (among SPI model) is more adequate;

10) Encryption and key management: how higher scalability via infrastruc-

ture sharing affects encryption and other mechanisms used for protecting resources and data;

11) Identity and access management: enabling authentication for cloud solutions while maintaining security levels and availability for customers and organizations;

12) Virtualization: risks related to multi-tenancy, isolation, virtual machine co-residence and hypervisor vulnerabilities, all introduced by virtualization technologies;

13) Security as a service: third party security mechanisms, delegating security responsibilities to a trusted third party provider;

CSA also published a document focusing on identifying top threats, aiming to aid risk management strategies when cloud solutions are adopted[12]. As a complete list of threats and pertinent issues is countless, the document targets those that are specific or intensified by fundamental characteristics of the cloud, such as shared infrastructures and greater flexibility. As a result, seven threats were selected:

1) Abuse and nefarious used of cloud computing: while providing flexible and powerful resources and tools, IaaS and PaaS solutions also unveil critical exploitation possibilities built on anonymity. This leads to abuse and misuse of the provided infrastructure for conducting distributed denial of service attacks, hosting malicious data, controlling botnets or sending spam;

2) Insecure application programming interfaces: cloud services provide APIs for management, storage, virtual machine allocation and other service-specific operations. The interfaces provided must implement security methods to identify, authenticate and protect against accidental or malicious use, which can introduce additional complexities to the system such as the need for third-party authorities and services;

3) Malicious insiders: although not specific to cloud computing, its effects are amplified by the concentration and interaction of services and

management domains;

4) Shared technology vulnerabilities: scalability provided by cloud solutions are based on hardware and software components which are not originally designed to provide isolation. Even though hypervisors offer an extra granularity layer, they still exhibit flaws which are exploited for privilege escalation;

5) Data loss and leakage: insufficient controls concerning user access and data security (including privacy and integrity), as well as disposal and even legal issues;

6) Account, service and traffic hijacking: phishing and related frauds are not a novelty to computing security. However, not only an attacker is able to manipulate data and transactions, but also to use stolen credentials to perform other attacks that compromise customer and provider reputation.

7) Unknown risk profile: delegation of control over data and infrastructure allows companies to better concentrate on their core business, possibly maximizing profit and efficiency. On the other hand, the consequent loss of governance leads to obscurity[66]: information about other customers sharing the same infrastructure or regarding patching and updating policies is limited. This situation creates uncertainty concerning the exact risk levels that are inherent to the cloud solution;

It is interesting to notice the choice for cloud-specific issues as it allows the identification of central points for further development. Moreover, this compilation of threats is closely related to CSA security guidance, composing a solid framework for security and risk analysis assessments while providing recommendations and best practices to achieve acceptable security levels.

Another approach adopted by CSA for organizing information related to cloud security and governance is the TCI Reference Architecture Model[64]. This document focuses on defining guidelines for enabling trust in the cloud while establishing open standards and capabilities for all cloud-based operations. The architecture defines different organization levels by combining frameworks like the SPI model, ISO 27002, COBIT, PCI, SOX and architectures such as SABSA,

TOGAF, ITIL and Jericho. A wide range of aspects are then covered: SABSA defines business operation support services, such as compliance, data governance, operational risk management, human resources security, security monitoring services, legal services and internal investigations; TOGAF defines the types of services covered (presentation, application, information and infrastructure; ITIL is used for information technology operation and support, from IT operation to service delivery, support and management of incidents, changes and resources; finally, Jericho covers security and risk management, including information security management, authorization, threat and vulnerability management, policies and standards. The result is a tri-dimensional relationship between cloud delivery, trust and operation that aims to be easily consumed and applied in a security-oriented design.

5.1.3. NIST

NIST has recently published a taxonomy for security in cloud computing[67] that is comparable to the taxonomy introduced in section "Cloud computing security taxonomy". This taxonomy's first level encompass typical roles in the cloud environment: cloud service provider, responsible for making the service itself available; cloud service consumer, who uses the service and maintains a business relationship with the provider; cloud carrier, which provides communication interfaces between providers and consumers; cloud broker, that manages use, performance and delivery of services and intermediates negotiations between providers and consumers; and cloud auditor, which performs assessment of services, operations and security. Each role is associated to their respective activities and decomposed on their components and subcomponents. The clearest difference from our taxonomy is the hierarchy adopted, as our proposal primarily focuses on security principles in its higher level perspective, while the cloud roles are explored in deeper levels. The concepts presented here extend NIST's initial definition for cloud computing[9], incorporating a division of roles and responsibilities that can be directly applied to security assessments. On the other hand, NIST's taxonomy incorporates concepts such as deployment models, service types and activities related to cloud management (portability, interoperability, provisioning), most of them largely employed in publications related to cloud computing—including this one.

5.1.4. Frameworks Summary

Table 1 and **Table 2** summarize the information about each framework.

5.2. Books, Papers and Other Publications

Rimal, Choi and Lumb[3] present a cloud taxonomy created from the perspective of the academia, developers and researchers, instead of the usual point of view related to vendors. Whilst they do provide definitions and concepts such as cloud architecture (based on SPI model), virtualization management, service types, fault tolerance policies and security, no further studies are developed focusing on cloud specific security aspects. This characteristic is also observed in other cloud taxonomies[68]–[70] whose efforts converge to the definition of service models and

Table 1. Summary of CSA security frameworks.

Framework	Objectives	Structure and comments
CSA Guidance	<ul><li>Recommendations for reducing risks</li><li>No restrictions regarding specific solutions or service types</li><li>Guidelines not necessarily applicable for all deployment models</li><li>Provide initial structure to divide efforts for researches</li></ul>	<ul><li>One architectural domain</li><li>Governance domains: risk management, legal concerns, compliance, auditing, information management, interoperability and portability</li><li>Operational domains: traditional and business security, disaster recovery, data center operations, encryption, application security, identification, authorization, virtualization, security outsourcing</li><li>Emphasis on the fact that cloud is not bound to virtualization technologies, though cloud services heavily depend on virtualized infrastructures to provide flexibility and scalability</li></ul>
CSA Top Threats	<ul><li>Provide context for risk management decisions and strategies</li><li>Focus on issues which are unique or highly influenced by cloud computing characteristics</li></ul>	<ul><li>Seven main threats: – Abuse and malicious use of cloud resources – Insecure APIs – Malicious insiders – Shared technology vulnerabilities – Data loss and leakage – Hijacking of accounts, services and traffic – Unknown risk profile (security obscurity)</li><li>Summarizes information on top threats and provide examples, remediation guidelines, impact caused and which service types (based on SPI model) are affected</li></ul>
CSA Architecture	<ul><li>Enable trust in the cloud based on well-known standards and certifications allied to security frameworks and other open references</li><li>Use widely adopted frameworks in order to achieve standardization of policies and best practices based on already accepted security principles</li></ul>	<ul><li>Four sets of frameworks (security, NIST SPI, IT audit and legislative) and four architectural domains (SABSA business architecture, ITIL for services management, Jericho for security and TOGAF for IT reference)</li><li>Tridimensional structure based on premises of cloud delivery, trust and operations</li><li>Concentrates a plethora of concepts and information related to services operation and security</li></ul>

Table summarizing information related to CSA security frameworks (guidance, top threats and TCI architecture).

Table 2. Summary of ENISA and NIST security frameworks.

Framework	Objectives	Structure and comments
ENISA Report	• Study on benefits and risks when adopting cloud solutions for business operations • Provide information for security assessments and decision making	• Three main categories of cloud specific risks (policy and organizational, technical, legal) plus one extra category for not specific ones • Offers basic guidelines and best practices for avoiding or mitigating their effects • Presents recommendations for further studies related to trust building (certifications, metrics and transparency), large scale data protection (privacy, integrity, incident handling and regulations) and technical aspects (isolation, portability and resilience) • Highlights the duality of scalability (fast, flexible and accessible resources versus concentrations of data attracting attackers and also providing infrastructure for aiding their operations) • Extensive study on risks considering their impact and probability

NIST Taxonomy	• Define what cloud services should provide rather than how to design and implement solutions
	• Ease the understanding of cloud internal operations and mechanisms
		• Taxonomy levels:
		  – First level: cloud roles (service provider, consumer, cloud broker, cloud carrier and cloud auditor)
		  – Second level: activities performed by each role (cloud management, service deployment, cloud access and service consumption)
		  – Third and following levels: elements which compose each activity (deployment models, service types and auditing elements)
		• Based on publication SP 500-292, highlighting the importance of security, privacy and levels of confidence and trust to increase technology acceptance
		• Concentrates many useful concepts, such as models for deploying or classifying services

Table summarizing information on ENISA and NIST security frameworks.

types rather than to more technical aspects such as security, privacy or compliance concerns—which are the focus of this paper.

In[7], Mather, Kumaraswamy and Latif discuss the current status of cloud security and what is predicted for the future. The result is a compilation of security-related subjects to be developed in topics like infrastructure, data security and storage, identity and access management, security management, privacy, audit and compliance. They also explore the unquestionable urge for more transparency regarding which party (customer or cloud provider) provides each security capability, as well as the need for standardization and for the creation of legal agreements reflecting operational SLAs. Other issues discussed are the inadequate encryption and key management capabilities currently offered, as well as the need for multi-entity key management.

Many publications also state the need for better security mechanisms for cloud environments. Doelitzscher *et al.*[71] emphasize security as a major research area in cloud computing. They also highlight the lack of flexibility of classic intrusion detection mechanisms to handle virtualized environments, suggesting the use of special security audit tools associated to business flow modeling through security SLAs. In addition, they identify abuse of cloud resources, lack of security monitoring in cloud infrastructure and defective isolation of shared resources as focal points to be managed. Their analysis of top security concerns is also based on publications from CSA, ENISA and others, but after a quick evaluation of issues their focus switch to their security auditing solution, without offering a deeper quantitative compilation of security risks and areas of concern.

Associations such as the Enterprise Strategy Group[72] emphasize the need for hypervisor security, shrinking hypervisor footprints, defining the security perimeter virtualization, and linking security and VM provisioning for better resource management. Aiming to address these requirements, they suggest the use of increased automation for security controls, VM identity management (built on top of Public Key Infrastructure and Open Virtualization Format) and data encryption (tightly connected to state-of-art key management practices). Wallom et al.[73] emphasize the need of guaranteeing virtual machines' trustworthiness (regarding origin and identity) to perform security-critical computations and to handle sensitive data, therefore presenting a solution which integrates Trusted Computing technologies and available cloud infrastructures. Dabrowski and Mills[74] used simulation to demonstrate virtual machine leakage and resource exhaustion scenarios leading to degraded performance and crashes; they also propose the addition of orphan controls to enable the virtualized cloud environment to offer higher availability levels while keeping overhead costs under control. Ristenpart et al.[44] also explore virtual machine exploitation focusing on information leakage, specially sensitive data at rest or in transit. Finally, Chadwick and Casenove[75] describe a security API for federated access to cloud resources and authority delegation while setting fine-grained controls and guaranteeing the required levels of assurance inside cloud environments. These publications highlight the need of security improvements related to virtual machines and virtualization techniques, concern that this paper demonstrates to be valid and urgent.

6. Discussion

Considering the points raised in the previous section, a straightforward conclusion is that cloud security includes old and well-known issues—such as network and other infrastructural vulnerabilities, user access, authentication and privacy—and also novel concerns derived from new technologies adopted to offer the adequate resources (mainly virtualized ones), services and auxiliary tools. These problems are summarized by isolation and hypervisor vulnerabilities (the main technical concerns according to the studies and graphics presented), data location and e-discovery (legal aspects), and loss of governance over data, security and even decision making (in which the cloud must be strategically and financially considered as a decisive factor).

Another point observed is that, even though adopting a cloud service or provider may be easy, migrating to another is not[76]. After moving local data and processes to the cloud, the lack of standards for protocols and formats directly affects attempts to migrate to a different provider even if this is motivated by legitimate reasons such as non-fulfillment of SLAs, outages or provider bankruptcy[77]. Consequently, the first choice must be carefully made, as SLAs are not perfect and services outages happen at the same pace that resource sharing, multi-tenancy and scalability are not fail proof. After a decision is made, future migrations between services can be extremely onerous in terms of time and costs; most likely, this task will require an extensive work for bringing all data and resources to a local infrastructure before redeploying them into the cloud.

Finally, the analysis of current trends for cloud computing reveals that there is a considerable number of well-studied security concerns, for which plenty solutions and best practices have been developed, such as those related to legal and administrative concerns. On the other hand, many issues still require further research effort, especially those related to secure virtualization.

7. Considerations and Future Work

Security is a crucial aspect for providing a reliable environment and then enable the use of applications in the cloud and for moving data and business processes to virtualized infrastructures. Many of the security issues identified are observed in other computing environments: authentication, network security and legal requirements, for example, are not a novelty. However, the impact of such issues is intensified in cloud computing due to characteristics such as multi-tenancy and resource sharing, since actions from a single customer can affect all other users that inevitably share the same resources and interfaces. On the other hand, efficient and secure virtualization represents a new challenge in such a context with high distribution of complex services and web-based applications, thus requiring more sophisticated approaches. At the same time, our quantitative analysis indicates that virtualization remains an underserved area regarding the number of solutions provided to identified concerns.

It is strategic to develop new mechanisms that provide the required security level by isolating virtual machines and the associated resources while following best practices in terms of legal regulations and compliance to SLAs. Among other requirements, such solutions should employ virtual machine identification, provide an adequate separation of dedicated resources combined with a constant observation of shared ones, and examine any attempt of exploiting cross-VM and data leakage.

A secure cloud computing environment depends on several security solutions working harmoniously together. However, in our studies we did not identify any security solutions provider owning the facilities necessary to get high levels of security conformity for clouds. Thus, cloud providers need to orchestrate/harmonize security solutions from different places in order to achieve the desired security level.

In order to verify these conclusions in practice, we deployed testbeds using OpenNebula (based on KVM and XEN) and analyzed its security aspects; we also analyzed virtualized servers based on VMWARE using our testbed networks. This investigation lead to a wide research of PaaS solutions, and allowed us to verify that most of them use virtual machines based on virtualization technologies such as VMWARE, XEN, and KVM, which often lack security aspects We also learned that Amazon changed the XEN source code in order to include security features, but unfortunately the modified code is not publicly available and there appears to be no article detailing the changes introduced. Given these limitations, a deeper study on current security solutions to manage cloud computing virtual machines inside the cloud providers should be a focus of future work in the area. We are also working on a testbed based on OpenStack for researches related to identity and credentials management in the cloud environment. This work should address basic needs for better security mechanisms in virtualized and distributed architectures, guiding other future researches in the security area.

Competing Interests

The authors declare that they have no competing interests.

Author's Contributions

NG carried out the security research, including the prospecting for information and references, categorization, results analysis, taxonomy creation and analysis of related work. CM participated in the drafting of the manuscript as well as in the analysis of references, creation of the taxonomy and revisions of the text. MS, FR, MN and MP participated in the critical and technical revisions of the paper including the final one, also helping with the details for preparing the paper to be published. TC coordinated the project related to the paper and also gave the final approval of the version to be published. All authors read and approved the final manuscript.

Acknowledgements

This work was supported by the Innovation Center, Ericsson Telecomunicações S.A., Brazil.

Source: Gonzalez N, Miers C, Redígolo F, et al. A quantitative analysis of current security concerns and solutions for cloud computing[C]//IEEE Third International Conference on Cloud Computing Technology and Science. IEEE Computer Society, 2011: 231–238.

References

[1] IDC (2009) Cloud Computing 2010—An IDC Update. slideshare.net/JorFigOr/cloud-computing-2010-an-idc-update.

[2] Armbrust M, Fox A, Griffith R, Joseph AD, Katz RH, Konwinski A, Lee G, Patterson DA, Rabkin A, Stoica I, Zaharia M (2009) Above the Clouds: A Berkeley View of Cloud Computing. Technical Report UCB/EECS-2009-28, University of California at Berkeley. eecs.berkeley.edu/Pubs/TechRpts/2009/EECS-2009-28.html.

[3] Rimal BP, Choi E, Lumb I (2009) A Taxonomy and, Survey of Cloud Computing Systems. In: Fifth International Joint Conference on INC, IMS and IDC, NCM'09, CPS. pp 44–51.

[4] Shankland S (2009) HP's Hurd dings cloud computing, IBM. CNET News.

[5] Catteddu D, Hogben G (2009) Benfits, risks and recommendations for information security. Tech. rep., European Network and Information Security Agency. enisa.europa.eu/act/rm/files/deliverables/cloud-computing-risk-assessment.

[6] CSA (2009) Security Guidance for Critical Areas of Focus in Cloud Computing. Tech. rep., Cloud Security Alliance.

[7] Mather T, Kumaraswamy S (2009) Cloud Security and privacy: An Enterprise Perspective on Risks and Compliance. 1st edition. O'Reilly Media.

[8] Chen Y, Paxson V, Katz RH (2010) What's New About Cloud Computing Security? Technical Report UCB/EECS-2010-5, University of California at Berkeley. eecs.berkeley.edu/Pubs/TechRpts/2010/EECS-2010-5.html.

[9] Mell P, Grance T (2009) The NIST Definition of Cloud Computing. Technical Report 15, National Institute of Standards and Technology. www.nist.gov/itl/cloud/upload/cloud-def-v15.pdf.

[10] Ibrahim AS, Hamlyn-Harris J, Grundy J (2010) Emerging Security Challenges of Cloud Virtual Infrastructure. In: Proceedings of APSEC 2010 Cloud Workshop, APSEC'10.

[11] Gonzalez N, Miers C, Red´ıgolo F, Carvalho T, Simpl´ıcio M, Naslund M, Pourzandi M (2011) A quantitative analysis of current security concerns and solutions for cloud computing. In: Proceedings of 3rd IEEE CloudCom. Athens/Greece: IEEE Computer Society.

[12] Hubbard D, Jr LJH, Sutton M (2010) Top Threats to Cloud Computing. Tech. rep., Cloud Security Alliance. cloudsecurityalliance.org/research/projects/top-threats-to-cloud-computing/.

[13] Tompkins D (2009) Security for Cloud-based Enterprise Applications. http://blog.dt.org/index.php/2009/02/security-for-cloud-based-enterprise-applications/.

[14] Jensen M, Schwenk J, Gruschka N, Iacono LL (2009) On Technical Security Issues in Cloud Computing. In: IEEE Internation Conference on Cloud Computing. pp. 109–116.

[15] TrendMicro (2010) Cloud Computing Security—Making Virtual Machines Cloud-Ready. Trend Micro White Paper.

[16] Genovese S (2009) Akamai Introduces Cloud-Based Firewall. http://cloudcomputing.sys-con.com/node/1219023.

[17] Hulme GV (2011) CloudPassage aims to ease cloud server security management. http://www.csoonline.com/article/658121/cloudpassage-aims-to-ease-cloud-server-security-management.

[18] Oleshchuk VA, Køien GM (2011) Security and Privacy in the Cloud—A Long-Term View. In: 2nd International Conference on Wireless Communications, Vehicular Technology, Information Theory and Aerospace and Electronic Systems Technology (Wireless VITAE), WIRELESS VITAE'11. pp 1–5. http://dx.doi.org/10.1109/WIRELESSVITAE.2011.5940876.

[19] Google (2011) Google App Engine. code.google.com/appengine/.

[20] Google (2011) Google Query Language (GQL). code.google.com/intl/en/appengine/docs/python/overview.html.

[21] StackOverflow (2011) Does using non-SQL databases obviate the need for guarding against SQL injection? stackoverflow.com/questions/1823536/does-using-non-sql-databases-obviate-the-need-for-guarding-against-sql-injection.

[22] Rose J (2011) Cloudy with a chance of zero day. www.owasp.org/images/1/12/Cloudy with a chance of 0 day Jon Rose-Tom Leavey.pdf.

[23] Balkan A (2011) Why Google App Engine is broken and what Google must do to fix it. aralbalkan.com/1504.

[24] Salesforce (2011) Salesforce Security Statement. salesforce.com/company/privacy/security.jsp.

[25] Espiner T (2007) Salesforce tight-lipped after phishing attack. zdnet.co.uk/news/security-threats/2007/11/07/salesforce-tight-lipped-after-phishing-attack-39290616/.

[26] Yee A (2007) Implications of Salesforce Phishing Incident. ebizq.net/blogs/security_insider/2007/11/-implications_of salesforc_e_phi.php.

[27] Salesforce (2011) Security Implementation Guide. login.salesforce.com/help/doc/en/salesforcesecurityimplguide.pdf.

[28] Li H, Dai Y, Tian L, Yang H (2009) Identity-Based Authentication for Cloud Computing. In: Proceedings of the 1st International Conference on Cloud Computing, CloudCom'09.

[29] Amazon (2011) Elastic Compute Cloud (EC2). aws.amazon.com/ec2/.

[30] Kaufman C, Venkatapathy R (2010) Windows Azure Security Overview. go.microsoft.com/?linkid=9740388.

[31] McMillan R (2010) Google Attack Part of Widespread Spying Effort. PCWorld.

[32] Mills E (2010) Behind the China attacks on Google. CNET News.

[33] Arrington M (2010) Google Defends Against Large Scale Chinese Cyber Attack: May Cease Chinese Operations. TechCrunch.

[34] Bosch J (2009) Google Accounts Attacked by Phishing Scam. BrickHouse Security Blog.

[35] Telegraph T (2009) Facebook Users Targeted By Phishing Attack. The Telegraph.

[36] Pearson S (2009) Taking account of privacy when designing cloud computing services. In: Proceedings of the 2009 ICSE Workshop on Software Engineering Challenges of Cloud Computing, CLOUD'09.

[37] Musthaler L (2009) Cost-effective data encryption in the cloud. Network World.

[38] Yan L, Rong C, Zhao G (2009) Strengthen Cloud Computing Security with Federal Identity Management Using Hierarchical Identity-Based Cryptography. In: Proceedings of the 1st International Conference on Cloud Computing, CloudCom'09.

[39] Tech C (2010) Examining Redundancy in the Data Center Powered by the Cloud and Disaster Recovery. Consonus Tech.

[40] Lyle M (2011) Redundancy in Data Storage. Define the Cloud.

[41] Dorion P (2010) Data destruction services: When data deletion is not enough. SearchDataBackup.com.

[42] Mogull R (2009) Cloud Data Security: Archive and Delete (Rough Cut). securosis.com/blog/cloud-data-security-archive-and-delete-rough-cut/.

[43] Messmer E (2011) Gartner: New security demands arising for virtualization, cloud computing. http://www.networkworld.com/news/2011/062311-security-summit.html.

[44] Ristenpart T, Tromer E, Shacham H, Savage S (2009) Hey, you, get off of my cloud: exploring information leakage in third-party compute clouds. In: Proceedings of the 16th ACM conference on Computer and communications security, CCS'09. New York, NY, USA, ACM, pp 199–212. doi.acm.org/10.1145/1653662.1653687.

[45] Chow R, Golle P, Jakobsson M, Shi E, Staddon J, Masuoka R, Molina J (2009) Controlling data in the cloud: outsourcing computation without outsourcing control. In: Proceedings of the 2009 ACM workshop on, Cloud computing security, CCSW'09. New York, NY, USA, ACM, pp 85–90. http://doi.acm.org/10.1145/1655008.1655020.

[46] Sadeghi AR, Schneider T, Winandy M (2010) Token-Based Cloud Computing—Secure Outsourcing of Data and Arbitrary Computations with Lower Latency. In: Proceedings of the 3rd international conference on Trust and trustworthy computing, TRUST'10.

[47] Brandic I, Dustdar S, Anstett T, Schumm D, Leymann F (2010) Compliant Cloud Computing (C3): Architecture and Language Support for User-driven Compliance Management in Clouds. In: 2010 IEEE 3rd International Conference on Cloud Computing. pp 244–251. http://dx.doi.org/10.1109/CLOUD.2010.42.

[48] Brodkin J (2008) Gartner: Seven cloud computing security risks. http://www.infoworld.com/d/security-central/gartner-seven-cloud-computing-security-risks-853.

[49] Kandukuri BR, Paturi R, Rakshit A (2009) Cloud Security Issues. In: Proceedings of the 2009 IEEE International Conference on Services Computing, SCC'09.

[50] Winterford B (2011) Amazon EC2 suffers huge outage. http://www.crn.com.au/News/255586,amazon-ec2-suffers-huge-outage.aspx.

[51] Clarke G (2011) Microsoft BPOS cloud outage burns Exchange converts. http://www.theregister.co.uk/2011/05/13/.

[52] Shankland S (2011) Amazon cloud outage derails Reddit, Quora.

[53] Young E (2009) Cloud Computing—The role of internal audit.

[54] CloudAudit (2011) A6—The automated audit, assertion, assessment and assurance API. http://cloudaudit.org/.

[55] Anand N (2010) The legal issues around cloud computing. http://www.labnol.org/internet/cloud-computing-legal-issues/14120/.

[56] Hunter S (2011) Ascending to the cloud creates negligible e-discovery risk. http://ediscovery.quarles.com/2011/07/articles/information-technology/ascending-to-the-cloud-creates-negligible-ediscovery-risk/.

[57] Sharon D, Nelson JWS (2011) Virtualization and Cloud Computing: benefits and e-discovery implications. http://www.slaw.ca/2011/07/19/virtualization-and-cloud-computing-benefits-and-e-discovery-implications/.

[58] Bentley L (2009) E-discovery in the cloud presents promise and problems. http://www.itbusinessedge.com/cm/community/features/interviews/blog/e-discovery-in-the-cloud-presents-promise-and-problems/?cs=31698.

[59] Zierick J (2011) The special case of privileged users in the cloud. http://blog.beyondtrust.com/bid/63894/The-Special-Case-of-Privileged-Users-in-the-Cloud.

[60] Dinoor S (2010) Got Privilege? Ten Steps to Securing a Cloud-Based Enterprise. http://cloudcomputing.sys-con.com/node/1571649.

[61] Pavolotsky J (2010) Top five legal issues for the cloud. http://www.forbes.com/2010/04/12/cloud-computing-enterprise-technology-cio-network-legal.html.

[62] ENISA (2011) About ENISA. http://www.enisa.europa.eu/about-enisa.

[63] CSA (2011) About. https://cloudsecurityalliance.org/about/.

[64] CSA (2011) CSA TCI Reference Architecture. https://cloudsecurityalliance.org/wp-content/uploads/2011/11/TCI-Reference-Architecture-1.1.pdf.

[65] CSA (2011) Security Guidance for Critical Areas of Focus in Cloud Computing V3.0. Tech. rep., Cloud Security Alliance. [Http://www.cloudsecurityalliance.org/guidance/csaguide.v3.0.pdf].

[66] Ramireddy S, Chakraborthy R, Raghu TS, Rao HR (2010) Privacy and Security Practices in the Arena of Cloud Computing—A Research in Progress. In: AMCIS 2010 Proceedings, AMCIS'10. http://aisel.aisnet.org/amcis2010/574.

[67] NIST (2011) NIST Cloud Computing Reference Architecture: SP 500-292. http://collaborate.nist.gov/twiki-cloud-computing/pub/CloudComputing/ReferenceArchitectureTaxonomy/NIST SP 500-292-090611.pdf.

[68] Youseff L, Butrico M, Silva DD (2008) Toward a Unfied Ontology of Cloud Computing. In: Grid Computing Environments Workshop, 2008. GCE'08. http://dx.doi.org/10.1109/GCE.2008.4738443.

[69] Johnston S (2008) Sam Johnston: taxonomy: the 6 layer cloud computing stack. http://samj.net/2008/09/taxonomy-6-layer-cloud-computing-stack.html.

[70] Linthicum D (2009) Defining the cloud computing framework. http://cloudcomputing.sys-con.com/node/811519.

[71] Doelitzscher F, Reich C, Knahl M, Clarke N (2011) An autonomous agent based incident detection system for cloud environments. In: Third IEEE International Conference on Cloud Computing Technology and Science, CloudCom 2011, CPS. pp 197–204. http://dx.doi.org/10.1109/CloudCom.

[72] Oltsik J (2010) Information security, virtualization, and the journey to the cloud. Tech. rep., Cloud Security Alliance.

[73] Wallom D, Turilli M, Taylor G, Hargreaves N, Martin A, Raun A, McMoran A (2011) myTrustedCloud: Trusted Cloud Infrastructure for Security-critical Computation and Data Managment. In: Third IEEE International Conference on Cloud Computing Technology and Science, CloudCom 2011, CPS. pp 247–254.

[74] Dabrowski C, Mills K (2011) VM Leakage and Orphan Control in Open-Source Clouds. In: Third IEEE International Conference on Cloud Computing Technology and Science, CloudCom 2011, CPS. pp 554–559.

[75] Chadwick DW, Casenove M (2011) Security APIs for My Private Cloud. In: Third IEEE International Conference on Cloud Computing Technology and Science, CloudCom 2011, CPS. pp 792–798.

[76] Claybrook B (2011) How providers affect cloud application migration.

http://searchcloudcomputing.techtarget.com/tutorial/How-providers-affect-cloud-application-migration.

[77] CSA (2011) Interoperability and portability.

Chapter 4
An Analysis of Security Issues for Cloud Computing

Keiko Hashizume[1], David G. Rosado[2], Eduardo Fernández-Medina[2], Eduardo B. Fernandez[1]

[1]Department of Computer Science and Engineering, Florida Atlantic University, Boca Raton, USA
[2]Department of Information Systems and Technologies GSyA Research Group, University of Castilla-La Mancha, Ciudad Real, Spain

Abstract: Cloud Computing is a flexible, cost-effective, and proven delivery platform for providing business or consumer IT services over the Internet. However, cloud Computing presents an added level of risk because essential services are often outsourced to a third party, which makes it harder to maintain data security and privacy, support data and service availability, and demonstrate compliance. Cloud Computing leverages many technologies (SOA, virtualization, Web 2.0); it also inherits their security issues, which we discuss here, identifying the main vulnerabilities in this kind of systems and the most important threats found in the literature related to Cloud Computing and its environment as well as to identify and relate vulnerabilities and threats with possible solutions.

Keywords: Cloud Computing, Security, SPI Model, Vulnerabilities, Threats, Countermeasures

1. Introduction

The importance of Cloud Computing is increasing and it is receiving a growing attention in the scientific and industrial communities. A study by Gartner[1] considered Cloud Computing as the first among the top 10 most important technologies and with a better prospect in successive years by companies and organizations.

Cloud Computing enables ubiquitous, convenient, on-demand network access to a shared pool of configurable computing resources (e.g., networks, servers, storage, applications, and services) that can be rapidly provisioned and released with minimal management effort or service provider interaction.

Cloud Computing appears as a computational paradigm as well as a distribution architecture and its main objective is to provide secure, quick, convenient data storage and net computing service, with all computing resources visualized as services and delivered over the Internet[2][3]. The cloud enhances collaboration, agility, scalability, availability, ability to adapt to fluctuations according to demand, accelerate development work, and provides potential for cost reduction through optimized and efficient computing[4]–[7].

Cloud Computing combines a number of computing concepts and technologies such as Service Oriented Architecture (SOA), Web 2.0, virtualization and other technologies with reliance on the Internet, providing common business applications online through web browsers to satisfy the computing needs of users, while their software and data are stored on the servers[5]. In some respects, Cloud Computing represents the maturing of these technologies and is a marketing term to represent that maturity and the services they provide[6].

Although there are many benefits to adopting Cloud Computing, there are also some significant barriers to adoption. One of the most significant barriers to adoption is security, followed by issues regarding compliance, privacy and legal matters[8]. Because Cloud Computing represents a relatively new computing model, there is a great deal of uncertainty about how security at all levels (e.g., network, host, application, and data levels) can be achieved and how applications security is moved to Cloud Computing[9]. That uncertainty has consistently led information executives to state that security is their number one concern with Cloud Computing[10].

Security concerns relate to risk areas such as external data storage, dependency on the "public" internet, lack of control, multi-tenancy and integration with internal security. Compared to traditional technologies, the cloud has many specific features, such as its large scale and the fact that resources belonging to cloud providers are completely distributed, heterogeneous and totally virtualized. Traditional security mechanisms such as identity, authentication, and authorization are no longer enough for clouds in their current form[11]. Security controls in Cloud Computing are, for the most part, no different than security controls in any IT environment. However, because of the cloud service models employed, the operational models, and the technologies used to enable cloud services, Cloud Computing may present different risks to an organization than traditional IT solutions. Unfortunately, integrating security into these solutions is often perceived as making them more rigid[4].

Moving critical applications and sensitive data to public cloud environments is of great concern for those corporations that are moving beyond their data center's network under their control. To alleviate these concerns, a cloud solution provider must ensure that customers will continue to have the same security and privacy controls over their applications and services, provide evidence to customers that their organization are secure and they can meet their service-level agreements, and that they can prove compliance to auditors[12].

We present here a categorization of security issues for Cloud Computing focused in the so-called SPI model (SaaS, PaaS and IaaS), identifying the main vulnerabilities in this kind of systems and the most important threats found in the literature related to Cloud Computing and its environment. A threat is a potential attack that may lead to a misuse of information or resources, and the term vulnerability refers to the flaws in a system that allows an attack to be successful. There are some surveys where they focus on one service model, or they focus on listing cloud security issues in general without distinguishing among vulnerabilities and threats. Here, we present a list of vulnerabilities and threats, and we also indicate what cloud service models can be affected by them. Furthermore, we describe the relationship between these vulnerabilities and threats; how these vulnerabilities can be exploited in order to perform an attack, and also present some countermeasures related to these threats which try to solve or improve the identified problems.

The remainder of the paper is organized as follows: Section 2 presents the results obtained from our systematic review. Next, in Section 3 we define in depth the most important security aspects for each layer of the Cloud model. Later, we will analyze the security issues in Cloud Computing identifying the main vulnerabilities for clouds, the most important threats in clouds, and all available countermeasures for these threats and vulnerabilities. Finally, we provide some conclusions.

1.1. Systematic Review of Security Issues for Cloud Computing

We have carried out a systematic review[13]-[15] of the existing literature regarding security in Cloud Computing, not only in order to summarize the existing vulnerabilities and threats concerning this topic but also to identify and analyze the current state and the most important security issues for Cloud Computing.

1.2. Question Formalization

The question focus was to identify the most relevant issues in Cloud Computing which consider vulnerabilities, threats, risks, requirements and solutions of security for Cloud Computing. This question had to be related with the aim of this work; that is to identify and relate vulnerabilities and threats with possible solutions. Therefore, the research question addressed by our research was the following: What security vulnerabilities and threats are the most important in Cloud Computing which have to be studied in depth with the purpose of handling them? The keywords and related concepts that make up this question and that were used during the review execution are: secure Cloud systems, Cloud security, delivery models security, SPI security, SaaS security, Paas security, IaaS security, Cloud threats, Cloud vulnerabilities, Cloud recommendations, best practices in Cloud.

1.3. Selection of Sources

The selection criteria through which we evaluated study sources was based on the research experience of the authors of this work, and in order to select these sources we have considered certain constraints: studies included in the selected sources must be written in English and these sources must be web-available. The following list of sources has been considered: ScienceDirect, ACM digital library, IEEE digital library, Scholar Google and DBLP. Later, the experts will refine the results and will include important works that had not been recovered in these sources and will update these work taking into account other constraints such as impact factor, received cites, important journals, renowned authors, etc.

Once the sources had been defined, it was necessary to describe the process and the criteria for study selection and evaluation. The inclusion and exclusion criteria of this study were based on the research question. We therefore established that the studies must contain issues and topics which consider security on Cloud Computing, and that these studies must describe threats, vulnerabilities, countermeasures, and risks.

1.4. Review Execution

During this phase, the search in the defined sources must be executed and the obtained studies must be evaluated according to the established criteria. After executing the search chain on the selected sources we obtained a set of about 120 results which were filtered with the inclusion criteria to give a set of about 40 relevant studies. This set of relevant studies was again filtered with the exclusion criteria to give a set of studies which corresponds with 15 primary proposals[4][6][10][16]–[27].

2. Results and Discussion

The results of the systematic review are summarized in **Table 1** which shows a summary of the topics and concepts considered for each approach.

As it is shown in **Table 1**, most of the approaches discussed identify, classify, analyze, and list a number of vulnerabilities and threats focused on Cloud Computing. The studies analyze the risks and threats, often give recommendations on how they can be avoided or covered, resulting in a direct relationship between

Table 1. Summary of the topics considered in each approach.

Topics/References	[4]	[6]	[10]	[16]	[17]	[18]	[19]	[20]	[21]	[22]	[23]	[24]	[25]	[26]	[27]
Vulnerabilities		X		X	X	X	X	X	X			X			X
Threats		X		X	X	X	X	X	X	X	X	X	X	X	X
Mechanisms/Recommendations	X			X		X		X				X	X	X	X
Security Standards							X				X				
Data Security	X	X					X		X			X		X		X
Trust		X										X		X	X	X
Security Requirements	X	X							X		X			X	X
SaaS, PaaS, IaaS Security					X				X			X			

vulnerability or threats and possible solutions and mechanisms to solve them. In addition, we can see that in our search, many of the approaches, in addition to speaking about threats and vulnerabilities, also discuss other issues related to security in the Cloud such as the data security, trust, or security recommendations and mechanisms for any of the problems encountered in these environments.

2.1. Security in the SPI Model

The cloud model provides three types of services[21][28][29]:

- Software as a Service (SaaS). The capability provided to the consumer is to use the provider's applications running on a cloud infrastructure. The applications are accessible from various client devices through a thin client interface such as a web browser (e.g., web-based email).

- Platform as a Service (PaaS). The capability provided to the consumer is to deploy onto the cloud infrastructure his own applications without installing any platform or tools on their local machines. PaaS refers to providing platform layer resources, including operating system support and software development frameworks that can be used to build higher-level services.

- Infrastructure as a Service (IaaS). The capability provided to the consumer is to provision processing, storage, networks, and other fundamental computing resources where the consumer is able to deploy and run arbitrary software, which can include operating systems and applications.

With SaaS, the burden of security lies with the cloud provider. In part, this is because of the degree of abstraction, the SaaS model is based on a high degree of integrated functionality with minimal customer control or extensibility. By contrast, the PaaS model offers greater extensibility and greater customer control. Largely because of the relatively lower degree of abstraction, IaaS offers greater tenant or customer control over security than do PaaS or SaaS[10].

Before analyzing security challenges in Cloud Computing, we need to understand the relationships and dependencies between these cloud service models[4]. PaaS as well as SaaS are hosted on top of IaaS; thus, any breach in IaaS will impact the security of both PaaS and SaaS services, but also it may be true on the other way around. However, we have to take into account that PaaS offers a platform to build and deploy SaaS applications, which increases the security dependency between them. As a consequence of these deep dependencies, any attack to any cloud service layer can compromise the upper layers. Each cloud service model comprises its own inherent security flaws; however, they also share some challenges that affect all of them. These relationships and dependencies between cloud models may also be a source of security risks. A SaaS provider may rent a development environment from a PaaS provider, which might also rent an infrastructure from an IaaS provider. Each provider is responsible for securing his own services, which may result in an inconsistent combination of security models. It also creates confusion over which service provider is responsible once an attack happens.

2.2. Software-as-a-Service (SaaS) Security Issues

SaaS provides application services on demand such as email, conferencing software, and business applications such as ERP, CRM, and SCM[30]. SaaS users have less control over security among the three fundamental delivery models in the cloud. The adoption of SaaS applications may raise some security concerns.

2.3. Application Security

These applications are typically delivered via the Internet through a Web browser[12][22]. However, flaws in web applications may create vulnerabilities for the SaaS applications. Attackers have been using the web to compromise user's computers and perform malicious activities such as steal sensitive data[31]. Security challenges in SaaS applications are not different from any web application technology, but traditional security solutions do not effectively protect it from attacks, so new approaches are necessary[21]. The Open Web Application Security Project (OWASP) has identified the ten most critical web applications security threats[32]. There are more security issues, but it is a good start for securing web applications.

2.4. Multi-Tenancy

SaaS applications can be grouped into maturity models that are determined by the following characteristics: scalability, configurability via metadata, and multi-tenancy[30][33]. In the first maturity model, each customer has his own customized instance of the software. This model has drawbacks, but security issues are not so bad compared with the other models. In the second model, the vendor also provides different instances of the applications for each customer, but all instances use the same application code. In this model, customers can change some configuration options to meet their needs. In the third maturity model multi-tenancy is added, so a single instance serves all customers[34]. This approach enables more efficient use of the resources but scalability is limited. Since data from multiple tenants is likely to be stored in the same database, the risk of data leakage between these tenants is high. Security policies are needed to ensure that customer's data are kept separate from other customers[35]. For the final model, applications can be scaled up by moving the application to a more powerful server if needed.

2.5. Data Security

Data security is a common concern for any technology, but it becomes a major challenge when SaaS users have to rely on their providers for proper security[12][21][36]. In SaaS, organizational data is often processed in plaintext and stored in the cloud. The SaaS provider is the one responsible for the security of the data while is being processed and stored[30]. Also, data backup is a critical aspect in order to facilitate recovery in case of disaster, but it introduces security concerns as well[21]. Also cloud providers can subcontract other services such as backup from third-party service providers, which may raise concerns. Moreover, most compliance standards do not envision compliance with regulations in a world of Cloud Computing[12]. In the world of SaaS, the process of compliance is complex because data is located in the provider's datacenters, which may introduce regulatory compliance issues such as data privacy, segregation, and security, that must be enforced by the provider.

2.6. Accessibility

Accessing applications over the internet via web browser makes access from any network device easier, including public computers and mobile devices. However, it also exposes the service to additional security risks. The Cloud Security Alliance[37] has released a document that describes the current state of mobile computing and the top threats in this area such as information stealing mobile malware, insecure networks (WiFi), vulnerabilities found in the device OS and official applications, insecure marketplaces, and proximity-based hacking.

2.7. Platform-as-a-Service (PaaS) Security Issues

PaaS facilitates deployment of cloud-based applications without the cost of buying and maintaining the underlying hardware and software layers[21]. As with SaaS and IaaS, PaaS depends on a secure and reliable network and secure web browser. PaaS application security comprises two software layers: Security of the PaaS platform itself (*i.e.*, runtime engine), and Security of customer applications deployed on a PaaS platform[10]. PaaS providers are responsible for securing the platform software stack that includes the runtime engine that runs the customer applications. Same as SaaS, PaaS also brings data security issues and other challenges that are described as follows:

2.7.1. Third-Party Relationships

Moreover, PaaS does not only provide traditional programming languages, but also does it offer third-party web services components such as mashups [10][38]. Mashups combine more than one source element into a single integrated unit. Thus,

PaaS models also inherit security issues related to mashups such as data and network security[39]. Also, PaaS users have to depend on both the security of web-hosted development tools and third-party services.

2.7.2. Development Life Cycle

From the perspective of the application development, developers face the complexity of building secure applications that may be hosted in the cloud. The speed at which applications will change in the cloud will affect both the System Development Life Cycle (SDLC) and security[12][24]. Developers have to keep in mind that PaaS applications should be upgraded frequently, so they have to ensure that their application development processes are flexible enough to keep up with changes[19]. However, developers also have to understand that any changes in PaaS components can compromise the security of their applications. Besides secure development techniques, developers need to be educated about data legal issues as well, so that data is not stored in inappropriate locations. Data may be stored on different places with different legal regimes that can compromise its privacy and security.

2.7.3. Underlying Infrastructure Security

In PaaS, developers do not usually have access to the underlying layers, so providers are responsible for securing the underlying infrastructure as well as the applications services[40]. Even when developers are in control of the security of their applications, they do not have the assurance that the development environment tools provided by a PaaS provider are secure.

In conclusion, there is less material in the literature about security issues in PaaS. SaaS provides software delivered over the web while PaaS offers development tools to create SaaS applications. However, both of them may use multi-tenant architecture so multiple concurrent users utilize the same software. Also, PaaS applications and user's data are also stored in cloud servers which can be a security concern as discussed on the previous section. In both SaaS and PaaS, data is associated with an application running in the cloud. The security of this data while it is being processed, transferred, and stored depends on the provider.

2.8. Infrastructure-as-a-Service (IaaS) Security Issues

IaaS provides a pool of resources such as servers, storage, networks, and other computing resources in the form of virtualized systems, which are accessed through the Internet[24]. Users are entitled to run any software with full control and management on the resources allocated to them[18]. With IaaS, cloud users have better control over the security compared to the other models as long there is no security hole in the virtual machine monitor[21]. They control the software running in their virtual machines, and they are responsible to configure security policies correctly[41]. However, the underlying compute, network, and storage infrastructure is controlled by cloud providers. IaaS providers must undertake a substantial effort to secure their systems in order to minimize these threats that result from creation, communication, monitoring, modification, and mobility[42]. Here are some of the security issues associated to IaaS.

2.9. Virtualization

Virtualization allows users to create, copy, share, migrate, and roll back virtual machines, which may allow them to run a variety of applications[43][44]. However, it also introduces new opportunities for attackers because of the extra layer that must be secured[31]. Virtual machine security becomes as important as physical machine security, and any flaw in either one may affect the other[19]. Virtualized environments are vulnerable to all types of attacks for normal infrastructures; however, security is a greater challenge as virtualization adds more points of entry and more interconnection complexity[45]. Unlike physical servers, VMs have two boundaries: physical and virtual[24].

2.10. Virtual Machine Monitor

The Virtual Machine Monitor (VMM) or hypervisor is responsible for virtual machines isolation; therefore, if the VMM is compromised, its virtual machines may potentially be compromised as well. The VMM is a low-level software that controls and monitors its virtual machines, so as any traditional software it entails security flaws[45]. Keeping the VMM as simple and small as possible reduces the risk of security vulnerabilities, since it will be easier to find and fix any vulnerability.

Moreover, virtualization introduces the ability to migrate virtual machines between physical servers for fault tolerance, load balancing or maintenance [16][46]. This useful feature can also raise security problems[42][43][47]. An attacker can compromise the migration module in the VMM and transfer a victim virtual machine to a malicious server. Also, it is clear that VM migration exposes the content of the VM to the network, which can compromise its data integrity and confidentiality. A malicious virtual machine can be migrated to another host (with another VMM) compromising it.

2.11. Shared Resource

VMs located on the same server can share CPU, memory, I/O, and others. Sharing resources between VMs may decrease the security of each VM. For example, a malicious VM can infer some information about other VMs through shared memory or other shared resources without need of compromising the hypervisor[46]. Using covert channels, two VMs can communicate bypassing all the rules defined by the security module of the VMM[48]. Thus, a malicious Virtual Machine can monitor shared resources without being noticed by its VMM, so the attacker can infer some information about other virtual machines.

2.12. Public VM Image Repository

In IaaS environments, a VM image is a prepackaged software template containing the configurations files that are used to create VMs. Thus, these images are fundamental for the overall security of the cloud[46][49]. One can either create her own VM image from scratch, or one can use any image stored in the provider's repository. For example, Amazon offers a public image repository where legitimate users can download or upload a VM image. Malicious users can store images containing malicious code into public repositories compromising other users or even the cloud system[20][24][25]. For example, an attacker with a valid account can create an image containing malicious code such as a Trojan horse. If another customer uses this image, the virtual machine that this customer creates will be infected with the hidden malware. Moreover, unintentionally data leakage can be introduced by VM replication[20]. Some confidential information such as passwords or cryptographic keys can be recorded while an image is being created. If the image is not "cleaned", this sensitive information can be exposed to other users. VM images are dormant artifacts that are hard to patch while they are offline[50].

2.13. Virtual Machine Rollback

Furthermore, virtual machines are able to be rolled back to their previous states if an error happens. But rolling back virtual machines can re-expose them to security vulnerabilities that were patched or re-enable previously disabled accounts or passwords. In order to provide rollbacks, we need to make a "copy" (snapshot) of the virtual machine, which can result in the propagation of configuration errors and other vulnerabilities[12][44].

2.14. Virtual Machine Life Cycle

Additionally, it is important to understand the lifecycle of the VMs and their changes in states as they move through the environment. VMs can be on, off, or suspended which makes it harder to detect malware. Also, even when virtual machines are offline, they can be vulnerable[24]; that is, a virtual machine can be instantiated using an image that may contain malicious code. These malicious images can be the starting point of the proliferation of malware by injecting malicious code within other virtual machines in the creation process.

2.15. Virtual Networks

Network components are shared by different tenants due to resource pooling. As mentioned before, sharing resources allows attackers to launch cross-tenant attacks[20]. Virtual Networks increase the VMs interconnectivity, an important security challenge in Cloud Computing[51]. The most secure way is to hook each VM with its host by using dedicated physical channels. However, most hypervisors use virtual networks to link VMs to communicate more directly and efficiently. For instance, most virtualization platforms such as Xen provide two ways to configure virtual networks: bridged and routed, but these techniques increase the possibility to perform some attacks such as sniffing and spoofing virtual network[45][52].

2.16. Analysis of Security Issues in Cloud Computing

We systematically analyze now existing security vulnerabilities and threats of Cloud Computing. For each vulnerability and threat, we identify what cloud service model or models are affected by these security problems.

Table 2 presents an analysis of vulnerabilities in Cloud Computing. This analysis offers a brief description of the vulnerabilities, and indicates what cloud service models (SPI) can be affected by them. For this analysis, we focus mainly on technology-based vulnerabilities; however, there are other vulnerabilities that are common to any organization, but they have to be taken in consideration since they can negatively impact the security of the cloud and its underlying platform. Some of these vulnerabilities are the following:

- Lack of employee screening and poor hiring practices[16]—some cloud providers may not perform background screening of their employees or providers. Privileged users such as cloud administrators usually have unlimited access to the cloud data.

- Lack of customer background checks—most cloud providers do not check their customer's background, and almost anyone can open an account with a valid credit card and email. Apocryphal accounts can let attackers perform any malicious activity without being identified[16].

- Lack of security education—people continue to be a weak point in information security[53]. This is true in any type of organization; however, in the cloud, it has a bigger impact because there are more people that interact with the cloud: cloud providers, third-party providers, suppliers, organizational customers, and end-users.

Cloud Computing leverages many existing technologies such as web services, web browsers, and virtualization, which contributes to the evolution of cloud

Table 2. Vulnerabilities in cloud computing.

ID	Vulnerabilities	Description	Layer
V01	Insecure interfaces and APIs	Cloud providers offer services that can be accessed through APIs (SOAP, REST, or HTTP with XML/JSON) [42]. The security of the cloud depends upon the security of these interfaces [16]. Some problems are:	SPI
		a) Weak credential	
		b) Insufficient authorization checks	
		c) Insufficient input-data validation	
		Also, cloud APIs are still immature which means that are frequently updated. A fixed bug can introduce another security hole in the application [54].	
V02	Unlimited allocation of resources	Inaccurate modeling of resource usage can lead to overbooking or over-provisioning [17].	SPI
V03	Data-related vulnerabilities	a) Data can be colocated with the data of unknown owners (competitors, or intruders) with a weak separation [36]	SPI
		b) Data may be located in different jurisdictions which have different laws [19,54,55]	
		c) Incomplete data deletion – data cannot be completely removed [19,20,25,56]	
		d) Data backup done by untrusted third-party providers [56,57]	
		e) Information about the location of the data usually is unavailable or not disclosed to users [25]	
		f) Data is often stored, processed, and transferred in clear plain text	
V04	Vulnerabilities in Virtual Machines	a) Possible covert channels in the colocation of VMs [48,58,59]	I
		b) Unrestricted allocation and deallocation of resources with VMs [57]	
		c) Uncontrolled Migration - VMs can be migrated from one server to another server due to fault tolerance, load balance, or hardware maintenance [42,44]	
		d) Uncontrolled snapshots – VMs can be copied in order to provide flexibility [12], which may lead to data leakage	
		e) Uncontrolled rollback could lead to reset vulnerabilities - VMs can be backed up to a previous state for restoration [44], but patches applied after the previous state disappear	
		f) VMs have IP addresses that are visible to anyone within the cloud - attackers can map where the target VM is located within the cloud (Cloud cartography [58])	
V05	Vulnerabilities in Virtual Machine Images	a) Uncontrolled placement of VM images in public repositories [24]	I
		b) VM images are not able to be patched since they are dormant artifacts [44]	
V06	Vulnerabilities in Hypervisors	a) Complex hypervisor code [60]	I
		b) Flexible configuration of VMs or hypervisors to meet organization needs can be exploited	
V07	Vulnerabilities in Virtual Networks	Sharing of virtual bridges by several virtual machines [51]	I

environments. Therefore, any vulnerability associated to these technologies also affects the cloud, and it can even have a significant impact.

From **Table 2**, we can conclude that data storage and virtualization are the most critical and an attack to them can do the most harm. Attacks to lower layers have more impact to the other layers. **Table 3** presents an overview of threats in Cloud Computing. Like **Table 2** it also describes the threats that are related to the technology used in cloud environments, and it indicates what cloud service models are exposed to these threats. We put more emphasis on threats that are associated with data being stored and processed remotely, sharing resources and the usage of virtualization.

Table 3. Threats in cloud computing.

ID	Threats	Description	Layer
T01	Account or service hijacking	An account theft can be performed by different ways such as social engineering and weak credentials. If an attacker gains access to a user's credential, he can perform malicious activities such as access sensitive data, manipulate data, and redirect any transaction [16].	SPI
T02	Data scavenging	Since data cannot be completely removed from unless the device is destroyed, attackers may be able to recover this data [10,17,25].	SPI
T03	Data leakage	Data leakage happens when the data gets into the wrong hands while it is being transferred, stored, audited or processed [16,17,20,58].	SPI
T04	Denial of Service	It is possible that a malicious user will take all the possible resources. Thus, the system cannot satisfy any request from other legitimate users due to resources being unavailable.	SPI
T05	Customer-data manipulation	Users attack web applications by manipulating data sent from their application component to the server's application [20,32]. For example, SQL injection, command injection, insecure direct object references, and cross-site scripting.	S
T06	VM escape	It is designed to exploit the hypervisor in order to take control of the underlying infrastructure [24,61].	I
T07	VM hopping	It happens when a VM is able to gain access to another VM (i.e. by exploting some hypervisor vulnerability) [17,43]	I
T08	Malicious VM creation	An attacker who creates a valid account can create a VM image containing malicious code such as a Trojan horse and store it in the provider repository [20].	I
T09	Insecure VM migration	Live migration of virtual machines exposes the contents of the VM state files to the network. An attacker can do the following actions: a) Access data illegally during migration [42] b) Transfer a VM to an untrusted host [44] c) Create and migrate several VM causing disruptions or DoS	I
T10	Sniffing/Spoofing virtual networks	A malicious VM can listen to the virtual network or even use ARP spoofing to redirect packets from/to other VMs [45,51].	I

The relationship between threats and vulnerabilities is illustrated in **Table 4**, which describes how a threat can take advantage of some vulnerability to compromise the system. The goal of this analysis is also to identify some existing defenses that can defeat these threats. This information can be expressed in a more detailed way using misuse patterns[62]. Misuse patterns describe how a misuse is performed from the point of view of the attacker. For instance, in threat T10, an attacker can read or tamper with the contents of the VM state files during live migration. This can be possible because VM migration transfer the data over network channels that are often insecure, such as the Internet. Insecure VM migration can be mitigated by the following proposed techniques: TCCP[63] provides confidential execution of VMs and secure migration operations as well. PALM[64] proposes a secure migration system that provides VM live migration capabilities under the condition that a VMM-protected system is present and active. Threat 11 is another cloud threat where an attacker creates malicious VM image containing any type of virus or malware. This threat is feasible because any legitimate user can create a VM image and publish it on the provider's repository where other users can retrieve them. If the malicious VM image contains malware, it will infect other VMs instantiated with this malicious VM image. In order to overcome this threat,

Table 4. Relationships between threats, vulnerabilities, and countermeasures.

Threat	Vulnerabilities	Incidents	Countermeasures
T01	V01	An attacker can use the victim's account to get access to the target's resources.	Identity and Access Management Guidance [65]
			Dynamic credential [66]
T02	V03a, V03c	Data from hard drives that are shared by several customers cannot be completely removed.	Specify destruction strategies on Service-level Agreements (SLAs)
T03	V03a, V03c, V03d, V03f, V04a-f, V05a, V07	Authors in [58] illustrated the steps necessary to gain confidential information from other VMs co-located in the same server as the attacker.	FRS techniques [67]
			Digital Signatures [68]
		Side channel [69]	Encryption [69]
			Homomorphic encryption [70]
T04	V01, V02	An attacker can request more computational resources, so other legal users are not able to get additional capacity.	Cloud providers can force policies to offer limited computational resources
T05	V01	Some examples are described in [32] such as SQL, command injection, and cross-site scripting	Web application scanners [71]
T06	V06a, V06b	A zero-day exploit in the HyperVM virtualization application that destroyed about 100,000 websites [72]	HyperSafe [60]
			TCCP (Trusted Cloud Computing Platform) [63]
			TVDc (Trusted Virtual Datacenter) [73,74]
T07	V04b, V06b	[75] presents a study that demonstrates security flaws in most virtual machines monitors	
T08	V05a, V05b	An attacker can create a VM image containing malware and publish it in a public repository.	Mirage [49]
T09	V04d	[76] has empirically showed attacks against the migration functionality of the latest version of the Xen and VMware virtualization products.	PALM [64]
			TCCP [63]
			VNSS [52]
T10	V07	Sniffing and spoofing virtual networks [51]	Virtual network framework based on Xen network modes: "bridged" and "routed" [51]

an image management system was proposed, Mirage[49]. It provides the following security management features: access control framework, image filters, provenance tracking system, and repository maintenance services.

2.17. Countermeasures

In this section, we provide a brief description of each countermeasure mentioned before, except for threats T02 and T07.

2.17.1. Countermeasures for T01: Account or Service Hijacking

1) Identity and access management guidance Cloud Security Alliance (CSA) is a non-profit organization that promotes the use of best practices in order to provide security in cloud environments. CSA has issued an Identity and Access Management Guidance[65] which provides a list of recommended best practiced to assure identities and secure access management. This report includes centralized directory, access management, identity management, role-based access control, user access certifications, privileged user and access management, separation of duties, and identity and access reporting.

2) Dynamic credentials[66] presents an algorithm to create dynamic credentials for mobile cloud computing systems. The dynamic credential changes its value once a user changes its location or when he has exchanged a certain number of data packets.

2.17.2. Countermeasures for T03: Data Leakage

1) Fragmentation-redundancy-scattering (FRS) technique[67] this technique aims to provide intrusion tolerance and, in consequence, secure storage. This technique consists in first breaking down sensitive data into insignificant fragments, so any fragment does not have any significant information by itself. Then, fragments are scattered in a redundant fashion across different sites of the distributed system.

2) Digital signatures[68] proposes to secure data using digital signature with RSA algorithm while data is being transferred over the Internet. They claimed that RSA is the most recognizable algorithm, and it can be used to protect data in cloud environments.

3) Homomorphic encryption The three basic operations for cloud data are transfer, store, and process. Encryption techniques can be used to secure data while it is being transferred in and out of the cloud or stored in the provider's premises. Cloud providers have to decrypt cipher data in order to process it, which raises privacy concerns. In[70], they propose a method based on the application of fully homomorphic encryption to the security of clouds. Fully homomorphic encryption allows performing arbitrary computation on ciphertexts without being decrypted. Current homomorphic encryption schemes support limited number of homomorphic operations such as addition and multiplication. The authors in[77] provided some real-world cloud applications where some basic homomorphic operations are needed. However, it requires a huge processing power which may impact on user response time and power consumption.

4) Encryption Encryption techniques have been used for long time to secure sensitive data. Sending or storing encrypted data in the cloud will ensure that data is secure. However, it is true assuming that the encryption algorithms are strong. There are some well-known encryption schemes such as AES (Advanced Encryption Standard). Also, SSL technology can be used to protect data while it is in transit. Moreover, [69]describes that encryption can be used to stop side channel attacks on cloud storage de-duplication, but it may lead to offline dictionary attacks reveling personal keys.

2.17.3. Countermeasures for T05: Customer Data Manipulation

Web application scanners Web applications can be an easy target because they are exposed to the public including potential attackers. Web application scanners[71] is a program which scans web applications through the web front-end in order to identify security vulnerabilities. There are also other web application security tools such as web application firewall. Web application firewall routes all web traffic through the web application firewall which inspects specific threats.

2.17.4. Countermeasures for T06: VM Escape

1) HyperSafe[60] It is an approach that provides hypervisor control-flow integrity. HyperSafe's goal is to protect type I hypervisors using two techniques: non-bypassable memory lockdown which protects write-protected memory pages from being modified, and restricted pointed indexing that converts control data into pointer indexes. In order to evaluate the effectiveness of this approach, they have conducted four types of attacks such as modify the hypervisor code, execute the injected code, modify the page table, and tamper from a return table. They concluded that HyperSafe successfully prevented all these attacks, and that the performance overhead is low.

2) Trusted cloud computing platform TCCP[63] enables providers to offer closed box execution environments, and allows users to determine if the environment is secure before launching their VMs. The TCCP adds two fundamental elements: a trusted virtual machine monitor (TVMM) and a trusted coordinator (TC). The TC manages a set of trusted nodes that run TVMMs, and it is maintained but a trusted third party. The TC participates in the process of launching or migrating a VM, which verifies that a VM is running in a trusted platform. The authors in[78] claimed that TCCP has a significant downside due to the fact that all the transactions have to verify with the TC which creates an overload. They proposed to use Direct Anonymous Attestation (DAA) and Privacy CA scheme to tackle this issue.

3) Trusted virtual datacenter TVDc[73][74] insures isolation and integrity in cloud environments. It groups virtual machines that have common objectives into workloads named Trusted Virtual Domains (TVDs). TVDc provides isolation between workloads by enforcing mandatory access control, hypervisor-based isolation, and protected communication channels such as VLANs. TVDc provides integrity by employing load-time attestation mechanism to verify the integrity of the system.

2.17.5. Countermeasures for T08: Malicious Virtual Machine Creation

Mirage In[49], the authors propose a virtual machine image management system in a cloud computing environments. This approach includes the following security features: access control framework, image filters, a provenance tracking, and repository maintenance services. However, one limitation of this approach is that filters may not be able to scan all malware or remove all the sensitive data from the images. Also, running these filters may raise privacy concerns because they have access to the content of the images which can contain customer's confidential data.

2.17.6. Countermeasures for T09: Insecure Virtual Machine Migration

1) Protection aegis for live migration of VMs (PALM)[64] proposes a secure live migration framework that preserves integrity and privacy protection during and after migration. The prototype of the system was implemented based on Xen and GNU Linux, and the results of the evaluation showed that this scheme only adds slight downtime and migration time due to encryption and decryption.

2) VNSS[52] proposes a security framework that customizes security policies for each virtual machine, and it provides continuous protection thorough virtual machine live migration. They implemented a prototype system based on Xen hypervisors using stateful firewall technologies and userspace tools such as iptables, xm commands program and conntrack-tools. The authors conducted some experiments to evaluate their framework, and the results revealed that the security policies are in place throughout live migration.

2.17.7. Countermeasures for T010: Sniffing/Spoofing Virtual Networks

Virtual network security Wu and *et al.*[51] presents a virtual network framework that secures the communication among virtual machines. This framework is based on Xen which offers two configuration modes for virtual networks: "bridged" and "routed". The virtual network model is composed of three layers: routing layers, firewall, and shared networks, which can prevent VMs from sniffing and spoofing. An evaluation of this approach was not performed when this publication was published.

Furthermore, web services are the largest implementation technology in cloud environments. However, web services also lead to several challenges that need to be addressed. Security web services standards describe how to secure communication between applications through integrity, confidentiality, authentication and authorization. There are several security standard specifications[79] such as Security Assertion Markup Language (SAML), WS-Security, Extensible Access Control Markup (XACML), XML Digital Signature, XML Encryption, Key Management Specification (XKMS), WS-Federation, WS-Secure Conversation, WS-Security Policy and WS-Trust. The NIST Cloud Computing Standards Roadmap Working Group has gathered high level standards that are relevant for Cloud Computing.

3. Conclusions

Cloud Computing is a relatively new concept that presents a good number of benefits for its users; however, it also raises some security problems which may slow down its use. Understanding what vulnerabilities exist in Cloud Computing will help organizations to make the shift towards the Cloud. Since Cloud Computing leverages many technologies, it also inherits their security issues. Traditional web applications, data hosting, and virtualization have been looked over, but some of the solutions offered are immature or inexistent. We have presented security issues for cloud models: IaaS, PaaS, and IaaS, which vary depending on the model. As described in this paper, storage, virtualization, and networks are the biggest security concerns in Cloud Computing. Virtualization which allows multiple users to share a physical server is one of the major concerns for cloud users. Also, another challenge is that there are different types of virtualization technologies, and each type may approach security mechanisms in different ways. Virtual networks are also target for some attacks especially when communicating with remote virtual machines.

Some surveys have discussed security issues about clouds without making any difference between vulnerabilities and threats. We have focused on this distinction, where we consider important to understand these issues. Enumerating these security issues was not enough; that is why we made a relationship between threats and vulnerabilities, so we can identify what vulnerabilities contribute to the execution of these threats and make the system more robust. Also, some current solutions were listed in order to mitigate these threats. However, new security techniques are needed as well as redesigned traditional solutions that can work with cloud architectures. Traditional security mechanisms may not work well in cloud environments because it is a complex architecture that is composed of a combination of different technologies.

We have expressed three of the items in **Table 4** as misuse patterns[46]. We intend to complete all the others in the future.

Competing Interests

The authors declare that they have no competing interests.

Authors' Contributions

KH, DGR, EFM and EBF made a substantial contribution to the systematic review, security analysis of Cloud Computing, and revised the final manuscript version. They all approved the final version to be published.

Acknowledgements

This work was supported in part by the NSF (grants OISE-0730065). Any opinions, findings, and conclusions or recommendations expressed in this material are those of the author(s) and do not necessarily reflect those of the NSF. We also want to thank the GSyA Research Group at the University of Castilla-La Mancha, in Ciudad Real, Spain for collaborating with us in this project.

Source: Hashizume K, Rosado D G, Fernández-Medina E, *et al.* An analysis of security issues for cloud computing [J]. Journal of Internet Services & Applications, 2013, 4(1): 1–13.

References

[1] Gartner Inc Gartner identifies the Top 10 strategic technologies for 2011. Online. Available: http://www.gartner.com/it/page.jsp?id=1454221. Accessed: 15-Jul-2011.

[2] Zhao G, Liu J, Tang Y, Sun W, Zhang F, Ye X, Tang N (2009) Cloud Computing: A Statistics Aspect of Users. In: First International Conference on Cloud Computing (CloudCom), Beijing, China. Springer Berlin, Heidelberg, pp 347–358.

[3] Zhang S, Zhang S, Chen X, Huo X (2010) Cloud Computing Research and Development Trend. In: Second International Conference on Future Networks (ICFN'10), Sanya, Hainan, China. IEEE Computer Society, Washington DC, USA, pp 93–97.

[4] Cloud Security Alliance (2011) Security guidance for critical areas of focus in Cloud Computing V3.0. Available: https://cloudsecurityalliance.org/guidance/csaguide.v3.0.pdf.

[5] Marinos A, Briscoe G (2009) Community Cloud Computing. In: 1st International Conference on Cloud Computing (CloudCom), Beijing, China. Springer-Verlag Berlin, Heidelberg.

[6] Centre for the Protection of National Infrastructure (2010) Information Security Briefing 01/2010 Cloud Computing. Available: http://www.cpni.gov.uk/Documents/Publications/2010/2010007-ISB_cloud_computing.pdf.

[7] Khalid A (2010) Cloud Computing: applying issues in Small Business. In: International Conference on Signal Acquisition and Processing (ICSAP'10), pp 278–281.

[8] KPMG (2010) From hype to future: KPMG's 2010 Cloud Computing survey. Available: http://www.techrepublic.com/whitepapers/from-hype-to-futurekpmgs-2010-cloud-computing-survey/2384291.

[9] Rosado DG, Gómez R, Mellado D, Fernández-Medina E (2012) Security analysis in the migration to cloud environments. Future Internet 4(2):469–487.

[10] Mather T, Kumaraswamy S, Latif S (2009) Cloud Security and Privacy. O'Reilly Media, Inc., Sebastopol, CA.

[11] Li W, Ping L (2009) Trust model to enhance Security and interoperability of Cloud environment. In: Proceedings of the 1st International conference on Cloud Computing. Springer Berlin Heidelberg, Beijing, China, pp 69–79.

[12] Rittinghouse JW, Ransome JF (2009) Security in the Cloud. In: Cloud Computing. Implementation, Management, and Security, CRC Press.

[13] Kitchenham B (2004) Procedures for perfoming systematic review, software engineering group. Department of Computer Scinece Keele University, United Kingdom and Empirical Software Engineering, National ICT Australia Ltd., Australia. TR/SE-0401.

[14] Kitchenham B, Charters S (2007) Guidelines for performing systematic literature reviews in software engineering. Version 2.3 University of keele (software engineering group, school of computer science and mathematics) and Durham. Department of Conputer Science, UK.

[15] Brereton P, Kitchenham BA, Budgen D, Turner M, Khalil M (2007) Lessons from applying the systematic literature review process within the software engineering domain. J Syst Softw 80(4):571–583.

[16] Cloud Security Alliance (2010) Top Threats to Cloud Computing V1.0. Available: https://cloudsecurityalliance.org/research/top-threats.

[17] ENISA (2009) Cloud Computing: benefits, risks and recommendations for information Security. Available: http://www.enisa.europa.eu/activities/riskmanagement/files/deliverables/cloud-computing-risk-assessment.

[18] Dahbur K, Mohammad B, Tarakji AB (2011) A survey of risks, threats and vulnerabilities in Cloud Computing. In: Proceedings of the 2011 International conference on intelligent semantic Web-services and applications. Amman, Jordan, pp 1–6.

[19] Ertaul L, Singhal S, Gökay S (2010) Security challenges in Cloud Computing. In: Proceedings of the 2010 International conference on Security and Management SAM' 10. CSREA Press, Las Vegas, US, pp 36–42.

[20] Grobauer B, Walloschek T, Stocker E (2011) Understanding Cloud Computing vulnerabilities. IEEE Security Privacy 9(2):50–57.

[21] Subashini S, Kavitha V (2011) A survey on Security issues in service delivery models of Cloud Computing. J Netw Comput Appl 34(1):1–11.

[22] Jensen M, Schwenk J, Gruschka N, Iacono LL (2009) On technical Security issues in Cloud Computing. In: IEEE International conference on Cloud Computing (CLOUD' 09). 116, 116, pp 109–116.

[23] Onwubiko C (2010) Security issues to Cloud Computing. In: Antonopoulos N, Gillam L (ed) Cloud Computing: principles, systems & applications. 2010, Springer-Verlag.

[24] Morsy MA, Grundy J, Müller I (2010) An analysis of the Cloud Computing Security problem. In: Proceedings of APSEC 2010 Cloud Workshop. APSEC, Sydney, Australia.

[25] Jansen WA (2011) Cloud Hooks: Security and Privacy Issues in Cloud Computing. In: Proceedings of the 44th Hawaii International Conference on System Sciences, Koloa, Kauai, HI. IEEE Computer Society, Washington DC, USA, pp 1–10.

[26] Zissis D, Lekkas D (2012) Addressing Cloud Computing Security issues. Futur Gener Comput Syst 28(3):583–592.

[27] Jansen W, Grance T (2011) Guidelines on Security and privacy in public Cloud Computing. NIST, Special Publication 800–144, Gaithersburg, MD.

[28] Mell P, Grance T (2011) The NIST definition of Cloud Computing. NIST, Special Publication 800–145, Gaithersburg, MD.

[29] Zhang Q, Cheng L, Boutaba R (2010) Cloud Computing: state-of-the-art and research challenges. Journal of Internet Services Applications 1(1):7–18.

[30] Ju J, Wang Y, Fu J, Wu J, Lin Z (2010) Research on Key Technology in SaaS. In: International Conference on Intelligent Computing and Cognitive Informatics (ICICCI), Hangzhou, China. IEEE Computer Society, Washington DC, USA, pp 384–387.

[31] Owens D (2010) Securing elasticity in the Cloud. Commun ACM 53(6):46–51.

[32] OWASP (2010) The Ten most critical Web application Security risks. Available: https://www.owasp.org/index.php/Category:OWASP_Top_Ten_Project.

[33] Zhang Y, Liu S, Meng X (2009) Towards high level SaaS maturity model: methods and case study. In: Services Computing conference. APSCC, IEEE Asia-Pacific, pp 273–278.

[34] Chong F, Carraro G, Wolter R (2006) Multi-tenant data architecture. Online. Available: http://msdn.microsoft.com/en-us/library/aa479086.aspx. Accessed: 05-Jun-2011.

[35] Bezemer C-P, Zaidman A (2010) Multi-tenant SaaS applications: maintenance dream or nightmare? In: Proceedings of the Joint ERCIM Workshop on Software Evolution (EVOL) and International Workshop on Principles of Software Evolution (IWPSE), Antwerp, Belgium. ACM New York, NY, USA, pp 88–92.

[36] Viega J (2009) Cloud Computing and the common Man. Computer 42 (8):106–108.

[37] Cloud Security Alliance (2012) Security guidance for critical areas of Mobile Computing. Available: https://downloads.cloudsecurityalliance.org/initiatives/mobile/Mobile_Guidance_v1.pdf.

[38] Keene C (2009) The Keene View on Cloud Computing. Online. Available: http://www.keeneview.com/2009/03/what-is-platform-as-service-paas.html. Accessed: 16-Jul-2011.

[39] Xu K, Zhang X, Song M, Song J (2009) Mobile Mashup: Architecture, Challenges and Suggestions. In: International Conference on Management and Service Science. MASS'09. IEEE Computer Society, Washington DC, USA, pp 1–4.

[40] Chandramouli R, Mell P (2010) State of Security readiness. Crossroads 16 (3):23–25.

[41] Jaeger T, Schiffman J (2010) Outlook: cloudy with a chance of Security challenges and improvements. IEEE Security Privacy 8(1):77–80.

[42] Dawoud W, Takouna I, Meinel C (2010) Infrastructure as a service security: Challenges and solutions. In: the 7th International Conference on Informatics and Systems (INFOS), Potsdam, Germany. IEEE Computer Society, Washington DC, USA, pp 1–8.

[43] Jasti A, Shah P, Nagaraj R, Pendse R (2010) Security in multi-tenancy cloud. In: IEEE International Carnahan Conference on Security Technology (ICCST), KS, USA. IEEE Computer Society, Washington DC, USA, pp 35–41.

[44] Garfinkel T, Rosenblum M (2005) When virtual is harder than real: Security challenges in virtual machine based computing environments. In: Proceedings of the 10th conference on Hot Topics in Operating Systems, Santa Fe, NM. volume 10. USENIX Association Berkeley, CA, USA, pp 227–229.

[45] Reuben JS (2007) A survey on virtual machine Security. Seminar on Network Security. http://www.tml.tkk.fi/Publications/C/25/papers/Reuben_final.pdf. Technical report, Helsinki University of Technology, October 2007.

[46] Hashizume K, Yoshioka N, Fernandez EB (2013) Three misuse patterns for Cloud Computing. In: Rosado DG, Mellado D, Fernandez-Medina E, Piattini M (ed) Security engineering for Cloud Computing: approaches and Tools. IGI Global, Pennsylvania, United States, pp 36–53.

[47] Venkatesha S, Sadhu S, Kintali S (2009) Survey of virtual machine migration techniques., Technical report, Dept. of Computer Science, University of California, Santa Barbara. Available: http://www.academia.edu/760613/Survey_of_Virtual_Machine_Migration_Techniques.

[48] Ranjith P, Chandran P, Kaleeswaran S (2012) On covert channels between virtual machines. Journal in Computer Virology Springer 8:85–97.

[49] Wei J, Zhang X, Ammons G, Bala V, Ning P (2009) Managing Security of virtual machine images in a Cloud environment. In: Proceedings of the 2009 ACM workshop on Cloud Computing Security. ACM New York, NY, USA, pp 91–96.

[50] Owens K Securing virtual compute infrastructure in the Cloud. SAVVIS. Available: http://www.savvis.com/en-us/info_center/documents/hoswhitepaper-securingvirutalcomputeinfrastructureinthecloud.pdf.

[51] Wu H, Ding Y, Winer C, Yao L (2010) Network Security for virtual machine in Cloud Computing. In: 5th International conference on computer sciences and convergence information technology (ICCIT). IEEE Computer Society Washington DC, USA, pp 18–21.

[52] Xiaopeng G, Sumei W, Xianqin C (2010) VNSS: a Network Security sandbox for virtual Computing environment. In: IEEE youth conference on information Computing and telecommunications (YC-ICT). IEEE Computer Society, Washington DC, USA, pp 395–398.

[53] Popovic K, Hocenski Z (2010) Cloud Computing Security issues and challenges. In: Proceedings of the 33rd International convention MIPRO. IEEE Computer Society Washington DC, USA, pp 344–349.

[54] Carlin S, Curran K (2011) Cloud Computing Security. International Journal of Ambient Computing and Intelligence 3(1):38–46.

[55] Bisong A, Rahman S (2011) An overview of the Security concerns in Enterprise Cloud Computing. International Journal of Network Security & Its Applications (IJNSA) 3(1):30–45.

[56] Townsend M (2009) Managing a security program in a cloud computing environment. In: Information Security Curriculum Development Conference, Kennesaw, Georgia. ACM New York, NY, USA, pp 128–133.

[57] Winkler V (2011) Securing the Cloud: Cloud computer Security techniques and tactics. Elsevier Inc, Waltham, MA.

[58] Ristenpart T, Tromer E, Shacham H, Savage S (2009) Hey, you, get off of my cloud: exploring information leakage in third-party compute clouds. In: Proceedings of the 16th ACM conference on Computer and communications security, Chicago, Illinois, USA. ACM New York, NY, USA, pp 199–212.

[59] Zhang Y, Juels A, Reiter MK, Ristenpart T (2012) Cross-VM side channels and their use to extract private keys. In: Proceedings of the 2012 ACM conference on Computer and communications security, New York, NY, USA. ACM New York, NY, USA, pp 305–316.

[60] Wang Z, Jiang X (2010) HyperSafe: a lightweight approach to provide lifetime hypervisor control-flow integrity. In: Proceedings of the IEEE symposium on Security and privacy. IEEE Computer Society, Washington DC, USA, pp 380–395.

[61] Wang C, Wang Q, Ren K, Lou W (2009) Ensuring data Storage Security in Cloud Computing. In: The 17th International workshop on quality of service. IEEE Computer Society, Washington DC, USA, pp 1–9.

[62] Fernandez EB, Yoshioka N, Washizaki H (2009) Modeling Misuse Patterns. In: Proceedings of the 4th Int. Workshop on Dependability Aspects of Data Warehousing and Mining Applications (DAWAM 2009), in conjunction with the 4th Int.Conf. on Availability, Reliability, and Security (ARES 2009), Fukuoka, Japan. IEEE Computer

Society, Washington DC, USA, pp 566–571.

[63] Santos N, Gummadi KP, Rodrigues R (2009) Towards Trusted Cloud Computing. In: Proceedings of the 2009 conference on Hot topics in cloud computing, San Diego, California. USENIX Association Berkeley, CA, USA.

[64] Zhang F, Huang Y, Wang H, Chen H, Zang B (2008) PALM: Security Preserving VM Live Migration for Systems with VMM-enforced Protection. In: Trusted Infrastructure Technologies Conference, 2008. APTC'08, Third Asia-Pacific. IEEE Computer Society, Washington DC, USA, pp 9–18.

[65] Cloud Security Alliance (2012) SecaaS implementation guidance, category 1: identity and Access managament. Available: https://downloads.cloudsecurityalliance.org/initiatives/secaas/SecaaS_Cat_1_IAM_Implementation_Guidance.pdf.

[66] Xiao S, Gong W (2010) Mobility Can help: protect user identity with dynamic credential. In: Eleventh International conference on Mobile data Management (MDM). IEEE Computer Society, Washington DC, USA, pp 378–380.

[67] Wylie J, Bakkaloglu M, Pandurangan V, Bigrigg M, Oguz S, Tew K, Williams C, Ganger G, Khosla P (2001) Selecting the right data distribution scheme for a survivable Storage system. CMU-CS-01-120, Pittsburgh, PA.

[68] Somani U, Lakhani K, Mundra M (2010) Implementing digital signature with RSA encryption algorithm to enhance the data Security of Cloud in Cloud Computing. In: 1st International conference on parallel distributed and grid Computing (PDGC). IEEE Computer Society Washington DC, USA, pp 211–216.

[69] Harnik D, Pinkas B, Shulman-Peleg A (2010) Side channels in Cloud services: deduplication in Cloud Storage. IEEE Security Privacy 8(6):40–47.

[70] Tebaa M, El Hajji S, El Ghazi A (2012) Homomorphic encryption method applied to Cloud Computing. In: National Days of Network Security and Systems (JNS2). IEEE Computer Society, Washington DC, USA, pp 86–89.

[71] Fong E, Okun V (2007) Web application scanners: definitions and functions. In: Proceedings of the 40th annual Hawaii International conference on system sciences. IEEE Computer Society, Washington DC, USA.

[72] Goodin D (2009) Webhost hack wipes out data for 100,000 sites. The Register, 08-Jun-2009. [Online]. Available: http://www.theregister.co.uk/2009/06/08/webhost_attack/. Accessed: 02-Aug-2011.

[73] Berger S, Cáceres R, Pendarakis D, Sailer R, Valdez E, Perez R, Schildhauer W, Srinivasan D (2008) TVDc: managing Security in the trusted virtual datacenter. SIGOPS Oper. Syst. Rev. 42(1):40–47.

[74] Berger S, Cáceres R, Goldman K, Pendarakis D, Perez R, Rao JR, Rom E, Sailer R, Schildhauer W, Srinivasan D, Tal S, Valdez E (2009) Security for the Cloud infrastructure: trusted virtual data center implementation. IBM J Res Dev 53 (4):560–571.

[75] Ormandy T (2007) An empirical study into the Security exposure to hosts of hostile virtualized environments. In: CanSecWest applied Security conference, Vancouver. Available: http://taviso.decsystem.org/virtsec.pdf.

[76] Oberheide J, Cooke E, Jahanian F (2008) Empirical exploitation of Live virtual achine migration. In: Proceedings of Black Hat Security Conference, Washington DC. Available: http://www.eecs.umich.edu/fjgroup/pubs/blackhat08-migration.pdf.

[77] Naehrig M, Lauter K, Vaikuntanathan V (2011) Can homomorphic encryption be practical? In: Proceedings of the 3rd ACM workshop on Cloud Computing Security workshop. ACM New York, NY, USA, pp 113–124.

[78] Han-zhang W, Liu-sheng H (2010) An improved trusted cloud computing platform model based on DAA and privacy CA scheme. In: International Conference on Computer Application and System Modeling (ICCASM), vol. 13, V13–39. IEEE Computer, Society, Washington DC, USA, pp V13–V33.

[79] Fernandez EB, Ajaj O, Buckley I, Delessy-Gassant N, Hashizume K, Larrondo-Petrie MM (2012) A survey of patterns for Web services Security and reliability standards. Future Internet 4(2):430–450.

Chapter 5

Estimating the Sentiment of Social Media Content for Security Informatics Applications

Kristin Glass[1], Richard Colbaugh[2]

[1]Institute for Complex Additive Systems Analysis, New Mexico Institute of Mining and Technology, Socorro, USA
[2]Analytics and Cryptography Department, Sandia National Laboratories, Albuquerque, USA

Abstract: Inferring the sentiment of social media content, for instance blog posts and forum threads, is both of great interest to security analysts and technically challenging to accomplish. This paper presents two computational methods for estimating social media sentiment which address the challenges associated with Web-based analysis. Each method formulates the task as one of text classification, models the data as a bipartite graph of documents and words, and assumes that only limited prior information is available regarding the sentiment orientation of any of the documents or words of interest. The first algorithm is a semi-supervised sentiment classifier which combines knowledge of the sentiment labels for a few documents and words with information present in unlabeled data, which is abundant online. The second algorithm assumes existence of a set of labeled documents in a domain related to the domain of interest, and leverages these data to estimate sentiment in the target domain. We demonstrate the utility of the proposed methods by showing they outperform several standard techniques for the task of inferring the sentiment of online movie and consumer product reviews. Additionally, we illustrate the potential of the methods for security informatics by estimating regional public opinion regarding two events: the 2009 Jakarta hotel bombings and 2011 Egyptian revolution.

Keywords: Sentiment Analysis, Social Media, Security Informatics, Machine Learning

1. Introduction

There is increasing recognition that the Web represents a valuable source of security-relevant intelligence and that computational analysis offers a promising way of dealing with the problem of collecting and analyzing data at Web scale (e.g., [1]–[4]). As a consequence, tools and algorithms have been developed which support various security informatics objectives[3][4]. To cite a specific example, we have recently shown that blog network dynamics can be exploited to provide reliable early warning for a class of extremist-related, real-world protest events[5].

Monitoring social media to spot emerging issues and trends and to assess public opinion concerning topics and events is of considerable interest to security professionals; however, performing such analysis is technically challenging. The opinions of individuals and groups are typically expressed as informal communications and are buried in the vast, and largely irrelevant, output of millions of bloggers and other online content producers. Consequently, effectively exploiting these data requires the development of new, automated methods of analysis[3][4]. Although helpful computational analytics have been derived for traditional forms of written content, less has been done to develop techniques that are well-suited to the particular characteristics of the content found in social media.

This paper considers one of the central problems in the new field of social media analytics: deciding whether a given document, such as a blog post or forum thread, expresses positive or negative opinion to-ward a particular topic. The informal nature of social media content poses a challenge for language-based sentiment analysis. While statistical learning-based methods often provide good performance in unstructured settings like this (e.g.,[6]–[13]), obtaining the required labeled instances of data, such as a collection of "exemplar" blog posts of known sentiment polarity, is usually an expensive and time-consuming undertaking.

We present two new computational methods for inferring sentiment orientation of social media content which address these challenges. Each method formulates the task as one of text classification, models the data as a bipartite graph of documents and words, and assumes that only limited prior information is available regarding the sentiment orientation of any of the documents or words of interest. The first algorithm adopts a semi-supervised approach to sentiment classification, combining knowledge of the sentiment polarity for a few documents and a small lexicon of words with information present in a corpus of unlabeled documents; note that such unlabeled data are readily obtainable in online applications. The second algorithm assumes existence of a set of labeled documents in a domain related to the domain of interest, and provides a procedure for transferring the sentiment knowledge contained in these data to the target domain. We demonstrate the utility of the proposed algorithms by showing they outperform several standard methods for the task of inferring the sentiment polarity of online reviews of movies and consumer products. Additionally, we illustrate the potential of the methods for security informatics through two case studies in which sentiment analysis of Arabic, Indonesian, and Danish (language) blogs is used to estimate regional public opinion regarding the 2009 Jakarta hotel bombings and 2011 Egyptian revolution.

2. Preliminaries

We approach the task of estimating the sentiment orientations of a collection of documents as a text classification problem. Each document of interest is represented as a "bag of words" feature vector $x \in \Re^{|V|}$, where the entries of × reflect some measure of the frequency with which the words in the vocabulary set V appear in the document. For example, the elements of × can be simple word-counts, or × can be normalized in various ways[6]. In this paper the elements x_i of × are defined to indicate the presence ($x_i = 1$) or absence ($x_i = 0$) of the corresponding words in the document; however, specifying × using word-counts yields similar results.) We wish to learn a vector $c \in \Re^{|V|}$ such that the classifier orient = $\text{sign}(c^T x)$ accurately estimates the sentiment orientation of document x, returning +1 (−1) for documents expressing positive (negative) sentiment about the topic of interest.

Knowledge-based classifiers leverage prior domain information to construct the vector c. One way to obtain such a classifier is to assemble lexicons of positive words $V^+ \subseteq V$ and negative words $V^- \subseteq V$, and then to set $c_i = +1$ if word $i \in V^+$, $c_i = -1$ if $i \in V^-$, and $c_i = 0$ if i is not in either lexicon; this classifier simply sums the positive and negative sentiment words in the document and assigns document orientation accordingly. While this scheme can provide acceptable performance in certain settings, it is unable to improve its performance or adapt to new domains, and it is usually labor-intensive to construct lexicons which are sufficiently complete to enable useful sentiment classification performance to be achieved.

Alternatively, learning-based methods attempt to generate the classifier vector c from examples of positive and negative sentiment. To obtain a learning-based classifier, one can begin by assembling a set of n_l labeled documents $\{(x_i, d_i)\}$, where $d_i \in \{+1, -1\}$ is the sentiment label for document i. The vector c then can be learned through "training" with the set $\{(x_i, d_i)\}$, for instance by solving the following set of equations for c:

$$\left[X^T X + \gamma I_{|V|}\right] c = X^T d, \quad (1)$$

where matrix $X \in \Re^{nl \times |V|}$ has document vectors for rows, $d \in \Re^{nl}$ is the vector of document labels, $I_{|V|}$ denotes the $|V| \times |V|$ identity matrix, and gγ0 is a constant; this corresponds to regularized least squares (RLS) learning[14]. Many other strategies can be used to compute c, including Naïve Bayes (NB) statistical inference[6]. Learning-based classifiers have the potential to improve their performance and to adapt to new situations, but standard methods for realizing these capabilities require that fairly large training sets of labeled documents be obtained and this is usually expensive.

Sentiment analysis of social media content for security informatics applications is often characterized by the existence of only modest levels of prior knowledge regarding the domain of interest, reflected in the availability of a few labeled documents and small lexicon of sentiment-laden words, and by the need to rapidly learn and adapt to new domains. As a consequence, standard knowledge-based and learning-based sentiment analysis methods are typically ill-suited for security informatics. In order to address this challenge, the sentiment analysis methods developed in this paper enable limited labeled data to be combined with readily available "auxiliary" information to produce accurate sentiment estimates. More specifically, the first proposed method is a semi-supervised algorithm (e.g., [9][10]) which leverages a source of supplementary data which is abundant online: unlabeled documents and words. Our second algorithm is a novel transfer learning method (e.g.,[11]) which permits the knowledge present in data that has been previously labeled in a related domain (say online movie reviews) to be transferred to a new domain (e.g., reviews of consumer products).

Each of the algorithms proposed in this paper assumes the availability of a modest lexicon of sentiment-laden words. This lexicon is encoded as a vector $w \in \Re^{V_1}$, where $V_l = V^+ \cup V^-$ is the sentiment lexicon and the entries of w are set to +1 or −1 according to the polarity of the corresponding words. The development of the algorithms begins by modeling the problem data as a bipartite graph G_b of documents and words (see **Figure 1**). It is easy to see that the adjacency matrix A for graph G_b is given by

$$A = \begin{bmatrix} 0 & X \\ X^T & 0 \end{bmatrix} \quad (2)$$

where the matrix $X \in \Re^{n \times |V|}$ is constructed by stacking the document vectors as rows, and each "0" is a matrix of zeros. In both the semi-supervised and transfer learning algorithms, integration of labeled and "auxiliary" data is accomplished by exploiting the relationships between documents and words encoded in the bipartite graph model. The basic idea is to assume that, in the bipartite graph G_b, positive

Figure 1. Cartoon of bipartite graph model G_b, in which documents (red vertices) are connected to the words (blue vertices) they contain and the link weights (black edges) can reflect word frequencies.

documents will tend to be connected to (contain) positive words, and analogously for negative documents/words.

3. Semi-Supervised Sentiment Analysis

We now derive our first sentiment estimation algorithm for social media content. Consider the common situation in which only limited prior knowledge is available about the way sentiment is expressed in the domain of interest, in the form of small sets of documents and words for which sentiment labels are known, but where abundant unlabeled documents can be easily collected (e.g., via Web crawling). In this setting it is natural to adopt a semi-supervised approach, in which labeled and unlabeled data are combined and leveraged in the analysis process. In what follows we present a novel bipartite graph-based approach to semi-supervised sentiment analysis.

Assume the initial problem data consists of a corpus of n documents, of which $n_l \ll n$ are labeled, and a modest lexicon V_1 of sentiment-laden words, and suppose that this label information is encoded as vectors $d \in \Re^{nl}$ and $w \in \Re^{|V1|}$, respectively. Let $d_{est} \in \Re^n$ be the vector of estimated sentiment orientations for the documents in the corpus, and define the "augmented" classifier $c_{aug} = \begin{bmatrix} d_{est}^T c^T \end{bmatrix}^T \in \Re^{n+|V|}$ which estimates the polarity of both documents and words. Note that the quantity c_{aug} is introduced for notational convenience in the subsequent development and is not directly employed for classification. More specifically, in the proposed methodology we learn c_{aug}, and therefore c, by solving an optimization problem involving the labeled and unlabeled training data, and then use c to estimate the sentiment of any new document of interest with the simple linear classifier orient = sign($c^T x$). We refer to this classifier as semi-supervised because it is learned using both labeled and unlabeled data. Assume for ease of notation that the documents and words are indexed so the first *nl* elements of d_{est} and $|V_1|$ elements of c correspond to the labeled data.

We wish to learn an augmented classifier c_{aug} with the following three properties: 1) if a document is labeled, then the corresponding entry of d_{est} should be close to this ±1 label; 2) if a word is in the sentiment lexicon, then the corresponding entry of c should be close to this ±1 sentiment polarity; and 3) if there is an edge X_{ij} of G_b that connects a document × and a word $v \in V$ and X_{ij} possesses significant weight, then the estimated polarities of × and v should be similar. These objectives are encoded in the following minimization problem:

$$\min_{c_{aug}} c_{aug}^T L c_{aug} + \beta_1 \sum_{i=1}^{n_1} \left(d_{est,i} - d_i \right)^2 + \beta_2 \sum_{i=1}^{|V_1|} \left(c_i - w_i \right)^2 \qquad (3)$$

where $L = D-A$ is the graph Laplacian matrix for G_b, with D the diagonal degree matrix for A (i.e., $D_{ii} = \sum_j A_{ij}$), and b_1, b_2 are nonnegative constants. Minimizing (3) enforces the three properties we seek for c_{aug}, with the second and third terms penalizing "errors" in the first two properties. To see that the first term enforces the third property, observe that this expression is a sum of components of the form $X_{ij}(d_{est,i}-c_j)^2$. The constants β_1, β_2 can be used to balance the relative importance of the three properties.

The c_{aug} which minimizes the objective function (3) can be obtained by solving the following set of linear equations:

$$\begin{bmatrix} L_{11} + \beta_1 I_{nl} & L_{12} & L_{13} & L_{14} \\ L_{21} & L_{22} & L_{23} & L_{24} \\ L_{31} & L_{32} & L_{33} + \beta_2 I_{|V_1|} & L_{34} \\ L_{41} & L_{42} & L_{43} & L_{44} \end{bmatrix} c_{aug} = \begin{bmatrix} \beta_1 d \\ 0 \\ \beta_2 w \\ 0 \end{bmatrix} \qquad (4)$$

where the L_{ij} are matrix blocks of L of appropriate dimension. The system (4) is sparse because the data matrix × is sparse, and therefore large-scale problems can be solved efficiently. Note that in situations where the set of available labeled documents and words is very limited, say less than a couple hundred documents, sentiment classifier performance can be improved by replacing L in (4) with the normalized Laplacian $L_n = D^{-1/2} L D^{-1/2}$, or with a power of this matrix L_n^k (for k a positive integer). The Appendix of this paper demonstrates that replacing L with L_n^k serves to "smooth" the polarity estimates assigned to the vertices of G_b, thereby reducing the possibility for over-fitting and increasing the capability for generalization.

We summarize this discussion by sketching an algorithm for learning the proposed semi-supervised classifier:

Algorithm SS

1) Construct the set of Equations (4), possibly by replacing the graph Laplacian L with L_n^k.

2) Solve equations (4) for $c_{aug} = [d_{est}{}^T c^T]^T$ (for instance using the Conjugate Gradient method).

3) Estimate the sentiment orientation of any new document x of interest as: orient = $\text{sign}(c^T x)$.

The utility of Algorithm SS is now examined through a case study involving a standard sentiment analysis task: estimation of the sentiment polarity of online movie reviews (an exercise which is known to be difficult).

4. Case Study One: Movie Reviews

This case study examines the performance of Algorithm SS for the problem of estimating sentiment of online movie reviews. The data used in this study is a publicly available set of 2000 movie reviews, 1000 positive and 1000 negative, collected from the Internet Movie Database and archived at the website[15]. The Lemur Toolkit[16] was employed to construct the data matrix x and vector of document labels d from these reviews. A lexicon of ~1400 domain-independent sentiment-laden words was obtained from[17] and employed to build the lexicon vector w.

This study compares the movie review orientation classification accuracy of Algorithm SS with that of three other schemes: 1.) lexicon-only, in which the lexicon vector w is used as the classifier as summarized in Section II, 2.) a classical NB classifier obtained from[18], and 3.) a well-tuned version of the RLS classifier (1). Algorithm SS is implemented with the following parameter values: $\beta_1 = 0.1$, $\beta_2 = 0.5$, and $k = 10$. A focus of the investigation is evaluating the extent to which good sentiment estimation performance can be achieved even if only relatively few labeled documents are available for training; thus we examine training sets which incorporate a range of numbers of labeled documents: $n_l = 50, 100, 150,$

200, 300, 400, 600, 800, 1000.

Sample results from this study are depicted in **Figure 2**. Each data point in the plots represents the average of ten trials. In each trial, the movie reviews are randomly partitioned into 1000 training and 1000 test documents, and a randomly selected subset of training documents of size n_l is "labeled" (*i.e.*, the labels for these reviews are made available to the learning algorithms). As shown in **Figure 2**, Algorithm SS outperforms the other three methods. Note that, in particular, the accuracy obtained with the proposed approach is significantly better than the other techniques when the number of labeled training documents is small. It is expected that this property of Algorithm SS will be of considerable value in security informatics applications that involve social media data.

5. Transfer Learning Sentiment Analysis

This section develops the second proposed sentiment estimation algorithm for social media content. Many security informatics applications are characterized by the presence of limited labeled data for the domain of interest but ample labeled

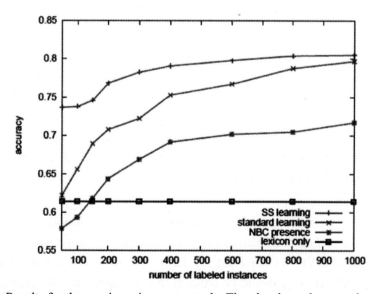

Figure 2. Results for the movie reviews case study. The plot shows how sentiment estimation accuracy (vertical axis) varies with number of available labeled movie reviews (horizontal axis) for four different classifiers: lexicon only (black), NB (magenta), RLS (blue), and Algorithm SS (red).

information for a related domain. For instance, an analyst may wish to ascertain the sentiment of online discussions about an emerging topic of interest, and may have in hand a large set of labeled examples of positive and negative posts regarding other topics (e.g., from previous studies). In this setting it is natural to adopt a transfer learning approach, in which knowledge concerning the way sentiment is expressed in one domain, the so-called source domain, is transferred to permit sentiment estimation in a new target domain. In what follows we present a new bipartite graph-based approach to transfer learning sentiment analysis.

Assume that the initial problem data consists of a corpus of $n = n_T + n_S$ documents, where n_T is the (small) number of labeled documents available for the target domain of interest and $n_S \gg n_T$ is the number of labeled documents from some related source domain; in addition, suppose that a modest lexicon V_1 of sentiment-laden words is known. Let this label data be encoded as vectors $d_T \in \Re^{n_T}$, $d_S \in \Re^{n_S}$, and $w \in \Re^{|V|}$, respectively. Denote by $d_{T,est} \in \Re^{n_T}$, $d_{S,est} \in \Re^{n_S}$, and $c \in \Re^{|V|}$ the vectors of estimated sentiment orientations for the target and source documents and the words, and define the augmented classifier as $c_{aug} = [d_{S,est}^T d_{T,est}^T c^T]^T \in \Re^{n+|V|}$. Note that the quantity c_{aug} is introduced for notational convenience in the subsequent development and is not directly employed for classification.

In what follows we derive an algorithm for learning c_{aug}, and therefore c, by solving an optimization problem involving the labeled source and target training data, and then use c to estimate the sentiment of any new document of interest via the simple linear classifier orient = sign($c^T x$). This classifier is referred to as transfer learning-based because c is learned, in part, by transferring knowledge about the way sentiment is expressed from a domain which is related to (but need not be identical to) the domain of interest.

We wish to learn an augmented classifier c_{aug} with the following four properties: 1.) if a source document is labeled, then the corresponding entry of $d_{S,est}$ should be close to this ±1 label; 2.) if a target document is labeled, then the corresponding entry of $d_{T,est}$ should be close to this ±1 label, and the information encoded in d_T should be emphasized relative to that in the source labels d_S; 3.) if a word is in the sentiment lexicon, then the corresponding entry of c should be close to this ±1 sentiment polarity; and 4.) if there is an edge X_{ij} of G_b that connects a document × and a word $v \in V$ and X_{ij} possesses significant weight, then the estimated polarities of × and v should be similar.

The four objectives listed above may be realized by solving the following minimization problem:

$$\min_{c_{aug}} c_{aug}^T L c_{aug} + \beta_1 \|d_{S,est} - k_S d_S\|^2 + \beta_2 \|d_{T,est} - k_T d_T\|^2 + \beta_3 \|c - w\|^2 \quad (5)$$

where $L = D - A$ is the graph Laplacian matrix for G_b, as before, and $\beta_1, \beta_2, \beta_3, k_S$, and k_T are nonnegative constants. Minimizing (5) enforces the four properties we seek for c_{aug}. More specifically, the second, third, and fourth terms penalize "errors" in the first three properties, and choosing $\beta_2 > \beta_1$ and $k_T > k_S$ favors target label data over source labels. To see that the first term enforces the fourth property, note that this expression is a sum of components of the form $X_{ij} (d_{T,est,i} - c_j)^2$ and $X_{ij} (d_{S,est,i} - c_j)^2$. The constants $\beta_1, \beta_2, \beta_3$ can be used to balance the relative importance of the four properties.

The c_{aug} which minimizes the objective function (5) can be obtained by solving the following set of linear equations:

$$\begin{bmatrix} L_{11} + \beta_1 I_{nS} & L_{12} & L_{13} \\ L_{21} & L_{22} + \beta_2 I_{nT} & L_{23} \\ L_{31} & L_{32} & L_{33} + \beta_3 I_{|V_1|} \end{bmatrix} c_{aug} = \begin{bmatrix} \beta_1 k_S d_S \\ \beta_2 k_T d_T \\ \beta_3 w \end{bmatrix} \quad (6)$$

where the L_{ij} are matrix blocks of L of appropriate dimension. The system (6) is sparse because the data matrix × is sparse, and therefore large-scale problems can be solved efficiently. In applications with very limited labeled data, sentiment classifier performance can be improved by replacing L in (6) with the nor-malized Laplacian L_n or with a power of this matrix L_n^k.

Note that developing systematic methods for characterizing how similar the source and target domains must be to enable useful transfer learning, or for selecting an appropriate source domain given a target set of interest, remain open research problems. Some helpful guidance for these tasks is provided in[12].

We summarize the above discussion by sketching an algorithm for constructing the proposed transfer learning classifier:

Algorithm TL:

1) Construct the set of Equations (6), possibly by replacing the graph Laplacian L with L_n^k

2) Solve equations (6) for $c_{aug} = \left[d_{S,est}^T d_{T,est}^T c^T \right]^T$.

3) Estimate the sentiment orientation of any new document × of interest as: orient = sign($c^T x$).

The utility of Algorithm TL is now examined through a case study involving online consumer product reviews.

6. Case Study Two: Product Reviews

This case study examines the performance of Algorithm TL for the problem of estimating sentiment of online product reviews. The data used in this study is a publicly available set of 1000 reviews of electronics products, 500 positive and 500 negative, and 1000 reviews of kitchen appliances, 500 positive and 500 negative, collected from Amazon and archived at the website[19]. The Lemur Toolkit[16] was employed to construct the data matrix × and vectors of document labels d_S and d_T from these reviews. A lexicon of 150 domain-independent sentiment-laden words was constructed manually and employed to form the lexicon vector w.

This study compares the product review sentiment classification accuracy of Algorithm TL with that of four other strategies: 1) lexicon-only, in which the lexicon vector w is used as the classifier as summarized in Section II, 2) a classical NB classifier obtained from[18], 3) a well-tuned version of the RLS classifier (1), and 4) Algorithm SS. Algorithm TL is implemented with the following parameter values: $\beta_1 = 1.0$, $\beta_2 = 3.0$, $\beta_3 = 5.0$, $k_S = 0.5$, $k_T = 1.0$, and $k = 5$. A focus of the investigation is evaluating the extent to which the knowledge present in labeled reviews from a related domain, here kitchen appliances, can be transferred to a new domain for which only limited labeled data is available, in this case electronics. Thus we assume that all 1000 labeled kitchen reviews are available to Algorithm TL (the only algorithm which is designed to exploit this information), and examine training sets which incorporate a range of numbers of labeled documents from the electronics domain: $n_T = 20, 50, 100, 200, 300, 400$.

Sample results from this study are depicted in **Figure 3**. Each data point in the plots represents the average of ten trials. In each trial, the electronics reviews are randomly partitioned into 500 training and 500 test documents, and a randomly selected subset of reviews of size n_T is extracted from the 500 labeled training instances and made available to the learning algorithms. As shown in **Figure 3**, Algorithm TL outperforms the other four methods. Note that, in particular, the accuracy obtained with the transfer learning approach is significantly better than the other techniques when the number of labeled training documents in the target domain is small. It is expected that the ability of Algorithm TL to exploit knowledge from a related domain to quickly learn an effective sentiment classifier for a new domain will be of considerable value in security informatics applications involving social media data.

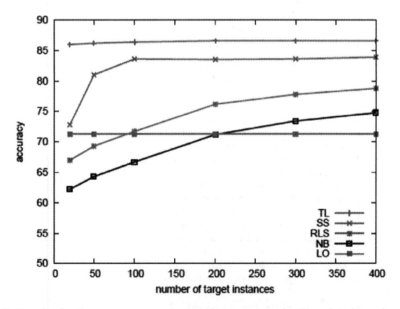

Figure 3. Results for the consumer product reviews case study. The plot shows how sentiment estimation accuracy (vertical axis) varies with number of available labeled electronics reviews (horizontal axis) for five different classifiers: lexicon only (orange), NB (black), RLS (magenta), Algorithm SS (blue), and Algorithm TL (red).

7. Case Study Three: Jakarta Hotel Bombings

On 17 July 2009 the JW Marriott and Ritz-Carlton Hotels in Jakarta, Indonesia were hit by suicide bombing attacks within five minutes of each other. A little over a week later, on 26 July 2009, a document claiming responsibility for the attacks and allegedly written by N.M. Top was posted on the blog[20]; see **Figure 4** for a screenshot of a portion of this blog post. In subsequent discussions we will refer to this post as the "Top post" for convenience, with the understanding that the authorship of the post is uncertain. At this time, senior U.S. intelligence and security officials expressed interest in understanding sentiment in the region regarding the bombings and the alleged claim of responsibility by a well-known extremist

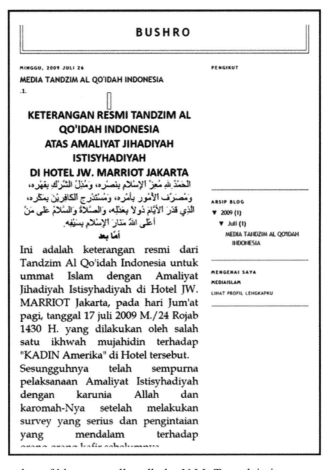

Figure 4. Screenshot of blog post, allegedly by N.M. Top, claiming responsibility for the July 2009 bombings of two hotels in Jakarta, Indonesia.

[personal communications, Senior U.S. Intelligence and Security Officials, July 2009]. Among other things, officials felt that characterizing this sentiment might provide insight into Indonesian public opinion concerning violent extremist organizations.

To enable a preliminary assessment along these lines, we collected two sets of social media data related to the Top post: 1) the ~3000 comments made to the post during the two week period immediately following the post, and 2) several hundred posts made to other Indonesian language blogs in which the Top post was discussed. We manually labeled the sentiment of a small subset of these documents, and also translated into Indonesian the generic sentiment lexicon used in Case Study One (in this paper all language translation was performed using the tool available at http://translate.google.com). Observe that this approach to constructing a sentiment lexicon is far from perfect. However, because our proposed algorithms employ several sources of information to estimate the sentiment of content, it is expected that they will exhibit robustness to imperfections in any single data source. This study therefore offers the opportunity to explore the utility of a very simple approach to multilingual sentiment analysis: translate a small lexicon of sentiment-laden words into the language of interest and then apply Algorithm SS or Algorithm TL directly within that language (treating words as tokens). The capability to perform automated, multilingual content analysis is of substantial interest in many security-related applications.

We implemented Algorithm SS to estimate the sentiment expressed in the corpus of comments made to the Top post[20] and in the set of related discussions posted at other blogs. Sample sentiment estimation results for the comments made to the Top post are shown at the top of **Figure 5**. These comments are almost universally negative, condemning both the bombings and the justification for the bombings given in the Top post. Manual examination of a subset of the comments confirms the results provided by Algorithm SS. For example, a typical negative comment reads, in part (translated from the Indonesian):

> "...a savage kind—Noordin M. Top does not deserve to live in this world, he only claims to serve Islam but actually he is in truth a disbeliever. You are all cowards, you terrorists reversing Islam, and Noordin Top is just a stupid terrorist who escaped Malaysia just to inflict violence and hatred..."

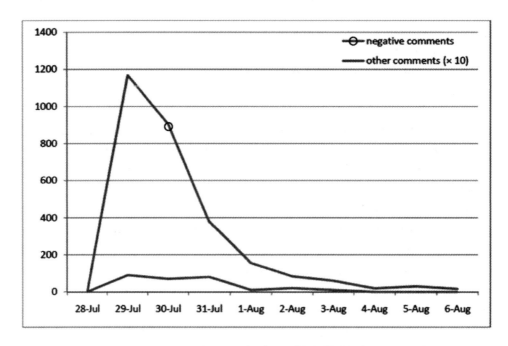

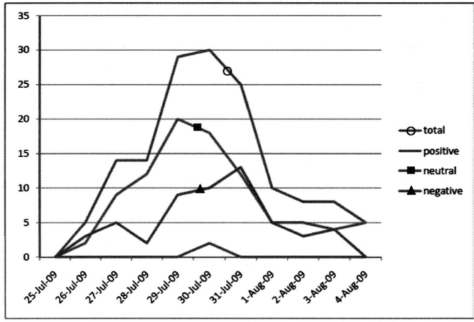

Figure 5. Sample results for blog analysis case study. The top plot shows the estimated sentiment of comments made directly to the blog[20] in response to the Top post (blue is negative, red is "conspiracy-related", with red data multiplied by a factor of 10 to be visible on the graph). The bottom plot gives estimated sentiment of blog posts discussing the Top post (total is blue, positive is red, olive is neutral, and purple is negative, respectively).

Interestingly, by slightly reformulating the classifier it is possible to discover that while almost all comments are negative, there is a thread of comments which puts forth various conspiracies associated with the origins of the bombings and/or the Top post. The following comment is illustrative of this theme:

"...the only explanation for the sophistication of the attacks is that the bombers were aided by the CIA or Mossad."

Sample results obtained through sentiment analysis of relevant posts made to other blogs are presented at the bottom of **Figure 5**. These posts also express largely negative sentiments about the Top post, although they are not as consistently negative as are the comments made directly on the Top post blog site. It should be noted that the "neutral" post sentiment depicted in the plot at the bottom of **Figure 5** are mainly news articles. Again, manual evaluation of a subset of these posts confirm the results obtained via auto-mated classification with Algorithm SS.

8. Case Study Four: Egyptian Revolution

Beginning on 25 January 2011, a popular uprising swept across Egypt in the form of massive demonstrations and rallies, labor strikes in various sectors, and violent clashes between protestors and security forces, ultimately leading to the resignation of Egyptian President Hosni Mubarak on 11 February. National security analysts and officials have expressed interest in understanding public sentiment regarding the Egyptian revolution generally and Mubarak specifically, especially 1) in the weeks before the protests and 2) for different regions of the globe.

To enable a preliminary assessment along these lines, we collected three sets of blog posts which are related to Egyptian unrest and Mubarak and were posted during the two week period immediately before the protests began on 25 January: 1) 100 Arabic posts, 2) 100 Indonesian posts, and 3) 100 Danish posts. We manually labeled the sentiment of a small subset of these documents, and translated into the appropriate language the generic sentiment lexicon used in Case Study One for implementation in this study (using the translation tool available at http://translate.google.com). Observe that this approach to constructing a sentiment lexicon is far from perfect. However, because lexical information is only one of the data sources used in the proposed approaches to sentiment estimation, it is expected that overall performance will exhibit robustness to imperfections in any one source of data; indeed, this study provides a simple test of this expected robustness.

We used Algorithm SS to estimate the sentiment expressed in the three sets of blog posts noted above, classifying the posts as either "negative" or "positive/neutral". The analysis reveals that, while the sentiment expressed by the bloggers in the sample is largely negative toward Mubarak, the fraction of negative posts varies by post language (and thus possibly by geographic region). In particular, as shown in **Figure 6**, Arabic language posts are the most negative, followed by Indonesian posts, with the Danish posts in our sample actually being slightly more positive/neutral than negative. Manual inspection of a subset of the comments confirms the results provided by Algorithm SS.

9. Summary

Sentiment analysis of social media content for security informatics applications is often characterized by the existence of only modest levels of prior knowledge regarding the domain of interest and the need to rapidly adapt to new domains; consequently, standard content analysis methods typically perform poorly

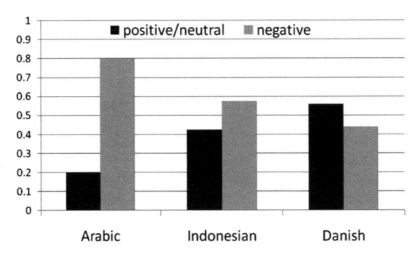

Figure 6. Results for Egyptian revolution case study. The bar chart shows that, while blog sentiment toward former Egyptian President Hosni Mubarak in the weeks leading up to the protests was largely negative, this sentiment varies with the language of posts. More specifically, the fraction of posts expressing negative sentiment regarding Mubarak is 0.80 for Arabic posts, 0.58 for Indonesian posts, and 0.44 for Danish posts.

in this setting. This paper presents two new computational methods for inferring the sentiment expressed in social media which address these challenges. Each method formulates the task as one of text classification, models the data as a bipartite graph of documents and words, and enables prior knowledge concerning the sentiment orientation of documents or words of interest to be effectively combined with "auxiliary" information to produce accurate sentiment estimates. The first proposed method is a semi-supervised algorithm that leverages a source of supplementary data which is abundant online: unlabeled documents and words. The second algorithm is a novel transfer learning method which permits the knowledge present in data that has been previously labeled in a related domain to be transferred to a new domain. We demonstrate the utility of the proposed algorithms by showing they outperform several standard methods for the task of inferring the sentiment polarity of online reviews of movies and consumer products. Additionally, we illustrate the potential of the methods for security informatics by estimating regional public opinion regarding two security-relevant events: the 2009 Jakarta hotel bombings and the 2011 Egyptian revolution.

The proposed algorithms are complementary in that they exploit different sources of auxiliary information (each of which is frequently available in security informatics applications). For instance, Algorithm SS is able to extract useful information from unlabeled documents and words, which are typically abundant online. This capability is particularly valuable in applications for which it is difficult or expensive to acquire any form of labeled data. Alternatively, if previous analysis has produced labeled information in a domain related to the current domain of interest, Algorithm TL can be used to effectively leverage this related data; observe that this situation is common in intelligence analysis settings.

Appendix

As summarized in Section 3, solving the optimization problem balances three objectives:

$$\min_{c_{aug}} c_{aug}^T L_n c_{aug} + \beta_1 \sum_{i=1}^{n_1} (d_{est,i} - d_i)^2 + \beta_2 \sum_{i=1}^{|V_1|} (c_i - w_i)^2$$

- clustering: reducing $c^T_{aug} L_n c_{aug}$ ensures that polarity estimates for documents,

$d_{est,i}$, and words, c_j, are assigned to reduce the magnitude of terms of the form

$$X_{ij}\left(\frac{d_{est,i}}{D_{1,ii}^{1/2}} - \frac{c_j}{D_{2,jj}^{1/2}}\right)^2$$

where $D_{1,ii}^{1/2} = \left(\sum_j X_{ij}\right)^{1/2}$ and $D_{2,jj}^{1/2} = \left(\sum_i X_{ij}\right)^{1/2}$;

- prior knowledge on documents: if document i is labeled d_i then the polarity estimate $d_{est,i}$ should be close to this label;

- prior knowledge on words: if word i is labeled w_i then the polarity estimate c_i should be close to this label.

It is shown in this paper that this procedure enables accurate sentiment classifiers to be learned. However, in situations in which very little labeled data is available, this approach can produce numerous isolated polarity clusters around labeled instances on the graph G_b, resulting in over-fitted solutions with little power for generalization. Here we show that replacing the term $\varphi = c^T_{aug} L_n c_{aug}$ with $\varphi_k = c^T_{aug} L^k_n c_{aug}$, where k a positive integer, smoothes the sentiment polarity estimates on G_b and resolves this difficulty.

Note first that L_n can be expressed as

$$L_n = \sum_{i=1}^{n+|V|} \lambda_i Z_i Z_i^T$$

where (λ_i, z_i) are eigenvalue-eigenvector pairs for L_n, and similarly that L_n^k can be written

$$L_n^k = \sum_{i=1}^{n+|V|} \lambda_i^k Z_i Z_i^T$$

Next observe that the quantity

$$\emptyset = c^T_{aug} L_n c_{aug} = \sum_i \sum_j X_{ij} \left(\frac{d_{est,i}}{D_{1,ii}^{1/2}} - \frac{c_j}{D_{2,jj}^{1/2}}\right)^2$$

measures the smoothness of the document-word polarity assignment specified by caug. If the eigenvalues λ_i are ordered so that $0 = \lambda_1 \leq \lambda_2 \leq ... \leq \lambda_{n+|V|}$ then, because $z_j^T L_n z_j = \lambda_j$, it is seen that the eigenvectors z_i of L_n are ordered by their smoothness. Then, since $\{z1, ..., z_{n+|V|}\}$ is an orthonormal basis (L_n is sym-metric), caug can be expanded as $c_{aug} = \sum_i \alpha_i z_i$ and φ becomes $\varphi = \sum_i \lambda_i \alpha_i^2$

Analogously, $\varphi_k = c_{aug}^T L_n^k c_{aug}$ can be written

$$\varnothing_k = \left(\sum_i \alpha_i z_i\right)\left(\sum_j \lambda_j^k z_j z_j^T\right)\left(\sum_m \alpha_m z_m\right) = \sum_i \lambda_i^k \alpha_i^2$$

It follows that minimizing φ_k instead of φ results in smoother polarity specifications on G_b, because the former imposes a more aggressive penalization of polarity assignments that include the larger (and less smooth) eigenvalue terms.

Acknowledgements

This work was supported by the U.S. Department of Defense and the Laboratory Directed Research and Development Program at Sandia National Laboratories. Sandia National Laboratories is a multi-program laboratory managed and operated by Sandia Corporation, a wholly owned subsidiary of Lockheed Martin Corporation, for the U.S. Department of Energy's National Nuclear Security Administration under contract DE-AC04-94AL85000.

Authors' Contributions

KG and RC designed the research, KG and RC developed the computational algorithms, KG conducted the empirical tests, and RC wrote the paper. All authors read and approved the final manuscript.

Competing Interests

The authors declare that they have no competing interests.

Source: Glass K, Colbaugh R. Estimating the sentiment of social media content for security informatics applications [J]. Security Informatics, 2011, 1(1): 65–70.

References

[1] US Committee on Homeland Security and Government Affairs, Violent Extremism, the Internet, and the Homegrown Terrorism Threat. 2008.

[2] Bergin A, Osman S, Ungerer C, Yasin N: Countering Internet Radicalization in Southeast Asia. ASPI Special Report 2009.

[3] In Intelligence and Security Informatics. Edited by: Chen H, Yang C, Chau M, Li S. Lecture Notes in Computer Science, Springer, Berlin; 2009.

[4] Proc 2010 IEEE International Conference on Intelligence and Security Informatics, Vancouver 2010.

[5] Colbaugh R, Glass K: Early warning analysis for social diffusion events. Proc 2010 IEEE International Conference on Intelligence and Security Informatics, Vancouver 2010.

[6] Pang B, Lee L: Opinion mining and sentiment analysis. Foundations and Trends in Information Retrieval 2008, 2:1–135.

[7] Dhillon I: Co-clustering documents and words using bipartite spectral graph partitioning. Proc ACM International Conference on Knowledge Discovery and Data Mining San Francisco; 2001.

[8] Kim S, Hovy E: Determining the sentiment of opinions. Proc International Conference on Computational Linguistics 2004.

[9] Sindhwani V, Melville P: Document-word co-regularization for semi-supervised sentiment analysis. Proc 2008 IEEE International Conference on Data Mining, Pisa 2008.

[10] Colbaugh R, Glass K: Estimating sentiment orientation in social media for intelligence monitoring and analysis. Proc 2010 IEEE International Conference on Intelligence and Security Informatics, Vancouver 2010.

[11] Pan S, Yang Q: A survey on transfer learning. IEEE Trans Knowledge and Data Engineering 2010, 22:1345–1359.

[12] Blitzer J, Dredze M, Perieia F: Biographies, bollywood, boom-boxes, and blenders: Domain adaptation for sentiment classification. Proc 45th Annual Meeting of the ACL, Prague 2007.

[13] He J, Liu Y, Lawrence R: Graph-based transfer learning. Proc 18th ACM Conference on Information and Knowledge Management Hong Kong; 2009.

[14] Hastie T, Tibshirani R, Friedman J: The Elements of Statistical Learning. Second edition. Springer, New York; 2009.

[15] http://www.cs.cornell.edu/People/pabo/movie-review-data/ accessed 2009.

[16] http://www.lemurproject.org/ accessed Dec 2009.

[17] Ramakrishnan G, Jadhav A, Joshi A, Chakrabarti S, Bhattacharyya P: Question answering via Bayesian inference on lexical relations. Proc 41st Annual Meeting of the ACL Sapporo; 2003.

[18] http://www.borgelt.net/bayes.html accessed Dec. 2009.

[19] http://www.cs.jhu.edu/~mdredze/ accessed Dec. 2010.

[20] www.mediaislam-bushro.blogspot.com accessed Aug. 2009.

Chapter 6

If You Want to Know About a Hunter, Study His Prey: Detection of Network Based Attacks on KVM Based Cloud Environments

Nikolaos Pitropakis[1], Dimitra Anastasopoulou[1], Aggelos Pikrakis[2], Costas Lambrinoudakis[1]

[1]Department of Digital Systems, University of Piraeus, Piraeus, Greece
[2]Department of Informatics, University of Piraeus, Piraeus, Greece

Abstract: Computational systems are gradually moving towards Cloud Computing Infrastructures, using the several advantages they have to offer and especially the economic advantages in the era of an economic crisis. In addition to this revolution, several security matters emerged and especially the confrontation of malicious insiders. This paper proposes a methodology for detecting the co-residency and network stressing attacks in the kernel layer of a Kvm-based cloud environment, using an implementation of the Smith-Waterman genetic algorithm. The proposed approach has been explored in a test bed environment, producing results that verify its effectiveness.

Keywords: Cloud Computing, Security, Co-Residency, Network Stressing, Malicious Insider, KVM, System Calls, Smith-Waterman

1. Introduction

Distributed systems have made a huge renovation in Information Technology (IT) infrastructures. Their continuation is the Cloud Computing. Despite a modern trend and a new economic model, Cloud Computing has made its statement turning into the technological model employed by the majority of large companies and organizations for facilitating their everyday needs. It is well known however that every novelty, despite offering a lot of advantages, also brings several disadvantages. The latter usually remains hidden, until a horror story appears. We refer to the security threats that the new technology has raised. They can be classified as: related to the service provider or to the infrastructure or to the host of the Cloud System.

Several of them are well known from conventional IT infrastructures: Distributed Denial of Service[1] came with distributed systems and still draws the attention of security experts, while social engineering attacks[2], malware and Trojan horses[3] are also popular for their impact on modern IT infrastructures. Despite the inherited threats, there are newly generated risks that need confrontation. The most important of them are Loss of governance[4], data interception[3] and replay attacks[3].

Our work focuses on the older and most unpredictable threat that existed before IT systems were born: the human factor. We refer to malicious insiders[4][5] of a Cloud Computing Infrastructure. Their activities can harm the confidentiality, integrity and availability of the data and services of a cloud system. The commonest role that a malicious insider has in a cloud infrastructure is that of the administrator; either the administrator of the host or one of the administrators of the virtual machines (VM). The privileges of an administrator allow several kinds of attacks to be launched. However, our work focuses on the network attacks and especially the stressing of the host network and the co-residency attack[6]. To be specific the stressing of the network is the basic component of DOS and DDOS attacks[7], where packets are continuously sent to the target in order to stop it from behaving properly and eventually deny its services to others. In the case of co-residency attack[6], we talk about the detection of neighbouring VMs and the retrieval of information about them such as their operating system. The leakage of so important information can seriously harm the cloud infrastructure.

There have been numerous attempts to counter networking stressing attacks[7][8] in their DOS and DDOS form. There are also attempts aiming to handle the activities of a malicious insider through the implementation of several different IDSs, connected through an event gatherer[9]. However, none of these attempts has managed to successfully prevent the actions of malicious insiders.

This paper, presents a novel method for identifying network based attacks in a cloud infrastructure. To this respect a KVM-based[10] system has been employed with its host OS Dom0 having direct access to all I/O functions of the system. This access is materialized by monitoring the system calls made by the kernel of the Dom0 operating systems. The proposed method has utilized the Smith-Wa- terman algorithm[11] to prove that by monitoring the system calls, the malicious actions of a potential cloud insider can be detected.

The rest of the paper is organised as follows: Section Related work and network attacks briefly describes the co-residency and the network stressing attacks. Section Detection method provides background information about the Smith-Waterman algorithm and detailed description of the proposed method. Section Test-bed environment and results of the experiments presents the test-bed environment, the applied automation methodology and the results of the tests conducted. Section Discussion contains annotations about the results, while section Conclusion and future work draws the conclusions giving some pointers to future work.

2. Related Work and Network Attacks

There are several approaches attempting to track, disable or even eliminate the malicious insider threat. Some of them focus on a specific aspect of the cloud such as the employees or the network, while others try to present a global solution. Few of them are able to differentiate themselves from existing solutions, inherited by conventional information systems.

Spring suggests that a firewall at the cloud border that blocks troublesome packets can limit, but cannot eliminate, access to known malicious entities[12]. Al-zain, Pardede, Soh and Thom suggest that moving from single cloud to multi-clouds will greatly reduce the malicious insider s threat as the information is spread among the interclouds and can t be retrieved from a single Cloud Infrastructure[13]. Another effort focuses on employing logistic regression models to estimate false positive/negatives in intrusion detection and identification of malicious insiders. Furthermore, it insists on developing new protocols that cope with denial of service and insider attacks and ensure predictable delivery of mission critical data[14].

Magklaras, Furnell and Papadaki[15] suggest an audit engine for logging user actions in relational mode (LUARM) that attempts to solve two fundamental problems of the insider IT misuse domain. Firstly, is the lack of insider misuse case data repositories that could be used by post-case forensic examiners to aid incident investigations and, secondly, how information security researchers can enhance their ability to accurately specify insider threats at system level.

Tripathi and Mishra[16] insist that cloud providers should provide controls to customer, which can detect and prevent malicious insiders threats. They add that malicious insider threats can be mitigated by specifying human resources requirements as part of legal contracts, conducting a comprehensive supplier assessment. This procedure would lead to reporting and determining security breach notification processes.

Fog computing[17] suggests an approach totally different from the others. The access operations of each cloud user are monitored, realising a sort of profiling for each user. This profiling facilitates the detection of abnormal behaviour. When unauthorized access is suspected and then verified, the method uses disinformation attacks by returning large amounts of decoy information to the malicious insiders, keeping this way the privacy of the real users data.

An approach, which is totally different from the latter, is that of Cuong Hoang H. Lee[18], which achieves security in a Xen based hypervisor[19] by trapping hypercalls, as they are fewer than system calls. The hypercalls are checked before their execution and thus malicious ones can be detected. A combination of the two latter methods takes advantage of the system calls, collecting them and classifying them in normal and abnormal through binary weighted cosine metric and k nearest neighbour classifier[20].

Paying special attention to access control mechanisms, Kollam and Sunnyvale[21] present a mechanism that generates immutable security policies for a client, propagates and enforces them at the provider s infrastructure. This is one of the few methods aiming directly at malicious insiders and especially   system administrators.

The reference to co-residence or (co-tenancy) implies that multiple independent customers share the same physical infrastructure[22]. This fact results in a scheme where Virtual Machines owned by different customers may be placed in the same physical machine. There are several methods that can achieve the discovery of neighbouring Virtual Machines in a Cloud infrastructure. There are also other methods who wish to counter this specific attack.

Adam Bates[23], claims that co-residency detection is also possible through network flow watermarking. To be specific, this is a type of network converting timing channel, capable of breaking anonymity by tracing the path of the network flow. It can also perform a variety of traffic analysis tasks. However, many drawbacks exist in this method, with the most important one being the introduction of a considerable delay in the network.

Ristenpart[6] presents the co-residency potential attacks on Amazon EC2, one of the largest Cloud Infrastructures. In his methodology he includes network tools such as nmap[24], hping[25] and wget[26], which are utilized in order to create network probes that will acquire the addresses of the potential targets. Additionally, the addresses are used to make a hypothetic map of the cloud network that will be tested in the third step. In the manifestation of the method he explores whether two instances are co-resident or not through a series of checks that depend on:

1) matching Dom0 IP address,

2) small packet round trip times, or,

3) numerically close internal IP address.

Project Silverline[27] aims to achieve both data and network isolation. Pseudo randomly-allocated IP address are used for each VM, hiding the actual IP addresses provided by the cloud provider. Then, in each Dom0, SilverLine replaces the pseudo IP addresses by the actual addresses before packets leave the machine. Since IP addresses are also discovered through DNS requests, the SilverLine also rewrites DNS responses to appropriate pseudo addresses.

Another approach, namely Homealone[28] allows the verification of the physical isolation of a Virtual Machine through the same tool that can launch co-residency attacks, performed through side channels that usually offer vulnerabilities. L2 memory cache is a popular way to reach the data of another VM. However, in the latter scenario L2 memory is silenced for the period of time needed by the system with upper purpose the residence information not to be acquired by another physical machine. In practice this is rather difficult as the L2 memory in a virtualized environment is never quiet and in most cases there is no physical isolation among the Virtual Machines.

There are numerous attempts to protect Cloud Infrastructures, not only from the co-residency attack but from other network stressing attacks too, by employing Intrusion Detection Systems (IDS). Most of them make use of multiple agents that are installed in different Virtual Machines and collect the data into a centralized point. The disadvantage is that they introduce considerable overhead to the Cloud infrastructure, since they consume significant amount of resources[29]–[34]. An interesting approach is that of Bakshi and Yogesh[7], who transfer the targeted applications to VMs hosted in another data center when they pick up grossly abnormal spike in inbound traffic.

It can be deduced that the majority of attacks that can be launched by insiders for detecting neighbouring virtual machines or just stressing the network of a Cloud Infrastructure, are based on simple network attacks. In a similar fashion the attacks that have been utilized in this paper for demonstrating the proposed detection method are very simple. Before explaining the attacks it should be stated that in order to launch them the attacker should know the ip address of the virtual machine. In our scenario the attacker is the administrator of a virtual machine with the Kali Linux Operating System[35], the ancestor of Backtrack Operating System[36], which offers to our hypothetic malicious insider a variety of tools.

In the case of the co-residecny attack, the attacker after obtaining the ip address of his virtual machine, is working on finding the Domain Name System (DNS) address. This can be easily retrieved through the command nslookup followed by the ip address of the Virtual Machine (VM). This command, executed in the Kali Linux kernel, will return the DNS address. After obtaining the DNS address, the attacker can use the nmap command to acquire the ip addresses of all virtual machines (including host) utilising the specific DNS. Specifically the command executed is nmap sP DNS_Adress/24. Having the ip addresses of all virtual machines that use the same DNS, the attacker can identify the Operating System of either the Host or of the other Virtual Machines, by executing the command nmap v O Ip_address. Through the aforementioned three distinct steps, all co- residents can be identified along with additional information about their operating systems, something that can allow the attacker to launch further attacks harming the Cloud Infrastructure.

Network stress is executed by launching a smurf attack[37] on a specially configured virtual network. In order to perform a smurf attack, the attacker needs the IPv6 address of the victim. The victim can be the Host or any other Virtual Machine on the same network. His IPv6 address can be obtained using two methods. The first one is via the ifconfig command, which can be executed on the Host. The second method is detecting IPv6-active hosts on the same network via the ping6 command[38]. The attacker can easily ping the link-local all-node multicast address ff02::1 from any virtual machine by executing the command "ping6 -I < interface > ff02::1". After obtaining the IPv6 address, the attacker can use the smurf6 tool to perform the attack, executing the command "smurf6 < interface > victim_ipv6_address". Through this method the attacker VM (or the Host) will flood the Virtual Network with spoofed ICMPv6 echo request packets, the source address of which is the IPv6 address of the victim machine and destination address is the link-local all-node multicast address ff02::1. Then the remaining machines on the same network will flood the victim with ICMPv6 echo replies, thus stressing the virtual network even more.

3. Detection Method

3.1. Algorithm

The proposed detection scheme has adopted the standard Smith-Waterman algorithm which was originally introduced in the context of molecular sequence analysis[9]. This was possible because the data streams under study consist of symbols drawn from a finite discrete alphabet. A minor modification introduced has to do with two parameters which refer to the number of horizontal and vertical predecessors which are allowed to be scanned in order to determine the accumulated cost at each node of the similarity grid. In other words, these two parameters define the maximum allowable gap length, both horizontally and vertically. This type of minor modification causes a significant improvement in response times and it is also in accordance with the nature of the data that are processed. The values of these two parameters, along with the gap penalty have been the result of extensive experimentation. Next the adopted Smith-Waterman algorithm is presented.

First of all, the pair wise (local) similarity between the individual elements of the two symbol sequences must be defined. To this end, let A and B be the two symbol sequences and $A(i), i = 1, M$, $B(j), j = 1, N$, be the i-th symbol of A and j-th symbol of B, respectively. The local similarity, $S(i,j)$, between $A(i)$ and $B(j)$ is then defined as

$$S(i;j) = 1; \text{ if } A(i) = B(j)$$

and

$$S(i;j) = -Gp; \text{ if } A(i) \neq B(j);$$

where Gp is the penalty for dissimilarity (a parameter to our method).

3.1.1. Initialization

Then a similarity grid, H, is created with its first row and column being initialized to zeros, *i.e.*,

$$H(0;j) = 0; \quad j = 0; N$$

and

$$H(i;0) = 0; \quad i = 0; M$$

As a result, the dimensions of the similarity grid are (M + 1) x (N + 1), its rows are indexed 0, ..., M and its columns are indexed 0, N.

3.1.2. Iteration

For each node, (i,j),i > =1, j > =1, of the grid, the accumulated similarity cost is computed according to the equation:

$$H(i,j) = \max \begin{cases} 0, \\ H(i-1, j-1) + S(i,j), \\ H(i-k, j-1) - k*Gp, k \in [1, Pv], \\ H(i, j-l-1) - l*Gpl, l \in [1, Ph], \end{cases}$$

$$i \in [1, M], j \in [1, N],$$

where Pv and Ph are the maximum allowable vertical and horizontal gaps (measured in number of symbols) respectively and Gp is the previously introduced dissimilarity penalty (which in this case also serves as a gap penalty). The above equation is repeated for all nodes of the grid, starting from the lowest row (i= 1) and moving from left to right (increasing index j). It can be seen that vertical and horizontal transitions (third and fourth branch of the equation) introduce a gap penalty, i.e., reduce the accumulated similarity by an amount which is proportional to the number of nodes that are being skipped (length of the gap).

In addition, if the accumulated similarity, H(i,j), is negative, then it is set to zero (first branch of the equation) and the fictitious node (0,0) becomes the predecessor of (i,j). If, on the other hand, the accumulated similarity is positive, the predecessor of (i,j) is the node which maximizes H(i,j). The coordinates of the best predecessor of each node are stored in a separate matrix. Concerning the first row and first column of the grid, the predecessor is always the fictitious node (0,0).

3.1.3. Backtracking

After the accumulated cost has been computed for all nodes, the node which corresponds to the maximum detected value is selected and the chain of predecessors is followed until a (0,0) node is encountered. This procedure is known as backtracking and the resulting chain of nodes is the best (optimal alignment) path.

In the experiments performed, different values of the parameters Pv, Ph and Gp have been used and finally the values that provided the most satisfactory performance have been selected.

3.2. Proposed Method

Fictional character David Rossi, inspired by John E. Douglas, one of the creators of criminal profiling program, once said *If you want to know about a hunter study his prey*[39]. The proposed methodology has been inspired by the above quote. The work of a malicious insider on a KVM-based cloud system, is performed with system calls of the host operating system. In order to investigate the type and sequence of system calls employed, the Linux Audit[40] tool has been used for capturing them.

The procedure that has been followed is the following:

- The system calls engaged during the execution of the *nslookup* command (first step of the co-residency attack), *nmap sP DNS_Adress/24* command (second step of the co-residency attack), *nmap v O Ip_address* (third step of the co-residency attack) and *smurf6 < interface > victim_ipv6_address* (smurf attack) are captured.

- The system calls engaged during the same time period of normal system operation (no attack is being launched) are captured.

- The above log files have been processed with the use of regular expressions and the "*sed*" command[41], leaving only the ID of each system call.

- Finally, the Smith-Waterman algorithm has been employed to compare the logs (every system call ID is being used by the algorithm as a DNA element).

Initially, the similarity between multiple executions of each attack step, at different time periods, was calculated with the use of an automated system that reduced the errors because of the human responsiveness. Then the similarity between an attack step and the respective time period of normal operation was derived. Ideally, this approach would facilitate the identification of specific system call patterns that will form the attack signature.

4. Test-Bed Environment and Results of the Experiments

4.1. Setup the Environment

In order to launch the attack and monitor the system logs, a minimal Cloud Infrastructure was built using one Dell PowerEdge T410 server with the following configuration: Intel Xeon E5607 as Central Processing Unit, 8 Gigabytes of memory running at 1333 MHz and 300 Gigabytes SAS HDD @10000rpms. The server was running OpenSuse Linux 12.1[42]. Also the Linux audit[40] tool was installed; this tool has a configuration file that stores a list of rules that specify which type of system calls will be logged. To avoid losing valuable information during our experiments all system calls were captured. Specifically the rule used was -a entry, always s all. Finally, two VMs with Kali Linux[35], containing the majority of the tools used for penetration testing and attacks, were set up on the server (see **Figure 1**).

4.2. Automating the Attack and System Calls Auditing Procedure

During our effort to automate the attack and the system call auditing procedure, a script was written in Expect[43]. Expect is an extension to the Tcl scripting language and it's used to automate interactions with programs that expose a text terminal interface. This feature can be installed through the expect package. Our script focuses on waiting for expected output with the use of the "expect" command, sending proper input with the use of the "send" command and eventually execute the necessary bash commands with the use of the "system" command. Initially, a directory in which the system calls are going to be saved, was created. Next, the "spawn" command to open the Virsh console[44] and connect to the virtual machine via a configured serial console, was executed. Virsh is a command

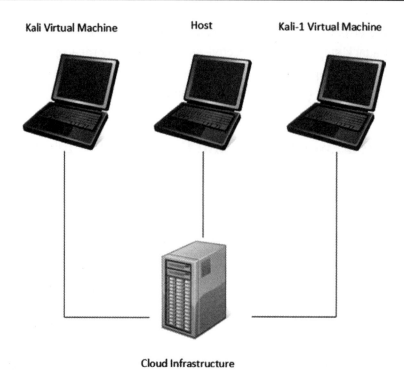

Figure 1. Test-bed environment.

line interface tool, used for the management of guests and the hypervisor. Then the Linux auditing system was enabled and the attack command was sent to the virtual machine that will be executed. Knowledge about when the attack is finished is acquired by waiting for a specific output of the expect command. Finally, the Linux auditing system is disabled and the saved system calls are extracted.

4.3. Launching the Attack

Having setup the environment, each one of the three steps of the co-residency attack (nslookup, nmap and nmap v O Ip_address commands; see section Proposed method) and the step of smurf attack (smurf6 < interface > victim_ipv6_address) were executed six times, each time capturing the system calls engaged.

After every single execution of a command (attack step), the system was left working in normal state for a time period equal to the execution time of the command, capturing again all the system calls engaged during that period. The time periods for the attack and the respective normal state periods are depicted in **Figure 2** and **Figure 3**.

Then by employing the Smith Waterman implementation (see Section Algorithm) in Matlab, using Gp equal to 1/3 and 1/5, Pv and Ph equal to 5 the following log sets were compared between them:

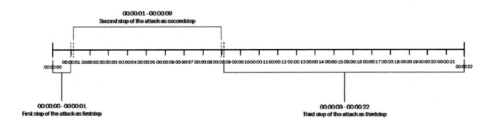

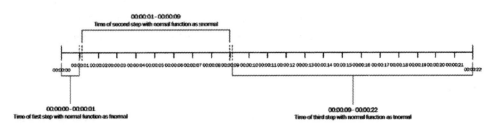

Figure 2. Time periods for the execution of the three attack steps and the respective time periods that the system was kept in normal state.

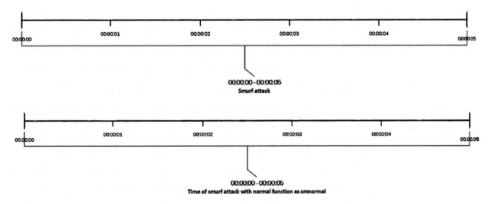

Figure 3. Time periods for the execution of the smurf attack and the respective time periods that the system was kept in normal state.

- The six log files (one for each execution round) of the first attack step; nslookup command.

- The six log files (one for each execution round) of the second attack step; nmap sP DNS_address/24 command.

- The six log files (one for each execution round) of the third attack step; nmap v O Ip_address command.

- The six log files (one for each execution round) of the smurf attack step.

- The twenty four log files of the attack (six log files for all executions of each attack step and smurf attack) with the respective log files for normal system operation.

As demonstrated in the next section, the results met our initial hypothesis. Greater similarity was found between the log files corresponding to the attack steps rather than between the attack logs and the logs of a normal system state.

4.4. Results

The results of the log files comparison are presented in the following **Table 1**, **Table 2**, **Table 3**, **Table 4**, **Table 5**, **Table 6**, **Table 7**, **Table 8** and **Table 9**. As illustrated in **Figure 2** and **Figure 3**, the logs of the first attack step are referred as firststep, the logs of the second attack step as secondstep, the logs of the third one as thirdstep and the logs of the smurf attack as smurfstep. Furthermore, the logs corresponding to normal system operation for a time period equal to that of the first attack step are referred as fnormal, of the second attack step are referred as snormal, of the third attack step are referred as tnormal and of the smurf attack as smnormal. The estimated similarity numbers that appear in the Gp columns represent the longest subseries of system calls that ware found similar using the Smith Waterman algorithm. It is expected from the training procedure that the similarity values will be larger when comparing the logs of the attack steps, and smaller when comparing the logs of an attack step and the respective log of normal system operation; i.e. it is expected that for the same Gp the firstep 1 2 will have larger similarity from the similarity of firstep1-fnormal1. This assumption is

Table 1. Comparison of the six log files (one for each execution round) of the first attack step for Gp equal to 1/3 and 1/5.

Log file comparison	Gp = 1/3	Gp = 1/5
firststep 1-2	1697.000000	1783.800000
firststep 2-3	2065.000000	2160.600000
firststep 3-4	2116.333333	2212.600000
firststep 4-5	1825.000000	1939.400000
firststep 5-6	1805.333333	1898.600000

Table 2. Comparison of the six log files (one for each execution round) of the first attack step for Gp equal to 1/3 and 1/5.

Log file comparison		Gp = 1/3	Gp = 1/5
firststep1	fnormal1	571.333333	630.800000
firststep2	fnormal2	1180.666667	1261.400000
firststep3	fnormal3	1162.666667	1227.800000
firststep4	fnormal4	1107.666667	1189.000000
firststep5	fnormal5	1198.000000	1261.200000
firststep6	fnormal6	144.000000	247.000000

Table 3. Comparison of the six log files (one for each execution round) of the second attack step for Gp equal to 1/3 and 1/5.

Log file comparison	Gp = 1/3	Gp = 1/5
secondstep 1-2	2419.333333	3103.000000
secondstep 2-3	1870.666667	2662.200000
secondstep 3-4	1907.666667	2816.600000
secondstep 4-5	2477.333333	3276.600000
secondstep 5-6	1668.000000	2351.200000

Table 4. Comparison of the six log files (one for each execution round) of the second attack step for Gp equal to 1/3 and 1/5.

Log file comparison		Gp = 1/3	Gp = 1/5
secondstep1	snormal1	171.333333	174.400000
secondstep2	snormal2	452.333333	889.200000
secondstep3	snormal3	1004.666667	1343.800000
secondstep4	snormal4	562.000000	977.600000
secondstep5	snormal5	787.000000	1123.400000
secondstep6	snormal6	595.000000	1051.800000

Table 5. Comparison of the six log files (one for each execution round) of the third attack step for Gp equal to 1/3 and 1/5.

Log file comparison	Gp = 1/3	Gp = 1/5
thirdstep 1-2	2024.000000	2776.000000
thirdstep 2-3	2739.666667	3691.000000
thirdstep 3-4	2486.666667	3447.000000
thirdstep 4-5	3226.000000	4222.800000
thirdstep 5-6	3129.333333	4140.600000

Table 6. Comparison of the six log files (one for each execution round) of the third attack step for Gp equal to 1/3 and 1/5.

Log file comparison		Gp = 1/3	Gp = 1/5
thirdstep1	tnormal1	536.666667	559.200000
thirdstep2	tnormal2	573.666667	1042.400000
thirdstep3	tnormal3	688.666667	1269.000000
thirdstep4	tnormal4	478.666667	970.600000
thirdstep5	tnormal5	878.000000	1323.400000
thirdstep6	tnormal6	562.333333	973.200000

Table 7. Comparison of the six log files (one for each execution round) of the smurf attack step for Gp equal to 1/3 and 1/5.

Log file comparison	Gp = 1/3	Gp = 1/5
smurfstep 1-2	3155.333333	3277.000000
smurfstep 2-3	2758.333333	2891.400000
smurfstep 3-4	3093.333333	3179.800000
smurfstep 4-5	3230.666667	3304.800000
smurfstep 5-6	2712.666667	2838.400000

Table 8. Comparison of the six log files (one for each execution round) of the smurf attack step for Gp equal to 1/3 and 1/5.

Log file comparison		Gp = 1/3	Gp = 1/5
smurfstep1	smnormal1	217.000000	443.600000
smurfstep2	smnormal2	176.666667	403.400000
smurfstep3	smnormal3	641.333333	791.600000
smurfstep4	smnormal4	695.666667	922.400000
smurfstep5	smnormal5	106.000000	265.000000
smurfstep6	smnormal6	738.333333	1052.800000

Table 9. Comparison of the two log files for each attack step with normal execution with a large amount of network operations for Gp equal to 1/3.

Log file comparison	Gp = 1/3
firststep1 fnormal1	422.000000
firststep2 fnormal2	449.000000
secondstep1 snormal1	529.666667
secondstep2 snormal2	556.333333
thirdstep1 snormal1	218.666667
thirdstep2 snormal2	259.666667
smurfstep1-smnormal1	126.333333
smurfstep2-smnormal2	211.666667

strengthened with the results of our last **Table 9** where we compare the logs of the execution of each step of the attack with the logs of a system that performs a large amount of network operations that greatly increases the number of system calls. All results are visualized in **Figure 4**.

5. Discussion

Recalling our main objective, that was to identify the existence of an attack through the sequences of the system calls. The results, which were presented in the previous section, have indeed verified that approach, since the comparison of the system calls triggered during the attack steps exhibits a much larger similarity than that produced when comparing the logs from some attack step and the respective logs for normal system operation. This assumption came true for all three steps of the co-residence attack and the smurf attack.

It would be a common query whether the results are accurate or not, and how can we verify their correctness. This question can be easily answered through the error parameter, Gp, which was used. To be specific, Gp is a variable that offers flexibility to the algorithm and defines how tolerant the algorithm will be during the comparison of the data sets. If we use the error value of 1/3, we have a less tolerant algorithm than when we use the value 1/5. This assumption leads to greater similarity figures being produced with a Gp of 1/5 than with a Gp of 1/3. Of course this is proved with our results, which were presented in the previous section.

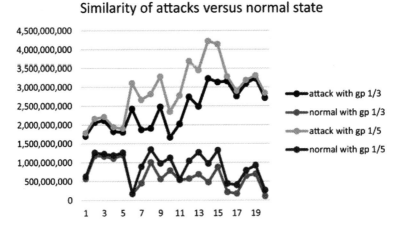

Figure 4. Graph depicting similarity between attacks and between attacks and normal system state for gp 1/3 and 1/5 respectively. Lower gp offers greater similarity.

In addition to that, we have to pay attention to the fact that the more tolerant the algorithm is, the better the similarity that we get among the logs of the attack steps. However, this is not the case for the comparison of logs produced during an attack step and the respective normal operation; specifically, even though the similarity is better for bigger values of Gp, the scaling is not the same.

Another important issue that should be considered is the workload of the system. During our experimentations we used three Virtual Machines and none of them had any permanent jobs other than those corresponding to the attack steps. In a real time environment, which has extra load on the virtual machines, the number of system calls would be much larger, with results on the time required for processing the log files (as described earlier in the paper). Furthermore, the tracking of the attack in this workload would be more difficult as the algorithm compares identities without being able to recognize whether or not a specific element is useful or not. Nevertheless, an initial set of experiments performed with increased workload indicate that the accuracy and effectiveness of the proposed detection method remains unaltered.

Conclusion and Future Work

In this paper a practical method for detecting malicious insider attacks from the system calls of the Host Operating System of a KVM based Cloud Infrastructure has been proposed. The approach has been evaluated by comparing the list of system calls produced during the different steps of the attack, not only with other executions of the same attack steps, but also with the normal system state during the same time the attack took place. The results have confirmed the initial assumption that the system calls can be utilized for the detection of an insider attack.

The focus of our current research work is the construction of system call patterns that will be used as 'attack signatures. The latter will help us build an IDS mechanism, which will be used for the generation of alerts and the prevention of many malicious actions.

Competing Interests

The authors declare that they have no competing interests.

Authors' Contributions

AP was the one who proposed the utilization of the Smith-Waterman algorithm, worked on its configuration and the specific implementation, while he wrote the section about the Smith-Waterman algorithm. DA was responsible for setting up the smurf attack and for conducting the experiments together with NP. She wrote the appropriate sections about the smurf attack and KVM hypervisor. NP was responsible for all technical issues and for setting up the test bed environment and the system calls recovery method. He also wrote the remaining sections of the paper. CL supervised the whole effort providing advice and guidelines on scientific issues, on the experimental methods adopted and on the writing process. All authors read and approved the final manuscript.

Acknowledgements

We sincerely thank all the researchers of the Systems Security Laboratory at University of Piraeus for the inspiration they provide and their useful comments throughout our research.

Source: Pitropakis N, Anastasopoulou D, Pikrakis A, *et al.* If you want to know about a hunter, study his prey: detection of network based attacks on KVM based cloud environments [J]. Journal of Cloud Computing, 2014, 3(1): 1–10.

References

[1] Douligeris C, Mitrokotsa A (2004) Ddos Attacks and Defense Mechanisms: Classification and State-Of-The-Art. In: Computer Networks, pp.643–666.

[2] Orgill GL, Romney GW, Bailey MG, Orgill PM (2004) The Urgency for Effective User Privacy-education to Counter Social Engineering Attacks on Secure Computer Systems. Proceedings of the Conference on Information Technology Education, CITC5.

[3] Krutz RL, Vines RD (2010) Cloud Security: A Comprehensive Guide To Secure Cloud Computing. Wiley Publishing Inc., Indianapolis.

[4] Enisa (2009) Cloud Computing Benefits, risks and recommendations for information security.

[5] Kandias M., Virvilis N., Gritzalis D. (2011) "The Insider Threat in Cloud Computing," in Proc. of the 6th International Conference on Critical Infrastructure Security (CRITIS-2011), Wolthusen S., *et al.* (Eds.), pp. 95–106, Springer, Switzerland, September 2011.

[6] Ristenpart T, Tromer E, Shacham H, Savage S (2009) Hey, You, Get Off of My Cloud: Exploring Information Leakage in Third-Party Compute Clouds. In ACM CCS, Chicago.

[7] Bakshi, Aman, and B. Yogesh (2010) "Securing cloud from ddos attacks using intrusion detection system in virtual machine." Communication Software and Networks, 2010. ICCSN'10. Second International Conference on. IEEE.

[8] Liu, Huan (2010) "A new form of DOS attack in a cloud and its avoidance mechanism." Proceedings of the 2010 ACM workshop on Cloud computing security workshop, ACM.

[9] Roschke S, Cheng F, Meinel C (2010) An Advanced IDS Management Architecture. J Inform Assur Security 5:246–255.

[10] KVM Hypervisor. http://www.linux-kvm.org/page/Main_Page.

[11] Smith TF, Waterman MS (1981) Identification of common molecular subsequences. J Mol Biol 147.1:195–197.

[12] Spring J (2011) "Monitoring Cloud Computing By Layer, Part 1." Security & Privacy, IEEE 9.2, pp.66–68.

[13] AlZain MA, Pardede E, Soh B, Thom JA (2012) Cloud computing security: from

single to multi-clouds. System Science (HICSS), 2012 45th Hawaii International Conference on. IEEE.

[14] Sandhu R, Boppana R, Krishnan R, Reich J, Wolff T, Zachry J (2010) "Towards A Discipline Of Mission-Aware Cloud Computing." Proceedings of The 2010 ACM Workshop On Cloud Computing Security Workshop. ACM, Chicago.

[15] Magklaras G, Furnell S, Papadaki M (2011) LUARM: An audit engine for insider misuse detection. Int J Digital Crime Forensics 3(3):37–49.

[16] Tripathi, Alok, and Abhinav Mishra (2011) "Cloud computing security considerations." Signal Processing, Communications and Computing (ICSPCC), 2011 IEEE International Conference on. IEEE.

[17] Stolfo, Salvatore J., Malek Ben Salem, and Angelos D. Keromytis (2012) "Fog computing: Mitigating insider data theft attacks in the cloud." Security and Privacy Workshops (SPW), 2012 IEEE Symposium on. IEEE.

[18] Hoang C (2009) Protecting Xen hypercalls, MSC thesis. University of British Columbia, Canada.

[19] XEN. http://www.xenproject.org/developers/teams/hypervisor.html.

[20] Rawat S, Gulati VP, Pujari AK, Vemuri VR (2006) Intrusion detection using text processing techniques with a binary-weighted cosine metric. J Inform Assur Security 1(1):43–50.

[21] Sundararajan S, Narayanan H, Pavithran V, Vorungati K, Achuthan K (2011) Preventing Insider attacks in the Cloud. In: Advances in Computing and Communications. Springer, Berlin Heidelberg, pp.488–500.

[22] Xiao Z., Xiao Y. (2012) Security and Privacy in Cloud Computing, Communications Sur- veys & Tutorials, IEEE, vol. PP no.99, pp.1–17.

[23] Bates A, Mood B, Pletcher J, Pruse H, Valafar M, Butler K (2012) Detecting co-residency with active traffic analysis techniques. In: Proceedings of the 2012 ACM Workshop on Cloud computing security workshop. ACM, NC, USA, pp.1–12.

[24] Nmap, http://nmap.org/.

[25] Hping, http://sectools.org/tool/hping/.

[26] Wget, http://www.gnu.org/software/wget/.

[27] Mundada Y, Ramachndran A, Feamster N (2011) SilverLine: Data and network isolation for cloud services , In Proceedings of the USENIX Workshop on Hot Topics in Cloud Computing (HotCloud).

[28] Zhang Y, Juels A, Oprea A, Reiter A (2011) HomeAlone: Co-Residency Detection in the Cloud via Side-Channel Analysis. Security and Privacy IEEE Symposium, Berkeley, CA.

[29] Mazzariello C, Bifulco R, Canonico R (2010) integrating a Network IDS into an

Open Source Cloud Computing Environment, Sixth International Conference on Information Assurance and Security.

[30] Schulter A, Vieira K, Westphal C, Westaphal C, Abderrrahim S (2008) Intrusion Detection For Computational Grids . In: Proc. 2nd Int l Conf. New Technologies Mobility, and Security. IEEE Press, Tangier, Morocco.

[31] Cheng F, Roschke S, Meinel C (2009) Implementing IDS Management on Lock-Keeper, Proceedings of 5th Information Security Practice and Experience Conference (ISPEC 09). Springer LNCS 5451:360–371.

[32] Cheng F, Roschke S, Meinel C (2010) An Advanced IDS Management Architecture , Journal of Information Assurance and Security, Dynamic Publishers Inc., vol. 51, Atlanta, GA 30362, USA. ISSN 1554-1010:246–255.

[33] Cheng F, Roschke S, Meinel C (2009) Intrusion Detection in the Cloud, Eighth IEEE International Conference on Dependable. Autonomic and Secure Computing, China.

[34] Bharadwaja S., Sun W., Niamat M., Shen F., Collabra: Axen Hypervisor based Collaborative Intrusion Detection System, Proceedings of the 8th International Conference on Information Technology: New Generations (ITNG 11), pp. 695–700, Las Vegas, Nev, USA, 2011.

[35] Kali Linux. http://www.kali.org/.

[36] Backtrack Linux. http://www.backtrack-linux.org/.

[37] Smurf Attack. http://www.ciscopress.com/articles/article.asp?p=1312796.

[38] IPv6 Ping. http://www.tldp.org/HOWTO/Linux%2BIPv6-HOWTO/x811.html.

[39] John E. Douglas. http://en.wikipedia.org/wiki/John_E._Douglas.

[40] Linux Audit. http://doc.opensuse.org/products/draft/SLES/SLES-security_sd_draft/cha.audit.comp.html.

[41] Sed command. http://linux.die.net/man/1/sed.

[42] OpenSuse. http://www.opensuse.org/.

[43] TCL Scripting. http://www.tcl.tk/man/expect5.31/expect.1.html.

[44] Virsh. https://access.redhat.com/site/documentation/en-US/Red_Hat_Enterprise_Linux/6/html/Virtualization_Administration_Guide/chapVirtualization_Administration_Guide-Managing_guests_with_virsh.html.

Chapter 7
Intelligent Feature Selection and Classification Techniques for Intrusion Detection in Networks: A Survey

Sannasi Ganapathy, Kanagasabai Kulothungan, Sannasy Muthurajkumar, Muthusamy Vijayalakshmi, Palanichamy Yogesh, Arputharaj Kannan

Department of Information Science and Technology, College of Engineering Guindy, Anna University, Chennai 25, Tamilnadu, India

Abstract: Rapid growth in the Internet usage and diverse military applications have led researchers to think of intelligent systems that can assist the users and applications in getting the services by delivering required quality of service in networks. Some kinds of intelligent techniques are appropriate for providing security in communication pertaining to distributed environments such as mobile computing, e-commerce, telecommunication, and network management. In this paper, a survey on intelligent techniques for feature selection and classification for intrusion detection in networks based on intelligent software agents, neural networks, genetic algorithms, neuro-genetic algorithms, fuzzy techniques, rough sets, and particle swarm intelligence has been proposed. These techniques have been useful for effectively identifying and preventing network intrusions in order to provide security to the Internet and to enhance the quality of service. In addition to the survey on existing intelligent techniques for intrusion detection systems, two new algorithms namely intelligent rule-based attribute selection algorithm for effective feature selection and intelligent rule-based enhanced multiclass support vector machine have been proposed in this paper.

Keywords: Survey, Intrusion Detection System, Neural Networks, Fuzzy Systems, Swarm Intelligence, Particle Swarm Intelligence

1. Review

1.1. Intrusion Detection Systems

Recently, Internet has become a part and parcel of daily life. The current internet-based information processing systems are prone to different kinds of threats which lead to various types of damages resulting in significant losses. Therefore, the importance of information security is evolving quickly. The most basic goal of information security is to develop defensive information systems which are secure from unauthorized access, use, disclosure, disruption, modification, or destruction. Moreover, information security minimizes the risks related to the three main security goals namely confidentiality, integrity, and availability.

Various systems have been designed in the past to identify and block the Internet-based attacks. The most important systems among them are intrusion detection systems (IDS) since they resist external attacks effectively. Moreover, IDSs provide a wall of defense which overcomes the attack of computer systems on the Internet. IDS could be used to detect different types of attacks on network communications and computer system usage where the traditional firewall cannot perform well. Intrusion detection is based on an assumption that the behavior of intruders differ from a legal user[1]. Generally, IDSs are broadly classified into two categories namely anomaly and misuse detection systems based on their detection approaches[2][3]. Anomaly intrusion detection determines whether deviation from the established normal usage patterns can be flagged as intrusions. On the other hand, misuse detection systems detect the violations of permissions effectively. Intrusion detection systems can be built by using intelligent agents and classification techniques. Most IDSs work in two phases namely preprocessing phase and intrusion detection phase. The intrusions identified by the IDSs can be prevented effectively by developing an intrusion prevention system. This paper mainly provides a survey on intelligent techniques proposed for developing IDSs. In addition, it explains about a new IDS which has been developed using two proposed algorithms namely intelligent rule-based attribute selection algorithm and intelligent rule-based enhanced multiclass support vector machine (IREMSVM).

1.2. Intelligent Intrusion Detection Systems

Intelligent IDSs are the ones considered to be intelligent computer programs situated in either a host or a network which analyzes the environment and acts flexibly to achieve higher detection accuracy[4][5]. These programs compute the actions to be performed on the environment both by learning the environment and by firing rules of inference[6]. Intelligent IDSs are capable of decision making and constraint checking. In most intelligent systems, either rules are fired or agents are used for decision making. Moreover, a set of static agents or a set of mobile and static agents have been used to achieve a single goal. Intelligent intrusion detection systems have been developed by proposing intelligent techniques for preprocessing and effective classification. Such IDSs have provided better detection rate in comparison with the other approaches.

1.3. Intelligent Preprocessing Techniques

Feature selection (or preprocessing) consists of detecting the relevant features and discarding the irrelevant ones, with the goal of obtaining a subset of features that describe the given problem properly with a minimum degradation of performance. It has several advantages[7], such as improving the performance of the machine learning algorithms, data understanding, gaining knowledge about the process and helping to visualize it, data reduction, limiting storage requirements, and helping in reducing processing costs.

There are two main models that deal with feature selection: filter methods and wrapper methods[8]. While wrapper models involve optimizing a predictor as part of the selection process, filter models rely on the general characteristics of the training data to select features with independence of any predictor. Wrapper models tend to give better results and this model is more precise than the filter model.

1.4. Intelligent Classification Techniques

Classification[9][10] is used to learn a model called classifier from a set of labeled data instances called training and then to classify a test instance into one of the classes using the learned model known as testing. Classification-based anomaly detection techniques operate in a similar two- phase fashion. The training phase learns a classifier using the available labeled training data. The testing phase classifies a test instance as normal or anomalous, using the classifier. Classification-based anomaly detection techniques operate under either one-class classifier or multi-class classifier.

One-class-classification-based anomaly detection techniques assume that all training instances have only one class label. Such techniques learn a discriminative boundary around the normal instances using a one-class classification algorithm. Any test instance that does not fall within the learned boundary is declared as anomalous.

Multi-class-classification-based anomaly detection techniques assume that the training data contains labeled instances belonging to multiple normal classes[11]. Such anomaly detection techniques teach a classifier to distinguish between each normal class and the rest of the classes. A test instance is considered anomalous if it is not classified as normal by any of the classifiers. Some techniques in this subcategory associate a confidence score with the prediction made by the classifier. If none of the classifiers are confident in classifying the test instance as normal, the instance is declared to be anomalous.

Many intelligent classification techniques namely decision trees, neural networks (NN), navie bayes and fuzzy set-based approach are available in the literature. This paper considers the most important intelligent classification techniques for comparison.

1.4.1. Decision Trees

A decision tree[12] is a tree where each non-terminal node represents a test or decision on the considered data item. Choice of a certain branch depends upon the outcome of the test. To classify a particular data item, the decision tree algorithms start at the root node and follow the assertions down until it reaches a terminal node (or leaf). A decision is made when a terminal node is approached. Decision trees can also be interpreted as a special form of a rule set, characterized by their hierarchical organization of rules.

1.4.2. Neural Networks

NN[13] are those systems modeled based on the human brain working. As the human brain consists of millions of neurons that are interconnected by synapses, a neural network is a set of connected input or output units in which each connection has a weight associated with it. The network learns in the learning phase by adjusting the weights so as to be able to predict the correct class label of the input. An artificial neural network consists of connected set of processing units. The connections have weights that determine how one unit will affect another. Subsets of such units act as input and output nodes, and the remaining nodes constitute the hidden layer. By assigning activation to each of the input node and allowing them to propagate through the hidden layer nodes to the output nodes, neural network performs a functional mapping from input values to output values.

1.4.3. Naive Bayesian Classifiers

Naive Bayesian classifiers[14] use Baye's theorem to classify the new instances of a data sample X. Each instance is a set of attribute values described by a vector, $X = (x_1, x_2, ..., x_n)$. Considering m classes, the sample X is assigned to the class Ci if and only if $P(X|C_i) P(Ci) > P(X|C_j) P(C_j)$ for all i and j in (1, m) such that j < > i. The sample belongs to the class with maximum posterior probability for the sample. For categorical data, $P(X_k|C_i)$ is calculated as the ratio of frequency of value X_k for attribute A_k and the total number of samples in the training set. For continuous valued attributes, Gaussian distribution can be assumed without loss of generality. In naive Bayesian approach, the attributes are assumed to be conditionally independent. In spite of this assumption, naive Bayesian classifiers give satisfactory results because focus is on identifying the classes for the instances, not the exact probabilities. Applications like spam mail classification and text classification can use naive Bayesian classifiers. Theoretically, Bayesian classifiers are least prone to errors. The limitation is the requirement of the prior probabilities. The amount of probability information required is exponential in terms of number of attributes, number of classes, and the maximum cardinality of attributes. With increase in number of classes or attributes, the space and computational complexity of Bayesian classifiers increase exponentially.

1.4.4. Fuzzy Sets

Fuzzy sets[15][16] form a key methodology for representing and processing uncertain information. Uncertainty arises in many forms in today's databases: imprecision, non-specificity, inconsistency, vagueness, etc. Fuzzy sets exploit uncertainty in an attempt to make system complexity manageable. As such, fuzzy sets constitute a powerful approach not only to deal with incomplete, noisy, or imprecise data but also to help in developing uncertain models of the data that provide smarter and smoother performance than traditional systems.

1.5. Dataset and Performance Metrics

1.5.1. Dataset

Since 1999, KDD'99[17][18] has been the most widely used data set for the evaluation of anomaly intrusion detection methods. The KDD'99 Cup data set was prepared by Stolfo et al[19] and was built based on the data captured in DARPA'98 IDS evaluation program[20][21]. This dataset was taken from the Third International Knowledge Discovery and Data Mining Tools Competition (KDD Cup 99). In this data set, each connection record is described by 41 attributes. The list of attributes consists of both continuous-type and discrete type variables, with statistical distributions varying drastically from each other, which makes the intrusion detection a very challenging task. The simulated attacks fall in one of the following four categories namely, denial of service (DoS), user to root (U2R), remote to local (R2L), and probe attacks. KDD'99 features are classified into three groups namely, basic features, traffic features, and content features. Traffic features are also classified into two types namely same host features and same service features.

Features present in KDD'99 Cup data set are grouped into three categories and are discussed below.

1) Basic Features: Basic features comprises of all the attributes that are extracted from a TCP/IP connection. These features are extracted from the packet header and includes src_bytes, dst_bytes, protocol etc.

2) Content Features: These features are used to evaluate the payload of the original TCP packet and looks for suspicious behavior in the payload portion. This includes features such as the number of failed login attempts, number of file creation operations etc. Moreover, most of the R2L and U2R attacks don't have any frequent sequential patterns. This is due to the fact that DoS and Probing attacks involve many connections to some host(s) in a very short duration of time but the R2L and U2R attacks are embedded in the data portions of the packets, and generally involves only a single connection. Hence, content based features are used to detect the attacks.

3) Traffic Features: These include features that are computed with respect to a window interval and are divided into two categories.

a) "Same host" features: These features are derived only by examining the connections in the past 2 seconds that have the same destination host as the current connection, and compute statistics related to protocol behavior, service etc.

b) "Same service" features: These features examine only the connections in the past 2 seconds that have the same service as the current connection. The above two types are called "time based traffic features".

Apart from these, there are various slow probing attacks that scan the hosts or ports using time interval greater than 2 seconds. As a result, these types of attacks do not generate intrusion patterns with a time window of 2 seconds. To overcome this problem, the "same host" and "same service" features are normally re-computed using a connection window of 100 connections.

1.5.2. Performance Metrics

By using its ability to make correct predictions, the effectiveness of the IDS is evaluated based on four possible metrics namely true negative rate (TNR), true positive rate (TPR), false positive rate (FPR), and false negative rate (FNR). If the actual class in the validation dataset is normal and is classified as normal, then TPR is incremented by 1 for each of the record. TNR is obtained if abnormal records are classified as abnormal records. FNR is obtained if normal record is classified as an anomaly record, and FPR is attained if abnormal record is classified as normal. Moreover, the most popular performance metrics for IDSs namely the detection rate (DR) and the false positive rate (FPR) are considered in this paper for effective analysis. DR is a ratio between number of anomaly (normal) correctly classified and total number of anomaly (normal). FPR is a ratio between number of anomalies incorrectly classified and total number of anomalies. An IDS should have a high DR and a low FPR. Other commonly used combinations include precision and recall, or sensitivity and specificity.

1.5.3. Testing Scenario

In this paper, the KDD cup data have been used for evaluating the most prominent algorithms available in the literature for feature selection and classification. A subset that consists of 10% of the records available from the KDD cup data set were used to evaluate the algorithms discussed in this paper due to the large number of records present in the data set. The data set were chosen in such a way this subset reflects all the properties necessary for distributing the four types of attacks and also the normal records. Moreover, tenfold cross validation was carried out on this subset of data by dividing this chosen data set into ten parts in which each tie nine parts are used for training and one part for testing. The experiments were carried out using Java programs written and tested for different algorithms presented in this paper. The results obtained are discussed in this paper based on the data analysis using KDD cup data set.

2. Works on Intelligent IDSs

In recent times, a lot of computational intelligence approaches were used for effective intrusion detection. The techniques include intelligent agent-based system, neural network-based IDSs, genetic algorithms, fuzzy and rough sets, particle swarm intelligence, and soft computing techniques.

2.1. Survey of Intelligent Agent-Based Systems

Intelligent agent-based systems are classified into four types namely simple agents, multi agents, mobile agents, and ant-based agents.

2.1.1. Static Agents

In the past, two types of static agents have been proposed namely simple and multi-agents. Simple agents have the capability to sense the environment and to act upon them. Bakar et al.[22] proposed a new agent-based approach for intrusion detection using rough set-based classification technique that uses simple agents. This technique generates rules from the data available on a large database and has mechanisms through rough sets to handle noise and uncertainty in data. However, provision of a rough classification model or rough classifier is computationally expensive, especially in its reduced computation phase.

Multi agent systems are employed to attain inherently robust solutions to many robotic applications like exploration, surveillance, patrolling, target tracking, and intelligent transportation. In these situations, agents could perform different and possibly independent tasks, but at the same time, they cooperate in order to guarantee the entire system's safety. Cooperation among agents is obtained through a shared set of rules according to which all agents are supposed to plan their actions. Adriano Fagiolini et al.[23] addressed the major problem where the uncooperative behavior in a team of hybrid agents are detected and proposed the architecture of a decentralized monitor to be embedded on the agents. By the path of this monitoring process, each agent was able to establish whether its neighbors are cooperative or not. The major advantages of this agent architecture are the scalability and decentralization. The disadvantage of this agent is not considering the implementation aspects of such monitors. Xiaodong Zhu et al.[24] presented a multi-agent-based intrusion detection system named multi-agent-based intelligent intrusion detection system. The learning agent module in that system is self-adjusting and learns the network-based audit data and the host-based audit data, with a capability of learning more than one technique of data mining, such as association rules and so on. The learning agent also could produce rules, and the detection agent could detect audit data according to these rules and respond to them. The experimental results show that their system has very high self-adapting ability, intelligence, and expansibility.

In the work of Gou Xiantai et al.[25], they have focused on the first-class automatic reaction of containment since it is quite practical and easy to set up a worm containment system for a metropolitan area networks (MAN) but not for the whole Internet. Hence, multi-agent system for worm detection and containment in MAN was given to limit the propagation of worms in MAN. The major advantage of the system are that it could prevent the whole MAN from being fallen down because of the worm scan and the worm attack such as distributed denial of service (DDoS), and it is very effective in blocking random scanning worms that most commonly have been encountered. The disadvantage of this system is that it is not appropriate for restricting other types of worms such as the flash worms, topological worms[26] and random scanning worms that infect networks faster than any other type of worms. Due to various drawbacks, they tend to be very noisy and hence waste a lot of network bandwidth and crash the routers. So, it is very important to have some automatic reaction mechanism to limit the propagation of such type of worms.

2.1.2. Mobile Agents

Mobile agents move from one host to another to carry out specific tasks. Ghenima Bourkache et al.[27] proposed a prototype of architecture of an anomaly distributed intrusion detection system for ad hoc networks that functions using a society of mobile and reactive agents carrying out intelligent and distributed intrusion detection. Their distributed model aimed at solving the problems faced by the hierarchical intrusion detection systems namely lower detection workload and great overhead. It proposes a technique for finding the main cause of the attack by the response engine in order to isolate the intruder from the network. Another work that uses mobile agents for intrusion detection was proposed by Wang et al.[28].

An integrated framework was proposed in[29] to guide the design of a mobile agent-based network management system namely the mobile agent-based framework for security-enhanced autonomous network and system management. This framework has offered two distinct advantages: 1) the provision of a secure agent-based management infrastructure and 2) the capability of achieving enhanced network management functionalities. They proposed two novel security schemes, namely the visibility protection scheme and the visa-based authentication scheme, for protection of management information and authentication and resource access control of management agents, respectively. Mobile agents could facilitate the implementation of robust, attack-resistant IDS architectures[30]. Agents were supposed to relocate when sensing danger or suspicious activity, clone for redundancy or replacement, operate autonomously and asynchronously from where created, collaborate and share knowledge, and be self organizing. Moreover, agents are amenable to genetic diversity, which also helped to avoid attacks aimed at circumventing the known and stable detection mechanisms of IDS.

2.1.3. Ant Based Agents

The ant colony optimization (ACO) algorithm depicted a probabilistic technique for solving computational problems which could be reduced to find good paths through graphs based on the strategies of real ants[31]. It was initially proposed in 1992 by Colorni, Dorigo, and Maniezzo[29][30]. In ACO, each artificial ant is considered as a simple agent, communicating with other ants only indirectly and by affecting changes to a common environment.

Chi-Ho Tsang *et al.*[32] presented a multi-agent IDS architecture for scalable intrusion detection and prevention in large switched networks in industrial plants. This unsupervised anomaly detection model based on ant colony was proposed and implemented in the decision agents in order to search heuristically for the near-optimal clustering with compact structure. The empirical results proved that this system could significantly improve the overall performance of existing ant-based clustering algorithms. On the other hand, this ant colony-based model could automatically determine the number of clusters that was critically required to be input in other clustering algorithms such as k-means, fuzzy c-means, and e-m clustering. They proved that ant agents are useful for reducing false positives in IDS.

2.2. Neural Networks Based IDSs

Neural networks are composed of basic units somewhat analogous to neurons. These units are linked with each other using the connection whose strength is modifiable as a result of a learning process or algorithm. Each of these units were integrated independently (in parallel) the information provided by its synapses in order to evaluate its state of activation. The unit response was then a linear or nonlinear function of its activation. Linear algebra concepts are used, in general, to analyze linear units, with eigenvectors and eigenvalues being the core concepts involved. This analysis made clear about the strong similarity between linear neural networks and the general linear model developed by statisticians. The linear models presented here are the perceptron and the linear associator. The behavior of nonlinear networks could be described within the framework of optimization and approximation techniques with dynamical systems.

2.2.1. Neural Networks for Preprocessing

A NN-based approach was introduced by Verikas and Bacauskiene[31] with salient features for classification. This is a feed-forward neural network. This approach involved neural network training with an augmented cross-entropy error function. A new feature selection algorithm[33] based on the wrapper approach using neural networks. The vital aspect of this algorithm is the automatic determination of neural network architectures during the feature selection process. According to this algorithm, it used a constructive approach involving correlation information in selecting features and determining neural network architectures. It would reduce the redundancy information resulting in compact neural network architecture.

2.2.2. Neural Networks for Classification

Jeich Mar et al.[34] proposed an IDS based on adaptive neuro-fuzzy inference system (ANFIS) rule to minimize the detection delay for the de-authentication attacks on the medium access control layer of a wireless local area network (WLAN). Both the average sequence number gap between the successive packets and the average statistical value of the de-authentication packets received by an access point are used to detect the de-authentication DoS attack. This ANFIS-IDS experimental platform was implemented and tested by the authors against real deauthentication DoS attack to empirically evaluate its average detection delay and average FAR. The performance of the IDS using the proposed ANFIS method was compared by them with non-parametric sequential change point detection algorithm in a practical WLAN environment.

Debar et al.[35] developed a NN model for IDS. The main advantage of this system is that the deviation to the normal behavior of the user could be easily diagnosed fairly and quickly by the NN. They used this capability as the goal of the IDS to detect potential intruders as quick as possible. Joo et al.[36] proposed a NN model to improve the performance of IDS using the asymmetric cost of false positive and false negative errors. Their approach differs from other approaches in measuring the system performance since it has considered asymmetric costs of errors rather than prediction accuracy in intrusion detection. Liu et al.[37] described a network IDS based on artificial neural networks (ANN). According to that system, the NNs are used to classify without consulting a domain expert; hence, this automation helped to detect both known and novel intrusions. The key part of the work was focused on the development of an adaptive resonance theory (ART) NN, and it is trained in real-time in an unsupervised way.

Moradi and Zulkernine[38] presented an ANN approach for intrusion detection. One of the limitations of that approach was the increase in training time, and also it does not provide a description on why certain network traffic was intrusive. Sarasamma et al.[39] proposed a novel multilevel hierarchical Kohonen net to detect intrusions in networks. In their work, randomly selected data points forming the KDD Cup 99 were used to train and test the classifier. The results obtained by them proved that the hierarchical Kohenen net in which each layer operates on a small subset of the feature space was superior to a Kohenen net operating on the entire feature space in detecting various kinds of attacks. Amini et al.[40] introduced an intelligent method for detecting known and unknown attacks. Unsupervised neural nets are used by them to detect intrusions in real time which can perform the analysis of new data over time without retraining. ART and self-organizing map NNs are evaluated using offline data in their work.

Koutsoutos et al.[41] presented a NN classifier ensemble system using a combination of NNs which is capable of detecting network attacks on web servers. Their system could identify unseen attacks and categorize them. The performance of the NN used by them for detecting attacks from audit dataset is fair with success rates of more than 78% in detecting novel attacks. However, it suffers from high false alarm rates; hence, it was necessary to propose suitable enhancements to the work. Shun and Malki[42] presented a NN-based IDS for detecting Internet-based attacks on a computer network. NNs are used by them to identify and predict current and future attacks in which the feed-forward NN with the back propagation training algorithm was employed to detect intrusions. They observed that the experimental results on KDD Cup 99 dataset show promising results for detection of intrusion when NNs are used for classification. Thomas and Balakrishnan[43] addressed the problem of optimizing the performance of IDS using fusion of multiple sensors. The trade-off between the detection rate and false alarm highlighted that the performance of the detector is better when the fusion threshold is less. In their work, NN-supervised learner has been designed and implemented to determine the weights of individual IDSs depending on their reliability in detecting a certain attack. The final stage of this data-dependent fusion architecture is a sensor fusion unit which computes the weighted aggregation in order to make an appropriate decision. The major limitation with this approach was that it required large computing power; hence, the training time was increased.

Linda et al.[44] presented an IDS using NN-based modeling for detection of anomalous activities. The major contributions of their approach are the use and analysis of real network data obtained from an existing critical infrastructure, the development of a specific window-based feature mining technique, construction of training dataset using randomly generated intrusion vectors, and the use of a combination of two NN learning algorithms, namely the error-back propagation and Levenberg-Marquardt algorithms, for normal behavior modeling. The major limitations of the approaches discussed in the literature using ANN for IDS is in training of NNs, i.e., computational load is very high. The time required for training is normally very high which is important for obtaining efficient NNs. Therefore, in the proposed work, a neuro-genetic algorithm has been designed and implemented by incorporating a genetic algorithm (GA) component into the ANN in the training phase. GA generates optimal weights by means of a special fitness function which has been designed specifically for weight adjustment in this research work; hence, it is used by ANN to learn the characteristics of normal pattern and attack types effectively.

2.3. Genetic Algorithm-Based IDSs

Genetic algorithm for simplified security audit trials analysis (GASSATA) proposed by Me[45], introduced a new genetic algorithm for the misuse intrusion detection. This GASSATA constructed a two-dimensional matrix. First, axis of the matrix specified different attacks which had been known already. Second, the axis represents different kinds of events derived from audit trails. Therefore, this matrix actually represented the patterns of intrusions. Given an audit record being monitored which had information about the number of occurrences of every event, this method applied genetic algorithms to find the potential attacks appearing in the audit record. However, the assumption that the attacks are dependent only on events in this method restricts its generality. There are two steps involved in genetic algorithm: one was coding a solution to the problem with a string of bits and the other was finding a fitness function to test each individual of the population against evaluation criteria.

2.3.1. Genetic-Based Feature Selection

Most of the real life problems definitely need an optimal and acceptable solution rather than calculating them precisely at the cost of degraded performance, time and space complexities. Therefore, it was necessary to carry out the analysis using selected features. The problem of selecting significant features from KDD Cup 99 dataset for intrusion detection could not be represented in terms of formula since it was too complex. Moreover, when all the features are used without feature selection, it took a very longer time to calculate a solution precisely. Therefore, the feasible approach is to use a heuristic method which has performed the feature selection effectively. GA[46] was a heuristic, which stated that it estimated a solution and generated the optimized results. Among various heuristic methods, GA[47] was supposed to be more promising since it has differed in many ways from other heuristics. First, GA works on population of possible solutions, while other heuristic methods use a single solution in their iterations. Second, most heuristics are probabilistic or stochastic in nature and hence were not deterministic. On the other hand, each individual in the GA population contributes well to obtain a possible solution to the problem. In GA, the algorithm starts with a set of possible solutions represented by chromosomes called population. A potential solution to a specific problem was encoded in the form of a chromosome. By using the solution of one population, a new population is formed. Solutions are selected to form new solutions called offspring and are selected according to their fitness value. Finally, GA is more suitable in reducing the search space. Therefore, the convergence of the algorithm is faster when GA is employed. Subset generation[48] is a method of heuristic search, in which each instance in the search space specifies a candidate solution for subset evaluation. The decision process of this method is determined by some basic issues. Initially, the search starting point must be decided since it has controlled the direction of search. Feature selection search has started either with null set where features were added one by one or it was started with a full set of features and was eliminated one by one. But these methods have the drawback of being trapped into local optima[49]. Sindhu et al.[12] introduced a new intrusion detection model which was the combination of the following: 1) removing redundant instances in order to make the learning algorithm to be unbiased, 2) identifying suitable subset of features by employing a wrapper-based feature selection algorithm, 3) realizing proposed IDS with neuro tree to achieve better detection accuracy. The lightweight IDS has been developed using a wrapper-based feature selection algorithm that maximizes the specificity and sensitivity of the IDS as well as by employing a neural ensemble decision tree iterative procedure to evolve optimal features. An extensive experimental evaluation of the proposed approach with a family of six decision tree classifiers namely decision stump, C4.5, naive Baye's tree, random forest, random tree, and representative tree model to perform the detection of anomalous network pattern has been introduced by them.

2.3.2. Neuro-Genetic Classification

Genetic paradigm is employed to choose the predominant features, which has revealed the occurance of intrusions. The neuro-genetic IDS (NGIDS) involves calculation of weightage value for each of the categorical attributes so that data of uniform representation could be processed by the neuro-genetic algorithm. In this system, unauthorized invasion of a user were identified and newer types of attacks were sensed and classified respectively by the neurogenetic algorithm. The experimental results obtained in this work shows that the system achieves improvement in terms of misclassification cost when compared with conventional IDS. The results of the experiments show that this system could be deployed based on a real network or database environment for effective prediction of both normal attacks and new attacks.

2.4. Fuzzy and Rough Sets

In this section, we discuss the topics namely fuzzy sets, neuro-fuzzy and rough-sets. Fuzzy logic would help to improve the detection accuracy when we are using different fuzzy logics. Rough sets could also be used to improve the detection accuracy.

2.4.1. Fuzzy Sets

A Fuzzy multi-class support vector machine (SVM) was proposed in literature for network intrusion detection and is a collaborative intrusion detection model. Four kinds of SVM detection agents are discussed in their work and these agents have different attributes. They are used to detect transmission control protocol (TCP) attacks, UDP attacks, ICMP attacks, and content-based detection separately. A TCP detection agent was used as an example to illustrate the construction process of detection agent. This multiagent collaborative detection method has increased the detection speed and accuracy. The intrusion detection based on fuzzy multi-class SVM has advantages in two aspects: 1) it selects the least attributes to build detection agents respectively and hence does not need all the attributes of the network packets; 2) it is a collaborative detection system which has improved not only the detection rate but also the detection accuracy. Du Hongle *et al.*[50] proposed an improved v-fuzzy support vector machine (FSVM) through introduction membership to each data point. They reformulate the improved v-FSVM so that different input points can make different contributions to decision hyper- plane. In order to verify the performance of the improved v-FSVM, they applied it to intrusion detection.

Yu-Ping Zhou *et al.*[51] presented a hierarchical neuro-fuzzy inference intrusion detection system (HFIS). In their proposed system, principal component analysis neural network was used to reduce the input data space. An enhanced fuzzy c-means clustering algorithm was applied to create and extract fuzzy rules. The adaptive neural fuzzy inference system was utilized repeatedly in their model. At last, the system was optimized by genetic algorithm. The main advantages of the HFIS model are its capability to perform not only misuse detection but also anomaly detection. Moreover, their method has higher speed and better performance.

A hybrid intrusion detection method based on hidden Markov model (HMM) and fuzzy logic has been proposed by Li *et al.*[52]. The experimental results showed that their method is efficient to classify the anomaly profile from the normal profile. Comparing with other methods based on the HMM only, this HMM- and fuzzy logic-based method has the following advantages: first, it needs only less storage without the profile database. With the processes being used by more and more users, the profile database will be greatly enlarged. So, the profile database would occupy much storage with the larger and larger profile database; second, it could reduce training time effectively, which needed less testing data. When the profile databases are very large, the detection speed is slower as the sequence must be compared with all the records in the profile database. Their approach detects network-based attacks only at high false- positive rates as the processes in those attack scenarios behave similar to the normal behavior.

2.4.2. Neuro-Fuzzy Algorithms

Neuro-fuzzy algorithms are useful for classifying large volume of data with uncertainty. A novel neuro-fuzzy network for pattern classification problem has been proposed[53]. This flexible classification system is able to determine all of the parameters from the training set without any prior knowledge. The proposed classification model has been used in calculating the initial weights from the training data. This model contains two networks: one is the feature extraction unit and the other, the inference unit. The feature extraction unit effectively reduces the dimension of the original feature variables. The inference determines the classification results according to the distributions of the new feature variables.

An evolutionary soft computing approach for intrusion detection has been introduced by Toosi and Kahani[54] and has successfully demonstrated its usefulness on the training and testing subset of KDD cup 99 dataset. The ANFIS network was used as a neuro-fuzzy classifier for intrusion detection since ANFIS is capable of producing fuzzy rules without the aid of human experts. Also, subtractive clustering has been utilized to determine the number of rules and membership functions with their initial locations for better classification. Moreover, a fuzzy decision-making engine was developed to make the system more powerful for attack detection, using the fuzzy inference approach. At last, they proposed a method to use genetic algorithms to optimize the fuzzy decision-making engine. Yu-Ping Zhou et al.[55] presented a hierarchical neuro-fuzzy inference intrusion detection system. In the proposed system, princi pal-component-analysis neural network has been used to reduce the input data space. An enhanced fuzzy c-means clustering algorithm has been applied to create and extract fuzzy rules. The adaptive neural fuzzy inference system was utilized repeatedly in their model. At last the system has been optimized by genetic algorithm. The main advantage of their model is the capability to detect not only misuse but also anomaly. Moreover, their proposed method has higher speed and better performance.

2.5. Fuzzy-Genetic Algorithms

A feature-extraction neuron-fuzzy classification model (FENFCM) has been proposed by Nai Ren Guo et al.[56] that enabled the extraction of feature variables and has provided the classification results. This classification model has been integrated with a standard fuzzy inference system and a neural network with supervised learning. The FENFCM automatically generated the fuzzy rules from the numerical data and triangular functions that were used as membership functions both in the feature extraction unit and in the inference unit. To adapt the proposed FENFCM, two modificatory algorithms are applied: first, they utilized evolutionary programming to determine the distribution of fuzzy sets for each feature variable of the feature extraction unit; second, the weight-revised algorithm is used to regulate the weight grade of the principal output node of the inference unit; finally, the FENFCM was validated using two benchmark data sets, the Wine database and the Iris database. Computer simulation results by them have demonstrated that the classification model provides a sufficiently high classification rate in comparison with that of other models proposed in the literature.

Rough Sets

Zihui Che Xueyun Ji[56] presented a new anomaly detection model based on rough set reduction and HMM on the basis of the analysis of shortcomings of other detection methods these days. Specifically, that method has the following advantages:

1) The method of rough set reduction provided an efficient way to reduce the number of attributes, as well as the complexity of the information expression system. It has decreased the training time of HMM after the reduction of redun-

dant information.

2) The rough set reduction process would also generate decision conditions, which could be applied to further detection after HMM evaluation. The strategy could revise the detection results to improve the accuracy of anomaly detection.

3) The HMM- and rough set-based approach could identify misuse and malicious intrusion by means of attribute reduction.

They could acquire a better HMM with a relatively small number of training data. Their method could promote the detection rate and decrease the false alarm rate stably.

A new feature selection algorithm combining a rough sets and genetic algorithms on the basis of clustering was proposed by Guo et al.[57]. Firstly, it uses the rough set theory to process selection and then uses the improved genetic algorithm based on clustering to find the optimal subset in the remaining subset. This algorithm has many advantages. In the end, it combines the results of the first two steps to get the final results. The results of the experiments show that this new method is better at detection accuracy and false rate than the other algorithms.

2.6. Particle Swarm Intelligence

2.6.1. Swarm Intelligence Approach

Honey bees exhibit many features that could be used as models for intelligent systems. These features include bee dance (communication), bee foraging, queen bee, task selection, collective decision making, nest site selection, mating, floral/pheromone laying and navigation systems.

Queen Bee Jung[58] proposed an evaluation method called queen-bee evolution simulating the queen bee role in the reproduction process. This method improved the optimization capability of genetic algorithms by enhancing exploitation and exploration processes. Xu et al.[59] developed a bee-swarm genetic algorithm for designing DNA sequences that satisfied some combinatorial and thermodynamic constraints, in which the optimum individual of population selected as a queen bee and a random population was introduced to reinforce the exploitation of genetic algorithm and increase the diversity of population.

Bee dance and communication Wedde *et al.*[60] presented a completely decentralized multi-agent approach on multiple layers where car or truck routing were handled through algorithms adapted from the BeeHive algorithms which in turn had been derived from honey bee behavior. They reported superior performance of over conventional approaches[61].

Task allocation Gupta and Koul[62] built an architecture named Swan based on the management of beehives by worker bees and the queen bee in the animal kingdom for network management of Internet protocol networks in order to overcome the shortcomings of traditional network management software. Similarity between honey bee and agents teamwork inspired Sadik *et al.*[63] proposed a system to develop a teamwork architecture to enhance the performance and task execution efficiency of software agents since a limited progress has been made towards efficient task execution mechanisms by group of agents in collaboration and coordination with each other. The authors named it Honey Bee teamwork architecture afterwards.

Collective decision and nest site selection Passino[64] established a mathematical model of the nest site selection process of honey bee swarms and highlighted the potential implications of the dynamics of swarm decision making. Gutierrez and Huhns[65] handled the quorum sensing during nest site selection in the area of design diversity of software fault tolerance.

2.6.2. Ant Colony Optimization

A novel approach for intrusion detection from the standpoint of feature selection was proposed by Gao *et al.*[66]. ACO was applied to select effective features for SVM classification. The simulation using KDD cup 99 dataset showed that SVM with obtained optimal feature subset could achieve better generalization performance than that without feature selection. This demonstrated the fact that dimension reduction could improve the generalization performance of intrusion detection and make the detection much more time efficient.

Rahul Karthik Sivagaminathan and Sreeram Ramakrishnan[67] presented a hybrid method based on ant colony optimization and ANNs to address feature selection. The proposed hybrid model was demonstrated by them using data sets from the domain of medical diagnosis, yielding promising results. An intrusion detection method based on ant colony fuzzy clustering has been proposed by Wei Song Li et al.[68]. The algorithm used the ant colony optimization algorithm which has a strong ability to deal with local minima since it is better than the random selected cluster centers that cause iterative process into a local optimal solution and dynamically determines the number and center of clusters. An efficient hybrid ant colony optimization-based feature selection algorithm has been presented by Md.Monirul Kabir et al.[69]. Since ants were the foremost strength of an ACO algorithm, guiding the ants in the correct directions was a critical requirement for high-quality solutions. Accordingly, this algorithm guided the ants during feature selection by determining the subset size. Furthermore, new sets of pheromone update and heuristic information measurement rules for individual features bring out the potential of the global search capability of this ACO-based feature selection algorithm.

2.6.3. Particle Swarm Optimization

A novel intrusion detection framework based on particle swarm optimization (PSO) was proposed by Jiang Tian and Gu[70] which had combined the idea of unsupervised learning method and the supervised strategy. Instead of calculating the accuracy, ROC analysis was utilized to evaluate the detection performance. This PSO algorithm has been executed for global optimal parameters of SVM. Best combination of TPR with FPR has been achieved after adjusting the offset of the detection function. The effectiveness of their method for anomaly detection was demonstrated on four benchmark datasets, and results have showed satisfactory performance.

3. Comparative Analysis

Over the past decade, intrusion detection based upon computational intelligence approaches has been a widely studied topic, being able to satisfy the growing demand of reliable and intelligent intrusion detection systems.

In our view, these approaches contribute to intrusion detection in different ways. Fuzzy sets have represented and processed numeric information in a linguistic format, so they could make the system complexity manageable by mapping a large numerical input space into a smaller search space. In addition, the use of linguistic variables is able to present normal or abnormal behavior patterns in a readable and easy to comprehend format. The uncertainty and imprecision of fuzzy sets smooth the abrupt separation of normal and abnormal data, thus enhanced the robustness of an IDS.

3.1. Feature Selection

3.1.1. Gradually Feature Removal Method

The gradually feature removal (GFR) method[71] decides the importance of the 41 features of the KDD Cup dataset and gradually removes the less important features. This algorithm well and selects 19 features namely 2, 4, 8, 10, 14, 15, 19, 25, 27, 29, 31, 32, 33, 34, 35, 36, 37, 38, and 40. Using these 19 features, 98.6249% accuracy was achieved with SVM in tenfold cross validation. In order to evaluate the advantage of the GFR method, the other three feature reduction algorithms were also undertaken by the authors. In the feature removal method, 10 important features are chosen. Similarly, the sole feature method chooses other 10 critical features. Moreover, by choosing the common features selected by the two algorithms, 10 critical features are derived in the hybrid method.

3.1.2. Modified Mutual Information-Based Feature Selection Algorithm

The modified mutual information-based feature selection algorithm (MMIFS) was proposed by Fatemeh Amiri *et al.*[72]. Moreover, the authors have analyzed the features selected by their proposed MMIFS method and their relationship with different attack types. In the KDD Cup 99, dataset was used for carrying out the experiments detecting attacks. This algorithm selects features for identifying DoS, Probe, R2L, and U2R attacks effectively by computing the mutual information.

Mutual information based feature selection algorithm was initially proposed by Battiti[73] to maximize the relevance between the input features and the output and to minimize the redundancy of the selected features. The algorithm selects one feature at a time which maximizes the information with outputs. In this mutual information based feature selection algorithm, the mutual information expression is adjusted by subtracting a quantity proportional to the average mutual information within the selected features. The main advantage of this algorithm is that it selects 13, 8, 15, and 10 features for Probe, DoS, R2L, and U2R which are optimal for classification.

3.1.3. CRF-Based Feature Selection

Conditional random field (CRF)-based feature selection is a statistical approach proposed by Gupta et al.[74] for effective feature selection. They proposed a layered approach in which each layer considers one type of attack. Therefore, the probability value for each relevant feature is measured, and for each type of attack, different features are selected.

They used domain knowledge along with the practical significance, and they performed feasibility analysis for each feature before selecting it for a particular layer. Thus, from the total 41 features, they selected only 5 features for the Probe layer, 9 features for the DoS layer, 14 features for the R2L layer, and 8 features for the U2R layer. Since each layer is independent of every other layer, the feature sets for the layers are not disjoint.

3.1.4. Wrapper Based Genetic Feature Selection

In this model, genetic feature selection algorithm follows a wrapper-based approach. Moreover, each iteration of this algorithm results in a decision tree. After n iterations, a series of trees are obtained, and the best have been used to generate rules. The tree with the highest sensitivity and specificity are identified as the best trees. Thus, best set of features are extracted[75][76] based on sensitivity and specificity values. The main advantage of this genetic- based feature selection algorithm is that it selects only the important and contributing features for classification.

3.1.5. Comparison

The gradual feature removal method first removes the repeated data from the KDD cup dataset and uses k-means clustering to remove the next important set of features. However, the number of clusters is predetermined. In the modified mutual information-based feature selection algorithm, mutual information is used to perform feature selection and hence more flexible. In the CRF-based feature selection method, conditional probability values are used to select the relevant features and hence can handle uncertainty effectively. Finally, the wrapper-based method uses a decision tree to remove redundant subsets of features. However, from the analysis of all these methods, it is observed that the combination of mutual information and information gain ratio values provide a better method for feature selection since it can be used to perform both attribute selection and tuple reduction.

3.2. Classification

There are many works on classification that are available in the literature. Among them, the most relevant works for IDS are discussed in this section.

3.2.1. Linear Programming System-Based Method for Detecting U2R Attacks

In this paper, a new approach for detecting U2R attacks has been investigated and evaluated. In their model, a behavior i belonging to an attack class j is represented by the variable x_{ij} and each class is represented by its feature vector as F_j. The distance between the behavior i and class j is represented as α_{ij}. The problem is represented for m attack classes and attack types contains β_j elements such that $\beta_j \geq 0$. Now, the problem is formulated using simplex model as

$$Z_{\min} = \sum_{i=1}^{n}\sum_{j=1}^{m} a_{ij} x_{ij}$$
$$x_{ij} \in \{0,1\}$$
$$x_{ij} = 1$$
$$\sum_{j=1}^{m} x_{ij} = 1$$
$$\sum_{i=1}^{m} x_j \geq 0$$

They proposed an optimal algorithm to solve this problem, and based on that, they classified the attacks effectively.

3.2.2. Layered Approach

We now describe the layer-based intrusion detection system (LIDS) proposed by Gupta et al.[74] in detail. According to them, the LIDS drew its motivation from the airport security model, where a number of security checks are performed one after the other in a sequence. Similar to this model, this LIDS represents a sequential layered approach was developed for ensuring availability, confidentiality, and integrity of data and (or) services over a network.

The major goal of using a layered model is to reduce computation time required to detect anomalous events and to improve the speed of operation of the system. In this approach, the algorithm uses the selected features and check whether there is a probe attack in the first layer called probe layer. Similarly, at each layer, it checks for the occurrence of the corresponding attacks. If there is an attack, it informs the prevention system. The main advantage of the layered approach is that it reduces the computation time by using separate feature selected by CRF-based feature selection algorithm.

3.2.3. Least Squares Support Vector Machine

SVM[77] is a supervised learning method used for solving classification and regression problems. An SVM can train with a large number of patterns. The least square support vector machine (LSSVM) is a modified algorithm[78] to the standard SVM. It solves a linear equation in the optimization stage and hence simplifies the process. Moreover, this LSSVM is effective since it avoids local minima in SVM problems used by LSSVM classifier is used by to detect normal and attacks data.

3.2.4. Neuro-Tree Classifier

In the neuro-tree classifier proposed by Sindhu et al.[12] for intrusion detection, the features selected by a genetic-based approach are used for classification. The major contributions of the neuro-tree classifier are the provisions of a new facility for the prevention of over fitting and the use of new fitness evaluation framework for maximizing the sensitivity and specificity. The main advantages of the neuro-tree classifier are that it reduces the false alarm rate and fast convergence.

3.2.5. Comparison

The linear programming system-based method used for classification is more efficient in detecting U2R attacks. The authors use the behavior distance between classes to find the similarity. However, it is necessary to focus on all types of attacks for providing effective security.

In the layered approach, each attack is analyzed in a separate layer and hence is effective in detecting all types of attacks. In the least square support vector machine-based classification uses an enhanced SVM to avoid local minima. This method detects all types of attacks with improved accuracy. The neuro-tree classifier provides effective classification when optimal features are provided. Hence, it reduces the false alarm rate effectively, and in addition, the algorithm converges fast.

4. Proposed Intelligent IDS

In this paper, an intelligent IDS developed by proposing a new feature selection algorithm and a new classification algorithm is also discussed.

Feature Selection

In this work, a new feature selection algorithm has been proposed by using an attribute selection and tuple selection. This algorithm has been proposed using rules and information gain ratio for attribute selection. In order to achieve this, the data set D is divided into n number of classes C_i. The attributes F_i having maximum number of nonzero values are chosen by the agent, and the information gain ratio is computed using Equations (1), (2), and (3), where F is the feature set.

$$\text{Info}(D) = -\sum_{j=1}^{m}\left[\frac{\text{freq}(C_j,D)}{|D|}\right]\log_2\left[\frac{\text{freq}(C_j,D)}{|D|}\right] \quad (1)$$

$$\text{Info}(F) = \sum_{i=1}^{n}\left[\frac{|F_i|}{|F|}\right] \times \text{info}(F_i) \quad (2)$$

$$\text{IGR}(A_i) = \left[\frac{\text{Info}(D) - \text{Info}(F)}{\text{Info}(D) + \text{Info}(F)}\right] \times 100 \quad (3)$$

In addition, tuple selection is also carried out using the rule-based approach.

5. Results and Discussion

The agent-based attribute selection algorithm has been selected with 19 important features in **Table 1**. This selection was based on the information gain ratio values of various attributes.

5.1. Classification

In this paper, a new classification algorithm called IREMSVM algorithm has been proposed from the existing intelligent agent-based enhanced multiclass SVM (IAEMSVM)[79].

5.2. Enhanced Multiclass Support Vector Machine

In the IREMSVM algorithm, the data set is first divided into R classes. Then the distance between any two classes of patterns are computed from the R classes using the Minkowski distance. According this method, the distance between two points

$$P = (x_1, x_2, x_3, \ldots, x_n) \text{ and}$$
$$Q = (y_1, y_2, y_3, \ldots, y_n) \in R^n \qquad (4)$$

is given by the formula given in Equation (5).

$$d_{ij} = \left(\sum_{i=1}^{n} |x_{ik} - x_{jk}|^p \right)^{\frac{1}{p}}, \qquad (5)$$

where p is the order and it also finds the centroids of each class, where j and k are the neighbors of i.

The centroid is computed using the formula given in Equation (6):

$$C_i = \sum_{m=1}^{nt} X_m^i / n_i, \qquad (6)$$

where C_i = centroid value of ith node, X = individual i^{th} lowest distance, and n = number of dimensions

The steps of this algorithm are as follows:

Algorithm 1. Intelligent rule-based attribute selection algorithm.

1. **Input:** Set of 41 features from KDD'99 Cup data set
2. **Output:** Reduced set of features R
3. **Steps of the algorithm:**
4. **Step 1:** Calculate the information gain ratio for each attribute $A_i \, \varepsilon \, D$ using Equation 3.
5. **Step 2:** Choose an attribute A_i from D with the maximum information gain ratio value.
6. **Step 3:** Split the data set D into subdatasets $\{D_1, D_2, \ldots, D_n\}$ depending on the attribute values of A_i where C_j stands for jth attribute of class C.
7. **Step 4:** Find all the attributes whose information gain ratio > the threshold.
8. **Step 5:** Store the selected attributes in the set R and output it.
9. **Step 6:** Compute the mutual information value for each tuple.
10. **Step 7:** Compare the key attribute values for each tuple with threshold.
11. **Step 8:** If it is less than the threshold, then exit.
12. **Step 9:** If key value is ≥ the threshold, then add the tuple into the ouput table.

Algorithm 2. Intelligent rule-based enhanced multiclass support vector machine.

1. Step 1: Rule select two initial cluster centers by applying the intelligent rules.
2. Step 2: Import a new class C from the dataset.
3. Step 3: Compute the Minkowski distance between two classes
4. Step 4: if (d (A, B) > d (A, C)) then
5.   B is assigned as Normal
6.   Else
7.   C is assigned as Attacker.
8. Step 5: Intelligent agent calculates the min and max of the distances.
9. Step 6: If (d (A, B) < threshold limit of the distance) then it creates a new cluster, and this is the center of the new cluster.
10.   Else
11.   B is assigned as a suspect.
12. Step 7: Now compute mutual information value, and check it with a threshold.
13. Step 8: If it is the mutual information value ≥ threshold then
14.   Accept the record
15.   Else
16.   Reject the record

5.3. Experimental Results

This work has been implemented using Java programs. Moreover, the experiments have been conducted to classify the KDD'99 Cup data set using both full features and selected features. So that comparative analysis can be performed.

Table 2 shows the comparison of SVM, IAEMSVM, and the proposed IREMSVM with respect to classification accuracy when the classification is proposed with the 19 selected features obtained from the proposed feature selection algorithms.

From this table, it is observed that the classification accuracy is increased in the proposed algorithm when it is compared with the existing algorithms for Probe, DoS, and others attacks. This is because the agents used in this proposed algorithm perform constraint checking for all types of experimental uses in the classification.

6. Conclusion

In this paper, a survey on intelligent techniques for feature selection and classification techniques used of Intrusion Detection has been presented and

Table 1. List of 19 selected features.

Selection number	Feature number	Feature name
1	2	protocol_type
2	4	src_byte
3	8	wrong_fragment
4	10	hot
5	14	root_shell
6	15	su_attempted
7	19	num_access_shells
8	25	rerror_rate
9	27	diff_srv_rate
10	29	srv_serror_rate
11	31	srv_diff_host_rate
12	32	dst_host_count
13	33	dst_host_srv_count
14	34	dst_host_same_srv_count
15	35	dst_host_diff_srv_count
16	36	dst_host_same_src_port_rate
17	37	dst_host_srv_diff_host_rate
18	38	dst_host_serror_rate
19	40	dst_host_rerror_rate

Table 2. Detection accuracy comparisons with 19 features.

Exp. No.	SVM			IAEMSVM			IREMSVM		
	Probe	DoS	Others	Probe	DoS	Others	Probe	DoS	Others
1	91.53	92.30	60.73	99.58	99.69	71.52	99.78	99.79	71.71
2	90.78	91.45	61.10	99.41	99.27	69.32	99.51	99.38	69.41
3	90.67	92.70	60.92	99.58	99.49	74.12	99.67	99.58	74.22
4	91.29	91.68	61.20	99.30	99.24	73.13	99.34	99.32	73.25
5	91.23	92.90	62.43	99.38	99.22	71.87	99.31	99.32	71.91

discussed. In addition, a new feature selection algorithm called Intelligent Rule based Attribute Selection algorithm and a novel classification algorithm named Intelligent Rule-based Enhanced Multiclass Support Vector Machine have been proposed. In this paper, intelligent algorithms for feature selection and classification have been proposed to design an effective intrusion detection system. The scope of this paper includes neural networks, fuzzy systems, genetic algorithm and particle swarm intelligence. The advantages and disadvantages of these intelligent techniques have been analyzed. The contributions of various research works in IDS are systematically summarized and compared, which allows us to clearly define existing research challenges and highlight promising new research directions. In addition the need for the new intelligent feature selection also called Intelligent Rule based Attribute Selection algorithm has been highlighted based on experimental results. In addition, the advantage of proposing the new classification also called Intelligent Rule-based Enhanced Multiclass Support Vector Machine has been discussed in detail so that the proposed system can be used to provide security to networks effectively.

Abbreviations

ACO: Ant colony optimization; ANFIS: Adaptive neuro-fuzzy inference system; ART: Adaptive resonance theory; CRF: Conditional random field; DDoS: Distributed denial of service; DR: Detection rate; FAR: False alarm rate; FENFCM: Feature-extraction neuron-fuzzy classification model; FNR: False negative rate; FPR: False positive rate; FSVM: Fuzzy support vector machine; GA: Genetic algorithm; GASSATA: Genetic algorithm for simplified security audit trials analysis; GFR: Gradually feature removal method; HFIS: Hierarchical neuro-fuzzy inference system; HMM: Hidden Markov model; IAEMSVM: Intelligent agent-based multiclass support vector machine; IDS: Intrusion detection system; IREMSVM: Intelligent rule-based multiclass support vector machine; LIDS: Layer-based intrusion detection system; LSSVM: Least square support vector machine; MAN: Metropolitan area networks; MMIFS: Modified mutual information-based feature selection; NN: Neural network; PSO: Particle swarm intelligence; SVM: Support vector machine; TNR: True negative rate; TPR: True positive rate; WLAN: Wireless local area network.

Competing Interests

The authors declare that they have no competing interests.

Source: Ganapathy S, Kulothungan K, Muthurajkumar S, *et al.* Intelligent feature selection and classification techniques for intrusion detection in networks: a survey [J]. Eurasip Journal on Wireless Communications & Networking, 2013, 2013(1): 1–16.

References

[1] W Stallings, Cryptography and Network Security Principles and Practices (Prentice Hall, Upper Saddle River, 2006).

[2] J Anderson, An Introduction to Neural Networks (MIT, Cambridge, 1995).

[3] B Rhodes, J Mahaffey, J Cannady, Multiple self-organizing maps for intrusion detection, Paper presented at the Proceedings of the 23rd National Information Systems Security Conference, Baltimore, 16–19, 2000.

[4] S Franklin, A Graser, Is it an agent or just a program? in ECAI'96 Proceedings of the Workshop on Intelligent Agents III, Agent Theories, Architectures, and Languages (Springer, London, 1996).

[5] N Jaisankar, SGP Yogesh, A Kannan, K Anand, Intelligent Agent Based Intrusion Detection System Using Fuzzy Rough Set Based Outlier Detection, Soft Computing Techniques in Vision Sci., SCI 395 (Springer, 2012), pp. 147–153.

[6] T Magedanz, K Rothermel, S Krause, Intelligent agents: an emerging technology for next generation telecommunications? In INFOCOM'96 Proceedings of the Fifteenth Annual Joint Conference of the IEEE Computer and Communications Societies, San Francisco, Mar 24–28, 1996.

[7] I Guyon, S Gunn, M Nikravesh, L Zadeh, Feature Extraction, Foundations and Applications (Springer, Berlin, 2006).

[8] R Kohavi, G John, Wrappers for feature subset selection. Artif Intell J Spec Issue Relevance 97(1–2), 273–324 (1997).

[9] P-N Tan, M Steinbach, V Kumar, Introduction to Data Mining (Addison-Wesley, Boston, 2005).

[10] RO Duda, PE Hart, DG Stork, Pattern Classification, 2nd edn. (Wiley, Hoboken, 2000).

[11] C Stefano, C Sansone, M Vento, To reject or not to reject: that is the question: An answer in the case of neural classifiers. IEEE Trans. Syst. Manag. Cyber 30(1), 84–94 (2000).

[12] SS Sivatha Sindhu, S Geetha, A Kannan, Decision tree based light weight intrusion detection using a wrapper approach. Expert Syst. Appl 39, 129–141 (2012).

[13] A Ghadiri, N Ghadiri, An adaptive hybrid architecture for intrusion detection based on fuzzy clustering and RBF neural networks, in Proceedings of the 2011 Ninth IEEE Conference on Annual Communication Networks and Services Research Conference, Otawa (IEEE Computer Society, Washington, 2011), pp. 123–129.

[14] S Chebrolu, A Abraham, P Johnson, Thomas Feature deduction and ensemble design of intrusion detection systems. Computers & Security 24(4), 295–307 (2005).

[15] W Zhang, S Teng, H Zhu, H Du, X Li, Fuzzy Multi-Class Support Vector Machines for Cooperative Network Intrusion detection. Proc. 9th IEEE Int. Conference on Cognitive Informatics (ICCI'10) (IEEE, Piscataway, 2010), pp. 811–818.

[16] L Zadeh, Role of soft computing and fuzzy logic in the conception, design and development of information/intelligent systems, in Computational Intelligence: Soft Computing and Fuzzy-neuro Integration with Applications, ed. by O Kaynak, L Zadeh, B Turksen, I Rudas. Proceedings of the NATO Advanced Study Institute on Soft Computing and its Applications held at Manavgat, Antalya, Turkey, 21–31 August 1996, volume 162 of NATO ASI Series (Springer, Berlin, 1998), pp. 1–9.

[17] M Tavallaee, E Bagheri, W Lu, AA Ghorbani, A Detailed Analysis of the KDD CUP 99 Data Set, in Proceedings of the IEEE Symposium on Computational Intelligence in Security and Defense Applications (Ottawa, 2009).

[18] KDD, KDD Cup. http://kdd.ics.uci.edu/databases/kddcup99/kddcup99.html. Accessed October 2007.

[19] SJ Stolfo, W Fan, W Lee, A Prodromidis, PK Chan, Cost-Based Modeling for Fraud and Intrusion Detection: Results From the JAM Project, in Proceedings of the 2000 DARPA Information Survivability Conference and Exposition (DISCEX '00), Hilton Head, SC, 2000.

[20] RP Lippmann, DJ Fried, I Graf, JW Haines, KR Kendall, D McClung, D Weber, SE Webster, D Wyschogrod, RK Cunningham, MA Zissman, Evaluating intrusion detection systems: the 1998 DARPA off-line intrusion detection evaluation, Hilton Head, 25–27 January 2000, vol 2 (IEEE, Amsterdam, 2000), pp. 10–12.

[21] MIT Lincoln Labs, DARPA Intrusion Detection Evaluation, 1998. http://www.ll.mit.edu/mission/communications/ist/corpora/ideval/index.html. Accessed February 2008.

[22] AA Bakar, ZA Othman, AR Hamdan, R Yusof, R Ismail, An Agent Based Rough Classifier for Data Mining. Eighth International Conference on Intelligent Systems Design and Applications, vol 1 (IEEE Computer Society, Washington, 2008), pp. 145–151.

[23] A Fagiolini, G Valenti, L Pallottino, G Dini, A Bic, Decentralized Intrusion Detection for Secure Cooperative Multi-Agent Systems, in Proceedings of the 46th IEEE Conference on Decision and Control (IEEE, Amsterdam, 2007), pp. 1553–1558.

[24] X Zhu, Z Huang, H Zhoul, Design of a Multi-agent Based Intelligent Intrusion Detection System. IEEE International Symposium on Pervasive Computing and Applications (IEEE, Amsterdam, 2006), pp. 290–295.

[25] G Xiantai, J Weidong, Z Dao, Multi Agent System for Detection and Containment in Metropolitan Area Networks. J. Electron. (China) 23(2), 259–265 (2006).

[26] N Joukov, C T-c, Internet worms as internet-wide threat. http://www.ecsl.cs.sunysb.edu/tr/TR143nikolaiRPE.pdf. Accessed Sept 2003.

[27] G Bourkache, M Mezghiche, K Tamine, A Distributed Intrusion Detection Model

Based on a Society of Intelligent Mobile Agents for Ad Hoc Network, in the 2011 Sixth IEEE International Conference on Availability, Reliability and Security, Vienna, August 2011 (IEEE, Amsterdam, 2011), pp. 569–572.

[28] Y Wang, S Behera, J Wang, G Helmer, V Honavar, L Miller, R Lutz, M Slagell, Towards the automatic generation of mobile agents for distributed intrusion detection system. J. Syst. Softw. 1(34), 1–14 (2006). Elsevier.

[29] C-h Fonk, GP Parr, PJ Morrow, Security schemes for Mobile Agent based Network and System Management Framework. J. Networks Syst. Manag. Springer 19, 232–256 (2011).

[30] P Mell, D Marks, M McLarnon, A Denial of service resistant intrusion detection architecture. Comput Networks J Elsevier, Amsterdam, 2000.

[31] A Verikas, M Bacauskiene, Feature selection with neural networks. Pattern Recognition Letters, Elsevier 23, 1323–1335 (2002).

[32] C-H Tsang, S Kwong, Multi-Agent Intrusion Detection System in Industrial Network using Ant Colony Clustering Approach and Unsupervised Feature Extraction, in the IEEE Conf. Proc. on Industrial Technology (IEEE, Amsterdam, 2005), pp. 51–56.

[33] MM Kabir, MM Islam, K Murase, A New Wrapper Feature selection approach using Neural Network. Neuro Computing 73, 3273–3283 (2010). Elsevier.

[34] J Mar, Y-C Yeh, I-F Hsiao, An ANFIS-IDS against Deauthentication DOS Attacks for a WLAN Taichung, 17–20 October 2010 (IEEE, Amsterdam, 2010), pp. 548–553.

[35] H Debar, M Becker, D Siboni, A neural network component for an intrusion detection system, in IEEE Symposium on Research in Computer Security and Privacy, Oakland, 4–6 May 1992 (IEEE, Amsterdam, 1992), pp. 240–250.

[36] D Joo, T Hong, I Han, The neural network models for IDS based on the asymmetric costs of false negative errors and false positive errors. Expert Syst. Appl 25, 69–75 (2003).

[37] Y Liu, D Tian, A Wang, ANNIDS: intrusion detection system based on artificial neural network, In Proceedings of the Second International Conference on Machine Learning and Cybernetics, vol. 3 (IEEE, Amsterdam, 2003), pp. 2–5.

[38] M Moradi, M Zulkernine, A neural network based system for intrusion detection and classification of attacks, in Proceedings of IEEE International Conference on Advances in Intelligent Systems—Theory and Applications, Luxembourg, vol. 148 (IEEE, Amsterdam, 2004), pp. 1–6.

[39] S Sarasamma, Q Zhu, J Huff, Hierarchical Kohonen net for anomaly detection in network security. IEEE Transactions on System, Man, Cybernetics, Part B, Cybernetics 35(2), 302–312 (2005).

[40] M Amini, R Jalili, HR Shahriari, RT-UNNID: A practical solution to real-time network-based intrusion detection using unsupervised neural networks. Computers and Security Science Direct 25(6), 459–468 (2006).

[41] S Koutsoutos, IT Christou, S Efremidis, A classifier ensemble approach to intrusion detection for network-initiated attacks, in Proceedings of the International Conference on Emerging Artificial Intelligence Applications in Computer Engineering: Real Word AI Systems with Applications in eHealth, HCI, Information Retrieval and Pervasive Technologies, vol. 160 (IOS, Amsterdam, 2007), pp. 307–319.

[42] J Shun, HA Malki, Network intrusion detection system using neural networks. Proc. Fourth IEEE Int Conf Nat Comput 5, 242–246 (2008). ICNC'08.

[43] C Thomas, N Balakrishnan, Improvement in intrusion detection with advances in sensor fusion. IEEE Trans. Inf Forensics Secur 4(3), 542–551 (2009).

[44] O Linda, T Vollmer, M Manic, Neural network based intrusion detection system for critical infrastructures, in Proceedings of IEEE International Joint Conference on Neural Networks, Georgia (IEEE, Amsterdam, 2009), pp. 102–109.

[45] L Me, GASSATA, a genetic algorithm as an alternative tool for security audit trials analysis, in Proceedings of 1st International workshop on Recent Advances in Intrusion Detection (Belgium, 1998).

[46] DE Goldberg, Genetic Algorithms in Search, Optimization, and Machine Learning (Addison-Wesley, Boston, 1989).

[47] G Stein, B Chen, AS Wu, KA Hua, Decision tree classifier for network intrusion detection with GA-based feature selection, Proceedings of the 43rd Annual Southeast Regional Conference, vol. 2 (ACM, Georgia, 2005), pp. 136–141.

[48] R Curry, P Lichodzijewski, MI Heywood, Scaling genetic programming to large datasets using hierarchical dynamic subset selection. IEEE Trans. Syst. Man Cybern. B Cybern. 37(4), 1065–1073 (2007).

[49] J Doak, An evaluation of feature selection methods and their application to computer security, Technical Report (University of California at Davis, Department of Computer Science, 1992).

[50] D Hongle, T Shaohua, Z Qingfang, Intrusion detection based on fuzzy support vector machines. International Conference on Networks Security, Wireless Communications and Trusted Computing, vol 2 (IEEE Computer Society, Washington, 2009), pp. 639–642.

[51] Y-P Zhou, J-A Fang, Y-P Zhou, Intrusion Detection Model Based on Hierarchical Fuzzy Inference System. Second IEEE International Conference on Information and Computing Science, vol 2 (IEEE Computer Society, Washington, 2009), pp. 144–147.

[52] Y Li, R Wang, J Xu, G Yang, B Zhao, Intrusion detection method based on fuzzy hidden Markov model. Sixth IEEE International Conference on Fuzzy Systems and Knowledge Discovery, vol 3 (IEEE, Piscataway, 2009), pp. 470–474.

[53] NR Guo, T-HS Li, Construction of a neuron-fuzzy classification model based on feature-extraction approach. Expert Syst. Appl 38, 682–691 (2011).

[54] AN Toosi, M Kahani, A new approach to intrusion detection based on an evolutionary

soft computing model using Neuro-fuzzy classifiers. Comput. Commun. 30, 2201–2212 (2007).

[55] Y-P Zhou, J-A Fang, Y-P Zhou, Research on Neuro-Fuzzy Inference System in Hierarchical Intrusion Detection. IEEE International Conference on Information Technology and Computer Science (IEEE Computer Society, Washington, 2009), pp. 253–256.

[56] ZCX Ji, An efficient intrusion detection approach based on hidden Markov model and rough set. IEEE International Conference on Machine Vision and Human-machine Interface (IEEE Computer Society, Washington, 2010), pp. 476–479.

[57] Y Guo, B Wang, X Zhao, X Xie, L Lin, Q Zhou, Feature Selection Based on Rough Set and Modified Genetic Algorithm for Intrusion Detection. IEEE International Conference on Computer Science & Education (IEEE, Piscataway, 2010), pp. 1441–1446.

[58] SH Jung, Queen-been evaluation for genetic algorithms. Electron. Lett. 36(6), 575–576 (2003).

[59] C Xu, Q Zhang, J Li, X Zhao, A bee swarm genetic algorithm for the optimization of dna encoding. 3rd International Conference on Innovative Computing Information and Control. 35 (IEEE, Piscataway, 2008).

[60] H Wedde, S Lehnohoff, B van Bonn, Z Bay, S Becker, S Bottcher, C Brunner, A Buscher, T Furst, A Lazagrescu, E Rotaru, S Senge, B Steinbach, F Yilmaz, T Zimmermann, A novel class of multi-agent algorithms for highly dynamic transport planning inspire by honey bee behavior. IEEE conference on emerging technologies and factory automation (IEEE, Piscataway, 2007), pp. 1157–1164.

[61] H Wedde, S Lehnohoff, B van Bonn, Z Bay, S Becker, S Bottcher, C Brunner, A Buscher, T Furst, A Lazagrescu, E Rotaru, S Senge, B Steinbach, F Yilmaz, T Zimmermann, Highly dynamic and adaptive traffic congestion avoidance in real time inspired by honey bee behavior. Mobilitat and Echtzeit, Informatik aktuell, 21–31 (Springer, Berlin, 2008).

[62] A Gupta, N Koul, SWAN: a swarm intelligence based framework for network management of ip networks. Int Conf Comput Intell Multimedia Appl 1, 114–118. IEEE Computer Society, Washington, 2007.

[63] S Sadik, A Ali, HF Ahmed, H Suguri, Honey bee teamwork architecture in multi-agent systems, Computer supported cooperative work in design III, Lecture notes in computer science, 4402/2007 (Springer, Berlin, 2007), pp. 428–437.

[64] K Passino, Systems biology of group decision making. 14th Mediterranean conference on control and automation (IEEE Computer Society, Washington, 2006).

[65] RLZ Gutierrez, M Huhns, Multi agent based fault tolerance management for robustness In Robust Intelligent Systems (Springer, Berlin, 2008), pp. 23–41.

[66] H-H Gao, H-H Yang, X-Y Wang, Ant Colony Optimization Based Network Intrusion Feature Selection And Detection. Proc. Fourth International Conference on Machine

Learning and Cybernetics (Springer, Berlin, 2005), pp. 3871–3875.

[67] RK Sivagaminathan, S Ramakrishnan, A hybrid approach for feature subset selection using neural networks and ant colony optimization. Expert Syst. Appl 33, 49–60 (2007).

[68] WS Li, XM Bai, LZ Duan, X Zhang, Intrusion Detection based on ant colony algorithm of Fuzzy clustering. International Conference on Computer Science and Network Technology (IEEE, Piscataway, 2011), pp. 1642–1645.

[69] MM Kabir, M Shahjahan, K Murase, A new hybrid ant colony optimization algorithm for feature selection. Elsevier-Expert Syst. Appl 39, 3747–3763 (2012).

[70] J Tian, H Gu, Anomaly detection combining one-class SVMs and Particle swarm optimization algorithms, Nonlinear Dynamics, vol. 61 (Springer, Berlin, 2010), pp. 303–310.

[71] Y Li, J Xia, S Zhang, J Yan, X Ai, K Dai, An efficient intrusion detection system based on support vector machines and gradually feature removal method. Expert Syst. Appl 39, 424–430 (2012).

[72] F Amiri, MMR Yousefi, C Lucas, A Shakery, N Yazdani, Mutual information-based feature selection for intrusion detection systems. J. Network Comput. Appl 34, 1184–1199 (2011).

[73] R Battiti, Using mutual information for selecting features in supervised neural net learning. IEEE Trans. Neural Netw. 5, 537–550 (1994).

[74] KK Gupta, B Nath, R Kotagiri, Layered Approach using Conditional Random Fields for Intrusion Detection. IEEE Trans. Dependable Secure Comput 7, 1 (2010).

[75] S Benferhat, K Tabia, On the combination of Naive Bayes and decision trees for intrusion detection, IEEE International Conference on Computational Intelligence for Modelling, Control and Automation, 2005 and International Conference on Intelligent Agents, Web Technologies and Internet Commerce, vol. 1 (Piscataway, IEEE, 2006), pp. 211–216.

[76] H Liu, L Yu, Toward integrating feature selection algorithms for classification and clustering. IEEE Trans. Knowl. Data Eng. 17, 491–502 (2005).

[77] C Cortes, V Vapnik, Support vector networks. Mach. Learn. 20, 1–25 (1995).

[78] JAK Suykens, L Lukas, DP Van, MB De, J Vandewalle, Least squares support vector machine classifiers: a large scale algorithm, in Proceedings of the European Conference on Circuit Theory and Design, 1999, pp. 839–842.

[79] S Ganapathy, P Yogesh, A Kannan, Intelligent Agent based Intrusion Detection using Enhanced Multiclass SVM. Comput. Intell. Neurosci. 2012, 10 (2012).

Chapter 8
Security in Cognitive Wireless Sensor Networks. Challenges and Open Problems

Alvaro Araujo, Javier Blesa, Elena Romero, Daniel Villanueva

Electronic Engineering Department, Universidad Politécnica de Madrid, Avda/Complutense 30, 28040 Madrid, Spain

Abstract: A cognitive wireless sensor network (CWSN) is an emerging technology with great potential to avoid traditional wireless problems such as reliability. One of the major challenges CWSNs face today is security. A CWSN is a special network which has many constraints compared to a traditional wireless network and many different features compared to a traditional wireless sensor network. While security challenges have been widely tackled in traditional networks, this is a novel area in CWSNs. This article discusses a wide variety of attacks on CWSNs, their taxonomy and different security measures available to handle the attacks. Also, future challenges to be faced are proposed.

Keywords: Cognitive, Security, Wireless Sensor Networks

1. Introduction

Global data traffic in telecommunications has an annual growth rate of over 50%. While the growth in traffic is stunning, both the rapid adoption of wireless technology over the globe and its penetration through all layers of society are even more amazing. Over the span of 20 years, wireless subscription has risen to 40% of the world population, and is expected to grow to 70% by 2015. Overall mobile data traffic is expected to grow to 6.3 exabytes per month by 2015, a 26 fold increase over 2010[1]. Over the recent years, wireless and mobile communications have increasingly become popular with consumers.

In regards to wireless networks, one of the fastest growing sectors in recent years was undoubtedly that of wireless sensor networks (WSNs). WSN consists of spatially distributed autonomous sensors that monitor a wide range of ambient conditions and cooperate to share data across the network. WSNs are introduced increasingly into our daily lives. Potential fields of applications can be found, ranging from the military to home control through commercial or industrial, to name a few. The emergence of new wireless technologies such as Zigbee and IEEE 802.15.4 has allowed for the development of interoperability of commercial products, which is important for ensuring scalability and low cost.

Most WSN solutions operate in unlicensed frequency bands. In general, they use ISM bands, like, the worldwide available 2.4 GHz band. This band is also used by a large number of popular wireless applications, for example, those that work over Wi-Fi or Bluetooth. For this reason, the unlicensed spectrum bands are becoming overcrowded with the increasing use of WSN-based systems. As a result, coexistence issues in unlicensed bands have been subject of extensive research[2][3], and in particular, it has been shown that IEEE 802.11 networks[4] can significantly degrade the performance of Zigbee/802.15.4 networks when operating in overlapping frequency bands[3].

The increasing demand for wireless communication presents an efficient spectrum utilization challenge. To address this challenge, cognitive radio (CR) has emerged as the key technology, which enables opportunistic access to the spectrum. A CR is an intelligent wireless communication system that is aware of its surrounding environment, and adapts its internal parameters to achieve reliable and efficient communication[5].

The main different between traditional WSN and new cognitive wireless sensor network (CWSN) paradigm is that in CWSN nodes change their transmission and reception parameters according to the radio environment. Cognitive capabilities are based in four technical components: sensing spectrum monitoring, analysis and environment characterization, optimization for the best communication strategy based on different constrains (reliability, power consumption, security, etc.) and adaptation and collaboration strategy.

Adding those cognition capabilities to the existing WSN infrastructure will bring about many benefits. In fact, WSN is one of the areas with the highest demand for cognitive networking. In WSN, node resources are constrained mainly in terms of battery and computation power but also in terms of spectrum availability.

Hence with cognitive capabilities, WSN could find a free channel in the unlicensed band to transmit or could find a free channel in the licensed band to communicate. CWSN could provide access not only to new spectrum (rather than the worldwide available 2.4 GHz band), but also to the spectrum with better propagation characteristics. A channel decision of lower frequency leads more advantages in a CWSN such us higher transmission range, fewer sensor nodes required to cover a specific area and lower energy consumption.

However, the cognitive technology will not only provide access to new spectrum but also provides better propagation characteristics. By adaptively changing system parameters like modulation schemes, transmit power, carrier frequency and constellation size, a wide variety of data rates can be achieved. This will certainly improve power consumption, network life and reliability in a WSN. Adding cognition to a WSN provides many advantages.

This way, CWSN is a new concept proposed in literature[6] with the following advantages.

- Higher transmission range.
- Fewer sensor nodes required to cover a specific area.
- Better use of the spectrum
- Lower energy consumption.
- Better communication quality.
- Lower delays.
- Better data reliability.

Despite the research interest in CWSN, security aspects have not yet been fully explored even though security will likely play a key role in the long-term commercial viability of the technology. The security paradigms are often inherited from WSN and do not fit with the specifications of CR networks. Looking at the literature related to CR, security researchers have seen that CR has special characteristics. This make CR security an interesting research field, since more chances are given to attackers by CR technology compared to general wireless networks. However, at present there are no specific secure protocols which integrate WSN and CR needs.

At this, still immature, point of CR, it is important to understand some fundamental issues such as potential threats, potential attacks and the consequences of these attacks.

As[7] says, the CR nature of the system introduces an entire new suite of threats and tactics that are not easily mitigated. The three main characteristics of CR are environment awareness, learning and acting capacity. At first, these characteristics should be an advantage against attacks but they can become in weaknesses. For example, CR nodes collaborate to make better decisions but these communications are ways to propagate the attack in the network.

Considering these characteristics since the attacker point of view, the fundamental differences between a traditional WSN and the CWSN network are

- The potential far reach and long-lasting nature of an attack.

- The ability to have a profound effect on network performance and behaviour through simple spectral manipulation.

The information sensed in a CRN is used to construct a perceived environment that will impact in a certain way in current and future behaviour s of all the nodes in the network. The induction of an incorrectly perceived environment will cause the wrong adaptation of the CRN, which could affect short-term behaviour but also because of their ability to learn, it will propagate the error to the new decisions. Thus, the malicious attacker has the opportunity for long-term impact on behaviour. Furthermore, CR collaborates with its fellow radios sharing information. Consequently, this provides an opportunity to propagate behaviour through the different networks.

Threats associated with each CRN features can be detected[7], such as

- Maintains awareness of surrounding environment and internal state. It could be an opportunity for spoofing that will send malicious data to the environment to provoke an erroneously perception.

- Adapts to its environment to meet requirements and goals. It is an opportunity to force desired changes in behaviour in the victim.

- Reasons on observations to adjust adaptation goals. It could be an opportunity to influence fundamental behaviour of CRN.

- Learns from previous experiences to recognize conditions and enables faster reaction times. This could an opportunity to affect long-lasting impact on CR behaviour.

- Anticipates events in support of future decisions. It could be an opportunity for long-lasting impact due to an erroneous prediction.

- Collaborates with other devices to make decisions based on collective observations and knowledge. This is an opportunity to propagate an attack through network.

- Wireless communication. Data might be eaves-dropped and altered without notice; and the channel might be jammed and overused by adversary. Access control, confidentiality, authentication and integrity must be guaranteed.

On the other hand, CRN features also help to mitigate malicious manipulation using:

- The ability to collaborate for authentication of local observations that are used to form perceived environments.

- The ability to learn from previous attacks.

- The ability to anticipate behaviours to prevent attacks.

- The ability to perform self-behaviour analysis.

Despite the extensive volume of research results on WSN[8], the considerable amount of ongoing research efforts on CR networks[9], and the new interest in CWSN[10], security in CWSN is vastly unexplored field. This is a new paradigm that offers many research opportunities.

The organization of this article is as follows. In Section 2, works in security are reviewed. In Section 3, a new taxonomy of attacks is proposed. In Section 4, countermeasures for CWSN attacks are analysed. Challenges and open works are shown in Section 5. Conclusions are offered in Section 6.

2. Related Work

First works about security in CR were developed specifically to analyse the effects produced by cognitive features and how they could be used to mitigate the negative effects. So, as we have said, in the article[7] each characteristic and the attacks that could take advantage of it are analysed. A different point of view is shown in the article of Zhang and Li[11]. They make a survey about the weaknesses introduced by the nature of CR. They base the security of the system in two tasks: protection and detection, and divide the attacks and countermeasures depending on which layer of the protocol stack affects. The article[12] studies threats that affect the ability to learn of cognitive networks and the dynamic spectrum access. To conclude the general references about security, it should be noted the article of Goergen and Clancy[9] where an attack classification in cognitive networks is done: DSA attacks, objective function attacks and malicious behaviour attacks.

In[13], two specific attacks against cognitive networks are analysed: primary user emulation (PUE), and sensing data falsification. It also provides some countermeasures well adapted to static scenarios such as TV system. In[14], a secure protocol spectrum sensing is presented. It bases its functionality on the generation and transmission of specific keys to each node. As a third example of safety sensing investigation, the research[15] proposes a collaborative algorithm based on energy detection and weighted combining (similar to a reputation system) to prevent malicious users.

Related to specifics attacks, the most studied against CR is the PUE, which was defined by Chen and Park[16] for the first time in 2006. Since then, research of the same authors[17] has focused on countermeasures against PUE. Also, in[18] a way to detect the PUs through an analytical model that does not require location information is shown. As well as the PUE attack, the community of researchers in CR has been studying other kind of attacks originate from different wireless networks, such as denial of service (DoS) attack or jamming attack. These attacks have special characteristics in cognitive networks, for example, article[19] studies these features for DoS, and[20] shows a countermeasure based on frequency hopping (technically possible in CR) to avoid jamming attacks.

Although previous articles help to understand the importance of securing CRNs[21]-[23] they do not take into account the specific characteristics of WSN.

On the other side, there are several articles related with security in WSNs, a topic very studied[8][24]-[27], but without using cognitive capabilities.

Summarizing the state of the art, there is still much to investigate in the area of security for CWSNs, because nowadays there is not any work focus on this topic.

3. Taxonomy of Attacks in CWSNs

As we shown in Section 1, CWSNs have special features that make security really interesting. However, security in CWSNs needs to be more studied by scientific community.

In this section, a complete taxonomy of attacks for CWSNs is shown. We are going to compare the differences in the scope between these attacks in a traditional WSN and in a cognitive one.

A taxonomy of attacks on CWSNs is very useful to design optimistic security mechanisms. There are several taxonomies of attacks on wireless networks[10] and focus on WSNs[6]. Moreover, some classifications of attacks in CR exist[3][9][11]. However, there is not a deep classification of attacks in CWSNs and study of attacks against cognitive WSNs does not exist.

We have analysed special network features that make CWSNs better against attacks: high transmission range, lower energy consumption, low delays and reliability of data. Their security is obviously endangered by the medium used, radio waves, but also by specific vulnerabilities of CWSNs like battery life or low computational resources.

Considering theses features, we propose a taxonomy which contains various attacks with different purposes, behaviours and targets. This will help researchers to better understand the principles of attacks in CWSNs, and further design more optimistic countermeasures for sensor networks. **Figure 1** shows an outline of this CWSN taxonomy of attacks. CWSN attacks are divided into communications, against privacy, node-targeted, power consumption, policy and cryptographic attacks.

3.1. Communication Attacks

First group is communication attacks. In this kind of attacks the attacker affects data transmissions between nodes with a concrete purpose. The goal could be from isolate a node to try to change the behaviour of whole network.

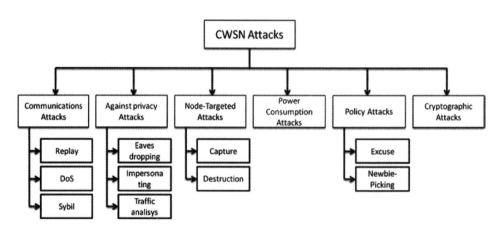

Figure 1. Taxonomy of attacks.

Communication attacks can be classified into three different types according to the attack behaviour: replay attack, DoS attack and Sybil attack. Replay attack[28] consists on the replay of messages from inside or outside the current run of communication. For example, message is directed to other than the intended node. This receiver node replays the message to the intended principal and this receives the delayed message. This delay is fundamental to calculate network characteristics (channel, topology, routing, etc.). CWSN could be affected in more degree that a regular WSN because nodes share information about the environment. If a node receives wrong information and also repeated, network behaviour could be affected deeply. If the PU packets are repeated, SU could have a wrong perspective of the spectrum too, avoiding the communications in frequencies or protocols used by the attacker.

DoS attack is characterized by an explicit attempt to present the legitimate use of a service. In this case, services are the spectrum or a special node. Different kinds of DoS attacks are

- Jamming attack, the transmission of a radio signal that interferes with the radio frequencies used by nodes. Jamming attack is one of the most studied attacks against WSN[29]. However, CWSN has great advantages to solve jamming but also can produce negative effects like energy consumption or communication failures. A typical jamming attack is a high power transmission using the PU frequency.

- Collision attack[30] consist of the intention of violate the communication protocol. This attack does not consume much energy of the attacker but can cause a lot of disruptions to the network operation. Due to the wireless broadcast nature, it is not trivial to identify the attacker. For example, the secondary users (SUs) have to share the spectrum. Therefore, the use of this type of attack is very efficient in order to disrupt the SU communication. Nodes, detecting collisions, will relay the information, making communication very difficult.

- Routing ill-directing attack. In this attack, a malicious node simply refuses to route messages. Examples of this kind of attacks are the grey hole and black hole ones. In these attacks, the nodes refuse all packets that arrive or a percentage thereof. Because of this misinformation, the network can change the routes, the topology or leaving isolated nodes.

- In flooding attack, a malicious node sends many connection request to a susceptible node, rendering the node or the resource useless. For instance, a joint network request to the coordinator node.

Sybil attack is defined as a malicious device illegitimately taking multiple identities. Sybil attack is effective against routing algorithms, voting, reputation systems and foiling misbehaviour detection. For instance, Sybil attack might utilize multiple identities to generate additional reputation to malicious nodes or to change the sensing spectrum information. The most studied attack against CR is the PUE.

3.2. Against Privacy Attacks

The other important attack class is attacks against privacy. CWSNs allow sharing resources to establish a communication and to be aware of environment. Attackers could use this access to take some of node information. The attacks against node privacy include eavesdropping, through taping the information; the attacker could easily discover the communication contents. Impersonating attack, where the attacker joins to the network and it can impersonate the original victim sensor node to receive packet, and traffic analysis, using wireless and cognitive features to listen in the entire spectrum. Traffic analysis attacks[31] try to deduce the context information of nodes analysing the traffic pattern from eavesdropping on wireless communication. Acquired information could be used to prepare a most harmful attack. For example, spectrum information can be used to know what the weakest spectrum zone is or where the PUs are emitting.

3.3. Node-Targeted Attacks

Node-targeted attacks need more attention that in a normal WSN because of the propagation of information is more important for the correct working of CWSN. A node can be captured[32][33] and attackers use reverse-engineered and become an instrument for mounting counterattacks. Other possibility is to destroy the nodes. This destruction not only affects to node functionality, but also affects whole network. Usually, node-targeted attacks ought to be less important for WSN. However, distributed information and co-operational behaviour in CWSN make a captured node a powerful weapon for attackers. Extracting a cryptographic key and modifying the internal device code are examples of node-targeted attacks.

3.4. Power Consumption Attacks

Battery life in WSN is a crucial factor. Small size of nodes and batteries makes CWSN very vulnerable to power consumption attacks. The attacker can inflict sleep torture on an energy constrained node by engaging in it unnecessary communication work to quickly drain its battery power. Depriving the power of a few crucial nodes (e.g. Access Point) may lead communication breakdown of the entire network. Attacker node can request a channel change every time, increasing power consumption.

3.5. Policy Attacks

The security and privacy policies are imperative since the policy basically influences the setup principles of a CWSN. Policy attacks can be classified as:

- Excuse attack, if the network policy is overly generous to recovering nodes that recently crashed or damaged by no requiring them to prove there are maintaining their quota, a malicious node may exploit this attack by repeatedly claiming to have been crashed/damaged. In this way, for instance, wrong spectrum information can be sent to the network very often to change the communications.

- Newbie-picking attack, if a CWSN requires that new nodes pay their dues by requiring them to give information to the net for some period of the time before they can consume any shared resource, therefore a veteran node could move from one newbie node to another, leeching their information without being required to give any information back.

3.6. Cryptographic Attacks

Concluding the taxonomy, the cryptographic attacks try to find the weaknesses in system analysing the information transmitted. Several cryptographic attacks exist but their objectives are the same: to acquire the cryptographic key, to identify weakness in the algorithms or in the node software. CWSN nodes do not have enough resources to implement a powerful cryptographic code and they are vulnerable to these attacks.

Apart from the above listed attacks that may hinder the key management of CWSNs, the following actions will also danger the key management within CWSNs: brute forces, dictionary attack and monitoring attack. One example of this kind of attack is Differential Power Analysis (DPA) attack. The DPA attack can be used to target an unsuspecting victim either by using special equipment that measures electromagnetic signals emitted by chips inside the device or by attaching a sensor to the device's power supply.

4. Countermeasures in CWSN

According to Section 3 is very clear that CWSN face a dangerous problem in security. Several attacks could be adapted from WSN to the new paradigm of cognitive networks. In the last 10 years, some researches related with security on CRN have appeared. They related specific attacks against these networks but a few counter-measures are proposed. In this section, we show three different groups of countermeasures according to the specific characteristics of CWSN.

4.1. Based on Geolocation

CR has its origin in United States where an important problem with the spectrum occupancy becomes real. The main reason is that the access to the radio spectrum is ruled by a restrictive regulatory regime that emerged when the Radio Act of 1927 declared the "ether" to be a publicly owned resource. The goal of CR was to use the radio spectrum when base stations did not transmit.

According to that, first real and simulated scenarios were static, with base stations making the role of PU and different devices like SU. If an attacker tries to emulate a PU the geolocation is an efficient method[18][34]. For example, in[34] the authors assume that the attacker is close to the victim and the real PU is much far from the SU and the attacker. Moreover, the position of each node, including the attacker, is fixed. Assuming that, SU can learn about the characteristics of the spectrum according to the received power.

Geolocation countermeasure does not work for most of cases in CWSN scenarios, almost with the same approach that previous mentioned papers. In a regular WSN, nodes can change their location, even attackers can change it. In fact, attackers have in the movement a great advantage to not be detected.

Another disadvantage of node mobility related to security is that if we would like to monitor PU we need to sense continuously the spectrum to detect new locations. The continuous sensing reduces node batteries. Moreover, if the PU could be in any spatial point, its location is irrelevant for security. For example, a mobile phone with Wi-Fi could be a PU and this device could stay in any location. Others parameters should be observed to differentiate between PU and an attacker.

To conclude, if we want to use a countermeasure based on geolocation, some restriction should be defined. For example, restricted areas for attackers or fixed number of PU in the scenario.

4.2. Based on Behaviour

In the same way that geolocation countermeasures, defences based on behaviour tries to modelling the PU[35]. The model is used to look for differences between a PU and attackers.

For example, in[17] authors use some radio parameters to decide if the transmitter is an incumbent transmitter or an attacker. These parameters are: signal characteristics, transmitted power and location. For a typical TV scenario on CR the PU model could be very precise. However as in geolocation countermeasures, the previous studies do not work for CWSN. Unfortunately it does not exist any model for PU in CWSN yet. PU usually are more unpredictable that in previous scenarios.

However, if we focus our CWSN in limited scenarios, for example intelligence ambient in a home or a building, the PU is defined specifically. Parameters like power transmission, time occupancy of spectrum and frequency used could be detected.

Genetic or Self Organizing Maps algorithms could be used to detect the PUs behaviour and to difference them against attackers. These algorithms can detect patterns and behaviour changes, so they are a good solution for this problem. However, computational cost and batteries life should be taking in account.

4.3. Based on Reputation and Trust of the CR Nodes

Two different groups of countermeasures related with the location and behaviour are proposed. Third group is a complement for the previous solutions that could improve the detection of attacks.

Reputation systems are very common in WSN[36]. Reputation takes advantage from the own characteristics of WSN: redundancy and adaptation. Usually several sensors form the networks and information is replied. Redundancy can be used to detect and isolate faulty or compromised nodes.

In CWSN where information is essential for the cognitive behaviour and sharing information is almost compulsory, reputation system can describe if the primary and SUs act like we expect. The big amount of information supplies the reputation system adjusting the reputation and trust of any node.

The best advantage of reputation systems is their versatility. The countermeasures could be implemented in any device, even small sensors with low resources, and could be used, in combination with others attacks, against most of attacks of Section 3.

5. Challenges and Open Problems

The nature of large, dynamic, adaptive, cognitive WSNs presents significant challenges in designing security schemes. A cognitive WSN is a special network which has many constraints compared to a traditional wireless network and many different features compared with a traditional WSN. While security challenges have been widely tackled in traditional networks, is a novel area in CWSN. In this section, most important challenges are discussed.

The wireless medium is inherently less secure because its broadcast nature makes eavesdropping simple. Any transmission can easily be intercepted, altered or replayed by an adversary. The wireless medium allows an attacker to easily intercept valid packets and easily inject malicious ones. Cognitive features allow a dynamic reconfiguration to avoid these attacks. However, malicious nodes can use the dynamic reconfiguration to create new attacks such us PUE. CWSNs have to adapt traditional wireless problems to cognitive networks and provide solutions to new problems.

The dynamic nature of sensor networks means no structure can be statically defined. Cognitive approach includes new dynamic issues: communication protocol, modulation, frequency, sensibility or emitted power. The attacker can use these powerful characteristics to affect the data transmissions between nodes with a concrete purpose. The goal could be from isolate a node to try to change the behaviour of entire network. Security schemes must be able to operate within this dynamic environment.

The next challenging factor is the hostile environment in which cognitive sensor nodes function. Nodes face the possibility of destruction or capture by attackers. Since nodes may be in a hostile environment, attackers can easily gain physical access to the devices. Attackers may capture a node, physically disassemble it, and extract from it valuable information (e.g. cryptographic keys). Because of the capacity of change the communication protocol, a capture node can affect to the whole network. For example, a malicious node can order to the network use a specific modulation or cryptographic algorithm and capture all the data. Also node can provide wrong information to the network causing a bad configuration. The ability to have a profound effect on network performance and behaviour through simple spectral manipulation is very dangerous. The highly hostile environment represents a serious challenge for security researchers.

The extreme resource limitations of CWSN devices pose considerable challenges to resource-hungry security mechanisms. The hardware constraints necessitate extremely efficient security algorithms in terms of bandwidth, computational complexity and memory. This is no trivial task. Energy is the most precious resource for these networks. Communication and cognitive algorithms are especially expensive in terms of power. Cognitive networks usually reduce power emission to save batteries. Attacker can isolate a node easily. Clearly, security mechanisms must give special effort to be communication efficient in order to be energy efficient.

The proposed scale of cognitive WSNs poses a significant challenge for security mechanisms. Cognitive networks are not only hundreds of sensors; they can also include different wireless interfaces and integrate a myriad of nodes in the same network. Providing security over such a network is equally challenging. Security mechanisms must be scalable to very large networks with different radio interfaces while maintaining high computation and communication efficiency.

One of the main goals of CWSNs is to allow a reliable communication. Certainly, unreliable communication is another threat to nodes security. The security of the network relies heavily on a defined protocol, which in turn depends on communication. Even if the channel is reliable, the communication may still be unreliable. The multi-hop routing, network congestion and node processing can lead to greater latency in the network, thus making it difficult to achieve synchronization among sensor nodes to change the communication scheme.

Depending on the function of the particular sensor network, the sensor nodes may be left unattended for long periods of time. There are two main cautions to unattended sensor nodes: exposure to physical attacks and managed remotely. Remote management of a sensor network makes it virtually impossible to detect physical tampering or DPA attack. Perhaps most importantly, the longer that a sensor is left unattended the more likely that an adversary has compromised the node.

CWSNs have a special feature for security mechanism: dynamic reconfiguration network scheme. Level security can adapts to a specific application, network topology, power and other constraints. Security level reconfiguration biased by different constraints has to be considered in order to improve network security.

6. Conclusions

CWSNs are increasingly being used in military, environmental, health and commercial applications. These networks are inherently different from traditional wireless networks as well as WSNs. Security is a mandatory feature for the deployment of CWSNs. This article summarizes the attacks and their taxonomy and also an attempt has been made to explore the security mechanisms widely used to handle those attacks. The challenges of WSNs are also briefly discussed. Security issues are a novel research area. This survey will hopefully motivate future researchers to design smarter and more robust security mechanisms and make their networks safer.

Acknowledgements

This study was funded by the Spanish Ministry of Science and Innovation, under Research Grant AMILCAR TEC2009-14595-C02-01, and the P8/08 within the National Plan for Scientific Research, Development and Technological Innovation 2008–2011.

Competing Interests

The authors declare that they have no competing interests.

Source: Araujo A, Blesa J, Romero E, *et al.* Security in cognitive wireless sensor networks. Challenges and open problems [J]. Eurasip Journal on Wireless Communications & Networking, 2012, 2012(1): 1–8.

References

[1] Cisco Systems Inc, Cisco Visual Networking Index: Global Mobile Data Traffic Forecast Update, 2010-2015. (2011) White Paper.

[2] I Howitt, J Gutierrez, IEEE 802.15.4 low rate–wireless personal area network coexistence issues, in Proceedings of IEEE Wireless Communications and Networking Conference (WCNC), vol. 3. New Orleans, Louisiana, USA, pp. 1481–1486 (March 2003).

[3] D Cavalcanti, R Schmitt, A Soomro, Achieving energy efficiency and QoS for low-rate applications with 802.11e, in IEEE Wireless Communications and Network-

ing Conference (WCNC), vol. 1. Hong Kong, pp. 2143–2148 (March 2007).

[4] IEEE 802.11 Standard, Wireless LAN medium access control (MAC) and physical layer (PHY) specifications, (Reaff 2003) Edition (1999).

[5] J Mitola, Cognitive radio: an integrated agent architecture for software defined RADIO, Ph.D. dissertation, (Royal Institute of Technology, Stockholm, Sweden, 2000).

[6] D Cavalcanti, S Das, J Wang, K Challapali, Cognitive radio based wireless sensor networks, in Proceedings of 17th International Conference on Computer Communications and Networks, vol. 1. St. Thomas, U.S. Virgin Islands, pp. 1–6 (August 2008).

[7] JL Burbank, Security in cognitive radio networks: the required evolution in approaches to wireless network security, in 3rd International Conference on Cognitive Radio Oriented Wireless Networks and Communications, (CrownCom), vol. 1. Singapore, pp. 1–7. 15-17 (May 2008).

[8] D Xiaojiang, C Hsiao-Hwa, Security in wireless sensor networks. IEEE Wirel Commun. 15(4), 60–66 (2008).

[9] TC Clancy, N Goergen, Security in cognitive radio networks: threats and mitigation, in 3rd International Conference on Cognitive Radio Oriented Wireless Networks and Communications, (CrownCom), vol. 1. Singapore, pp. 1–8 (May 2008).

[10] AS Zahmati, S Hussain, X Fernando, A Grami, Cognitive wireless sensor networks: emerging topics and recent challenges, in IEEE International Conference Science and Technology for Humanity (TIC-STH), vol. 1. Toronto, Canada, pp. 593–596 (September 2009).

[11] X Zhang, C Li, The security in cognitive radio networks: a survey, in Proceedings of the International Conference on Wireless Communications and Mobile Computing: Connecting the World Wirelessly (IWCMC), ACM, New York, NY 1, pp. 309–313 (2009).

[12] Y Zhang, G Xu, X Geng, Security threats in cognitive radio networks, in Proceedings of the 10th IEEE international Conference on High Performance Computing and Communications (HPCC), vol. 1. Dalian, China, pp. 1036–1041 (September 2008).

[13] C Ruiliang, P Jung-Min, YT Hou, JH Reed, Toward secure distributed spectrum sensing in cognitive radio networks. IEEE Commun Mag. 46, 50–55 (2008).

[14] G Jakimoski, KP Subbalakshmi, Towards secure spectrum decision, in Proceedings of IEEE Intl. Conference on Communications. (ICC), vol. 1. Piscataway, NJ, USA, pp. 2759–2763 (June 2009).

[15] T Zhao, Y Zhao, A new cooperative detection technique with malicious user suppression, in Proceedings of IEEE Intl. Conference on Communications (ICC), vol. 1. Dresden, Germany, pp. 14–18 (June 2009).

[16] R Chen, JM Park, Ensuring trustworthy spectrum sensing in cognitive radio networks, in 1st IEEE Workshop on Networking Technologies for Software Defined Radio Networks, (SDR), vol. 1. Orlando, Florida, USA, pp. 110–119 (September 2006).

[17] R Chen, JM Park, JH Reed, Defense against primary user emulation attacks in cognitive radio networks. IEEE J Sel Areas Commun. 26(1), 25–37 (2008).

[18] Z Jin, S Anand, KP Subbalakshmi, Detecting primary user emulation attacks in dynamic spectrum access networks, in IEEE International Conference on Communications, (ICC), vol. 1. Dresden, Germany, pp. 14–18 (June 2009).

[19] TX Brown, A Sethi, Potential cognitive radio denial-of-service vulnerailities and protection countermeasures: a multi-dimensional analysis and assessment, in 2nd International Conference on Cognitive Radio Oriented Wireless Networks and Communications, (CrownCom), vol. 1. Orlando, Florida, USA, pp. 456–464 (July 2007).

[20] L Zhang, J Ren, T Li, Spectrally efficient anti-jamming system design using message-driven frequency hopping, in IEEE International Conference on Communications (ICC), vol. 1. Dresden, Germany, pp. 1–5 (June 2009).

[21] G Baldini, T Sturman, A Biswas, R Leschhorn, G Godor, M Street, Security aspects in software defined radio and cognitive radio networks: a survey and a way ahead. IEEE Commun Surv Tutor. 99, 1–25 (2011).

[22] A Sethi, TX Brown, Hammer model threat assessment of cognitive radio denial of service attacks, in 3rd IEEE Symposium on New Frontiers in Dynamic Spectrum Access Networks, (DySPAN), vol. 1. Chicago, USA, pp. 1–12 (October 2008).

[23] S Arkoulis, L Kazatzopoulos, C Delakouridis, GF Marias, Cognitive spectrum and its security issues, in Proceedings of The Second International Conference on Next Generation Mobile Applications, Services, and Technologies (NGMAST), vol. 1. Cardiff, Wales, UK, pp. 565–570 (September 2008).

[24] P Walters, Z Liang, W Shi, V Chaudhary, Wireless sensor network security: a survey, (Security in Distributed, Grid, and Pervasive Computing, Auerbach Publications, CRC Press, New York, USA, 2006).

[25] W Yong, G Attebury, B Ramamurthy, A survey of security issues in wireless sensor networks. IEEE Commun Surv Tutor. 8(2), 2–23 (2006).

[26] Y Zhou, Y Fang, Y Zhang, Securing wireless sensor networks: a survey. IEEE Commun Surv Tutor. 10(3), 6–28 (2008).

[27] D Martins, H Guyennet, Wireless sensor network attacks and security mechanisms: a short survey, in 13th International Conference on Network-Based Information Systems (NBiS), vol. 1. Takayama, Gifu, Japan, pp. 313–320 (September 2010).

[28] DR Raymond, RC Marchany, SF Midkiff, Scalable, Cluster-Based Anti-Replay Protection for Wireless Sensor Networks, in IEEE Information Assurance and Security Workshop (IAW), vol. 1. West Point, NY, USA, pp. 127–134 (June 2007).

[29] H Sun, S Hsu, C Chen, Mobile jamming attack and its countermeasure in wireless sensor networks, in 21st International Conference on Advanced Information Networking and Applications Workshops (AINAW), vol. 1. Washington, DC, USA, pp. 457–462 (May 2007).

[30] P Reindl, K Nygard, D Xiaojiang, Defending malicious collision attacks in wireless sensor networks, in IEEE/IFIP 8th International Conference on Embedded and Ubiquitous Computing (EUC), pp. 771–776 (2010).

[31] X Luo, X Ji, MS Park, Location privacy against traffic analysis attacks in wireless sensor networks, in International Conference on Information Science and Applications (ICISA), vol. 1. Seoul, Korea, pp. 1–6 (April 2010).

[32] T Bonaci, L Bushnell, R Poovendran, Node capture attacks in wireless sensor networks: a system theoretic approach, in 49th IEEE Conference on Decision and Control (CDC), vol. 1. Atlanta, Georgia, USA, pp. 6765–6772 (December 2010).

[33] P Tague, R Poovendran, Modeling node capture attacks in wireless sensor networks. in 46th Annual Allerton Conference on Communication, Control, and Computing 1221–1224 (2008).

[34] Z Chen, T Cooklev, C Chen, C Pomalaza-Raez, Modeling primary user emulation attacks and defenses in cognitive radio networks, in IEEE 28th International Performance Computing and Communications Conference (IPCCC), vol. 1. Phoenix, Arizona, USA, pp. 208–215 (December 2009).

[35] TW Wu, YE Lin, HY Hsieh, Modeling and comparison of primary user detection techniques in cognitive radio networks, in IEEE Global Telecommunications Conference (GLOBECOM), vol. 1. New Orleans, LA, USA, pp. 1–5 (December 2008).

[36] S Ganeriwal, MB Srivastava, Reputation-based framework for high integrity sensor networks, in Proceedings of the 2nd ACM workshop on Security of ad hoc and sensor networks (SASN), vol. 1. ACM, New York, NY, pp. 66–77 (2004).

Chapter 9

Anticipating Complex Network Vulnerabilities through Abstraction-Based Analysis

Richard Colbaugh[1], Kristin Glass[2]

[1]Sandia National Laboratories, Albuquerque, NM 87111, USA
[2]New Mexico Institute of Mining and Technology, Socorro, NM 87801, USA

Abstract: Large, complex networks are ubiquitous in nature and society, and there is great interest in developing rigorous, scalable methods for identifying and characterizing their vulnerabilities. This paper presents an approach for analyzing the dynamics of complex networks in which the network of interest is first abstracted to a much simpler, but mathematically equivalent, representation, the required analysis is performed on the abstraction, and analytic conclusions are then mapped back to the original network and interpreted there. We begin by identifying a broad and important class of complex networks which admit vulnerability-preserving, finite state abstractions, and develop efficient algorithms for computing these abstractions. We then propose a vulnerability analysis methodology which combines these finite state abstractions with formal analytics from theoretical computer science to yield a comprehensive vulnerability analysis process for networks of real-world scale and complexity. The potential of the proposed approach is illustrated via case studies involving a realistic electric power grid, a gene regulatory network, and a general class of social network dynamics.

Keywords: Complex Networks, Finite State Abstraction, Vulnerability Analysis, Electric Power Grids, Biological Networks, Social Movements

1. Introduction

It is widely recognized that technological, biological, and social networks, while impressively robust in most circumstances, can fail catastrophically in response to focused attacks. Indeed, this combination of robustness and fragility appears to be an inherent property of complex, evolving networks ranging from the Internet and electric power grids to gene regulatory networks and financial markets e.g.,[1]-[5]. As a consequence, there is significant interest in developing methods for reliably detecting and characterizing the vulnerabilities of these networks e.g.,[6][7].

The challenges of vulnerability analysis are particularly daunting in the case of complex networks. Most such networks are large-scale "systems of systems", so that analysis methods must be computationally efficient. Additionally, because these networks perform reliably almost all of the time, standard techniques for finding vulnerabilities (e.g., computer simulations, "red teaming") can be ineffective and, in any case, are not guaranteed to identify all vulnerabilities. These observations suggest that, in order to be practically useful, any method for analyzing vulnerabilities of complex networks should be scalable, to enable analysis of networks of real-world complexity, and rigorous, so that for instance it is guaranteed to find all vulnerabilities of a given class.

This paper presents a new approach to vulnerability analysis which possesses these properties. The proposed methodology is based upon aggressive abstraction—dramatically simplifying, property preserving abstraction of the network of interest[4]. Once an aggressive abstraction is derived, all required analysis is performed using the abstraction. Analytic conclusions are then mapped back to the original network and interpreted there; this mapping is possible because of the property preserving nature of the abstraction procedure.

Our focus is on dynamical systems with uncountable state spaces, as many complex networks are of this type. We begin by identifying a large and important class of dynamical networks which admit vulnerability-preserving, finite state abstractions, and develop efficient algorithms for recognizing such networks and for computing their abstraction. We then offer a methodology which combines these finite state models with formal analytics from theoretical computer science[8] to provide a comprehensive vulnerability analysis process for large-scale networks. The potential utility of the proposed approach to vulnerability analysis is illustrated through case studies involving a realistic electric power grid, a gene regulatory network, and a general class of social network dynamics.

2. Preliminaries

This section introduces the class of network models to be considered in the paper and briefly summarizes some technical background that will be useful in our development.

The evolution to ensure robust performance in complex networks typically leads to systems that possess a "hybrid" structure, exhibiting both continuous and discrete dynamics[4]. More precisely, these networks often evolve to become hybrid dynamical systems—feedback interconnections of switching systems, which have discrete state sets, with systems whose dynamics evolve on continuous state spaces[9].

More quantitatively, consider the following definitions for hybrid dynamical system (HDS) models:

Definition 2.1: A continuous-time HDS is a control system

$$\sum\nolimits_{HDSct} \quad \begin{aligned} q+ &= h(q,k), \\ dx/dt &= f_q(x,u), \\ k &= p(x), \end{aligned}$$

where $q \in Q$ (with $|Q|$ finite) and $x \in X \subseteq \Re^n$ (with X bounded) are the states of the discrete and continuous systems that make up the HDS, $u \in \Re^m$ is the control input, h defines the discrete system dynamics, $\{f_q\}$ is a family of vector fields characterizing the continuous system dynamics, and p defines a partition of state space X into subsets with labels $k \in \{1,...,K\}$.

Definition 2.2: A discrete-time HDS is a control system

$$\sum_{HDSdt} \quad \begin{array}{c} q+ = h(q,k), \\ x+ = f_q(x,u), \\ k = p(x). \end{array}$$

We sometimes refer to an HDS using the symbol ΣHDS if the nature of the continuous system (continuous- or discrete-time) is either unimportant or clear from the context.

The concept of finite state abstraction for an infinite state system is illustrated in **Figure 1**. Consider a complex network with states that evolve on a continuous space and an analysis question of interest. Such a situation is depicted at the bottom of **Figure 1**, where the continuous dynamics are shown as curves on a continuous state space (blue region), and the analysis question involves deciding whether states in the green region can evolve to the red region. Reachability questions of this sort are difficult to answer for generic complex networks. However, if it is possible to construct a finite state abstraction of the network which possesses equivalent dynamics, then the analysis task becomes much easier. To see this, observe from **Figure 1** that a finite state abstraction of the original dynamics takes

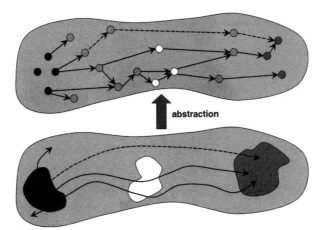

Figure 1. Finite state abstraction. Cartoon illustrates that the abstraction preserves network dynamics: trajectories of the infinite state system (curves in blue region at bottom) are mapped to equivalent finite state trajectories (sequences of state transitions at top).

the form of a graph, where the states are graph vertices (nodes within the blue region at top) and feasible state transitions define the graph's directed edges. Reachability analysis is straightforward with a graph, and if the complex network and its abstraction have equivalent reachability properties then the much simpler graph analysis also characterizes the reachability of the original system.

Reachability assessment, while valuable, is typically not sufficient to answer real-world vulnerability analysis questions. For instance, suppose that the red region in **Figure 1** is the set of failure states. It may be of interest to determine if all system trajectories which reach the red region first pass through the white "alerting" region, so there is warning of impending failure, or whether all trajectories which reach the red region subsequently return to the blue "normal" region, and thereby recover from failure. Addressing these more intricate questions requires that the analysis be conducted using a language which allows a nuanced description of, and reasoning about, network dynamics. We show in[4] that linear temporal logic (LTL) provides such a language, enabling quantitative specification of a broad range of complex network vulnerability problems. LTL extends propositional logic by including temporal operators, thereby allowing dynamical phenomena to be analyzed, and is similar to natural language and thus is easy to use[10].

As we wish to use LTL to analyze the dynamics of complex networks and we model these networks as HDS, we tailor our definition of LTL to be compatible with this setting:

Definition 2.3: The syntax of LTL consists of

- atomic propositions (q, k), where $q \in Q$ is an HDS discrete state and $k \in K$ is a label for a subset in the continuous system state space partition;

- formulas composed from atomic propositions using a grammar of Boolean ($\phi \vee \theta, \neg \phi$) and temporal ($\phi U \theta, \circ \phi$) operators.

The semantics of LTL follows from interpreting formulas on trajectories of HDS, that is, on sequences of (q, k) pairs: $(q, k) = (q_0, k_0), (q_1, k_1), \ldots, (q_T, k_T)$.

The Boolean operators $\vee$ and $\neg$ are disjunction and negation, as usual. The temporal operators U and ○ are read "until" and "next", respectively, with $\phi U \theta$ specifying that ϕ must hold until θ holds and ○ ϕ signifying that ϕ will be true at the next time instant (see[10] for a more thorough description).

Abstractions which preserve LTL also preserve all vulnerabilities expressible as LTL formulas; this property is stated more precisely in Section 4, where a formal definition is given for system vulnerability, and is also explained in[4]. Thus we seek an abstraction procedure which preserves LTL: given a system representation Σ_1, the procedure should generate a system abstraction Σ_2 which is such that $\{\Sigma_1 \models \phi\}$, $\{\Sigma_2 \models \phi\}$ for all LTL formulas ϕ (where $\models$ denotes formula satisfaction). Bisimulation is a powerful method for abstracting finite state systems to yield simpler finite state systems that are equivalent from the perspective of LTL[10]. However, the problem of constructing finite state bisimulations for continuous state systems is largely unexplored (but see the seminal work[11][12]). Indeed, one of the contributions of this paper is to develop a theoretically sound, practically implementable approach to obtaining finite state bisimulations for complex network models.

Bisimulation is typically defined for transition systems, so we first introduce this notion (see[10] for details):

Definition 2.4: A transition system is a four-tuple T = (S, $\rightarrow$, Y, h) with state set S, transition relation $\rightarrow \subseteq S \times S$, output set Y, and output map h: $S \rightarrow Y$. T is finite if |S| is finite.

Transition relation $\rightarrow$ defines admissible state transitions, so $(q, q') \in \rightarrow$, denoted $q \rightarrow q'$, if T can transition from q to q'.

Bisimilar transition systems share a common output set and have dynamics which are equivalent from the perspective of these outputs:

Definition 2.5: Transition systems $T_s = (S, \rightarrow_s, Y, h_s)$ and $T_p = (P, \rightarrow_p, Y, h_p)$ are bisimilar via relation $R \subseteq S \times P$ if:

- $s \sim p \Rightarrow h_s(s) = h_p(p)$ (R respects observations);

- $s \sim p, s \rightarrow_s s' \Rightarrow \exists p' \sim s'$ such that $p \rightarrow_p p'$ (T_P simulates T_S, denoted

$T_S < T_P$);

- $p \sim s, p \rightarrow_p p' \Rightarrow \exists s' \sim p'$ such that $s \rightarrow_s s'$ ($T_P < T_S$).

where ~ denotes equivalence under relation R.

A standard result from theoretical computer science e.g.,[10] shows that bisimulation preserves LTL:

Proposition 2.1: If T1 and T2 are bisimilar transition systems and φ is an LTL formula then $\{T_1 \mid = \varphi\}$, $\{T_2 \mid = \varphi\}$.

The following statements offer an alternative definition for bisimulation, which is easily shown to be equivalent to the one presented in Definition 2.5 and is useful in the subsequent development:

Definition 2.6: A finite partition $\Phi : S \rightarrow P$ of the state space S of transition system $T = (S, \rightarrow, Y, h)$ naturally induces a quotient transition system $T/\sim = (P, \rightarrow_\sim, Y, h_\sim)$ of T provided that

- $\Phi(s) = \Phi(s')$ (denoted $s \sim s'$) $\Rightarrow h(s) = h(s')$;

- $h_\sim(P) = h(s)$ if $P = \Phi(s)$;

- $\rightarrow_\sim$ is defined so that $\Phi(s) \mid \rightarrow_\sim \Phi(s')$ iff $s \rightarrow s'$

Transition system T and its quotient T/~ are bisimilar if an additional condition holds:

Proposition 2.2: Suppose T/~ is defined as in Definition 2.6 and, in addition, $\Phi(s) \rightarrow_\sim \Phi(s') \Rightarrow \forall s'' \sim s \exists s''' \sim s'$ such that $s'' \rightarrow s'''$. Then T and T/~ are bisimilar.

Finally, we introduce a class of continuous state (control) systems which is important in applications.

Definition 2.7: The continuous-time system dx/dt = f (x, u), with f:

$\Re^n \times \Re^m \to \Re^n$, is differentially flat if there exists (flat) outputs $z \in \Re^m$ such that $z = H(x), x = F_1(z, dz/dt, ..., d^r z/dt^r)$, and $u = F_2(z, dz/dt, ..., d^r z/dt^r)$ for some integer r and maps H, F_1, F_2.

Definition 2.8: The discrete-time system x + = f(x, u) is difference flat (with memory k) if there exists (flat) outputs $z \in \Re^m$ such that
$z = H(x), x(t) = F_1(z(t), z(t+1), ..., z(t+k-1))$, and
$u(t) = F_2(z(t), z(t+1), ..., z(t+k-1))$ for some maps H, F_1, F_2.

Background on flat systems may be found in[13]. Many real-world control systems are flat, including all controllable linear systems as well as all feedback linearizable systems. Perhaps more importantly, the complex, evolving networks underlying so much of advanced technology, biology, and social processes frequently possess flat subsystems.

3. Finite State Abstraction

In this section we demonstrate that hybrid systems with (differentially or difference) flat continuous systems admit finite state bisimulations and present algorithms for constructing the bisimilar abstractions.

Consider an HDS of the form given in Definition 2.1 or 2.2. The following provides a transition system representation for the continuous system dynamics of HDS:

Definition 3.1: The transition system model T_{HDSc} for the continuous system portion of ΣHDS is the collection $T_{HDSc} = \{T_q^k\}$, with one transition $T_q^k = (X_q^k, \to_q^k, Y_q^k, h_q^k)$ specified for each (q, k) pair. Each T_q^k has bounded state space X_q^k, finite output set Y_q^k, an output map $h_q^k : X_q^k \to Y_q^k$ that defines a finite partition of X_q^k with labels $y \in Y_q^k$, and transition relation $\to_q^k$ reflecting the discrete- or continuous-time dynamics:

- For discrete-time continuous systems, $x \to_q^k x'$ iff $\exists u$ such that $x' = f_q(x, u)$ on subset k;

- For continuous-time continuous systems, $x \to_q^k x'$ iff there is a trajectory

$x:[0,T] \to X_q^k$ of $dx/dt = f_q(x,u)$, a time $t' \in (0,T)$, and adjacent partitions of X_q^k labeled $y, y' \in Y_q^k$ such that $x(0) = x$, $x(T) = x'$, $x([0,t')) \subseteq y$, and $x((t',T]) \subseteq y'$.

We make the standard assumption that $k: X \to K$ partitions the HDS continuous system state space X into polytopes and that all HDS discrete system transitions are triggered by k transitions[9] (see Definitions 2.1 and 2.2).

Definition 3.1 allows Σ_{HDS} to be modeled as a feedback interconnection of two transition systems, one with continuous state space and one with finite state set:

Definition 3.2: The transition system T_{HDS} associated with the HDS given in Definition 2.1 or 2.2 is a feedback interconnection of 1) the continuous system transition system $T_{HDSc} = \{T_q^k\}$ given in Definition 3.1 and 2) the transition system associated with the HDS discrete system, given by $T_{HDSd} = (Q, \to_d, Q, id)$, where id is the identity map and $q \to_d q'$ iff $\exists k$ such that $q' = h(q,k)$. Thus $T_{HDS} = (Q \times X, \to_{HDS}, Q \times Y, h_{HDS})$, where $Q \times X = \bigcup_q (\bigcup_k \{q\} \times X_q^k)$, $Q \times X = \bigcup_q (\bigcup_k \{q\} \times Y_q^k)$, and the definitions for $\to_{HDS}$ and h_{HDS} follow immediately from the transition relation and output map definitions specified for T_{HDSc} and T_{HDSd}.

Because the transition system T_{HDSd} corresponding to the HDS discrete system is already a finite state system, the main challenge in abstracting HDS to finite state systems is associated with finding finite state bisimulations for the continuous systems $T_{HDSc} = \{T_q^k\}$. This is made explicit in the following

Theorem 1: If each transition system T_q^k associated with T_{HDS} is bisimilar to some finite quotient transition system $T_q^k/\sim = (Y_q^k, \to_\sim, Y_q^k, id)$ and the state space quotient partitions defined by the h_q^k satisfy a mild compatibility condition then T_{HDS} admits a finite bisimulation.

Proof: The proof is straightforward and is given in[4].

Theorem 1 shows that the key step in obtaining a finite state bisimulation for HDS T_{HDS}, and thus for Σ_{HDS}, is constructing bisimulations for the continuous state transition systems T_q^k. We therefore focus on this latter problem for the remainder of the section. Our first main result along these lines is for difference flat continuous systems:

Theorem 2: Given any finite partition $\pi : Z \to Y$ of the flat output space Z of a difference flat system, the associated transition system $T_F = (X, \to, Y, \pi \circ H)$ admits a bisimilar quotient $T_F/\sim$.

Proof: Consider the equivalence relation R that identifies state pairs (x, x') which generate identical sets of k- length output symbol sequences y = y_0 y_1... y_{k-1}, and the quotient system $T_F/\sim$ induced by R. R defines a finite partition of X (both $|Y|$ and k are finite), and $x \sim x' \Rightarrow \pi \circ H(x) = \pi \circ H(x')$ so that R respects observations. $T_F \angle T_F/\sim$ follows immediately from the definition of quotient systems. To see that TF/~ $\angle$ TF, note that flatness ensures any symbol string y = y_k y_{k+1} ... is realizable by transition system TF; thus x ~ x0 at time t implies that x and x0 can transition to equivalentstates at time t + 1. Therefore, from Definition 2.5, TF and TF/~ are bisimilar.

Remark 3.1: Efficient algorithms exist for checking if a given system is difference flat, so Theorem 2 provides a practically implementable means of identifying discrete-time continuous state systems which admit finite bisimulation[4].

Remark 3.2: The flat output trajectory completely defines the evolution of a difference flat system. As a consequence, because any finite partition of flat outputs pace induces a finite bisimilar quotient for the flat system, this partition can be refined to yield any desired level of detail in the abstraction.

An analogous result holds for differentially flat HDS continuous systems. Our development of this result requires the following lemmas.

Lemma 3.1: A control system is differentially flat iff it is dynamic feedback linearizable.

Proof: The proof is given in[14].

Lemma 3.2: Control system Σ admits a finite bisimulation iff any representation of Σ obtained through coordinate transformation and/or invertible feedback also admits a finite bisimulation.

Proof: The proof is given in[12].

Lemmas 3.1 and 3.2 suggest the following procedure for constructing finite bisimulations for differentially flat systems: 1) transform the flat system into a linear control system via feedback linearization, 2) compute a finite bisimulation for the linear system, and 3) map the bisimilar model back to the original system representation. As a result, we focus on building finite bisimulations for linear control systems.

In particular, the control system of interest is one "chain" of a Brunovsky normal form (BNF) system Σ_{BNF}[4]:

$$dx_1/dt = x_2,$$
$$dx_2/dt = x_3,$$
$$\ldots$$
$$dx_n/dt = u.$$

Concentrating on this system entails no loss of generality, as any controllable linear system can be modeled as a collection of these single chain systems, one for each input, and the decoupled nature of the chains ensures we can abstract each one independently and then "patch" the abstractions together to obtain an abstraction for the full system.

Consider the following partition of the (assumed bounded) state space $X \subseteq \mathfrak{R}^n$ of $\sum_{BNF}$:

Definition 3.3: Partition π_ϵ is the map $\pi_\epsilon : X \to Y$ that partitions X into subsets $y_{is} = \{x \in X | x_1 \in [i\epsilon, (i+1)\epsilon), \text{sign}(x_2) = s_1, \ldots, \text{sign}(x_n) = s_{n-1}\}$, where i is an integer and s is an (n-1)-vector of "signs" specifying a particular orthant of X.

Note that π_ϵ partitions X into "slices" orthogonal to the x_1-axis.

We are now in a position to state

Theorem 3: The transition system $T_{BNF} = (X, \rightarrow, Y, \pi_\varepsilon)$ associated with system Σ_{BNF} and partition π_ε admits a finite bisimilar quotient $T_{BNF}/{\sim} = (Y, \rightarrow_\sim, Y, id)$.

Proof: $T_{BNF}/{\sim}$ is finite because $|Y|$ is finite. Assume $\rightarrow_\sim$ is constructed so that $\pi_\varepsilon(x) \rightarrow_\sim \pi_\varepsilon(x') \Leftrightarrow x \rightarrow x'$, with the latter specified as in Definition 3.1. Then all the conditions of Definition 2.6 are satisfied, and from Proposition 2.2 we need only show $\pi_\varepsilon(x) \rightarrow_\sim \pi_\varepsilon(x') \Rightarrow \forall x'' \sim x \exists x''' \sim x'$ such that $x'' \sim x'''$. This amounts to demonstrating that if x can be driven through face F of slice $\pi_\varepsilon(x)$ then any x'' in that slice can be driven through F as well, which can be shown by checking this property for the system $x_1^{(n)} = u$ on each or thant of X (see e.g.,[13] for a proof that system $x_1^{(n)} = u$ possesses this property).

Remark 3.3: The result given in Theorem 3 is most useful in situations where the control input u can be chosen large relative to the system "drift". Abstraction methods for applications in which control authority is limited are given in[4].

Next we turn to the task of computing finite bisimulations for HDS. We focus on constructing bisimulations for HDS continuous systems, as HDS discrete systems already possess finite state representations, and in particular on abstracting differentially flat continuous systems; the derivation of algorithms for difference flat continuous systems is analogous but simpler and is therefore omitted (see[4]).

Consider, without loss of generality (see Lemma 3.1), the problem of computing a finite state abstraction for continuous-time linear control system Σ_{lc}: $dx/dt =$ Ax + Bu (where A, B are matrices). The transition system associated with Σ_{lc} is $T_{lc} = (X, \rightarrow_{lc}, Y, h)$, with h: $X \rightarrow Y$ any finite, hypercubic partition of X and $\rightarrow_{lc}$ specified as in Definition 3.1. The finite state abstraction of interest is quotient system $T_{lc}/{\sim} = (Y, \rightarrow_{lc\sim}, Y, id)$. Observe that in order to obtain $T_{lc}/{\sim}$ it is only necessary to determine the set of admissible transition relations $\rightarrow_{lc\sim}$.

As most applications of interest involve large-scale systems, it is desirable to develop efficient algorithms for computing $\rightarrow_{lc\sim}$. We now introduce such a procedure. The algorithm decides whether a transition $y \rightarrow_{lc\sim} y'$ between two adjacent cells of the lattice y, y' is allowed, and is repeated for all candidate transitions of interest. We begin by summarizing a simple algorithm, based on computational linear system results given in[15], for deciding whether $y \to_{lc\sim} y'$ is admissible. Let k be the number of the coordinate axis orthogonal to the common face between y and y', V be the set of vertices shared by y and y', and a_k^T represent row k of A. Define $\prod^k(w)$ to be the pro-jection of vector w onto axis k, and suppose $y < y'$. Then $y \to_{lc\sim} y'$ iff $\prod^k(Av_i + Bu) > 0$ for some $v_i \in V$ and $u \in U$ An algorithm which "operationalizes" this observation is

Algorithm 3.1:

If $y < y'$:

- If any element of row k of B is nonzero, $y \to_{lc\sim} y'$ is true. STOP.

- Repeat until $y \to_{lc\sim} y'$ is determined to be true or all vertices have been checked:

    o Select a vertex $v_1 \in V$

    o Compute the inner product $p = a_k^T v_i$

    o If $p > 0$ then $y \to_{lc\sim} y'$ is true. STOP.

- If $y \to_{lc\sim} y'$ has not been found to be true it is false.

If $y > y'$: Algorithm is the same except that the comparison $p > 0$ is replaced by $p < 0$.

A difficulty with Algorithm 3.1 is that the number of vertices shared by two adjacent cells is 2n−1, so that checking them becomes unmanageable even for moderately-sized systems. Interestingly, the algorithm can be modified so that feasibility of a transition can be tested by considering only a single well-chosen vertex, independent of the size of the model. The new algorithm is therefore extremely efficient and can be applied to very large systems. Let v_0 be the lowest vertex (in a component-wise sense) shared by y, y' and let $a_k^+ (a_k^-)$ be the sum of positive (negative) elements of row k of A, excluding the diagonal. We can now state

Algorithm 3.2[4][16]:

If $y < y'$:

- If any element of row k of B is nonzero, $y \rightarrow_{lc\sim} y'$ is true. STOP.

- Compute the inner product $p = a_k^T v_0$

- If $p + a_k^+ > 0$ then $y \rightarrow_{lc\sim} y'$ is ture. STOP

- Otherwise $y \rightarrow_{lc\sim} y'$ is false

If $y > y'$: Algorithm is the same except that the comparison $p + a_k^+ > 0$ is replaced by $p + a_k^- < 0$

A Matlab program which implements Algorithm 3.2 is presented in[4]. This program has been applied to systems with n = 10000 state variables using desktop computers. Additionally, standard timing studies reveal that the computational cost of Algorithm 3.2 scales quadratically in the system state space dimension n (while Algorithm 3.1 scale exponentially in n), providing further support for the argument that this approach to abstraction can be applied to problems of real-world scale and complexity.

4. Vulnerability Analysis

This section considers the vulnerability assessment problem: given a complex network and a class of failures of interest, does there exist an attack which causes the system to experience such failure? Other important vulnerability analysis tasks, including vulnerability exploitation and mitigation, are investigated in[4]. The proposed approach to vulnerability assessment leverages the finite state abstraction results derived in the preceding section. The basic idea is straightforward: given an HDS model for a network of interest and a class of failures of concern: 1) construct a finite bisimulation for the HDS network model, 2) conduct the vulnerability analysis on the system abstraction, and 3) map the analysis results back to the original system model.

Observe that the proposed approach possesses desirable characteristics. For instance, the analytic process is scalable, because both the abstraction methodology[4] and the tools available for detecting vulnerabilities in finite state systems (e.g., [8]) are scalable. Additionally, the analysis is rigorous. Because HDS vulnerabilities are expressible as LTL formulas, and bisimulation preserves LTL, the original complex network and its abstraction have identical vulnerabilities. Formal analysis tools such as model checking[8] can be structured to identify all vulnerabilities of the finite state abstraction, and bisimilarity then implies that the approach is guaranteed to find all vulnerabilities of the original network as well.

We now quantify the proposed approach to vulnerability assessment. It is supposed that the complex network of interest can be modeled as an HDS, Σ_{HDS}, and that the network's desired or "normal" behavior can be characterized with an LTL formula φ; generalizing the situation to a set of LTL formulas $\{\varphi_i\}$ is straightforward. Consider the following

Definition 4.1: Given an HDS Σ_{HDS} and an LTL encoding φ of the desired network behavior, the vulnerability assessment problem involves determining whether Σ_{HDS} can be made to violate φ.

The proposed vulnerability assessment method employs bounded model checking (BMC), a powerful technique for deciding whether a given finite state transition system satisfies a particular LTL specification over a finite, user-specified time horizon[8]. Briefly, BMC checks whether a finite transition system T satisfies an LTL specification φ on a time interval [0, k], denoted T $|= k\ \varphi$, in two steps: 1) translate T $|= k\ \varphi$ to a proposition $[T, \varphi]_k$ which is satisfied by, and only by, transition system trajectories that violate φ (this is always possible), and 2) check if $[T, \varphi]_k$ is satisfiable using a modern SAT solver[8]. Note that because modern SAT solvers are extremely powerful, this approach to model checking can be implemented with problems of real-world scale.

We are now in a position to state our vulnerability assessment algorithm. Let T_{HDS} denote the transition system associated with Σ_{HDS}, and consider the vulnerability assessment problem given in Definition 4.1. We have

Algorithm 4.1: Vulnerability assessment

1) Construct a finite bisimilar abstraction T for T_{HDS} using the results of Section 3.

2) Check satisfiability of $[T, \varphi]_k$ using BMC:

- if $[T, \varphi]_k$ is not satisfiable then T is not vulnerable and thus Σ_{HDS} is not vulnerable (on time horizon k);

- if $[T, \varphi]_k$ is satisfiable then T, and therefore Σ_{HDS}, is vulnerable, and the SAT solver "witness" is an exploitation of the vulnerability.

To illustrate the utility of the proposed approach to vulnerability assessment we apply the analytic method to an important complex network: an electric power (EP) grid. EP grids are naturally represented as HDS, with the continuous system modeling the generator and load dynamics as well as power flow constraints and the discrete system capturing protection logic switching and other "supervisory" behavior:

Definition 4.2: The HDS power grid model Σ_{EP} takes the form

$$\Sigma_{EP} \quad \begin{aligned} q+ &= h(q,k,v), \\ dx/dt &= f_q(x,y,u), \\ 0 &= g_q(x,y), \\ k &= p(x,y), \end{aligned}$$

where q and x are the discrete and continuous system states, v and u denote exogenous inputs, y is the vector of "algebraic variables", and all other terms are analogous to those introduced in Definition 2.1.

The continuous system portion of grid model Σ_{EP} is feedback linearizable[17], which implies the continuous system is differentially flat and consequently that Σ_{EP} admits a finite abstraction. Additionally, it can be shown that grid vulnerabilities are expressible as LTL formulas composed of atomic propositions which depend only on q and k[18]. Thus Algorithm 4.1 is directly applicable to power grids.

We now summarize the results of a vulnerability assessment for the 20bus grid shown in **Figure 2**. This grid provides a simple but useful representation of a real, national-scale EP system for which (proprietary) data are available to us[4]. The grid can be modeled as an HDS Σ_{EP} of the form given in Definition 4.2. The report[4] gives a Matlab encoding of the specific HDS model used in this study. Because the model Σ_{EP} corresponds to a real-world grid, the behaviors of the model and the actual grid can be compared. For example, the real grid recently experienced a large cascading voltage collapse, and data was collected for this event. We simulated this cascading outage (see **Figure 3**, top plot) and found close agreement between the behavior of the actual grid and the model ΣEP. Observe that this result is encouraging given the well-known difficulties associated with reproducing such cascading dynamics with computer models (see, e.g.,[17][18]).

Vulnerability assessment was performed using Algorithm 4.1. It was assumed that the grid's attacker wishes to drive the voltage at bus 11 to unacceptably low levels, so that the loads at this bus would not be served, and that sess to the generator at bus 2 via cyber means[18]. Note that this class of vulnerabilities is interesting because the access point—the generator at bus 2—is geographically remote from the target of the attack—the loads at bus 11.

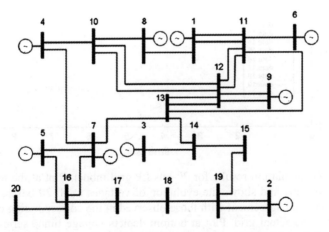

Figure 2. One-line diagram for the 20-bus EP grid model used in the vulnerability assessment case study.

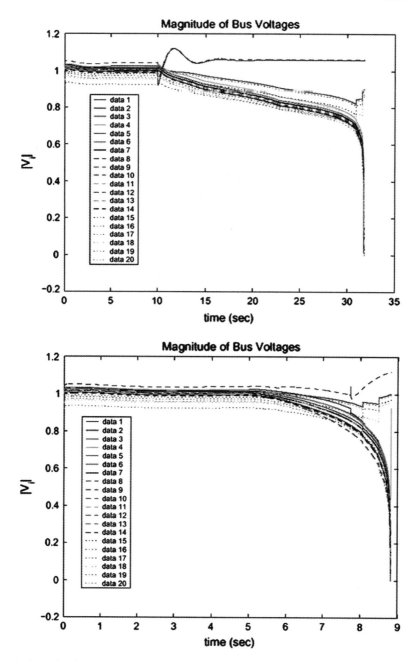

Figure 3. Sample simulation results for 20-bus EP grid model. Plot at the top is from the model validation study and shows the evolution of voltages at all 20 buses; these voltage time series are in good agreement with those observed in the corresponding cascading voltage collapse for the actual grid. Plot at bottom depicts voltage timec series which result from applying the vulnerability exploitation procedure designed using the proposed finite state abstraction methodology.

The first step in the vulnerability assessment procedure specified in Algorithm 4.1 involves constructing a finite state bisimulation T for Σ_{EP}; this abstraction is computed using Algorithm 3.2. The second step in Algorithm 4.1 is to apply BMC to T to determine if it is possible to realize the attack objective, *i.e.*, low voltage at bus 11, through admissible manipulation of the generator at bus 2. We employed NuSMV, an open source software tool for formal verification of finite state systems, for this analysis[19]. This vulnerability assessment reveals that it is possible for the attacker to realize the given objective via the assumed grid access, and gives a finite state "trace" of one means of exploiting the vulnerability. Using this trace, we synthesize an exploitation attack which is directly implementable with the HDS model Σ_{EP}. Sample simulation results are shown in **Figure 3** (bottom plot). It can be seen from the bus voltage time series in **Figure 3** that the attacker's goals can indeed be realized, in this case by initiating a cascading voltage collapse which takes down bus 11 as well as most of the rest of the grid.

5. Discussion

This paper presents an approach for analyzing complex networks in which the network of interest is first abstracted to a much simpler, but mathematically equivalent, representation, the required analysis is performed using the abstraction, and analytic conclusions are then mapped back to the original network and interpreted there. We identify an important class of complex networks which admit vulnerability-preserving, finite state abstractions, provide efficient algorithms for computing these abstractions, and offer a vulnerability analysis methodology that combines finite state network representations with formal analytics to enable rigorous vulnerability analysis for networks of real-world scale and complexity. The considerable potential of the method is demonstrated through a case study involving a realistic electric power grid model.

We now demonstrate that the proposed approach to analyzing complex network dynamics can also be applied to biological and social systems. Consider first a biological example. Many aspects of the physiology of living organisms oscillate with a period of approximately 24 hours, corresponding to the duration of a day, and the molecular basis for this circadian rhythm has been quantified in several organisms. For instance, a useful model for the gene regulatory network responsible for circadian rhythm in Drosophila melanogaster (fruit fly) is[20]:

$$\frac{dM_P}{dt} = v_{sP}\frac{K_{IP}^n}{K_{IP}^n + C_N^n} - v_{mP}\frac{M_P}{K_{mP} + M_P} - k_d M_P,$$

$$\frac{dP_0}{dt} = k_{sP}M_P - V_{1P}\frac{P_0}{K_{1P} + P_0} + V_{2P}\frac{P_1}{K_{2P} + P_1} - k_d P_0,$$

$$\frac{dP_1}{dt} = V_{1P}\frac{P_0}{K_{1P} + P_0} - V_{2P}\frac{P_1}{K_{2P} + P_1} - V_{3P}\frac{P_0}{K_{3P} + P_1} + V_{4P}\frac{P_2}{K_{4P} + P_2} - k_d P_1,$$

$$\frac{dP_2}{dt} = V_{3P}\frac{P_1}{K_{3P} + P_1} - V_{4P}\frac{P_2}{K_{4P} + P_2} - k_3 P_2^2 + k_4 C - v_{dP}\frac{P_2}{K_{dP} + P_2} - k_d P_2,$$

$$\frac{dC}{dt} = k_3 P_2^2 - k_4 C - k_1 C + k_2 C_N - k_{dC} C,$$

$$\frac{dC_N}{dt} = k_1 C - k_2 C_N - k_{dC} C,$$

where M_P, P_0, P_1, P_2, C, and C_N are state variables corresponding to the concentrations of the constituents of the circadian rhythm gene network, v_{sP} is an exogenous (control) input signal associated with the light-dark cycle of the environment, and all other terms are constant model parameters.

As is evident from Definition 2.7, a differentially flat system possesses (flat) outputs, equal in number to the number of inputs, which permit the system states and inputs to be recovered through algebraic manipulation of these outputs and their time derivatives. In the case of Drosophila circadian rhythm, C_N is one such flat output. To see this, note that C and its time derivatives can be obtained from the sixth equation through manipulation of C_N and its derivatives. These terms, in turn, permit P_2 (and its derivatives) to be obtained from the fifth equation, and continuing in this way up the "chain" of equations gives all of the states and the input v_{sP}. Thus the system states and input can be obtained from knowledge of C_N and its derivatives, proving that the above gene network model for Drosophila circadian rhythm is differentially flat. This, in turn, implies that the model admits a finite state bisimulation (Theorem 3).

We have constructed a finite state model for Drosophila using Algorithm 3.2, and then applied Algorithm 4.1 to this finite state representation to identify gene network vulnerabilities. More specifically, this methodology was employed to identify gene network parameters whose manipulation would quickly and efficiently reset the phase of Circadian rhythm. The parameters nominated by this analysis are:

- mRNA transcription rate;

- mRNA degradation rate;

- protein translation rate.

It is worth noting that these Circadian phase vulnerabilities are identical to the "control targets" obtained in[21] through a substantially more involved, computationally-intensive sensitivity analysis.

Consider next the phenomenon of social movements, that is, large, informal groupings of individuals and/or organizations focused on a particular issue, for instance of political, social, economic, or religious significance e.g.,[22]. Given the importance of social movements and the desire to understand their emergence and growth, numerous mathematical representations have been proposed to characterize their dynamics. For example, [23]suggests a model in which each individual in a population of interest can be in one of three states—member (of the movement), potential member, and ex-member—and interactions between individuals can lead to transitions between these affiliation states (e.g., potential members can be persuaded to become members). In particular, [23]proposes the following model for social movement dynamics:

$$\Sigma_{sm} \quad \begin{aligned} dP/dt &= \lambda - \beta PM + \delta_1 E, \\ dM/dt &= \beta PM - \delta_2 ME - \delta_3 M, \\ dE/dt &= \delta_2 ME + \delta_3 M - \delta_1 E, \end{aligned}$$

where P, M, and E denote the fractions of potential members, members, and ex-members in the population, Λ can be interpreted to be the system's input, and β, δ_1, δ_2, δ_3 are nonnegative constants related to the probabilities of individuals undergoing the various state transitions. It is worth noting that this model is shown in[23] to provide a good description for the growth of real-world social movements.

This model for social dynamics is differentially flat with flat output E. To see this, observe that M and its time derivatives can be obtained from the third equation through manipulation of E and its derivatives. These terms, in turn, permit P (and its derivatives) to be obtained from the second equation. Finally, knowledge of P, M, E, and their derivatives allows the input Λ to be recovered from the first equation. Thus all of the system states as well as the input can be obtained from knowledge of E and its derivatives, which shows that the social movement model is differentially flat. This, in turn, implies that the model admits a finite state bisimilar abstraction (see Theorem 3).

We now consider the vulnerability of social movement dynamics. In order to make this analysis more interesting, the dynamics Σ_{sm} is extended to enable modeling of social movements propagation on networks with realistic topologies. More specifically, it is known that real-world social networks possess community structure, that is, the presence of densely connected groupings of individuals which have only relatively few links to other groups e.g.,[22][23], and we are interested in modeling and analyzing social movement dynamics on these networks. One way to construct such a representation is to model movement dynamics as consisting of two components: 1) intra-community dynamics, involving frequent interactions between individuals within the same community and the resulting gradual change in the concentrations of movement members, and 2) inter-community dynamics, in which the movement jumps from one community to another, for instance because a movement member "visits" a new community.

It is natural to model these dynamics via HDS, with the continuous system representing intra-community dynamics via Σ_{sm}, or its finite state abstraction, the discrete system capturing the inter-community dynamics (e.g., using a simple switching rule), and the interplay between these dynamics being represented by the HDS feedback structure. A detailed description of the manner in which HDS models can be used to capture general social dynamics on networks with realistic topologies is given in[22], and the basic idea is illustrated in **Figure 4**.

We have constructed a finite state model for Σ_{sm} using Algorithm 3.2, and then connected these intra-community dynamics models together to represent the complete intra- and inter-community dynamics. We applied Algorithm 4.1 to this finite state representation to identify social movement vulnerabilities. More specifically, this methodology was employed to identify the characteristics of social

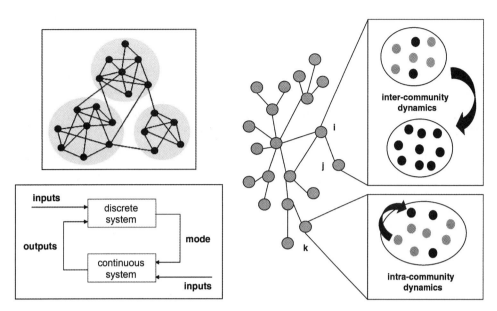

Figure 4. Modeling social dynamics on networks with community structure via HDS. The cartoon at top left depicts a network with three communities. The cartoon at right illustrates dynamics within a community k and between communities i and j. The schematic at bottom left shows the basic HDS feedback structure, in which the HDS discrete and continuous systems model the inter-community and intra-community dynamics, respectively.

movement dynamics which are most important for movement success. This analysis indicates that a key indicator of the ultimate success of a social movement is significant early dispersion of a movement across network communities; interestingly, this measure should be more predictive than the early volume of movement activity (which is a standard metric for predictive analysis of social dynamics).

Competing Interests

The authors declare that they have no competing interests.

Authors' Contributions

RC and KG designed the research, RC developed the theoretical results, RC and KG developed the computational algorithms, and RC wrote the paper. All authors read and approved the final manuscript.

Acknowledgements

This work was supported by the U.S. Department of Defense and the Laboratory Directed Research and Development Program at Sandia National Laboratories. Sandia National Laboratories is a multi-program laboratory managed and operated by Sandia Corporation, a wholly owned subsidiary of Lockheed Martin Corporation, for the U.S. Department of Energy's National Nuclear Security Administration under contract DE-AC04-94AL85000.

Source: Colbaugh R, Glass K. Anticipating complex network vulnerabilities through abstraction-based analysis [J]. Security Informatics, 2012, 1(1): 1–11.

References

[1] J. Carlson, J. Doyle, Complexity and Robustness. Proc. National Academy of Sciences USA 99, 2538–2545 (2002).

[2] J. Doyle, D. Alderson, L. Li, S. Low, M. Roughan, S. Shalunov, R. Tanaka, W. Willinger, The 'Robust Yet Fragile' Nature of the Internet. Proc. National Academy of Sciences USA 102, 14497–14502 (2005).

[3] P. Ormerod, R. Colbaugh, Cascades of Failure and Extinction in Evolving Complex Systems. J Artificial Societies and Social Simulation 9, 4 (2006).

[4] R. Colbaugh, K. Glass, G. Willard, Analysis of Complex Networks Using Aggressive Abstraction. SAND2008-7327 (Sandia National Laboratories, Albuquerque, NM, 2008). Additional file 1.

[5] R. LaViolette, K. Glass, R. Colbaugh, Deep Information from Limited Observation of 'Robust Yet Fragile' Systems. Physica A 388, 3283–3287 (2009).

[6] National Infrastructure Protection Plan (U.S. Department of Homeland Security, Washington DC, 2009).

[7] H. Chen, C. Yang, M. Chau, S. Li (eds.), Intelligence and Security Informatics (Lecture Notes in Computer Science, Springer, Berlin, 2009).

[8] E. Clarke, O. Grumberg, D. Peled, Model Checking (MIT Press, Cambridge, MA, 1999).

[9] R. Majumdar, P. Tabuada (eds.), Hybrid Systems: Computation and Control (Lecture Notes in Computer Science, Springer, Berlin, 2009).

[10] R. Milner, Communication and Concurrency (Prentice-Hall, NJ, 1989).

[11] R. Alur, T. Henzinger, G. Lafferriere, G. Pappas, Discrete Abstractions of Hybrid Systems. Proc. IEEE 88, 971–984 (2000).

[12] P. Tabuada, G. Pappas, Linear Time Logic Control of Discrete-Time Linear Systems. IEEE Trans. Automatic Control 51, 1862–1877 (2006).

[13] P. Martin, R. Murray, P. Rouchon, (California Institute of Technology, Pasadena, CA, 2003).

[14] E. Aranda-Bricaire, C. Moog, J. Pomet, A Linear Algebraic Framework for Dynamic Feedback Linearization. IEEE Trans. Automatic Control 40, 127–132 (1995).

[15] L. Habets, J. van Schuppen, A Control Problem for Affine Dynamical Systems on a Full-dimensional Polytope. Automatica 40, 21–35 (2004).

[16] J. Gardiner, R. Colbaugh, Development of a Scalable Bisimulation Algorithm, ICASA Technical Report (New Mexico Institute of Mining and Technology, Socorro, NM, 2006).

[17] M. Ilic, J. Zaborszky, Dynamics and Control of Large Electric Power Systems (Wiley, NY, 2000).

[18] J. Stamp, R. Colbaugh, R. Laviolette, A. McIntyre, B. Richardson, Impacts Analysis for Cyber Attack on Electric Power Systems, SAND2008-7300 (Sandia National Laboratories, Albuquerque, NM, 2008).

[19] NuSMV: a new symbolic model checker. http://nusmv.fbk.eu downloaded July 2007.

[20] J. Goncalves, T. Yi, (MTNS, Leuven, Belgium, 2004).

[21] N. Bagheri, J. Stelling, F. Doyle, Circadian Phase Resetting via Single and Multiple Control Targets. PLoS Comput. Biol. 4, e1000104 (2008).

[22] R. Colbaugh, K. Glass, Proc. IEEE Multi-Conference on Systems and Control (Saint Petersburg, Russia, 2009).

[23] P. Hedstrom, Explaining the Growth Patterns of Social Movements. Understanding Choice, Explaining Behavior (Oslo University Press, Oslo, Norway, 2006).

Chapter 10

Increasing Virtual Machine Security in Cloud Environments

Roland Schwarzkopf, Matthias Schmidt, Christian Strack, Simon Martin, Bernd Freisleben

Department of Mathematics and Computer Science, University of Marburg, Hans-Meerwein-Str. 3, D-35032 Marburg, Germany

Abstract: A common approach in Infrastructure-as-a-Service Clouds or virtualized Grid computing is to provide virtual machines to customers to execute their software on remote resources. Giving full superuser permissions to customers eases the installation and use of user software, but it may lead to security issues. The providers usually delegate the task of keeping virtual machines up to date to the customers, while the customers expect the providers to perform this task. Consequently, a large number of virtual machines (either running or dormant) are not patched against the latest software vulnerabilities. The approach presented in this article deals with these problems by helping users as well as providers to keep virtual machines up to date. Prior to the update step, it is crucial to know which software is actually outdated or affected by remote security vulnerabilities. While these tasks seem to be straightforward, developing a solution that handles multiple software repositories from different vendors and identifies the correct packages is a challenging task. The Update Checker presented in this article identifies outdated software packages in virtual machines, regardless if the virtual machine is running or dormant on disk. The proposed Online Penetration Suite performs pre-rollout scans of virtual machines for security vulnerabilities using established techniques and prevents execution of flawed virtual machines. The article presents the design, the implementation and an experimental evaluation of the two components.

1. Introduction

Infrastructure-as-a-Service (IaaS) Clouds[1] and virtualized Grid computing are based on the idea that users build individual virtual machines as execution environments for their tasks, allowing them to provide the required software stack without having to deal with Cloud or (multiple) Grid site administrators[2].

While the use of virtual machines is beneficial for service and infrastructure providers (users and providers in the Cloud nomenclature), by lowering the costs for the former and improving utilization and management capabilities for the latter, there are also some drawbacks. Since virtual machines are cheap and easy to create, users tend to create distinct virtual machines for different tasks. Users can branch new virtual machines based on old ones, snapshot machines or even rollback machines to a previous state. While these features provide great flexibility for users, they pose an enormous security risk for providers. A machine rollback, for example, could reveal an already fixed security vulnerability[3]. What makes the task of keeping the software stack up-to-date even more time-consuming is the increasing number of virtual machines, a phenomenon called virtual machine sprawl[4].

More problems arise because some of the virtual machines are likely to be dormant (not running) at some point in time. These virtual machines cannot be easily kept up-to-date, because typically this would require the virtual machines to be started, updated and shut down again, which is not only time-consuming, but may also be a tedious process. Different solutions[4]–[6] have been developed to solve the maintenance problem of (dormant) virtual machines. While these solutions can be used to update dormant machines, they suffer from a potential compatibility problem. They "forcibly" install updates, either by changing an underlying layer[5] or by replacing files[4][6], and there is no guarantee that the updates can be safely applied and that they are compatible to the software stack and the configuration of all affected virtual machines.

Moreover, all of these solutions lack the ability to properly identify which applications are truly outdated. Since this information is a prerequisite for the actual update process, it is a crucial step in the process of keeping (dormant) virtual machines in a Cloud or a virtualized Grid computing environment up-to-date. While such a check is easy to perform for running virtual machines, because of the commonly used package management systems on Linux platforms and automatic update facilities on Windows platforms, it is again a problem with dormant virtual machines. Even if virtual machines are kept up to date, the installed software might still contain design flaws or software vulnerabilities not fixed with the latest update. Thus, only checking for updates alone is not sufficient. Furthermore, machines used in a public IaaS environment are subject to external attacks, *i.e.*, they might be a selected or random target chosen by scripts. Therefore, it is indispensable to continuously analyze the used virtual machines and take proactive countermeasures such as patching the revealed flaws.

In this article a combined approach that checks for software updates and scans virtual machines for known security vulnerabilities is presented. The first component called *Update Checker* is proposed to check a potentially huge number of Linux-based virtual machines for the necessity of updates. Since the Update Checker copies the information about installed packages to a central database, the check can be executed on the central instance without booting the virtual machine beforehand and shutting it down afterwards, which is the most time-consuming part of checking for updates of a virtual machine. Thus, the check is independent of the status of the virtual machine (running or dormant). Both apt/dpkg and yum/rpm are supported and therefore all major Linux distributions. The solution allows easy checking of all registered virtual machines, returning either the number of available updates or details about each of the available updates. The second component called *Online Penetration Suite* (OPS) is proposed to perform periodic or pre-rollout online-scanning of virtual machines. While periodic scans can be done in idle times, pre-rollout scans are executed before machines go live, delaying the start of a machine but using the latest version of the scanners for up-to-date results. Virtual machines are scanned for software vulnerabilities, using a combination of well-known security products.

Furthermore, the proposed solutions can inform the owners about relevant findings via e-mail. Using an API, other management tools can utilize the results. To leverage existing software, our proposal is based on the Xen Grid Engine

(XGE)[2] and the Image Creation Station (ICS)[7] introduced in previous publications. The XGE is a software tool to create either virtualized Grid environments on-demand or to act as a Cloud IaaS middleware. The ICS offers an easy way for users to create, maintain and use virtual machines in the previously mentioned environments. An exemplary integration into the ICS, marking virtual machines that contain obsolete packages in virtual machine lists and providing details about available updates in detail views, and the XGE, preventing virtual machines containing obsolete packages from being started, is provided. The OPS scan process is triggered either by the ICS as a periodic maintenance operation or, if the additional overhead is acceptable, by the XGE as a pre-rollout check that might prevent a virtual machine from being started. As an alternative to preventing virtual machines from being started, those virtual machines can be started as usual and the owner is informed that his/her running machine is potentially unsafe. This can help administrators by giving them an overview of their dormant virtual machines, but also users without experience in the area of system maintenance (e.g. scientists that build custom virtual machines to execute their jobs), by making them aware of the problem.

The article is organized as follows. The next section presents the proposed design. Then, its implementation is discussed, followed by the presentation of experimental results. Afterwards, related work is discussed. The final section concludes the article and outlines areas for future research.

2. Design

The following sections present the design of the proposed approach. The first section outlines the Update Checker, a solution for checking for updates in virtual machines. The second section describes the Online Penetration Suite, an approach for online-scanning virtual machines for known software vulnerabilities.

2.1. Update Checker

Since the primary goal of the Update Checker is detecting obsolete software in (dormant) virtual machines, the term virtual machine is used throughout this article. Nevertheless, the solution is applicable to physical machines as well.

The concept of the Update Checker is to build a central database that contains all the information required for the task of checking for updates. This includes the list of installed packages, including the exact version of the installed package as well as the list of repositories that are used for each virtual machine. This information has to be imported into the central database when the virtual machine is first registered, and updated after each change of the virtual ma- chine, *i.e.*, after the installation of new software or the update of already installed software.

Since the Update Checker is not targeted at a single Linux distribution (compared to, e.g., Landscape for Ubuntu[8]), at least the two prevalent software management solutions are supported: apt/dpkg, used for example in Debian and Ubuntu, as well as yum/rpm, used for example in Red Hat and Fedora as well as SuSE. Both solutions use a specific package database format as well as a specific repository format. While apt/dpkg uses the same plaintext file format both as package database and as repository database, yum/rpm uses a Berkeley database as package database and an XML file as repository database. Nevertheless, this has no influence on the structure of the database used to store the required information, since both systems have the concept of distinct package names and a consistent versioning scheme in common.

The design of the solution is shown in **Figure 1**. There are specific importers for the package databases and for the repository databases of the different software management solutions. This makes the Update Checker easily adaptable to other software management solutions. Information about the installed packages of a virtual machine is stored in the Package DB. Metadata about the VM, *i.e.*, the time stamp of the import, the repositories used, etc., is stored in the Metadata DB. Information about the available packages on the different repositories is stored in the Repository Cache. When invoked, the Update Checker takes the information from these databases and the Repository Cache and matches installed and available packages to detect obsolete software and stores the results in the Result Cache.

When a query for the state of one or more virtual machines is issued, the Update Checker first checks to see if the result of that query is already available in the Result Cache and returns the cached result if it is not obsolete. Cached results are considered obsolete after a configurable amount of time, depending on factors

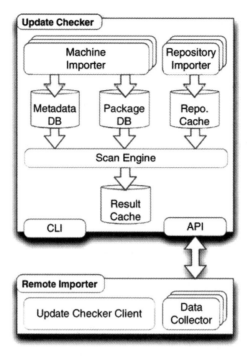

Figure 1. Update Checker architecture. The architecture of the Update Checker.

such as the frequency of updates or the need for security. Otherwise, it checks if the package lists of all repositories assigned to the virtual machine are available in the Repository Cache and not obsolete, *i.e.*, the configured validity period has not yet expired. If this is not the case, the package lists are downloaded from the software vendor's repository, parsed and stored in the Repository Cache for future use. When using the Repository Cache instead of the real repositories, there is the chance that the Update Checker fails to identify an outdated package. Nevertheless, the Repository Cache is very useful for checking many virtual machines and by using a small validity period, the risk can be minimized. Finally, the actual check of the virtual machine is started, comparing the version of each installed package with the version available at the repository. Information about outdated packages is then stored in the Result Cache, so that subsequent queries regarding the same virtual machine can be answered faster.

To help the user to judge whether the identified outdated software poses a risk to the virtual machine, the Update Checker infers information about the priority of an update. Unfortunately, there is no common way to do this for multiple distributions. As a first approach, the source repository of the updated packages is evaluated, since distributions like Debian or Ubuntu use special repositories for security updates. The source of an update can therefore be used as a hint of its significance.

The Update Checker allows to query for the number of available updates for a single or multiple virtual machines as well as for details about the outdated packages and available updates for a single virtual machine. The former query allows a good estimation of the state of the virtual machine, where zero means the virtual machine is up to date, while a number greater than zero means that there are updates available. If significance information is available, individual numbers for each level of significance as well as the sum of the numbers are returned. This can either be used in situations where an overview over a number of virtual machines is required, e.g., a list of virtual machines in a management tool like the ICS, or as a status check for a specific virtual machine, e.g., before it is started by the XGE.

Since the availability of updates itself allows no judgment about the threat resulting from the outdated packages, even when significance information is available, the latter query allows a detailed examination of the status of a virtual machine, by giving a list of outdated packages. This allows the user of the virtual machine to do a threat analysis based on the outdated packages and decide whether immediate action is required or not. The described functionality is used as an example of the integration of the Update Checker with other components. The complete solution is shown in **Figure 2**.

Two different interfaces are provided by the Update Checker: a command line interface (CLI) and an API for use by other software. The former can be used, when an administrator manually wants to execute an update check or register a virtual machine. The latter is provided for other tools like the ICS or XGE, allowing them to easily access the status information. This interface is provided using the language-independent protocol XML-RPC[9], to be available to tools written in any language.

The Update Checker can also be configured to run the checks at regular intervals, e.g., daily or weekly. This speeds up queries by other tools, because the

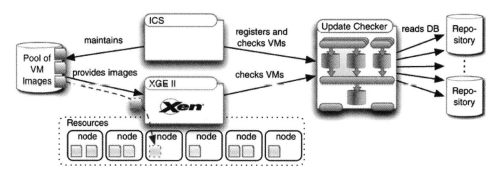

Figure 2. Usage scenario. The architecture of a complete system for virtualized Grid computing, consisting of the ICS, the XGE and the Update Checker. The figure shows the XGE deploying and starting a virtual machine, after the Update Checker has attested the virtual machine as being up-to-date.

information is already available. Users can be informed about obsolete software in their virtual machines via email. Additionally, administrators can also be informed about all virtual machines using obsolete software, to get an overview of the security of all virtual machines running on their infrastructure.

To ease the registration of virtual machines, the remote importer is provided (see **Figure 1**). It uses software management solution specific Data Collectors to gather the information required for the Update Checker, sends it to the machine the Update Checker is running on and triggers the registration process.

It might seem cumbersome to manually re-register virtual machines after every change, but with the remote importer it is merely a single command. Furthermore, it can be easily automated when software for management and maintenance of virtual machines is used.

2.2. Online Penetration Suite

This section presents the Online Penetration Suite (OPS) to scan an arbitrary number of virtual machines for security vulnerabilities utilizing multiple security scanners. The OPS combines and interprets the different results and generates a machine-readable and a human-readable report. Furthermore, the OPS is able to manage (start, stop, migrate, etc.) virtual machines if necessary. This allows automatic testing of virtual machines in a virtualized infrastructure to detect known security vulnerabilities. Once the vulnerabilities are known, the administrators and users can fix them to protect their systems with respect to unwanted attacks.

2.2.1. Architecture

The OPS is divided into two parts: the *logic* part, containing the flow control and the report generator, and the *backend* part, operating the registered vulnerability scanners and the virtual machines. The architecture of the OPS is shown in **Figure 3**, containing two adapters for OpenVAS[10] and Nessus[11].

The *OPS Logic* module controls the processes of the OPS. It configures the security scanners, boots the virtual machines to test (if required) and starts the actual scans. Since the vulnerability scanners are basically third-party products with individual characteristics and modes of operation, they are abstracted by *Adapters* that hide the differences and provide an unified interface to start and monitor the vulnerability scanners. They allow the OPS not only to start the actual scans, but also to watch the scanners during the execution to detect any failures and react accordingly.

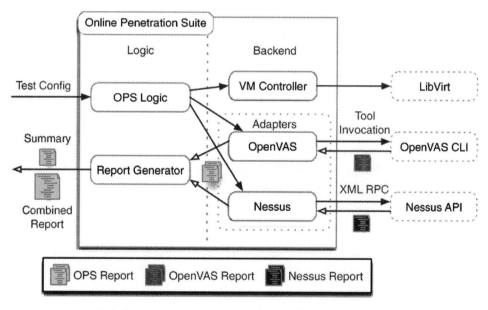

Figure 3. Online Penetration Suite architecture. The architecture of the Online Penetration Suite.

For a scan, the OPS needs two input parameters: the names of the target virtual machines and the name(s) of one or more vulnerability scanners. If no scanners are provided, the OPS chooses all scanners by default. A name uniquely identifies a virtual machine and allows the OPS to obtain further information like the IP and MAC address, path to the disk image(s), etc.

The *Report Generator* module collects the reports from the different scanners and generates the final result: a summary, containing the number of detected vulnerabilities categorized by a risk factor, and a combined report, containing the results from the security scanners in a unified format. To enable the Report Generator to analyze and understand the reports, the adapters have to convert the reports from the native format of the scanner to the unified OPS format.

The *backend* part of OPS consists of adapters to the required tools and libraries. It provides a module to control virtual machines using the libvirt[12] library as well as the vulnerability scanner adapters. Currently, the OPS supports two different scanners: OpenVAS[10] and Nessus[11], both well-known and established security-products.

2.2.2. Running Vulnerability Scans

OpenVAS is built as a client-server-architecture. The server is divided into three parts: administrator, manager and scanner. All clients communicate with either the manager or the administrator that both call the scanner. The OPS uses omp, a tool from the OpenVAS command line client for interaction. In order to guarantee a seamless scan, some of the countless options of OpenVAS are preset by the OpenVAS adapter module using a configuration file. This prevents the user from choosing wrong options that could possibly lead to false results. Nevertheless, by modifying the adapter configuration file it is possible for an administrator to enable/disable tests or set/unset options.

Nessus, being the ancestor of OpenVAS, is also built as a client-server-architecture. To control it, an XML-RPC interface is used. Nessus needs a number of parameters to start the scan process: the IP address of the server, authentication data and a scan configuration. Similar to the OpenVAS adapter, the Nessus adapter module presets a number of options to guarantee a seamless scan process.

2.2.3. Structure of the Reports

The combined report generated by the *Report Generator* is hierarchically divided into several parts. It starts with a summary of all reports and contains the results of each scanner structured by each tested virtual machine. Finally, the machine-specific report contains the vulnerabilities of this host. This includes a detailed description of the vulnerability, the severity level and if applicable, port number and transport protocol. The following paragraph shows an excerpt of a report:

```
<vulnerability>
<title>Microsoft Outlook SMB Attachment
Remote Code Execution Vulnerability (978212)</title>
<port>general/tcp</port>
<risk factor>HIGH</risk factor>
<description>
Overview: This host has critical security
update missing according to Microsoft Bulletin MS10-045.
[...]
CVE : CVE-2010-0266 BID : 41446
</description>
</vulnerability>
```

3. Implementation

In this section, the implementation of the Update Checker and the OPS is outlined.

3.1. Update Checker

This section describes important parts of the implementation of the Update Checker, working from the top to the bottom of **Figure 1**. First, the machine and repository importers and their sources of information are described using the Debian Package Manager (dpkg) and the Advanced Packaging Tool (apt) of Debian and its derivates as an example. Afterwards, the internal databases and caches, the

Scan Engine and the different interfaces are described. This section is concluded with details about the remote importer and the integration with other components. Further implementation details can be found in a previously published paper[13] of the authors. The implementation of the Update Checker has been done using the Ruby programming language.

3.1.1. Machine Importer

A machine importer is responsible for importing the list of installed packages and enabled repositories of a machine into the Package DB and Metadata DB, respectively. This information is collected from the package database, that keeps track of installed packages, versions, files belonging to each package, etc., and from the configuration files of the software management solution.

The package database of dpkg is stored in /var/lib/dpkg and consists of several text files, of which the file status is of particular interest, because it contains the metadata for each package that has ever been installed on the system. For each package it contains about a dozen key-value-pairs, of which three are required to extract the information: Package, which contains the package name, Status, which contains the state of the package (installed or not installed), and Version, which contains the exact version of the package. The following snippet shows the parsed parts of a dpkg package management database entry:

```
Package: openssh-server
Status: install ok installed
Version: 1:5.1p1-5
```

The repositories used by apt are stored in /etc/apt/ sources.list. This file contains multiple definitions, one per line, in the following format:

```
deb ROOT ARCHIVE COMPONENT
(COMPONENT...)
```

The meaning of these fields is explained in the next section. They are required to build the URL for the actual repository that is required to load the list of available packages.

3.1.2. Repository Importers

A repository importer is responsible for importing the list of available packages in a repository into the Repository Cache. This information is gathered from the repository database of the software management solution. The repository database of an apt repository can be found using the following URL that is built using information from the fields in the config file.

```
ROOT/dists/ARCHIVE/COMPONENT/'
    binary-ARCHITECTURE/Packages.TYPE
```

The *ROOT* field contains the root URL of the repository or mirror. The next two fields partition the repository: Debian and Ubuntu use *ARCHIVE* to divide the repository by the release (e.g. *stable* or *testing*) and *COMPO-NENT* to divide by license type and level of support (e.g. *main*, *contrib* or *non-free*). The last two fields specify the system architecture and the compression format of the repository database.

The repository database uses the same format as the package database of dpkg. Thus, parsing can be done using the same technique.

3.1.3. Internal Databases and Caches

The Package DB is used to store a name-version-pair for each installed package on every machine. Its counterpart is the Repository Cache that stores a name-version-pair for each available package on every repository. Initially, it was planned to store this information in a database. Unfortunately, importing a virtual machine or updating the list of available packages of a repository was very slow using this technique. As a faster alternative, a hash encoded in JSON[14] was chosen, written to an individual file per virtual machine or repository, respectively. This was faster by a factor of more than 23 when measured for the import of two Debian repositories (2.16 sec using the hash versus 50.02 sec using the database). The equivalent to the database snippets shown above in the internal format is the following:

```
...,"openssh-server :"1:5.1p1-5 ,...
```

Information about outdated packages is stored in the Result Cache. It stores name-old version-new version-priority-quadruplets in a JSON encoded list, written to an individual file per virtual machine.

The Metadata DB stores a list of all registered virtual machines and repositories as well as the mapping between them. Furthermore, it stores the names of all files that build the Package DB, Repository Cache and Result Cache, together with an expiration date for each file of the two caches.

3.1.4. Scan Engine

In this component, the actual identification of outdated packages takes place. Whenever a query for available updates of a virtual machine is submitted and there is no current result in the result cache, the Update Checker first determines the required repositories using the Metadata DB. If the repository cache does not contain current versions of the required repositories, a repository importer is used to update the cache. Afterwards, the list of installed packages is retrieved from the Package DB and the version of each package is compared with the version of that package stored in the repository cache. Outdated packages are stored in the result cache with installed and available version, so that subsequent queries can be handled faster. Finally, the number of outdated packages or the list of outdated packages is returned to the issuer of the query.

One particular problem discovered during the implementation of the Update Checker is the format of the version numbers used by the different package management systems or distributions, respectively. While most of the distributions use versions composed of the fields epoch, version and release, there are subtle differences between the distributions, e.g., separators, format of the release field, etc. Even the versiono my gem, a Ruby library especially designed for version comparisons, failed to correctly compare Debian version numbers.

One possibility is the use of the dpgk binary which provides an option to compare versions. This is very slow, since each comparison requires forking a new process. A Ruby library named dpkg-ruby implements version comparison using a native library. An old version of this library contains a Ruby-only version of the version comparison. Although slower, this solution is preferred to be independent of native libraries. By using an additional string comparison beforehand, performance losses can be cut down. Except for some minor tweaks, this version comparison library worked with all version numbers that were encountered in Debian and Fedora.

A daemon is used to provide some automation. All virtual machines can be checked for updates automatically at regular intervals. As described above, this frequently updates the cached repository databases and caches the results for all virtual machines. Queries using the API or the command line interface can then be served from the cache, requiring almost no time (only a file has to be read). The daemon also allows to notify users by email about outdated packages in their virtual machines. Additionally, the daemon can be configured to send emails about the status of all virtual machines to administrators.

3.2. Online Penetration Suite

The Online Penetration Suite is implemented in the Java programming language. Virtual machines are controlled using the Java binding of the libvirt library, the Nessus scanner is invoked using the Apache XML-RPC library and the reports of the vulnerability scanners are processed and converted using the Java API for XML Processing (JAXP).

Depending on the test configuration specified via the command line, the OPS frontend selects the required vulnerability scanners, starts their server components (if required), boots the virtual machines to scan (if they are not running already) and finally initiates and monitors the actual scan processes. All of these operations are hidden behind an interface that is implemented by the adapters, making the OPS easily extensible with new scanners. Since the report generation process is based entirely on reports in the unified OPS format, no vulnerability scanner dependent code is required for this step in the frontend.

The adapters use different techniques to control and monitor the actual vulnerability scanners. OpenVAS provides a command line interface, so its adapter needs to create a test configuration in the form of an XML file and pass it as an argument to the omp binary. Monitoring of OpenVAS requires analyzing the output of its client. For Nessus, the provided XMLRPC API is used. It contains methods to start and monitor the actual scan process. Both adapters contain code to convert the proprietary report formats into the unified OPS format.

4. Experimental Results

The following section presents an evaluation of the presented components.

4.1. Update Checker

Measurements have been conducted to evaluate the Update Checker on an Intel Xeon E5220 machine with 1 GB memory. The first measurement is a local measurement testing all components of the Update Checker, *i.e.*, machine import, repository import and update checking. Three Debian and three Fedora virtual machines have been used in this test, with varying numbers of installed packages and enabled repositories. Each test has been executed 20 times and average values have been calculated. The results are shown in **Table 1**.

In the first part of this evaluation, the different machine importers were tested. All required files were copied to the machine the test was executed on prior to the evaluation, thus no network communication is involved. Furthermore, before the measurement rpm-qa was executed on the source machine to generate a list of installed packages including their version. This is required to work around incompatibilities (*i.e.*, the rpm binary on Debian squeeze could not read the rpm database of a Fedora 15 installation).

Table 1. Update Checker component benchmark.

Distribution	Installed packages	Machine import	Repository import	Update import
Debian	563	0.04 secs	2.39 secs	0.44 secs
Debian	867	0.06 secs	2.80 secs	0.44 secs
Debian	1493	0.07 secs	2.68 secs	0.78 secs
Fedora	591	0.03 secs	13.59 secs	0.38 secs
Fedora	1063	0.04 secs	14.84 secs	1.00 secs
Fedora	2159	0.05 secs	15.38 secs	2.10 secs

Benchmark of all individual components of the Update Checker.

The growing import times can be explained with the growing number of installed packages that must be parsed.

The second part of the test measured the time required to download and parse all repository databases for the virtual machines (each machine had between 2 and 4 repositories configured) without using the repository cache. The times measured are thus artificial and are only of little relevance for actual usage, but allow evaluating the repository import and update checking. While the times for the Debian machines are quite stable, the increase of the time for Fedora is caused by the number of repositories used (2, 3 and 4, respectively). The very bad performance of the Fedora repository import is caused by the use of XML in the repository database.

The last part of the test evaluates the algorithm that actually checks for updates. Again, the increase in the times is caused by the growing number of packages. The reason for the worse results for Fedora are probably the longer and more complex version numbers used in Fedora, making the comparison harder and more time-consuming.

The measured values are promising. Checking for updates is a very fast process with the Update Checker. Because of the individual files used for the Package DB and Repository Cache, we do not expect performance degradation when the number of virtual machines increases. The relatively long time required for importing yum repositories is compensated by the repository cache, that results in every repository being downloaded and parsed only once during the configurable validity period of the cache.

To evaluate the influence of the repository cache, another measurement has been conducted that represents a more realistic scenario: checking all imported virtual machines for updates. The six machines from the last measurement were checked at once, taking advantage of the repository cache. The experiment was repeated 20 times and the average times are shown in **Figure 4**. The results indicate that the repository cache is very effective in cutting down the time required to check multiple virtual machines for updates.

To evaluate the scalability (and applicability for physical machines) of the

Update Checker, 115 physical nodes from our compute cluster were imported. All machines were checked at once using the repository cache. The experiment was repeated 20 times and the time required to check all virtual machines was calculated. The results shown in **Figure 5** provide evidence for the scalability of the Update Checker. The average check time was 34.53 seconds for all 115 machines, that is 0.30 seconds per machine.

Another measurement was conducted to evaluate the import time of the virtual machines, when the remote importer is used. This involves gathering all required files, executing rpm-qa in the case of rpm based distributions, sending everything to the Update Checker and starting the import process. For each virtual machine, 10 imports were executed. The results are shown in **Figure 6**.

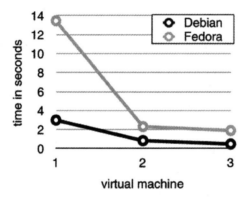

Figure 4. Update checking performance. Benchmark of the update checking process for multiple virtual machines using the repository cache.

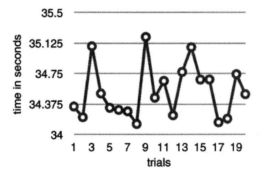

Figure 5. Results of the scalability evaluation. Benchmark of the update checking process for 115 machines using the repository cache.

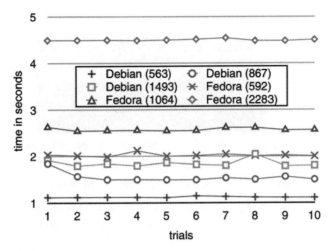

Figure 6. Remote import performance. Total time required to import each of the 6 virtual machines using the remote importer.

As expected, the amount of time the import process requires grows with the number of packages in the database. Generally, the import process is faster for apt/dpgk based virtual machines than for yum/rpm based virtual machines. The source of this problem seems to be the use of the rpm binary to extract the information from the database.

4.2. Online Penetration Suite

The following section presents measurements related to the OPS. All tested systems are Xen domainU virtual machines running Debian Squeeze and located on Pentium IV systems with 1 GB memory. The OPS node is an Intel Xeon E5220 machine and 1 GB memory. All systems are interconnected with switched fast Ethernet.

The first experiment measures the total runtime of the OPS depending on the number of virtual machines. **Figure 7** shows the results. The OPS used both vulnerability scanners in parallel while the number of target virtual machines was increased with every run. To get a robust mean, 100 trials were performed. Testing one virtual machine took 684 seconds on average, testing two machines took 859 seconds, testing three machines 1056 seconds, and it took 1279 seconds to test all

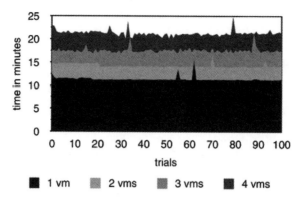

Figure 7. Results of the OPS evaluation. Benchmark of the runtime of the OPS depending on the number of tested systems.

four machines. Obviously, the measurement reveals that the runtime increases linearly with the number of tested systems. Furthermore, it reveals that it is more efficient to test multiple targets in parallel instead of scanning one after another.

In order to test the efficiency of the OPS, multiple tests against virtual machines running different versions of the Debian operating systems were conducted. The unpatched release version of Debian *Etch* (released April 2007), *Lenny* (released February 2009), *Squeeze* (released February 2011) and *Wheezy* (current unstable version) were used. The results of the tests are shown in **Table 2**.

The OPS successfully revealed a number of security vulnerabilities in all tested versions, including two high-risk flaws in each version. Debian *Etch* is the oldest release and contains the lowest number of vulnerabilities because it contains less features (in terms of installed services) than all other versions. Other flaws are related to the installed kernel version. The flaws appeared with newer kernel versions and thus, only in newer Debian versions.

5. Related Work

The Cloud computing risk report written by ENISA[15] mentions the failure of customer hardening procedures as one of the research problems needed to be solved. Customers failing to secure the computing environment may pose a vulnerability to the Cloud infrastructure.

Table 2. OPS results for Debian.

Distribution	Risk level none	Risk level Low	Risk level medium	Risk level high
Debian Etch	14	2	0	2
Debian Lenny	43	2	3	2
Debian Squeeze	44	2	3	2
Debian Wheezy	43	2	3	2

Number of security vulnerabilities the OPS detected in different versions of Debian Linux.

Automation of system administration, including system administration and updating systems is one of the relevant research topics mentioned in the Expert Group Report[16] created by the European Commission.

An image management system, called Mirage, is presented by Wei et al.[6]. Mirage addresses security concerns of a virtual machine image publisher, customer and administrator. To reduce the publisher's risk, an access control framework regulates the sharing of virtual machines images. Image filters remove unwanted information (e.g., logs, sensitive information, etc) from images prior to publishing. The authors also present a mechanism to update dormant images and apply security updates. While Mirage offers a complete solution for virtual disk image maintenance, it lacks the features presented in this article. Mirage cannot show whether the packages in a system are outdated and work with multiple package management systems.

Based on Mirage, Reimer et al.[4] present the Mirage image format (MIF), a new storage format for virtual machine disk images. MIF solves the problem of *virtual machine image sprawl*, i.e., the complexity of maintaining disk image content that changes continuously due to cloning or snapshotting. MIF stores the disk image content in a central repository and supports searching, installing and updating applications in all images. By using a special storage device, disk images share common blocks and thus take up only a fraction of the actual disk space. Using MIF it is also possible to update packages on a system although the update procedure is quite complex. At first, it is quite unclear how the system determines whether there is a need for an update. Furthermore, the system needs a modified version of dpkg, thus, it is not usable with off-the-shelf installations or other package management solutions. The authors state that "the optimized Dpkg does not support some of Dpkg's features".

A system for unscheduled system updates, called Auto- Pod, was presented by Potter et al.[17]. AutoPod is based on system call interposition and the chroot utility and is able to create file system namespaces, called pods. Every process in a pod can be offline-migrated to another physical machine by using a checkpoint mechanism. Unfortunately, AutoPod is bound to Debian Linux and cannot be used with other package managers. Furthermore, it also updates a system automatically, which could lead to problems in case of an incomplete update. In contrast to the presented solution, AutoPod is based on chroot, which is known for having several major security flaws in the past.

Sapuntzakis et al.[18] developed a utility, called the Collective, which assigns virtual appliances to hardware dynamically and automatically. By keeping software up to date, their approach prevents security break-ins due to fixed vulnerabilities. While their approach allow updating whole virtual machine appliances, it does not allow the update of certain packages within the appliance. Furthermore, it is not possible to determine whether certain packages are outdated.

Layered virtual machines[5] can be used to solve the maintenance problem of dormant virtual machines. These machines are split up in different layers, such as a common base layer, containing a base system with some commonly required libraries and tools, a user layer containing specific applications required by the user and potentially other layers. Besides benefits when it comes to storage and transfer of those virtual machines, considering shared layers that need to be stored and transferred only once and reused by many virtual machines, this architecture also helps with the problem of keeping machines up-to-date. Because a base layer is shared by many virtual machines, updating the base layer will affect all virtual machines built on top. Although not the complete software stack is affected by those updates, some of the most important parts of the system (e.g., the SSH libraries, which were affected by a serious bug in the Debian implementation back in 2008[19]) can be fixed this way.

Canonical, the company behind Ubuntu Linux, offers a commercial product called Landscape[8]. Landscape can be used to manage Ubuntu (virtual) machines, including package management and monitoring. While Landscape is able to detect and update outdated applications within virtual machines, it can only handle the Debian package format and is not able to update dormant machines. However, Landscape can update outdated machines once they are live the next time.

SAVEly, a tool to check Amazon Machine Images (AMIs) for vulnerabilities was presented by Bleikertz et al.[20]. The authors construct an attack graph based on the security polices used in EC2. These policies are used to group machines while restricting the communication between them. Based on the graph, the authors use the OpenVAS scanner to check the AMI for remote vulnerabilities. Their approach is tightly coupled to Amazon's EC2 and cannot be used with other IaaS implementations or in virtualized Grid environments.

Yoon and Sim[21] present an automated network vulnerability assessment framework. It uses a combination of a scan manager, message relay server and scanners to check the hosts in a network for vulnerabilities. Their approach uses similar techniques as the ones presented, but it lacks the ability to work in a Cloud computing environment. It is neither able to control virtual machines, nor to instrument an IaaS solution like the XGE.

6. Conclusions

In this article, a new approach to increase the security of virtual machines in either virtualized Grid or Cloud computing environments has been presented. It is based on two components: a first component called Update Checker to identify outdated packages can check either running or dormant virtual machine images efficiently. It supports the two major Linux software management solutions, namely apt/dpkg and yum/rpm, and thus all major Linux distributions currently used in Grid or Cloud environments. Due to its flexible design, plugins for other software management solutions can be easily added. The use of multiple caches speeds up the check process, resulting in a time less than a second for a complete check of an average virtual machine. A second component called Online Penetration Suite scans virtual machines for software vulnerabilities using established security techniques. It can identify flaws in software components listening on the network. Both components are integrated into two already existing solutions (XGE and ICS) that leverage their capabilities to deny running too outdated machines or provide the user with the ability to update his or her machines.

There are several areas for future work. For example, the current implementation of the Update Checker only supports software installed using the package management systems of current Linux distributions. Nevertheless, there are cases where software is installed in other ways, either by compiling it manually or by installing software from binary packages that are not available in repositories. The idea of a generic framework with software specific plugins that can determine the installed version seems to be promising. Problems to solve are binaries without a version parameter and even more locating the software that was installed without using the package management system. Furthermore, the current approach to infer the significance of updates is a very basic approach. Comparing the list of outdated packages to the security advisories of the distribution, if available, seems to be promising. This would require distribution specific parsers for the advisories, since there is no unified advisory format, and manual configuration of the advisory sources for each distribution. The OPS currently controls two vulnerability scanners. In the future, it would be desirable to support a larger number of scanners.

Competing Interests

The authors declare that they have no competing interests.

Authors' Contributions

All authors contributed equally. All authors read and approved the final manuscript.

Acknowledgements

This work is partly supported by the German Ministry of Education and Research (BMBF) (D-Grid Initiative and HPC-Call) and the Hessian Ministry of Science and Art (HMWK).

Source: Schwarzkopf R, Schmidt M, Strack C, *et al.* Increasing virtual machine security in cloud environments [J]. Journal of Cloud Computing Advances Systems & Applications, 2012, 1(1): 1–12.

References

[1] Armbrust M, Fox A, Griffith R, Joseph A (2009) Above the Clouds: A Berkeley View of Cloud Computing, Technical Report UCBEECS200928 53(UCB/EECS-2009-28). EECS Department University of California Berkeley.

[2] Smith M, Schmidt M, Fallenbeck N, Dörnemann T, Schridde C, Freisleben B (2009) Secure On-demand Grid Computing. J Future Generation Comput Syst 25(3): 315–325.

[3] Garfinkel T, Rosenblum M (2005) When Virtual is Harder than Real: Security Challenges in Virtual Machine Based Computing. In 10th Workshop on Hot Topics in Operating Systems 121–126.

[4] Reimer D, Thomas A, Ammons G, Mummert T, Alpern B, Bala V (2008) Opening Black, Boxes: Using Semantic Information to Combat Virtual Machine Image Sprawl. In Proceedings of the Fourth ACM SIGPLAN/SIGOPS International Conference on Virtual Execution Environments 111–120. Seattle: ACM.

[5] Schwarzkopf R, Schmidt M, Fallenbeck N, Freisleben B (2009) Multi-Layered Virtual Machines for Security Updates in Grid Environments. In Proceedings of 35th Euromicro Conference on Internet Technologies, Quality of Service and Applications (ITQSA) 563–570. Patras: IEEE Press.

[6] Wei J, Zhang X, Ammons G, Bala V, Ning P (2009) Managing Security of Virtual Machine Images in a Cloud Environment. In Proceedings of the 2009 ACM Workshop on, Cloud Computing Security, CCSW'09 91–96. New York: ACM.

[7] Fallenbeck N, Schmidt M, Schwarzkopf R, Freisleben B (2010) Inter-Site Virtual Machine Image Transfer in Grids and Clouds. In Proceedings of the 2nd International ICST Conference on Cloud Computing (CloudComp 2010) 1–19. Barcelona: Springer, LNICST.

[8] Canonical Inc (2011) Ubuntu Advantage Landscape. http://www.canonical.com/enterprise-services/ubuntu-advantage/landscape.

[9] Winer D (2003) XML-RPC Specification. http://www.xml-rpc.com/spec.

[10] OpenVAS Developers (2012) The Open Vulnerability Assessment System (OpenVAS). http://www.openvas.org/.

[11] Tenable Network Security (2012) Nessus Security Scanner. http://www.nessus.org/products/nessus.

[12] Libvirt Developers (2012) Libvirt—The Virtualization API. http://libvirt.org/.

[13] Schwarzkopf R, Schmidt M, Strack C, Freisleben B (2011) Checking Running and Dormant Virtual Machines for the Necessity of Security Updates in Cloud Environments. In Proceedings of the 3rd IEEE International Conference on Cloud Computing Technology and Science (CloudCom) 239–246. Athens: IEEE Press.

[14] Crockford D (2006) The application/json Media Type for JavaScript Object Notation (JSON). http://www.ietf.org/rfc/rfc4627.

[15] ENISA European Network and Information Security Agency (2009) Cloud Computing Risk Assessment. http://www.enisa.europa.eu/act/rm/files/deliverables/cloud-computing-risk-assessment.

[16] Lillard TV, Garrison CP, Schiller CA, Steele J (2010) The Future of Cloud Computing. In Digital Forensics for Network, Internet, and Cloud Computing 319–339. Boston: Syngress.

[17] Potter S, Nieh J (2005) AutoPod: Unscheduled System Updates with Zero Data Loss. In Autonomic Computing, International Conference on 367–368.

[18] Sapuntzakis C, Brumley D, Chandra R, Zeldovich N, Chow J, Lam MS, Rosenblum M (2003) Virtual Appliances for Deploying and Maintaining Software. In Proceedings of the 17th USENIX Conference on System Administration 181–194. Berkeley: USENIX Association.

[19] Debian Security Advisory 1576-1 OpenSSH (2008) Predictable Random Number Generator. http://www.debian.org/security/2008/dsa-1576.

[20] Bleikertz S, Schunter M, Probst CW, Pendarakis D, Eriksson K (2010) Security Audits of Multi-tier Virtual Infrastructures in Public Infrastructure Clouds. In Proceedings of the 2010 ACM Workshop on Cloud Computing Security, CCSW'10 93–102. Chicago.

[21] Yoon J, Sim W (2007) Implementation of the, Automated Network Vulnerability Assessment Framework. In Proceedings of the 4th International Conference on Innovations in Information Technology 153–157. Dubai: IEEE.

Chapter 11

Scaling of Wireless Sensor Network Intrusion Detection Probability: 3D Sensors, 3D Intruders, and 3D Environments

Omar Said[1,2], Alaa Elnashar[2,3]

[1]Department of Computer Science, College of Science, Menofia University, Shebin El Kom, Gamal Abdelnaser St., 32512, Menofia, Egypt
[2]College of Computers and Information Technology, Taif University, Al-Hawyia, 21974, Taif, Saudi Arabia
[3]Department of Computer Science, College of Science, Minia University, Ibrahimia St., 61519, Minia, Egypt

Abstract: In this paper, a new model that deploys heterogeneous sensors in 3D wireless sensor networks (WSNs) is proposed. The model handles the two sensing scenarios, single sensing and multiple sensing. The probabilities of intrusion detection in a 3D environment with sensors distributed using Gaussian, uniform, beta, and chi-square are compared. The resultant probabilities values help the WSN security designers in selecting the most suitable sensors deployment regarding some critical network parameters such as quality-of-service (QoS). WSN efficiency under different probabilistic distributions is also demonstrated. To evaluate the proposed model, a simulation environment is constructed using OPNET and NS2. The simulation results showed that Gaussian distribution provides the best efficiency and performance.

Keywords: WSN, WSN Security, Intrusion Detection, QoS, 3D Sensor Applications

1. Introduction

Wireless sensor networks (WSNs) consist of small main components which have limited power devices such as sensors and can be installed in open environments[1]-[3]. So, WSNs may be attacked by intruders. Since WSNs applications are used in different fields such as environmental sensing, industrial monitoring, and military process management, intrusion detection became an extremely important issue[4][5]. In WSN, a huge number of sensors should be deployed for intruder detection. However, the high cost of this solution makes it impractical. Furthermore, using a huge number of sensors does not guarantee a successful detection of a moving intruder within a certain distance since void area may be found in the WSN. There are two main categories for intrusion detection problem. The first one uses a component to monitor WSN security. This component may be software, hardware, or human. The target of this component is accomplished by using some sensors to ensure that the security level in WSN is acceptable[6]-[8]. The second one detects the intruder when it tries to storm unauthorized area[9]-[13]. The time consumed for intruder detection process is an important parameter that should be considered. Accordingly, the intruder should be detected at the same time of its entrance. So, raising the probability of intruder detection in WSNs is concerned to sensor deployment plan more than the number of sensors. Also, wired network contains many intrusion detection systems which are not accommodating the nature of WSNs[14]-[17]. Consequently, developing an innovative technique that deals with WSNs nature is an important issue. Furthermore, finding the optimal representation of sensor deployment that provides the best intruder detection is another important issue. In this paper, a model for intrusion detection in 3D environments is introduced. The model uses various probability distributions to deploy sensors within the entire WSN. A simulator is created to simulate the intrusion detection process and evaluate its efficiency based on the probability distribution used. This will guide the WSN designer to select the optimal sensors distribution that yields the best intrusion detection efficiency. The paper is organized as follows: in Section 2, the problem definition is introduced. In Section 3, the related works are demonstrated. In Section 4, the proposed model is presented. In Section 5, the simulation environment is constructed and the results are discussed. Finally, conclusion and future work are introduced in Sections 6 and 7, respectively.

2. Problem Definition

WSNs consist of large number of inter communicated sensors via wireless interfaces. These sensors may have various processing capabilities, computing resources, or coverage scenarios. Many WSN applications contain different types of sensors which are used in different types of tasks. Also, there are many applications that are not only used in 2D environment but also in 3D space. The problem of intrusion detection has not been studied extensively in the case of heterogeneous WSNs sensors and also in the case of 3D environments because of its complexity. Also, the effect of mobile sensors on the heterogeneous WSNs is not studied. So, till now, there is no standard model that helps security designers in selecting a suitable sensors distribution that accommodates the nature of WSN 3D applications and available network specifications.

The problem is defined as follows: find the best distribution of network sensors that minimizes the probability value $\Pr\left(V_s \subset \left(\cup_{m-1}^{n} V_{lm}\right)\right)$ or maximizes $\Pr\left(\left(\cup_{m-1}^{n} V_{lm}\right) \cap V_s = \emptyset\right)$, where *VS* is the space volume that is covered by a network sensor S, and *VI* is the space volume that is covered by an intruder I, see **Figure 1**.

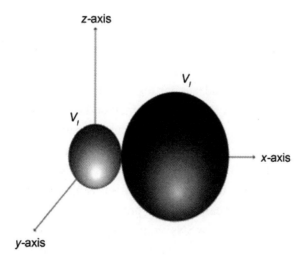

Figure 1. Sensor/intruder spaces in 3D.

For simplification, suppose one sensor S and one intruder I. The sensor S covers the volume V_S that is represented by a sphere of radius R_S centered at V_S center $= (x_1, y_1, z_1)$. The intruder I covers the volume V_I that is represented by a sphere of radius R_I centered at V_I center $= (x_2, y_2, z_2)$.

3. Related Works

WSN intrusion detection is a challenging process since it can be affected by several parameters such as number of sensors, sensor types, quality-of-service (QoS), detection cost, and WSN environment. Few studies have been introduced to solve the intrusion detection problem in different environments with different types of sensors. These studies can be classified into two categories. The first one manipulates the detection problem with different factors rather than the factors that are presented in this paper. The second category is close related work, which tried to study this problem under nearby parameters of the proposed idea but with different problem definitions and models.

3.1. General Related Work

Several models and studies were presented to handle intrusion detection problem. Anomaly-detection-based model with statistical analysis[18] was introduced for intrusion detection in ad hoc network. The model takes a long time to detect the intruder due to huge transmitted data and traffic. In[19], a framework, which enables the network designer to select the optimized intruder detection system according to his particular needs, is proposed. In[20], a method for detecting packet modification attacks and supporting target node location privacy is proposed. Moreover, accuracy of this proposed method is evaluated using small-scale WSN. Detection of a moving intruder in a curved path was studied in[21]. The technique uses sine function to define the moving curve of the intruder with detection probability in WSN. Regarding heterogeneous WSNs with the Gaussian distribution, it was proved that the probability of intrusion detection (PID) increases as the number of sensors increases[22]. The probability of intruder detection in homogeneous and heterogeneous WSNs with consideration of density and sensing range for each WSN node was studied in[23]. A comparison between the performance of WSN with Gaussian and Poisson probability distribution under different network settings was presented in[24]. The area covered by WSN at the exact time or within certain time interval in addition to calculating the time that have been consumed to detect a randomly sited inactive object was proposed in[25]. In[26], the relation between the intrusion detection time and the distance, which the intruder cuts down from the start point until hacking the field of interest, is determined. Furthermore, [27]showed that lack of sensors can be compensated by sensor mobility which improved network coverage. Some other ideas such as furtive wireless communication have been demonstrated in[28].

3.2. Closely Related Work

There are three close researches to the proposed idea. The first one[29] implemented Gaussian and uniform distributions to determine the probability of intruder detection in WSN. The proposed technique in distinguished between the detection probability as regards the requirements of each WSN application and the network parameters. In addition, the proposed system was studied under two scenarios: single sensing detection and multiple-sensing detection. But this proposed technique did not study the 3D environment under heterogeneous WSN. The second close research proposed a technique that combines Gaussian and Poisson probability distributions[30]. Gaussian distribution was applied to the central area covered by WSN. Poisson distribution was applied to the remaining area. Therefore, this technique is considered as a mixture of two probability distributions in one test area. Also, this technique supposed that the WSN contains heterogeneous sensors. But the tested environment was not in 3D space. Furthermore, it should test each probability distribution independently on the WSN area before proposing a mixture of them. The last one proposed an analysis for intrusion detection problem in a 3D environment that is represented by a cube with two types of sensors[31]. This technique has some drawbacks such as the probability density function is not calculated and a cube is not sufficient to prove the model results. In addition, this proposed model is not well defined.

4. The Proposed Model

Since many applications are applied in 3D environment, optimizing sensor distribution to enhance intrusion detection probability should be considered. In addition, sensor heterogeneity is also another important parameter that should be well studied since most of the intrusion detection researches focus on using homogenous sensors. The main purpose of the proposed model is to analyze the intrusion detection probability, which helps in sensor deploying, to gain a satisfied WSN security level. The effect of the used probability distribution such as Gaussian, uniform, beta, and chi-square are also studied. **Figure 2(a)**, **Figure 2(b)** shows samples from sensors distribution views using uniform and normal. Reducing communication redundancy can be achieved by intelligent sensor deployment in WSN. The proposed model deals with four components, a 3D environment within which the WSN should be installed, 3D sensors that are used to detect an intruder, 3D intruder(s) which is (are) predicted to hack WSN, and a target resource that is assumed to be protected by WSN sensors. The 3D environment is represented by a cubic space with predefined dimensions. Sensors, intruders, and targets are presented as sphere centers as shown in **Figure 3(a)**. Sensors are heterogonous since each sensor is assumed to be located at a sphere center with a predefined radius that may differ from other sensors radii as shown in **Figure 3(b)**. Each sensor covers its entire sphere volume space and is allocated dynamically. Each intruder is also represented as in case of sensors, and its sphere space represents the intrusion space as shown in **Figure 3(c)**. Target space is the sphere volume that is required to be protected. The target is hacked by a specific intruder if it is located outside any of sensors coverage spaces, and its distance from the intruder is less than or equal the sum of the intruder sphere and its sphere radii as shown in **Figure 3(d)**; otherwise, it is considered to be safe as shown in **Figure 3(e)**, **Figure 3(f)**. The steps of the proposed model are described in **Algorithm 1** shown below.

5. Simulation and Evaluation

A simulator has been developed to simulate the intrusion detection process with multiple intruders in 3D environment with sensors that are distributed by using various probability distributions. The simulator is also designed to evaluate the proposed model. The evaluation is based on the effect of the used probability distributions on efficiency of WSN and also the network parameters such as packet loss, end-to-end delay, and throughput. The simulation parameters and their values are listed in **Table 1**.

Algorithm 1

D_m is the minimum distance
I: Intruder and T: Target
M: The number of intruders
N: The number of sensors
D_{jT}: Intruder/Target distance
D_{TSi}: Nearest Sensor/Target distance
Suppose Dm = 0, F = 0.
I position is random.
T position is random.
Let D_m = X
The nearest sensor from target is in the same direction of intended intruder.
For each Gaussian, uniform, beta, and chi-square
Begin
While (F ≺ M)
Begin
For j = 1 TO M
Begin
For i = 1 to N
    Begin
    If (D_m ≻ D_{iT})
    D_m = D_{iT}
    End
S_i = 1 (Flaged)
If (D_{jT} = = D_{TSi})
Print 'Middle State.'
ElseIF D_{jT} ≻ D_{TSi}
Print 'Save State.'
ElseIF D_{jT} ≺ D_{TSi}
Print 'Hacking State.'
End
For K = 1 to M
Begin
If (($I_k \cap S_j \neq \Phi$), j=2 to n)
    Print 'Detection process is distributed over multi sensors
    depending on the overlapped volume spaces.'
End
F = F+1.
End
End

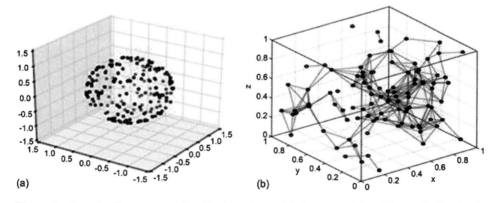

Figure 2. Samples from sensor distribution views. (a) Sensors with uniform distribution in 3D[32]. (b) Sensors with normal distribution in 3D[33].

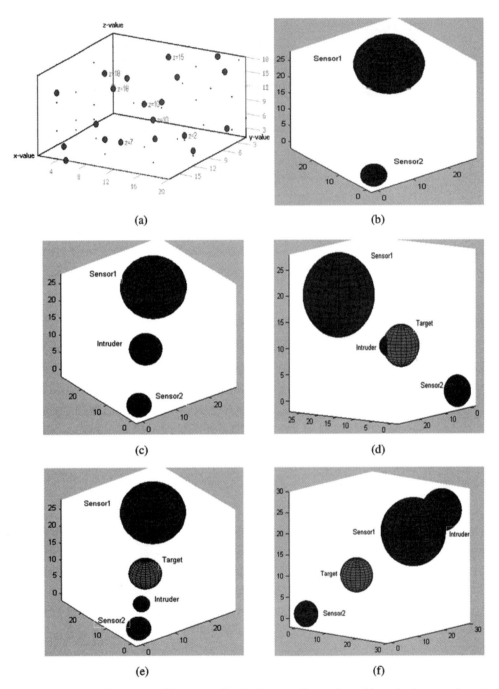

Figure 3. 3D environment with sensor distributions and samples of intruder/sensors/target states. (a) 3D environment with sensors spheres centers. (b) Two heterogeneous sensors coverage spaces. (c) Two heterogeneous sensor and an intruder representation. (d) Hacked target. (e) Safe target—no intruders detected. (f) Safe target—an intruder detected.

Table 1. Simulation parameters and their values.

Simulation parameter	Value
Packet size	1,024 Bytes
Destination	Random
Packet inter-arrival time	Constant (1.0)
Simulation time	2 h
Simulation time	2 h
Battery life	3 to 4 h [active sensing] and 1,000+ h [standby]
Start time	Uniform (20, 21)
Transmission range	1 to 100 m
ACK mechanism	Enabled
3D camera	170 (W) × 54 (H) × 49 (D) mm
MAC layer	802.15.4
Transmit power	0.05 mw/m^3
Transmit band	2.4 GHz
Environment space volume	10 × 10 × 10 km^3

OPNET 14.5 and NS2[34][35] are used to simulate a 3D environment that contains ZigBee with its three layers: application, media access control (MAC), and network. In the MAC layer, the 802.15.4 MAC protocol is used in addition to the ZigBee (CSMA/CA) model[34]. The network layer is implemented by the ZigBee network model, which is used in the routing and request handling.

In the simulated environment, each 3D camera is attached to one sensor to reflect the 3D nature. The simulator is built on large scale as shown in **Table 1** and includes different numbers of Zigbee end devices, coordinators, and routers. WSN sensors are communicated via infrared media using star topology and distributed using Gaussian, uniform, beta, and chi-square. Configuring WSN model components and selecting the scalable parameters for the entire simulated WSN are accomplished.

5.1. Intruder Detection Efficiency Evaluation

Suppose that the volume space covered by a network sensor is Vs and the volume space covered by intruder is Vi. $Pr(Vi \cap Vs)$ is the coverage spaces intersection probability of an intruder and a sensor. If this probability equals ϕ, then the system is safe and there is a distance between the intruder and its target. So, this probability should be maximized. *Pr (Vi ∩ Vs)* is the worst probability because the intruder is around to reach its target and hack the security system. So, this probability should be minimized. The distance between the target and an intruder and also its velocity are two factors that are used to determine the time consumed by the intruder to hack the target. Knowing this time value enables the security system manager to protect the target against hacking. In the proposed simulation, the values of intruders' velocities and their distances from the target are randomly generated. Equations 11 to 4 are for Gaussian, uniform, beta, and chi-square probability distributions, respectively[29][36][37]. Equation 5 describes how to extract the time from the intruder velocity and the current distance between the intruder and his target[38]. Equation 6 describes the relationship between the intruder and the sensors spheres in the security domain[39].

To evaluate the efficiency of intruder detection process, a simulator with 800 3D sensors covering 3D environment and two 3D intruders is used. The positions of both the target and intruders are created randomly in the simulated environment. The distance between the intruder and the target in addition to the distance between the intruder and the nearest sensor to the target are considered[26].

Two experiments were carried out using this simulated environment. In the first experiment, the values of the distance between the intruder and the target and also the distance between the intruder and the nearest sensor to the target are ignored, in contrast to the second one that considered these values.

Figure 4 shows the simulation results of the first experiment. The results show that Gaussian sensor distribution provides the best average performance for all the numbers of sensors except for 400 sensors. A lower performance is provided in case of using uniform sensor distribution especially for 100 and 700 sensors. Beta and chi-square distributions have the lowest average performance values. **Figure 5** shows the simulation results of experiment 2. The results show that Gaussian sensors distribution provides the best average performance for all the numbers of sensors. A lower performance is provided in the case of using uniform sensor distribution. Beta and chi-square distributions have the lowest average performance values.

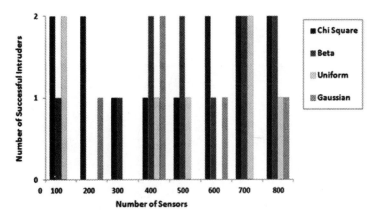

Figure 4. The simulation results of the first experiment. Number of successful intruders for each distribution without considering distance and velocity parameters.

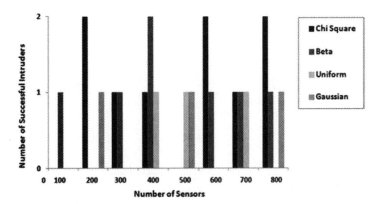

Figure 5. The simulation results of the second experiment. Number of successful intruders for each distribution with considering distance and velocity parameters.

Gaussian[29].

$$f(x,\mu,\sigma) = \frac{1}{\sigma\sqrt{2\Pi}} e^{\frac{(x-\mu)^2}{2\sigma^2}}$$

Uniform[29]

$$f(x,k) = \begin{cases} \dfrac{1}{u-1} & \text{if } \in R_x \\ 0 & \text{if } \notin R_x \end{cases}$$

Beta[36]

$$f(x,\alpha,\beta) = \frac{1}{B(\alpha,\beta)} x^{\alpha-1}(1-x)^{\beta-1}$$

Chi-square[37]

$$f(x,k) = \begin{cases} \dfrac{x^{\frac{k}{2}-1} e^{\frac{-\Pi}{2}}}{2^{\frac{k}{2}} \Gamma\left(\dfrac{k}{2}\right)} & x \geq 0 \\ 0 & \text{otherwise} \end{cases}$$

Time extraction[38]

$$T = \frac{D}{V}$$

Spheres intersection[39]

$$\frac{4d^2 R^2 - \left(d^2 - r^2 - R^2\right)^2}{4d^2}$$

5.2. Evaluation of WSN Sensor Distribution

It is obvious that if the WSN efficiency is not acceptable, the intruder may be detected but the security system may fail to recover this problem. To complete a test of the proposed model efficiency, the cycle after the intruder detection process should be evaluated. This cycle means that data, which are sent to inform the security manager with intruder, should be transmitted successfully. So, the WSN efficiency should be evaluated. The evaluation parameters are the average number of control bits, the number of hops, the average number of lost packets, the end-to-end delay, the throughput, and the general efficiency. In the following subsections, the simulation environment is constructed and each parameter results is showed and discussed.

5.2.1. Average Number of Control Bits

During simulation execution, trace files containing the number of control bits are created instantly for the number of used sensors. The control bits are the retransmitted bits, the ACK bits and the bits which are used to initiate the sessions between WSN sensors in case of sudden event occurrence such as intruder detection. In general, as the number of management bits decreases, the energy consumption decreases and sensors battery live increases.

Figure 6 shows the transmission control data for different WSNs sensors distributed using Gaussian, uniform, beta, and chi-square. The simulation results indicate that Gaussian WSN has the minimum number of management bits compared with other WSNs in contrast to chi-square WSNs that have the maximum number of control bits.

5.2.2. The Number of Hops

The routing path is determined by the number of hops, which are used in data transmission. So, the number of transmission hops is an important factor in determining the best routing paths and their alternatives.

Figure 7 shows that the number of hops used in the Gaussian WSN is less than the number of hops used in the other WSNs (uniform, beta, and chi-square). When the number of sensors equals 600, the Gaussian WSN and the uniform WSN have the same number of transmission hops. The hesitations in curve plots are owed to the randomly chosen sources and destinations in the WSNs. In addition, failure of multiple sensor nodes may cause data collision which requires alternative nodes to complete the transmission process that increases the number of hops. Furthermore, when the number of sensors is greater than 300, the plots' values are decreased. This occurs because the data, which are transmitted within the WSN at this number of sensors, use the best routing path. The simulation results indicate that the Gaussian WSN uses the minimum number of hops compared with the other WSNs. In contrast, the beta and the chi-square WSNs have the maximum number of used hops.

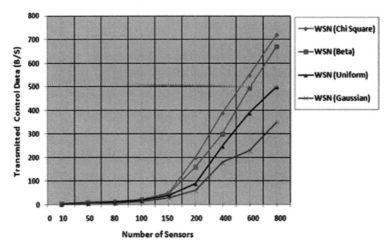

Figure 6. Transmission control data for different WSNs (Gaussian, uniform, beta, and chi-square).

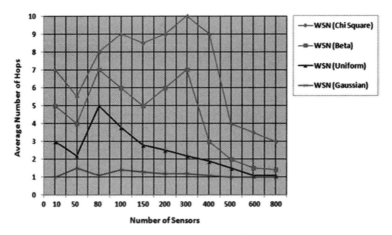

Figure 7. Average number of hops for different WSNs (Gaussian, uniform, beta, and chi-square).

5.2.3. The Average Number of Lost Packets

In the general networks, it is well known that the transmitted data increases when the number of nodes increases which may cause more packet loss[40]. In WSNs, there is a slight difference; the amount of transmitted data depends on the sensor-acquiring information when urgent events occurred[40]. Also, the number of lost packet affects the network efficiency. So, the number of packet loss should be monitored. **Figure 8** shows the average number of lost packets for different

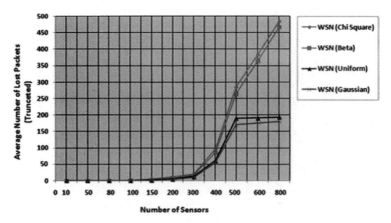

Figure 8. Average number of packet loss for different WSNs (Gaussian, uniform, beta, and chi-square).

WSNs sensors distributed by using Gaussian, uniform, beta, and chi-square distributions. The simulation results indicate that the number of sensors does not negatively affect the number of lost packets for a medium number of sensor (≤300) for all simulated WSNs. For a large number of sensors (>−300), the average number of lost packets exponentially increases as the number of sensors increases in case of beta and chi-square WSNs in contrast to Gaussian and uniform WSNs, in which the average number of lost packets increases slowly as the number of sensors increases.

5.2.4. Average End-to-End Delay

The end-to-end delay is defined as the total time, which is consumed in packet transmission from source to destination[40]. The end-to-end delay is measured by a time difference between message buffering time at the source and the time of receiving the last bit at the destination. **Figure 9** shows the average end-to-end delay for different WSNs sensors distributed by using Gaussian, uniform, beta, and chi-square distributions. The simulation results indicate that uniform and Gaussian WSNs have the minimum average delay, respectively, compared with the other WSNs. Chi-square and beta WSNs have the maximum delay. The interpretation for these results is that in case of the uniform and the Gaussian WSNs, little number of hops are used to transmit the data from source to destinations; this decreases the total time of hops data handling which leads to less end-to-end delay. On the other hand, the beta and the chi-square WSNs routing

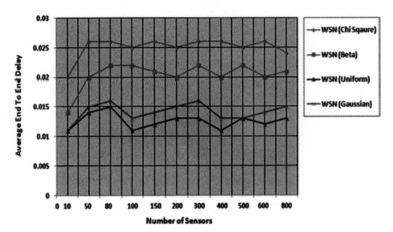

Figure 9. Average end-to-end delay for different WSNs (Gaussian, uniform, beta, and chi-square).

paths use more hops to transmit the required data from source to destinations (including sinks) which leads to an increased end-to-end delay.

5.2.5. Throughput

Throughput is defined as the correct transmitted data with a specific quality from source to destination within a time interval[40]. Increased number of sensors used in WSNs makes analysis of QoS parameter an important issue. There are many factors affecting the WSN through-put such as packet loss, end-to-end delay, energy consumption, long distance, and obstacles. **Figure 10** shows the throughput for different WSNs sensors distributed by using Gaussian, uniform, beta, and chi-square distributions. The simulation results indicate that Gaussian WSN achieves the maximum throughput, the second highest throughput is achieved by uniform WSN, and beta and chi-square WSNs have the lowest throughput. The interpretation for these results is that Gaussian and uniform WSN have direct sensor communication and high QoS paths. Furthermore, the WSN task may be shared among multiple sensors due to sensor overlaps as a result of less collisions and packet drops. On the other hand, beta and chi-square WSNs have complex paths as a result of higher collisions and packet drops take place as a result of which the throughput is minimum. Also, energy consumption makes some sensors not to provide their services; so other alternative routes should be generated to complete the routing process.

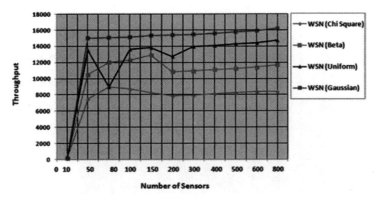

Figure 10. Throughput for different WSNs (Gaussian, uniform, beta, and chi-square).

5.2.6. General Efficiency

The general efficiency of the entire system is defined by a relationship between the numbers of events which are handled by sensors in one WSN in proportion to other WSNs regarding some parameters such as end-to-end delay, packet loss, throughput, and path complexity. The simulation results in **Figure 11** show that Gaussian WSN has the best efficiency and then uniform WSN. Beta and chi-square WSNs have the lowest efficiency. The general efficiency parameter is considered as cumulative result which summarizes all the previous results. Also, all previous results and final recommendation are summarized in **Table 2**.

6. Conclusions

In this paper, a new model for WSN intruder detection in 3D environment is proposed. Each element in the security system is represented by a center of various radii spheres. Sensors are distributed in the 3D environment using Gaussian, uniform, beta, and chi-square. The security status (hacked or safe) is determined by three parameters, the intruder/target and nearest sensor/target distances, the intruder velocity, and the relationship between spheres (intruder and sensors). The proposed model handles both detection scenarios, single and multiple. Two approaches of evaluations are presented. The first one concerns with WSN security issue. The second approach concerns with network parameters such as average number of control bits, the number of hops, the average number of lost packets, the end-to-end delay, the throughput, and the general efficiency.

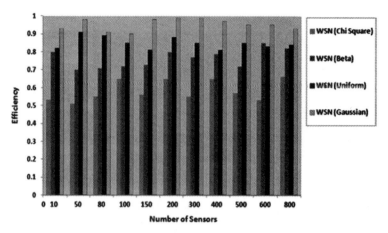

Figure 11. General efficiency for different WSNs (Gaussian, uniform, beta, and chi-square).

Table 2. Results summarization and final recommendation.

Distribution	WSN efficiency evaluation		Intruder detection evaluation		Final recommendation
Gaussian	Control data size	HP			
	Number of hops	HP	Without considering distant	HP	
	Packet loss	HP			
	End-to-end delay	LPR2			
	Throughput	HP	With considering distant	HP	
	General efficiency	HP			
Uniform	Control data size	LPR2			Using Gaussian
	Number of hops	LPR2	Without considering distant	LPR2	to distribute the
	Packet loss	LPR2			sensors in WSN
	End-to-end delay	HP			gives the best
	Throughput	LPR2	With considering distant	LPR2	performance as
	General efficiency	LPR2			regards intruder detection
Beta	Control data size	LPR3			probability and overall
	Number of hops	LPR3	Without considering distant	LPR3	WSN efficiency,
	Packet loss	LPR3			then uniform, beta,
	End-to-end delay	LPR3			and chi-square
	Throughput	LPR3	With considering distant	LPR3	
	General efficiency	LP			
Chi-square	Control data size	LP			
	Number of hops	LP	Without considering distant	LP	
	Packet loss	LP			
	End-to-end delay	LP			
	Throughput	LP	With considering distant	LP	
	General efficiency	LP			

HP: highest performance. LPR2: lower performance-rank2. LPR3: lower performance-rank3. LP: lowest performance.

The results showed that Gaussian WSN have the best performance in both evaluation approaches, then uniform WSN. The beta and the chi-square WSNs have the lowest performance. Also, the end-to-end delay is the only parameter where the uniform WSN provides better performance than the Gaussian WSN. Simulation results of the proposed model show that Gaussian sensors distribution in WSN is recommended in 3D environments.

Future Work

Mixed probability distributions of sensors in WSN (*i.e.*, Gaussian/uniform, Gaussian/beta, or more mixtures) should be studied and tested. Accordingly, the results should be compared with those of this paper. This will provide a standard sensor distribution. Also, the number of sensors, which are used in the simulation environment, should be larger. In addition, a tracer system for an intruder may be added to the proposed model for system security integrity.

Competing Interests

The authors declare that they have no competing interests.

Acknowledgements

I would like to pay special thanks and appreciation to the persons who made our research successful.

Source: Said O, Elnashar A. Scaling of wireless sensor network intrusion detection probability: 3D sensors, 3D intruders, and 3D environments [J]. Eurasip Journal on Wireless Communications & Networking, 2015, 2015(1): 1–12.

References

[1] Y Luo, S Morgera, R Sankar, in proceedings of the 2nd IEEE International Conference on Information Science and Engineering (ICISE). A survey on intrusion detec-

tion of wireless sensor network (China, 4–6 Dec 2010), pp. 1798–1802.

[2] Y Akyildiz, S Weilian, Y Sankarasubramaniam, E Cayirci, A survey on wireless sensor networks. IEEE Commun. Mag. 40(8), 102–114 (2002).

[3] K Sohraby, D Minoli, T Znati, Wireless sensor networks: technology, protocols, and applications, 12–68 (2007).

[4] J Al-Karaki, A Kamal, E Cayirci, Routing techniques in wireless sensor networks: a survey. IEEE Wireless Commun. 11(6), 6–28 (2004).

[5] S Tilak, N Abu-Ghazaleh, A taxonomy of wireless micro-sensor network models. ACM Mobile Comput. Commun. Rev. 6(2), 28–36 (2002).

[6] A Agah, S Das, K Basu, M Asadi, in proceedings of the Third IEEE International Symposium on Network Computing and Applications (NCA). Intrusion detection in sensor networks: a non-cooperative game approach (USA, January 30 Aug), pp. 343–346.

[7] A Agah, S Das, K Basu, A game theory based approach for security in wireless sensor networks, (USA, Unknown Month 15), pp. 259–263.

[8] V Giruka, M Singhal, J Royalty, S Varanasi, Security in wireless sensor networks. Wiley Wireless Commun. Mobile Comput. 8(1), 1–24 (2008).

[9] A Arora, P Dutta, S Bapat, V Kulathumani, H Zhang, V Naik, V Mittal, H Cao, M Demirbas, M Gouda, A line in the sand: a wireless sensor network for target detection, classification, and tracking. Int. J. Comput. Telecommun. Networking-Special Issue: Mil. Commun. Syst. Technol. 46(5), 605–634 (2004).

[10] H Solimana, N Hikalb, N Sakrb, A comparative performance evaluation of intrusion detection techniques for hierarchical wireless sensor networks. Elsevier Egypt. Inform. J. 13(3), 225–238 (2012).

[11] S Sachan, M Wazid, P Singh, H Goudar, A cluster based intrusion detection and prevention technique for misdirection attack inside WSN, (India, May 3), pp. 795–801.

[12] H Kung, D Vlah, Efficient location tracking using sensor networks. Proceedings of IEEE International Conference on Wireless Communications and Networking (WCNC), 1954–1961 (Unknown Month 20).

[13] C Lin, W Peng, Y Tseng, Efficient in-network moving object tracking in wireless sensor networks. IEEE Trans. Mobile Comput. 5(8), 1044–1056 (2006).

[14] Y Albagory, O Said, Performance enhancement of high-altitude platforms wireless sensor networks using concentric circular arrays. Elsevier AEU-Int. J. Electronics Commun. 69(1), 382–388 (2015).

[15] T Yang, D Mu, W Hu, H Zhang, Energy-efficient border intrusion detection using wireless sensors network. Springer EURASIP J. Wireless Commun. Netw. 1, 1–12 (2014).

[16] R Mitchell, I Chen, A survey of intrusion detection in wireless network applications.

Elsevier Comput. Commun. J. 42, 1–23 (2014).

[17] C Lin, W Chen, Y Tseng, Efficient in-network moving object tracking in wireless sensor networks. IEEE Trans. Mobile Comput. 5(8), 1044–1056 (2006).

[18] Y Zhang, W Lee, Intrusion detection in wireless ad-hoc networks, (USA, November 6), pp. 275–283.

[19] A Stetsko, T Smolka, V Matyas, Improving intrusion detection systems for wireless sensor networks. Springer Appl. Cryptography Netw. Secur. Lect. Notes Comput. Sci. 8479, 343–360 (2014).

[20] J Kur, V Matyas, A protocol for intrusion detection in location privacy-aware wireless sensor networks. Springer Trust Privacy Secur. Digital Bus. Lecture Notes Comput. Sci. 8647, 180–190 (2014).

[21] Y Wang, Y Leow, J Yin, Is straight-line path always the best for intrusion detection in wireless sensor networks, (China, November 8), pp. 564–571.

[22] Y Wang, W Fu, D Agrawal, Intrusion detection in Gaussian distributed heterogeneous wireless sensor networks, (USA, Summer 12), pp. 313–321.

[23] Y Wang, X Wang, B Wang, D Agrawal, Intrusion detection in homogeneous and heterogeneous wireless sensor networks. IEEE Trans. Mobile Comput. 7(6), 698–711 (2008).

[24] Y Wang, F Li, F Fang, Poisson versus Gaussian distribution for object tracking in wireless sensor networks, (China, Unknown Month 22), pp. 1–4.

[25] B Liu, P Brass, O Dousse, P Nain, D Towsley, Mobility improves coverage of sensor networks, (USA, Unknown Month 25), pp. 300–308.

[26] O Dousse, C Tavoularis, P Thiran, Delay of intrusion detection in wireless sensor networks, (Italy, Unknown Month 22), pp. 155–165.

[27] S Qiong, C Comaniciu, Efficient cooperative detection for wireless sentinel networks, (USA, Unknown Month 17), pp. 1–6.

[28] D Turgut, B Turgut, L Boloni, Stealthy dissemination in intruder tracking sensor networks, (Germany, Unknown Month 20), pp. 22–29.

[29] Y Wang, W Fu, D Agrawal, Gaussian versus uniform distribution for intrusion detection in wireless sensor networks. IEEE Trans. Parallel Distributed Syst. 24(2), 342–355 (2013).

[30] M Chopde, K Ramteke, S Kamble, Probabilistic model for intrusion detection in wireless sensor network. Int. J. Commun. Netw. Secur. (IJCNS). 1(3), 19–23 (2011).

[31] M Mubarak, S Sattar, A Sajitha, Intrusion detection: a probability model for 3D heterogeneous WSN. Int. J. Comput. Appl. 6(12), 15–20 (2010).

[32] C Ortiz, J Puig, C Palau, M Esteve, Wireless sensor network modeling and simulation,

(Valencia, Spain, Unknown Month 14), pp. 307-312.

[33] G Marsaglia, Choosing a point from the surface of a sphere. Ann. Math. Stat. 43(2), 645–646 (1972).

[34] R Kaparti, OPNET IT GURU: a tool for networking education. MSCIT Practicum Paper. http://staff.ustc.edu.cn/~bhua/experiments/ITGAE_Tool_Ntwrk_Ed.pdf. Accessed 3 Mar 2015.

[35] The network simulator - ns-2. http://www.isi.edu/nsnam/ns/. Accessed 10 Dec 2014.

[36] T Xuan, S Choi, I Koo, A novel blind event detection method for wireless sensor networks. Hindawi J. Sensors. 2014, 1–6 (2014).

[37] D Cohen, M Kelly, X Huang, N Srinath, Trustability based on beta distribution detecting abnormal behaviour nodes in WSN, (Indonesia, Unknown Month 29), pp. 339–344.

[38] M Islam, Motion analysis using distance and velocity-time function. Int. J. Sci. Knowl. 2(1), 1–6 (2013).

[39] Sphere-sphere intersection. http://mathworld.wolfram.com/Sphere-SphereIntersection.html. Accessed 10 Dec 2014.

[40] I Akyildiz, W Sankarasubramaniam, E Cayirci, Wireless sensor networks: a survey. Elsevier J. Comput. Netw. **38**(4), 393–422 (2002).

Chapter 12
Security Informatics Research Challenges for Mitigating Cyber Friendly Fire

Thomas E. Carroll[1], Frank L. Greitzer[2], Adam D. Roberts[1]

[1]Pacific Northwest National Laboratory, P.O. Box 999, 99352 Richland, Washington, USA
[2]PsyberAnalytix LLC, 99352 Richland, Washington, USA

Abstract: This paper addresses cognitive implications and research needs surrounding the problem of cyber friendly fire (FF). We define cyber FF as intentional offensive or defensive cyber/electronic actions intended to protect cyber systems against enemy forces or to attack enemy cyber systems, which unintentionally harms the mission effectiveness of friendly or neutral forces. We describe examples of cyber FF and discuss how it fits within a general conceptual framework for cyber security failures. Because it involves human failure, cyber FF may be considered to belong to a sub-class of cyber security failures characterized as unintentional insider threats. Cyber FF is closely related to combat friendly fire in that maintaining situation awareness (SA) is paramount to avoiding unintended consequences. Cyber SA concerns knowledge of a system's topology (connectedness and relationships of the nodes in a system), and critical knowledge elements such as the characteristics and vulnerabilities of the components that comprise the system and its nodes, the nature of the activities or work performed, and the available defensive and offensive countermeasures that may be applied to thwart network attacks. We describe a test bed designed to support empirical research on factors affecting cyber FF. Finally, we discuss mitigation strategies to combat cyber FF, including both training concepts and suggestions for decision aids and visualization approaches.

1. Introduction

Computer and network security are among the greatest challenges to maintaining effective information systems in public, private, and military organizations. In defining computer security, Landwehr[1] described three threats to Information systems: 1) the unauthorized disclosure of information, 2) the unauthorized modification of information, and 3) the unauthorized withholding of information (e.g., denial of service or DoS). Denning[2] referred to information warfare as a struggle between an offensive and a defensive player over an information resource, with outcomes that may affect availability or integrity of the resource. While much attention has been devoted to combating external threats such as worms, viruses, and DoS attacks, actions by insiders pose a significant threat to computer and network security. This insider threat, however, is not confined to "bad actors" that intentionally perform malicious acts against an information system. Just as unintended actions by friendly forces may impact physical resources and security of friendly forces in military engagements, the actions of well-intentioned cyber defenders may result in harm to information resources and security. The focus of the present paper is on understanding and mitigating threats to intelligence and security informatics posed by cyber friendly fire.

While friendly fire (FF) is a familiar term, cyber FF is a relatively new concept for the information security community. An initial proposed definition of cyber FF, from Greitzer et al.[3], emphasizes three key characteristics:

- Cyber/electronic actions are performed intentionally,

- Actions are offensive or defensive,

- Actions result in inhibiting, damaging, or destroying friendly or neutral infrastructure or operations.

Andrews and Jabbour[4] provide the second:

The employment of friendly cyber defenses and weapons with the intent of either defending the blue cyber systems from attack from red or gray forces, or attacking the enemy to destroy or damage their people, equipment, or facilities, which results in unforeseen and unintentional damage to friendly cyber systems.

These definitions have many similarities: cyber FF is a consequence of offensive or defensive actions, the actions were performed with purpose, and the damage occurs to friendly or neutral cyber assets. Both definitions imply or overtly identify consequences of the action as unintentional. Furthermore, incidents that are born from accidents, negligence, carelessness, or malicious insiders are not friendly fire. From there, the definitions diverge. Greitzer *et al.* consider harm to both cyber systems and mission effectiveness, while Andrews and Jabbour focus only on systems. A recent Air Force chief scientist's report on technology horizons mentions the need for "a fundamental shift in emphases from 'cyber protection' to 'maintaining mission effectiveness' in the presence of cyber threats"[5]. Thus, mission effectiveness, and not only systems, is an appropriate focus for friendly fire incidents. In addition, we argue that cyber FF consequences may be felt well beyond cyber space[6]. Consider cyber physical systems that closely integrate physical, computational, and communication components to sense and effect changes in the real world. These systems are heavily employed in critical infrastructure to control and monitor processes. Adversely impacting the operation of these systems may result in large-scale power failures, toxic waste releases, or explosions that can have catastrophic consequences on the environment and life.

Given these considerations, a revised definition of cyber FF[6] is:

Cyber friendly fire is intentional offensive or defensive cyber/electronic actions intended to protect cyber systems against enemy forces or to attack enemy cyber systems, which unintentionally harms the mission effectiveness of friendly or neutral forces.

The following two examples illustrate cyber FF incidents that derive from defensive actions that unintentionally harm the organization's missions:

Illustrative Example 1. As a cost saving measure, Company XYZ has outsourced their corporate website and email to a hosting company. A hacker who has compromised and is now on the hosting company's infrastructure disrupts services by attempting to break into Company XYZ's resources. An administrator at Company XYZ notes the activity and quickly takes actions to protect company resources by blocking traffic from network addresses that are the source of the attack. As a direct consequence of these actions, Company XYZ employees lose access to their corporate website and email.

Illustrative Example 2. A current vulnerability to widely-deployed web serving software is being actively exploited. The vendor for the software has issued a security patch. Company XYZ, who relies on the software as a critical component of their e-business platform, rapidly deploys the fix on their infrastructure. The patch introduces a problem into the software, causing transactions to fail and frustrating potential customers who are attempting to purchase the company's products.

The next examples illustrate defensive actions that unintentionally harm friendly assets, but do not constitute FF:

Illustrative Example 3. Company XYZ stores client personally identifiable information in a central database. The database is compromised by an adversary, who then actively engages in exfiltrating the sensitive data. Company XYZ administrators detect the extrusion of data and take action to stem the flow of data by severing the Internet connection until they can remediate and recover from the attack. The administrators fully comprehend that no client is able to access the company's services while disconnected, but the induced harm is far less than harm of continued data exfiltration.

Illustrative Example 4. A network administrator hastily writes a new firewall rule to block suspected malicious network traffic. He errors in composing the rule, but before he catches his mistake, he publishes the errant rule to production. The rule disrupts the operations of the company's web servers, which inhibits purchases, harming sales.

2. Cognitive Approaches to Cyber Friendly Fire Research

The concept of cyber FF is similar in many respects to combat friendly fire[3], and like combat friendly fire, a fundamental cognitive issue lies in maintaining situation awareness (SA). In addition, cyber FF is closely related to some aspects of insider threat, especially when viewed within the broad framework of cyber security failures. This section provides some background and perspective on the cognitive foundations for cyber FF.

2.1. Cyber Security Failures and the Unintentional Insider Threat

The domain of cyber security spans a broad spectrum of research and operational policies to address outsider cyber threats, insider threats, and other failures such as accidents or mishaps. **Figure 1** provides a conceptual view of where cyber FF fits within this broader framework of cyber security failures.

Included in the framework is the familiar branch (shown in the black boxes) representing cyber attacks and exploits by malicious insiders, the latter representing the most highly studied insider threat research topic. Engineering failures that may be attributed to system/hardware/software vulnerabilities are represented in unfilled boxes. Of most interest for the present discussion are the branches of the hierarchy that relate to human failures that may be attributed to actions of insiders. The topic of unintentional insider threat (UIT) has been largely ignored until recent research by[7] that has provided a working definition of UIT:

> An unintentional insider threat is 1) a current or former employee, contractor, or business partner 2) who has or had authorized access to an organization's network, system, or data and who, 3) through action or inaction without malicious intent, 4) unwittingly causes harm or substantially increases the probability of future serious harm to the confidentiality, integrity, or availability of the organization's resources or assets, including information, information systems, or financial systems.

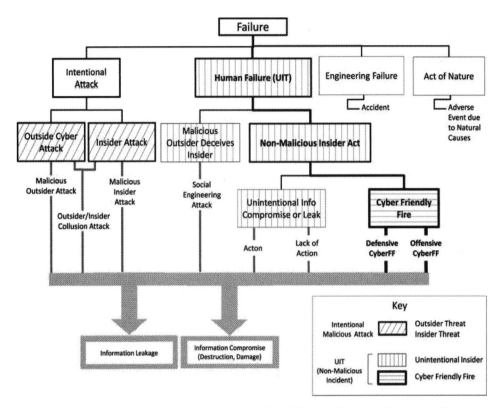

Figure 1. Conceptual framework of cyber security failures. The embolden path illustrates that cyber FF incidents is a subclass of unintentional insider threat incidents.

As pointed out in[7], UIT incidents share a common characteristic in which an organizational insider facilitates the actual or potential threat event. However, there is a distinction between the UIT cases that originate with actions performed by the internal, non-malicious member of the organization, versus UIT events that originate with an outside malicious agent (outside agent may recruit a malicious insider to participate in a collusion attack, or outside agent may deceive a non-malicious insider to take actions that enable an attack). These various cases are depicted in separate branches of **Figure 1**. The research reported in[7] relates primarily to the light blue boxes in **Figure 1**, especially social engineering exploits, information leakage and information compromise. For these UIT cases, there are four main types of incidents, which are referred to in[7] as UIT threat vectors: (a) accidental disclosure (DISC), (b) attack enabled through use of malicious code such as malware or spyware (UIT-HACK), (c) improper disposal of physical records (PHYS), and (d) lost or stolen portable equipment (PORT).

We suggest that the cyber FF definition clearly fits within the above broad definition of UIT. However, UIT research to date has not considered the case of cyber friendly fire; for example, cyber friendly fire is not mentioned in[7] nor are any cases included in their discussion or taxonomic descriptions. This is not surprising since there is no repository of such cases to draw from. By the same token, our original work on cyber FF emphasized the differences between cyber FF and insider threat, maintaining that they are distinct cyber threats. This is true if one considers only malicious insider threats, but as we now argue, cyber FF should legitimately be considered as a special case of UIT.

Therefore, it is useful to examine UIT research and associated UIT mitigation strategies to identify possible approaches for addressing cyber FF. The area of greatest commonality between UIT and cyber FF is in human performance failures. As noted in[7]: "A major part of the UIT definition is the failure in human performance. While human errors can never be eliminated completely, they can be dramatically reduced through human error mitigation techniques. Such techniques should focus on system conditions that contributed to… the resulting errors and adverse outcomes." These remarks and associated suggestions for enhancing the decision maker's situation awareness and reducing human errors pertain just as strongly to cyber FF mitigation strategies as they do to UIT threat mitigation. For this reason, we may use the arguments and suggestions provided in[7] to provide high level organizational and human factors strategies for combating cyber FF.

Problems associated with organizational factors, such as work setting, management systems, and work planning, may impact employee performance. For example, job stress[8] and time pressure[9] negatively affect performance; heavy and prolonged workload can cause fatigue, which adversely affects performance[10]. Moreover, organizational factors that increase stress may in turn lead to human factors/cognitive impacts such as narrowing of attention (attending to fewer cues)[11][12] and reduced working memory capacity[13]–[15]. Cognitive factors associated with UIT susceptibility include attention deficits and poor situation awareness[16][17], lack of knowledge and memory failures[18]–[20], and high workload or stress that impairs performance or judgment[10][21]. Finally, external or organizational factors may affect an individual's emotional states, both normal and abnormal, which in turn can affect the human error rate and lead to UIT occurrences.

2.2. Cognitive Systems Perspective

The traditional approach in accounting for performance failures such as combat friendly fire/fratricide and the lesser-examined cyber FF is to regard these events as aberrations—failures of an individual or a system. As with most performance failures (errors), assigning blame to the individual(s) responsible for a cyber FF incident is not a sufficient mitigation strategy: there is typically no single cause of these errors that occur in the "fog of war". To understand the causes (and persistence) of cyber FF, it is necessary to consider the human factors, and it seems particularly relevant to address the problem from a cognitive systems/naturalistic decision making perspective (e.g.,[22][23]). Thus, we should ask: *How did the individual perceive the situation? Why did the individual see the event that way? Why did the individual act in a way that turned out to be erroneous?*

A cognitive systems perspective leads us to consider research on SA and mental models. The SA scientific literature is substantial and no attempt is made here to report exhaustively on this topic. In short, the most accepted definition of SA is given by Endsley[16]: SA is the *perception* of the elements in the environment within a volume of time and space (Level 1 SA), the *comprehension* of their meaning (Level 2 SA), and the *projection* of their status into the future (Level 3 SA). Later work by McGuiness and Foy[24] added the *resolution* of the situation (Level 4 SA), which is deciding on a single course of action from a set of possible actions to achieve the required outcome to the situation.

SA depends on an accurate mental model[25]. Mental models have been described as well-defined, highly organized, and dynamic knowledge structures that are developed over time from experience (e.g.,[26]). By representing organized "chunks" of information in the environment, mental models serve to reduce the information load that would otherwise overwhelm the ability of decision makers to attend, process, and integrate the large amount of information that is inherent in complex operational environments. Cues in the environment activate these mental models, which in turn guide the decision- making process. Appropriate and effective mental models enable experienced decision makers to correctly assess and interpret the current situation (Level 1 and Level 2 SA) as well as to select an appropriate action based on patterns (mental models) stored in their long-term memory[27].

Considering that a lack of SA is often a contributing factor to human errors in decision making, it is clear that a study of cyber FF should focus on factors that affect the cyber security officer's/system administrator's SA. What constitutes cyber SA?

Tadda and Salerno[28] mapped constructs of SA to more cyber-relevant network environments. A SA process model was constructed that has general applicability as well as specific relevance to cyber SA. The paper also suggested a set of metrics that may be useful in assessing the effectiveness of tools for supporting SA. Consistent with Tadda and Salerno's characterization of SA, our notion of cyber SA focuses on knowledge of a system's topology (connectedness and relationships of the nodes in a system), the characteristics and vulnerabilities of the components that comprise the system (and populate the nodes), the nature of the activities or work performed, and the available defensive (and offensive) countermeasures that may be applied to thwart network attacks. SA must also include an understanding of *why* each node exists, *what* it is doing, and the harm associated with disrupting that function as a response to attack. The trade-offs between accepting the ongoing risks of attack must be properly balanced against the damage done to the overall organization's mission, and the process of balancing those elements should motivate and guide the defender to select responses that minimize the total amount of harm.

More specifically, we may speculate on implications for cyber defense and cyber SA based on the notion of "digital SA"[a]. Given the complexity of cyber structures (particularly at the national scale of critical infrastructures such as the Internet or the electric power grid), it is necessary to take a "system of systems" perspective. In this view, there is never 100 percent certainty or complete knowledge, and it must be assumed that systems will be attacked (*i.e.*, it is not possible to prevent all attacks with certainty). Thus, an appropriate cyber security strategy is *resiliency*, *i.e.*, the ability to anticipate, avoid, withstand, minimize, and recover from the effects of attacks (or for that matter, from the effects of natural disasters). To anticipate and avoid the effects of attacks or other adverse circumstances, a high level of SA is required. In particular, there is a critical need for operators to *anticipate* and *apply protocols* to avoid *cascade effects* in the network, thereby avoiding unintended consequences of defensive or offensive actions. The following types of knowledge (*critical knowledge units*) are required to invoke this anticipatory process:

- Knowledge of each enterprise, enterprise's network structure, and network component.

- Knowledge of each computer system of interest in each enterprise/component.

- Knowledge of each I/O port on each computer and how it is being used.

- Record of traffic flow and volume on every I/O port.

- Knowledge of the results of computing expected during the normal operation of each of the components in the network based on the current traffic flow and volume.

- Knowledge of operating limits for each component, enabling the decision maker to project "faults" that may lead to shut-downs and cascade failures.

- Knowledge of alternative corrective actions for such faults.

An additional consideration regarding the role of SA and cognitive models in cyber FF is the importance of Team SA: the degree to which each team member possesses the SA required for his or her responsibilities[16] and in particular, the extent to which team members possess the same SA on *shared* SA requirements[29][30]. Conflicts between goals and/or failures to coordinate goals among different members of the team are major underlying/root causes of many cyber FF incidents.

Given these considerations, a recommended approach is to capture the mental models that constitute the above types of knowledge, and then to tailor training approaches and tools to address associated cognitive factors.

2.3. Trends That Make Digital SA Harder

Current trends challenge the abilities of individuals and teams to maintain digital SA; more particularly, changes in the roles and communications among cyber security professionals, as well as paradigmatic changes brought about by cloud and utility computing, have increased the difficulty of acquiring, understanding, and maintaining critical knowledge units necessary for effective defense and operation of the information and communications infrastructure. One trend is a growing separation between the roles of individuals responsible for cyber defense mission planning and the roles of individuals responsible for operating and defending the information and communications infrastructure. Missions are defined in terms of abstract resources and quality of service attributes, rather than actual systems and devices. For example, a mission in support of a business-to-business portal is defined in terms of number of concurrent users and user experience attributes, such as page response time. The requirements are translated into resource and location requirements (e.g., "ten web servers in the East Coast data center will be tasked"). In many cases, external third parties provide and operate the infrastructure. Under these circumstances, the mission planner may not be aware of what resources are allocated, the underlying network topology, or the geographical location of the resources. The operators and defenders are compartmentalized—and often isolated—from the mission planners, understand the resources and infrastructure but are unaware of the missions that the infrastructure is serving. Communication between mission planners, operators, and defenders is complicated—if it can occur at all (e.g., "need to know" restrictions)—because planners focus on the mission, while operators and defenders focus on resources. A second trend, offered by cloud and utility computing, employs dynamic resource allocations in response to changing demands and requirements. Dynamic resource management can quickly revise and relocate allocations, as well as change the purpose or criticality of systems—this changing operational landscape challenges the ability of cyber defenders to maintain an accurate accounting of system resources, assets, and vulnerabilities. The ability to acquire and maintain critical knowledge units that are needed to support effective SA for defending these types of enterprises exceeds the efforts of an individual or even a small team—it demands the support of an entire enterprise and its complement of third parties, an extremely difficult goal. Research is needed to flesh out requirements for information sharing and for automated tools, visualization support, and decision aids to ensure that defenders have the necessary knowledge and SA to protect their enterprise.

3. Mitigation Approaches

In this section, we describe approaches and tools to mitigate cyber FF. Following research on organizational factors underlying human error, we first discuss organizational best practices (Section "Organizational best practices") that are recommended to foster productive work environments, relieve stress, and reduce cognitive load. Next, in Section "Training", we discuss relevant research on learning and cognition that may be applied to improve the effectiveness of training approaches in reducing the likelihood of cyber FF outcomes. In Section "Effects of stress on performance", we discuss implications of this research for the design and development of tools to help enhance SA and decision making (for example, by promoting the acquisition and use of critical knowledge units elaborated in Section "Cognitive systems perspective").

3.1. Organizational Best Practices

Following their discussion of possible organizational and human factors that contribute UIT, [7]described possible UIT mitigation strategies. Organizational best practices suggested in[7] that are most relevant to mitigating the incidence of cyber FF are those that help to reduce cognitive load, stress, and ultimately lead to lower risks of human errors:

- Review and improve management practices to align resources with tasks.

- Improve data flow by enhancing communication and maintaining accurate procedures.

- Maintain productive work setting by minimizing distractions.

- Implement effective work planning and control to reduce job pressure and manage time.

- Maintain employee readiness.

- Maintain staff values and attitudes that align with organizational mission and ethics.

- Implement security best practices throughout the organization.

3.2. Training

Conventional training, simulation-based training, and war gaming can each be utilized as parts of an integrated strategy to educate, raise awareness about cognitive biases and limitations, develop coping skills, and exercise skills designed to mitigate the environmental and situational factors that increase the likelihood of cyber FF. The goal of such training approaches is to provide the learner with experiences and instruction on cues, mental models, and actions that, with practice, will help establish a repertoire of well-learned concepts that can be executed under stressful or in novel, uncertain conditions. To address training requirements and approaches to reduce cyber FF, it is useful to examine factors that impact cognition and human performance, particularly with regard to SA. Research has demonstrated a number of factors that impact performance; in the present context, effects of stress, overlearning, and issues relating to cognitive bias are particularly relevant. Greitzer and Andrews[31] review cognitive foundations and implications for training to mitigate combat friendly fire. Here we describe aspects of this research that are pertinent to training requirements for cyber FF.

3.2.1. Effects of Stress on Performance

Stress has strong effects on every aspect of cognition from attention to memory to judgment and decision making. Under stress, attention appears to channel or tunnel, reducing focus on peripheral information and centralizing focus on main tasks[32]. Originally observed by Kohn[33], this finding has been replicated often, first by seminal work from Easterbrook[34] demonstrating a restriction in the range of cues attended to under stress conditions (tunneling) and many other studies (see[35]). Research by Janis and Mann[36] suggests that peripheral stimuli are likely to be the first to be screened out or ignored, and that under stress, individuals may make decisions based on incomplete information. Similarly, Friedman and Mann[37] note that individuals under stress may fail to consider the full range of alternatives available, ignore long-term consequences, and make decisions based on oversimplifying assumptions—often referred to as heuristics. Research on the effects of stress on vigilance and sustained attention, particularly regarding effects of fatigue and sleep deprivation, shows that vigilance tends to be enhanced by moderate levels of arousal (stress), but sustained attention appears to decrease with fatigue and loss of sleep[38].

3.2.2. Overlearning

Several investigations have shown that tasks that are well- learned tend to be more resistant to the effects of stress than those that are less-well-learned. Extended practice leads to commitment of the knowledge to long term memory and easier retrieval, as well as automaticity and the proceduralization of tasks. These over-learned behaviors tend to require less attentional control and fewer mental resources[39][40], which facilitates enhanced performance and yields greater resistance to the negative effects of stress—*i.e.*, overlearned behaviors are less likely to be forgotten and more easily recalled under stress. Van Overschelde and Healy[41] found that linking new facts learned under stress with preexisting knowledge sets helps to diminish the negative effect of stress. On the other hand, there is also a tendency for people under stress to "fall-back" to early-learned behavior[42]-[44] —even less efficient or more error prone behavior than more recently-learned strategies—possibly because the previously learned strategies or knowledge are more well-learned and more available than recently acquired knowledge.

3.2.3. Effects of Stress on Learning

Research suggests that high stress during instruction tends to degrade an individual's ability to learn. The research literature consistently demonstrates that elements of working memory are impaired, although the mechanisms behind these effects are poorly understood[35]. Stress appears to differentially affect working memory phases[45][46]. One instructional strategy to address stress effects is to use a phased approach with an initial learning phase under minimum stress, followed by gradual increasing exposure to stress more consistent with real-world conditions[31]. Similarly, stress inoculation training attempts to immunize an individual from reacting negatively to stress exposure. The method provides increasingly realistic pre-exposure to stress through training simulation; through successive approximations, the learner builds a sense of positive expectancy and outcome and a greater sense of mastery and confidence. This approach also helps to habituate the individual to anxiety-producing stimuli.

3.2.4. Team Performance

Finally, it is important to consider group processes in this context. Research on team decision making indicates that effective teams are able to adapt and shift strategies under stress; therefore, team training procedures should teach teams to adapt to high stress conditions by improving their coordination strategies. Driskell, Salas, and Johnston[47] observed the common phenomenon of Easterbrook's attentional narrowing is also applicable to group processes. They demonstrated that stress can reduce group focus necessary to maintain proper coordination and SA—*i.e.*, team members were more likely to shift to individualistic focus than maintaining a team focus.

3.2.5. Implications

Based on the brief foregoing discussion, we can summarize the challenges and needs for more effective training in general terms as well as more specifically focused on cyber defense and mitigation of cyber FF: training should incorporate stress situations and stress management techniques, development of realistic scenarios that systematically vary stress (e.g., as produced by varying cognitive workload through tempo of operations and density of attacks), and addressing challenges in preparing cyber warriors to overcome cognitive biases. The following factors should be included in designing training approaches:

- Training should provide extended practice, promoting more persistent memory and easier retrieval, and to encourage automaticity and the proceduralization of tasks to make them more resistant to the effects of stress.

- Training scenarios should include complex/dynamic threats that reflect the uncertainties of the real world—scenarios that force trainees to operate without perfect information and that incorporate surprises that challenge preconceptions or assumptions.

- Training scenarios should be designed to encourage the habit of testing one's assumptions to produce more adaptive, resilient cyber defense performance in the face of uncertainty.

- Training should enhance awareness of the effects of stress on cognitive performance—such as tunneling and flawed decision making strategies that ignore information—and coping strategies to moderate these effects. The training should be designed to make as explicit as possible what might happen to skill and knowledge under stress.

- Train awareness of cognitive biases and practices for managing these biases.

- Team training should focus on strategies for maintaining group cohesion and coordination, mitigating the tendency for team members to revert to an individual perspective and lose shared SA.

- Training should exercise the execution of cognitive tasks by both individuals and groups.

3.3. Tools

A key objective in the study of factors influencing cyber FF and mitigation strategies is to identify features of decision support tools with potential to reduce the occurrence of cyber FF. Our review of relevant research, as summarized in the foregoing discussion, strongly suggests that tools and visualizations to improve cyber SA are key ingredients of desired solutions. Important functions should include decision aids to support memory limitations, to counteract the negative effects of stress on performance (e.g., perceptual narrowing), and to avoid the negative consequences of cognitive biases on decisions.

3.3.1. Supporting Memory Limitations that Reduce Situation Awareness

As stated earlier, support for the cyber analyst should strive to encourage proactive decision making processes that anticipate and apply protocols to avoid cascade effects in the network, and concurrently avoid unintended consequences of defensive or offensive actions. We identified a set of critical knowledge units required for enhanced SA and anticipatory decision making, including knowledge of components of the network, details of each computer system, I/O ports, traffic flow/volumes, and ability to project impacts of possible courses of action. Decision aids and/or visualization support is needed to alleviate memory lapses and limitations by providing readily accessible information on network topology and component assets/vulnerabilities—typically referred to as external representations or external memory by researchers advocating the study of "distributed cognition" in the broader context of the social and physical environment that must be interwoven with the decision maker's internal representations (also referred to as "situated cognition"[48][49]). Thus, a decision aid that displays critical knowledge units for components that are being considered for application of remedial actions may help to avoid cyber FF effects that impair system effectiveness. This concept is similar to what Tadda and Salerno[28] refer to as "Knowledge of Us" (data relevant to the importance of assets or capabilities of the enterprise)—hence, a process that identifies to the decision maker whether there is a potential or current impact to capabilities or assets used to perform a mission. Similarly, a tool may be envisioned that helps the decision maker understand and prioritize risks that may be computed for various possible alternative actions.

3.3.2. Mitigating Cognitive Biases

Gestalt psychology tells us that we tend to see what we expect to see. Expectancy effects can lead to such selective perception as well as biased decisions or responses to situations in the form of other cognitive biases like confirmation bias (the tendency to search for or interpret information in a way that confirms one's preconceptions) or irrational escalation (the tendency to make irrational decisions based upon rational decisions in the past). The impact of cognitive biases on decision performance—particularly response selection—is to foster decisions by individuals and teams that are based on prejudices or expectations that they have gained from information learned before they are in the response situation. Decision aids and visualizations are needed that help to reduce confirmation bias, irrational escalation, and other forms of impaired decision making. One possible form of decision support designed to counteract these biases is the use of the analysis of competing hypotheses (e.g.,[50]). Other concepts that may serve as sources of ideas and strategies for the design of decision aids may be derived from problem solving techniques discussed by Jones in The Thinker's Toolkit[51].

3.4. Recommendations

Based on the foregoing discussion, we summarize the challenges and needs for more effective training and decision support to improve cyber defense and mitigate cyber FF:

- Training recommendations

  - Develop realistic cyber war gaming scenarios that systematically vary stress (e.g., as produced by varying cognitive workload through tempo of operations and density of attacks).

  - Incorporate stress management techniques.

  - Address challenges in preparing cyber warriors to overcome cognitive biases.

  - Conduct experiments to assess effectiveness of different training approaches.

- Information analysis and decision support recommendations

  - Conduct experiments to help identify effective features of decision support and information visualization tools. Will conventional training approaches to improve analytic process (e.g., analysis of alternative hypotheses, other decision making tools and strategies) be effective in the cyber domain? Our intuition suggests that the answer is "no" because of the massive data, extreme time constraints requiring near real-time responses, and the largely data-driven nature of the problem. New types of data preprocessing (triage) and visualization solutions will likely be needed to improve SA.

  - Perform cognitive engineering research to develop prospective information analysis and visual analytics solutions to enhance SA and decrease cyber FF.

4. Research Test Bed and Preliminary Studies

As concluded in Section "Mitigation approaches", more research is needed to enhance our understanding of the factors underlying cyber FF and to explore and validate possible mitigation approaches and tools. Cognitive engineering research is needed to focus on determinants of SA deficiencies and human errors in working with tools aimed to support cyber security analyst perception and decision making processes.

Research in cyber FF should be founded upon scientific principles and empirical studies in human factors and cognitive engineering, such as seminal human factors work on SA by Endsley[16] and later by Tadda and Salerno[28], who mapped constructs of SA to more cyber-relevant network environments. The present paper has sought to define research questions and to lay a foundation for empirical investigations of factors contributing to the cyber FF phenomenon and impacts on performance of proposed mitigations that can be in the form of training/awareness or decision aids.

Along these lines, we conducted a preliminary study at PNNL to help address these research questions using the simulation capabilities of PNNL's Unclassified Security Test Range test bed. The purpose of the pilot study was to demonstrate feasibility of an experimental methodology to assess effectiveness of decision aids and visualizations for cyber security analysis. Because the experiment was limited to a very small number of participants, interpretation of results was speculative, but the design and implementation of the testbed itself serve to advance the research goals described here.

4.1. Unclassified Security Test Range

The Unclassified Security Test Range consists of a combination of virtual and physical devices for testing, simulation, and evaluation. This closed network offers services found on a production network without the costs associated with duplicating a real environment. The idea is to duplicate enough of a real network to allow the test bed to appear realistic. In order to achieve this lofty goal, the virtual and physical environment is flexible and can be customized to represent different configurations based on requirements. With the proper configuration and orchestration of components, it is possible to create simulated environments that model Fortune 500 enterprises, and application and infrastructure service providers. The test range also has a room that is mocked up as an advanced "network operation center". Besides workstations provisioned with several large monitors, there were two large over head displays, allowing the projection of visualization such as network health and status. Observers can watch subjects from a vantage point, which is partially obscured from the participants view.

The test range creates virtual machines for user workstations and servers that interconnected using real networking switches, routers, and firewalls. Every virtual machine has at least two network interfaces, one for management and observation, and one or more for experiment network traffic. The test range features a unique simulation capability called ANTS. This software package simulates user behaviors: agents that are deployed on the virtual machines network have models or profiles of operator's use of real applications such as Microsoft Word, Outlook, and Internet Explorer. Application usage then generates the traffic found in normal networks. The advantage of this approach over others is its ability to create higher fidelity.

The test range has a network monitoring feature that provides the capability to monitor, log, and analyze all the traffic flowing through the network. Additionally, remote researchers and observers are provided a capability to view into the range.

4.2. Procedure

The test bed was configured to appear as an e-commerce website, a payment processor, and an "Internet" to ferret communication between e-commerce site and payment processor, and customers and malcontents to the e-commerce site. Participants were tasked with the role of network and security operations and were responsible for maintaining the operation of the e-commerce site. They were confronted with two types of events that interfered with customers assessing the website. The first event type, which manifested several times during the scenario, was a fault in the order-processing system that triggered the abnormal execution termination of the order system. The second event was a Denial-of-Service (DoS) that originated from the payment processor partner. While a partner attacking appears exceptional, there have been cases in which attackers have exploited partner relationships and used compromised partners as stepping stones to further their compromise towards reaching their goals. The DoS attack consumed large quantities of

Chapter 12

resources, slowing customer access. Both events appear, at least at first glance, to be similar.

Participants were furnished with four widely available tools. The first is Big Brother (BB) system and network monitor. BB was configured to monitor various aspects of the system and network object attributes (e.g., CPU utilization, data rate, system event logs) and alerts when these object attributes exceed defined thresholds. BB supplies alert notifications in an easily understood panel. **Figure 2** is a screen shot of the simulation's Big Brother network overview that is displaying "all conditions clear". The single alert informs the administrator that the system is unable to download updated malware/virus signatures. By design, the Test Range is isolated and constituent systems are unable to communicate with systems on the Internet. The second tool for monitoring is the Cisco ASA's ASDM panel (shown in **Figure 3**). The overview panel displays current network conditions, such as data rate and connection volume. Other ASDM panels display detailed network traffic traces and assist in traffic inspection. Half the participants was also furnished with

Figure 2. The Big Brother (BB) network and system monitor overview page. Each row is a network resource; each column is an indicator of a test result of the resource status. A green orb indicates healthy, while a red orb denotes a failed test.

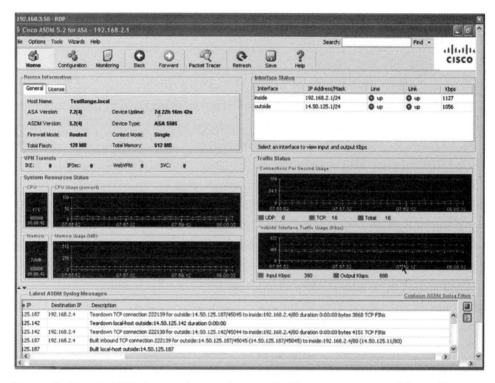

Figure 3. Cisco's ADSM network overview panel. The current status of firewall's interfaces are shown in the top right panel. The middle panels graph in real time resource utilization, connection rate, and data rate. Lastly, the bottom panel shows the device's system log messages.

Ether-Ape, a network monitor that displays network activity graphically. As depicted in **Figure 4**, the interface colorfully renders communication between systems by drawing a link between systems. The width of the link changes in proportion to the volume of traffic, *i.e.*, the link width expands as traffic increases.

4.3. Participants

Participants were four PNNL network operations staff, solicited via email as study volunteers. The invitation stated that they were invited to participate in several simulated scenarios as part of a study on network monitoring and security. The participants originate from different parts of PNNL and perform different job functions. Summarizing their jobs, two participants provide IT support for cyber security and national security research and development groups, another participant

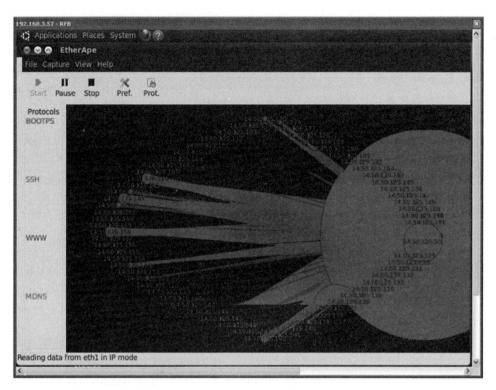

Figure 4. EtherApe network visualization tool. Rays indicate communicating systems; ray widths is proportional to data rates.

provides IT support for PNNL's various business applications, and the last performs a variety of tasks in support of general scientific computing. While every participant understood the concepts and skills necessary in performing the tasks at hand, only one had previous experience in operating in a small business climate as illustrated by the pilot study scenario. Examining the remaining three participants, two can be ranked as having intermediate level of experience and the last having little experience. All participants were familiar with the use of Big Brother network/system monitoring tool; in fact, all use it daily as part of their jobs. The experienced participant had some minimal level of exposure to the Cisco ASA and its ASDM overview page. No other participant had any prior experience. Finally, none of the participants had any exposure to EtherApe.

All participants were provided with BB and Cisco ASDM monitoring tools. Two participants were randomly selected (here identified as Participant 2 and Participant 4) and were furnished with EtherApe, which represented the "enhanced visualization" condition.

4.4. Results

To review, all participants had ready access to BB and Cisco ASDM monitoring tools; two were also provided with EtherApe as an additional visualization aid.

Participant 1 (without EtheApe) From our perch, it was not evident if the participant was choosing to use either Big Brother or the ASDM panel. During the first phase of the exercise, he relied on the alerts provided by the help desk before remediating problems. Due to a technical difficulty, the attack never reached a point to harm the ordering system. He did note the attack on the ASDM general overview panel and chose to ignore it.

Participant 2 (with EtherApe) This participant was hyper vigilant. Unfortunately, his choices and actions lead to cyber friendly fire. Nevertheless, his actions constitute cyber FF. He relied heavily on the ASDM overview display and noted problems nearly instantaneously. During the first phase, he reacted to the information by disabling the external interface of the ASA firewall device—in effect he chose to cut his network off from the Internet, thus committing textbook definition of cyber friendly fire. After we witnessed the participant act in this way twice, we informed the participant of the consequences. Unfortunately this was to no avail, as the participant disabled the external interface and was attempting to disable the internal interface during the DoS attack. If successful, he would no longer have remote access to administer the firewall.

Participant 3 (without EtheApe) The least experienced of the four participants, he preferred ASDM overview display over the Big Brother status display. He was methodical. In the first phase, he gathered available information before deciding on the course of restarting the ordering system. The deliberate approach was slow. While the participants 1 and 2 took under two minutes to note and correct the problem, it took him at least two minutes before taking corrective action. Four minutes passed before he realized the advent of the DoS attack and it took another three minutes before he decided to take any action.

Participant 4 (with Etherape) He was the most experienced of all the participants. While not part of his daily job function, he has served as a network operator in a part time job. During the exercise he was leaning back in his chair watching the EtherApe visualization or hunched over staring at the ASDM overview display. His response was rapid and in most cases remediated problems in under thirty seconds. Not once did we need to announce the occurrence of an event. He noted the DoS attack immediately from both the EtherApe and ASDM displays. He performed a packet trace and identified the source as the payment processor and that the attacking system was a critical component in processing payment transactions. He recognized that making changes to the firewall may cause harm later on; he would prefer contacting the partner first.

4.5. Discussion

It is not possible to draw generalizations from the small number of participants, particularly because of technical difficulties that affected performance of Participant 1 (and perhaps also because of existence of doubts about whether or not Participant 2 understood the instructions sufficiently to follow directions). Focus on results obtained for Participants 3 and 4 yields precious little data upon which to draw conclusions.

At a shallow level of analysis, we note that Participant 4 (who received the enhanced visualization condition) performed much better than Participant 3 (who did not receive the enhanced visualization condition). Besides the obvious conclusion that the experimental manipulation was effective, there are other possible explanations due to uncontrolled confounding factors: For example, Participant 4 had more experience than Participant 3. Because of time and budget constraints, we were unable to conduct a somewhat larger pilot study that could incorporate appropriate controls (such as a pretest-post-test design).

While we did not identify objectives related to training, some observations from the pilot study suggest training implications. Even when informed of the consequences of his choices, Participant 2 continued to engage in actions that resulted in cyber FF. This could have been a result of lack of experience with network and firewall operations, or possibly he missed the "message" communicated during the orientation session about the importance of maintaining business operations; or perhaps he believed that he was, in fact, taking the best course of action. In any case, this observation suggests that training approaches should be considered.

Because this was a pilot study, limitations and difficulties were not unexpected. Nevertheless, we still may conclude that the results at least suggest that one can demonstrate cyber FF performance differences that possibly can be related to the independent variable studied (visualization support); and perhaps more importantly for the present purposes, the Unclassified Cyber Test Range that we utilized at PNNL appears to be capable of supporting experimental studies of cyber FF. This point is important going forward, since it reinforces the recommended research strategy of conducting more controlled scientific studies of cyber SA and cyber FF in a high fidelity simulated environment.

Conclusions and recommendations that serve to inform future design of such experimental studies are:

- Access to a larger pool of participants is needed to allow for the possibility of statistically significant results. The observational methods employed, interview procedures, and the performance measures collected would readily apply to an expanded study with more participants.

- Participants should have a more relevant background and experience with the type of enterprise and network represented in the scenario. Participants should be fluent in the technical skills required to perform necessary actions. Experience ought to be controlled as a factor in the study.

Much more specific recommendations about the design and human factors of the Cyber Test Range were derived from observing the participants and obtaining feedback from the participants including deficiencies in software packages, monitor placement, choice of keyboard and mice, and the height of the overhead displays.

5. Conclusions and Future Direction

Cyber FF is one class of cyber security failures. Since it is based in human failure, we have argued that cyber FF belongs to a sub-class of cyber security failures that is characterized by unintentional insider threats. There is a tendency to regard such failures as aberrations that can be fixed with technological advances, such as technology solutions to improve SA, to increase accuracy in identifying targets, or to improve the precision of defensive or offensive actions. We have argued that a sound mitigation strategy, whether or not incorporating technological advancements, must be focused on identifying and accommodating the realities of human performance. Contributing factors attributed to cyber FF must be empirically studied along with the benefits of training, awareness, and tools meant to reduce the number and severity of incidents. To this end, we have described a preliminary study that we conducted that demonstrates cyber FF research in a controlled, isolated testbed environment. While results where speculative due to the small number of participants, it does demonstrate that experimentation in a testbed can advance the research goals described within.

Related research is ongoing: An advanced concept that is currently being pursued by PNNL cyber security research programs is the notion of Asymmetric Resilient Cybersecurity (ARC)[b], which is characterized by goals of standing up resilient and robust cyber infrastructure and network architectures that present a "moving target" to potential attackers in an attempt to overcome and hopefully reverse the current asymmetric state of affairs that favors the adversary. The goals and challenges of this program align with issues that we have articulated in our research on cyber FF, particularly in ways that can be seen as amplifying the cyber FF challenge: e.g., maintaining enterprise-wide SA when the network, systems, and components "move" continuously and dynamically. Moreover, ongoing research at PNNL seeks to develop and assess visualization and decision support tools that address cognitive limitations. Current research is developing Kritikos, a network resource identification, resource dependency discovery, and criticality assessment tool. Dependencies are identified by Self-Organizing Maps (SOM), a neural network machine learning algorithm, discovering repeated spatio-temporal patterns in IP Information Flow eXport (IPFIX) record sets. Patterns in time and space indicate usage; repeated observations of a pattern suggest a dependent relationship. The patterns allow a dependency-based network model to be generated. This model is a (disconnected) graph where vertices are resources and edges indicate dependent relationships. A business process/operations model annotated with indication of network resources and business criticality, assumptions not unusual for today's enterprises, can be fused with the network model, illuminating indirect resources. Furthermore, the relationship between business process and network resources can be used to assess the criticality of resources in terms of business objectives and requirements. Cyber FF research also directly meets essential needs of DOE cyber security as well as cyber security programs within the DoD and the intelligence community.

The fundamental research goal is to develop a scientific understanding of the behavioral implications of cyber FF. Research is needed to extend our current understanding of cyber SA and to develop metrics and measures for cyber FF. The principal scientific research questions include: What are root causes of cyber FF? What are possible mitigating solutions, both human factors and technical/automated? We have examined relevant research and cognitive theory, and we have taken some initial steps toward investigating these research questions in empirical laboratory studies using realistic test scenarios in a cyber SA/FF testbed facility[52]. Continued empirical research is required to investigate the phenomenon and relevant contributing factors as well as mitigation strategies. A major objective should be to investigate approaches to and assessment of effectiveness of cyber FF mitigation strategies, such as training and decision aids/tools. Such research promises to advance the general field of cyber SA and inform other ongoing cyber security research. In addition, it is hoped that this research will facilitate the design and prototyping of automated or semi-automated systems (or decision aids) to increase cyber SA and eliminate or decrease cyber FF; this provides a foundation for development of commercial products that enhance system effectiveness and resiliency.

Endnotes

[a]The following discussion is based in part on an essay on situation awareness in Wikipedia: http://en.wikipedia.org/wiki/Situation_awareness.

ᵇInformation about PNNL's Asymmetric Resilient Cybersecurity (ARC) Lab Direct Research & Development Initiative can be found at: http://cybersecurity.pnnl.gov/arc.stm.

Acknowledgements

Portions of the research were also funded by PNNL's Asymmetric Resilient Cybersecurity (ARC) Laboratory Research & Development Initiative. The views expressed in this report are the opinions of the Authors, and do not represent official positions of the Pacific Northwest National Laboratory, the Department of Energy, or the Air Force Research Laboratory.

Source: Carroll T E, Greitzer F L, Roberts A D. Security informatics research challenges for mitigating cyber friendly fire [J]. 2014, 3(13): 1–14.

References

[1] CE Landwehr, Formal models for computer security. ACM Comput. Surv. 13(3), 247–278 (1981).

[2] DE Denning, Information Warfare and Security. (ACM Press, New York, 1999).

[3] FL Greitzer, SL Clements, TE Carroll, JD Fluckiger, Towards a research agenda for cyber friendly fire. Technical Report PNNL-18995, Pacific Northwest National Laboratory (2009).

[4] DH Andrews, KT Jabbour, Mitigating cyber friendly fire: a sub-category of cyber mishaps. High Front. 7(3), 5–8 (2011).

[5] United States Air Force Chief Scientist (AF/ST), Report on Technology Horizons: A Vision for Air Force Science & Technology During 2010–2030 (2010). http://www.au.af.mil/au/awc/awcgate/af/tech_horizons_vol-1_may2010.pdf.

[6] FL Greitzer, TE Carroll, AD Roberts, Cyber friendly fire: Research challenges for security informatics, in Proc. of the IEEE International Conference on Intelligence and Security Informatics (ISI 2013) (IEEE, Piscataway, NJ, 2013), pp. 94–99.

[7] FL Greitzer, J Strozer, S Cohen, J Bergey, J Cowley, A Moore, D Mundie, Unintentional insider threat: contributing factors, observables, and mitigation strategies, in Proc. of the 47th Hawai'i International Conference on System Sciences (HICCS-47) (IEEE, Piscataway, NJ, 2014).

[8] S Leka, A Griffiths, T Cox, Work Organization & Stress: Systematic Problem Approaches for Employers, Managers and Trade Union Representatives, Protecting Workers' Health Series No. 3. (World Health Organization, 2004). http://www.who.int/occupational_health/publications/pwh3rev.pdf. Accessed 7 Dec 2013.

[9] P Lehner, M-M Seyed-Solorforough, MF O'Connor, S Sak, T Mullin, Cognitive biases and time stress in team decision making. 27(5), 698–703 (1997).

[10] E Soetens, J Hueting, F Wauters, Traces of fatigue in an attention task. Bull. Psychonomic Soc. 30(2), 97–100 (1992).

[11] BK Houston, Noise, task difficulty, and Stroop color-word performance. J. Exp. Psychol. 82(2), 403–404 (1969).

[12] AF Stokes, K Kite, Flight Stress: Stress, Fatigue, and Performance in Aviation. (Gower Technical, Aldershot, 1994).

[13] DR Davies, R Parasuraman, The Psychology of Vigilance. (Academic Press, London, 1982).

[14] GRJ Hockey, Changes in operator efficiency as a function of environmental stress, fatigue and circadian rhythms, in Handbook of Perception and Performance, ed. by K Boff, L Kaufman, and JP Thomas, vol. 2 (Wiley, New York, 1986), pp. 1–44.

[15] PL Wachtel, Anxiety, attention, and coping with threat. J. Abnorm. Psychol. 73(2), 137–143 (1968).

[16] MR Endsley, Towards a theory of situation awareness in dynamic systems. Hum. Factors. 37(1), 32–64 (1995).

[17] MD Rodgers, RH Mogfor, B Strauch, Post hoc assessment of situation awareness in air traffic control incidents and major aircraft accidents, in Situation Awareness Analysis and Measurement, ed. by MR Endsley, DJ Garland (Lawrenece Erlbaum Associates, Mahway, 2000). pp. 73–112.

[18] R Dhamija, JD Tygar, M Hearst, Why phishing works, in Proc. of the SIGCHI Conference on Human Factors in Computer Systems (CHI '06) (ACM, New York, 2006), pp. 581–590.

[19] D Sharek, C Swofford, M Wogalter, Failure to recognize fake Internet popup warning messages, in Proc. of the Human Factors and Ergonomics Society 52nd Anual Meeting (Human Factors and Ergonomics Society, Santa Monica, CA, 2008), pp. 557–560.

[20] J-P Erkkilä, Why we fall for phishing, in Proc. of the SIGCHI Conference on Human Factors in Computer Systems (CHI'11) (ACM, New York, 2011).

[21] SG Hart, CD Wickens, Workload assessment and predictions, in MANPRINT: An Approach to Systems Integration, ed. by HR Booher (Van Nostrand Reihold, New York, 1990), pp. 257–296.

[22] S Dekker, M Lützhöft, Correspondence, cognition and sensemaking: a radical empi-

ricist view of situation awareness, in A Cognitive Approach to Situation Awareness: Theory and Application, ed. by Banbury S, Tremblay S (Ashgate, Burlington, 2004), pp. 22–41.

[23] KE Weick, Sensemaking in Organizations, Foundations for Organiztional Science. (SAGE Publications, Thousand Oaks, 1995).

[24] B McGuiness, L Foy, A subjective measure of SA: the crew awareness rating scale (CARS), in Proc. of the Human Performance, Situation Awareness and Automation Conference (SA Technologies, Marietta, GA, 2000).

[25] N Sarter, D Woods, Situation awareness: a critical ill-defined phenomenon. Int. J. Aviat. Psychol. 1(1), 45–57 (1991).

[26] SWJ Kozlowski, Training and developing adaptive teams: Theory, principles, and research, in Making Decisions Under Stress: Implications for Individual and Team Training, ed. by JA Cannon-Bowers, E Salas (America Psychological Association, Washington, DC, 1988).

[27] D Serfaty, J MacMillan, EE Entin, EB Entin, The decision-making expertise of battle commanders, in Naturalistic Decision-Making, ed. by CE Zsambok, G Klein (Lawrence Erlbaum, New York, 1997).

[28] GP Tadda, JS Salerno, Overview of cyber situation awareness, in Cyber Situational Awareness: Issues and Research, ed. by S Jajodia, P Liu, V Swarup, and C Wang (Springer, New York, 2010), pp. 15–35.

[29] MR Endsley, WM Jones, Situation awareness, information dominance, & information warfare. Technical Report AL/CF-TR-1997-0156, US Air Force Armstrong Laboratory (1997). http://www.dtic.mil/dtic/tr/fulltext/u2/a347166.pdf.

[30] MR Endsley, WM Jones, A model of inter- and intrateam situation awareness: implications for design, training, and measurement, in New Trends in Cooperative Activities: Understanding System Dynamics in Complex Environments, ed. by M McNeese, E Salas, and M Endsley (Human Factors and Ergonomics Society, Santa Monica, 2001).

[31] FL Greitzer, DH Andrews, Training strategies to mitigate expectancy-induced response bias in combat identification: a research agenda, in Human Factors Issues in Combat Identification, ed. by DH Andrews, RP Herz, and MB Wolf (Ashgate, Farnham, UK, 2010).

[32] J Kavanagh, Stress and performance: a review of the literature and its applicability to the military. Technical Report TR-192, RAND (2005).

[33] H Kohn, Effects of variations of intensity of experimentally induced stress situations upon certain aspects of perception and performance. J. Genet. Psychol. 85, 289–304 (1954).

[34] JA Easterbrook, The effect of emotion on cue utilization and the organization of behavior. Psychol. Rev. 66, 183–201 (1959).

[35] MA Staal, Stress, cognition, and human performance: a literature review and conceptual framework. Technical Report NASA/TM-2004-212824, National Aeronautics and Space Administration (2004).

[36] IL Janis, L Mann, Decision Making. (The Free Press, New York, 1977).

[37] IA Friedman, L Mann, Coping patterns in adolescent decision-making: an Israeli-Australian comparison. J. Adolesc. 16, 187–199 (1993).

[38] DR Davies, GS Tune, Human Vigilance Performance. (Staples Press, London, 1970).

[39] J Leavitt, Cognitive demands of skating and stick handling in ice hockey. Can. J. Appl. Sport. Sci. 4, 46–55 (1979).

[40] MD Smith, CJ Chamberlin, Effect of adding cognitively demanding tasks on soccer skill performance. Percept. Mot. Skill. 75, 955–961 (1992).

[41] JP Van Overschelde, AF Healy, Learning of nondomain facts in high- and low-knowledge domains. J. Exp. Psychol. Learn. Mem. Cognit. 27, 1160–1171 (2001).

[42] M Allnutt, Human factors: Basic principles, in Pilot Error, ed. by R Hurst, LR Hurst (Aronson, New York, 1982), pp. 1–22.

[43] RP Barthol, ND Ku, Regression under stress to first learned behavior. J. Abnorm. Soc. Psychol. 59(1), 134–136 (1959).

[44] RB Zajonc, Social facilitation. Science. 149, 269–274 (1965).

[45] S Kuhlmann, M Piel, OT Wolf, Impaired memory retrieval after psychosocial stress in healthy young men. J. Neurosci. 25(11), 2977–2982 (2005).

[46] S Kuhlmann, OT Wolf, Arousal and cortisol interact in modulating memory consolidation in healthy yound men. Behav. Neurosci. 120(1), 217–223 (2006).

[47] JE Driskell, E Salas, J Jonston, Does stress lead to a loss of team perspective? Group Dynam.: Theor. Res. Pract. 3, 291–302 (1999).

[48] DA Norman, The Psychology of Everyday Things. (Basic Books, New York, 1988).

[49] J Hollan, E Hutchins, D Kirsh, Distributed cognition: toward a new foundation for human-computer interaction research. ACM Trans. Comput. Hum. Interact. 7(2), 174–196 (2000).

[50] RJ Heuer Jr, Analysis of competing hypotheses, in Psychology of Intelligence Analysis (Center for the Study of Intelligence, Central Intelligence Agency, Washington, D.C., 1999).

[51] MD Jones, The Thinker's Toolkit: Fourteen Powerful Techniques for Problem Solving. (Three Rivers Press, New York, 1998).

[52] FL Greitzer, TE Carroll, AD Roberts, Cyber friendly fire. Technical Report PNNL-20821, Pacific Northwest National Laboratory (2011).

Chapter 13
Security-by-Experiment: Lessons from Responsible Deployment in Cyberspace

Wolter Pieters[1,2], Dina Hadžiosmanović[1], Francien Dechesne[3]

[1]Delft University of Technology, CyberSecurity@TUDelft, P.O. Box 5015, 2600 GA Delft, The Netherlands

[2]University of Twente, Services, Cybersecurity and Safety, P.O. Box 217, 7500 AE Enschede, The Netherlands

[3]3TU.Ethics@Eindhoven, Eindhoven University of Technology, P.O. Box 513, 5600 MB Eindhoven, The Netherlands

Abstract: Conceiving new technologies as social experiments is a means to discuss responsible deployment of technologies that may have unknown and potentially harmful side-effects. Thus far, the uncertain outcomes addressed in the paradigm of new technologies as social experiments have been mostly safety-related, meaning that potential harm is caused by the design plus accidental events in the environment. In some domains, such as cyberspace, adversarial agents (attackers) may be at least as important when it comes to undesirable effects of deployed technologies. In such cases, conditions for responsible experimentation may need to be implemented differently, as attackers behave strategically rather than probabilistically. In this contribution, we outline how adversarial aspects are already taken into account in technology deployment in the field of cyber security, and what the paradigm of new technologies as social experiments can learn from this.

In particular, we show the importance of adversarial roles in social experiments with new technologies.

Keywords: Adversarial Experiments, Cyber Security, Empirical Security, Responsible Experimentation, Security-by-Experiment, Social Experiments

1. Introduction

1.1. Safety and Security

The paradigm of new technologies as social experiments (henceforth NTaSE) deals with responsible deployment of technologies that may have undesirable side-effects. The idea is that it is impossible to identify all potential problems in the design stage, and that it may therefore be necessary to subject society to a deployment experiment with uncertain outcomes.

Thus far, the uncertain outcomes addressed in the NTaSE-paradigm have been mostly safety-related (nuclear waste and accidents, health effects of nanoparticles, genetically modified crops etc.). Safety implies that potential harm is caused by the design plus accidental events in the environment, such as natural disasters and human mistakes. What the paradigm has not covered so far are so-called adversarial risks. These are risks that are not caused by probabilistic natural events or accidents, even human failures, but rather by the determined, strategic behaviour of an adversarial agent. This is often labelled security as opposed to safety.

The most obvious example in the NTaSE literature of a technology with a security component is nuclear technology: adversaries who get hold of nuclear material may use it for weapons. Recently, Lehtveer and Hedenus (2015) discussed this extensively in terms of nuclear proliferation. Where the NTaSE-literature has discussed nuclear technology, it has focused on accidents (safety), and has not addressed this adversarial perspective (security). For example, Krohn and Weingart (1987) explicitly focus on the "accident as implicit experiment", analysing the Chernobyl Meltdown, and in the same vein, Van de Poel (2011) analyses the Fukushima accident. Finally, Taebi et al. (2012) address four characteristics of nuclear energy that complicate traditional risk assessment (low probabilities with large consequences, uncertainty and ignorance, long term effects, intense emotions by both proponents and opponents), but do not include the potential effects of intentional adverse behaviour.

In general, the possible use of technologies for military purposes is termed dual use. However, security can be broader than that when considering use by enemies or adversaries, with opposing and conflicting goals, who need not always be of military type. New technologies may be used for weapons, but also for fraud, other criminal activities, or morally problematic behaviour that is not in itself illegal (yet). The NTaSE literature has not yet addressed security aspects for other new technologies with potential dual use-issues and other relevant adversarial risks either. The work on biotechnology and nanotechnology so far mainly focuses on responsibility: Robaey (2013) address distribution of responsibility for GM- risks through the philosophical notion of ownership; Jacobs *et al.* (2010) propose conditions for the responsible marketing of nano(bio)tech products. Thus, we conclude that the security perspective is currently underrepresented within the paradigm.

1.2. Cyber Security

Adversarial risks are also key in the domain of cyber security. Cyber security is a relatively new field of research and technology, which aims at protecting information technology and connected infrastructures against malicious actions by adversaries (hackers, terrorists, nation states engaged in cyber warfare). In this domain, many technologies have been deployed that allowed adversaries to take advantage. In some cases, benevolent people or groups have pointed developers at weaknesses, which have or have not been fixed subsequently.

We believe that the deployment of many new technologies may induce adversarial risks, next to accidental risks, that are not to be neglected. Dealing with these adversarial components has not yet been highlighted as a systematic part of the NTaSE framework, and it requires an extension of the framework to accommodate the adversarial perspective. In that respect, we think that the domain of cyber security contains important lessons for the NTaSE paradigm. In order to demonstrate this, we provide an overview of selected cyber security problems and solutions from the NTaSE perspective. In particular, we discuss the issue of adversarial roles in different interpretations of cyber security as social experiment.

This paper is a further elaboration and adaptation of ideas initiated in Pieters *et al.* (2014a). In that paper, we highlighted the relevance of the NTaSE paradigm for the cyber security community, by discussing how the conditions for responsible experimentation can be interpreted for experiments in the context of cyber security. In contrast to the previous paper, the present paper aims at identifying lessons that NTaSE can learn from cyber security. Therefore, the focus in this paper is on developments in cyber security that can be seen as social experiments, what techniques have been used to improve deployment processes in the light of adversarial risk, and also what hasn't worked, in terms of problematic cases. The sections on experimental practices ("Adversarial Experiments in Cyber Security" section) and deployment techniques ("Techniques for Responsible Deployment" section) are entirely new.

1.3. Structure of This Paper

In section "Deployment Problems", we will discuss several occasions in which the deployment of security-sensitive information technologies had the characteristics of a social experiment, but generally was not designed as such. These examples point to several possible "adversarial roles" in such histories. In section "Adversarial Experiments in Cyber Security", we discuss how adversarial roles have been embedded in more formal experimental practices related to cyber security, such as software testing but also manipulative experiments. In section "Techniques for Responsible Deployment", we discuss techniques in the cyber security domain that have been used to implement responsible deployment, in particular related to monitoring and feedback, from which we may be able to draw lessons for other technology domains in which adversarial risk plays a role. In section "Conditions for Responsible Deployment" we discuss relevant conditions for responsible deployment in the light of adversarial risks, and we conclude in section "Conclusions and Discussion".

2. Deployment Problems

To illustrate the issue of responsible deployment in relation to cyber security, we discuss three examples. These examples point to undesirable effects that the deployment of new technologies in the cyber domain may cause, and how potential problems have been identified in the course of the deployment experiments. The examples are electronic voting, smart electricity meters, and a public transport chipcard, all in the Dutch context.[1]

2.1. Electronic Voting

One example of the deployment of a security-sensitive new information technology is the introduction of electronic voting in the Netherlands (Jacobs and Pieters 2009; Pieters 2008; Pieters and Van Haren 2007). In the Netherlands, electronic voting machines had been introduced since the early 1990s. After the introduction of the machines, regulations were not revisited or updated, nor was there any renewed evaluation of the risks.

When—as one of the last municipalities—Amsterdam introduced electronic voting in 2006, a pressure group was founded to fight electronic voting. Their main argument in favour of a return to paper voting was the ability of citizens to observe and verify the procedure. In electronic voting, one cannot see or deduce that "what goes in is what comes out", unless one trusts the workings of the machines, which is precisely what is at stake here. The pressure group had an excellent media strategy, and achieved coverage of their manipulation of machines they had bought. Basically, they showed that it would be easy for anyone having access to the machines to replace the chips containing the counting programs with fraudulent ones. Such access is available both to manufacturer personnel as well as to those involved in transport and storage. Besides, storage security was shown to be low, giving external attackers the opportunity to gain access as well. Also, and more or less by accident, they demonstrated that it was possible to eavesdrop on the choice of the voter by means of a radio antenna, capturing signals emitted by the device while showing the voter's choice [TEMPEST attack, Gonggrijp and Hengeveld (2007)].

Because of the information-leak-by-radiation, the certification of the touch-

---

[1]The electronic voting and smart metering examples are the same as in the earlier conference paper (Pieters *et al.* 2014a).

screen machines was suspended before the 2006 elections. Commissions were installed to study both the past and the future of electronic voting. The "past" commission concluded that the government had too easily outsourced their responsibility for the electoral process, and requested that the government take control again (Hermans and Twist 2007). The "future" commission proposed a combination of ballot printer and vote counter, where the intermediate votes would be readably printed on paper (Election Process Advisory Commission 2007). However, because an expert group concluded that the radio signal problems could not be solved for any device on which the voter chooses a candidate, also the ballot printer was deemed infeasible. It was therefore decided to abandon all forms of electronic voting. Currently, future electronic voting possibilities are discussed, again based on the framework of ballot printer plus vote counter, based on a report by another commission (Commissie onderzoek elektronisch stemmen in het stemlokaal 2013).

2.2. Smart Meters

The EU Directive on energy efficiency (2006/32/EG) prescribed the installation of smart energy meters that provide end users with information on their actual use, so that they can contribute to energy savings. In the Netherlands, a combination of two separate legal bills was proposed in 2008, amounting to the mandatory roll-out of smart meters, which were still to be developed. These meters were to send measurements for gas (hourly), and electricity (quarter-hourly) to the network operators, who would forward this information to the energy providers, who could then inform the consumers about their consumption. The initial proposal also included signalling functions (to detect energy quality), switching functions (to remotely switch off in case of non-payment or disasters) and regulatory functions (to add options to the meter) for the network operators. In fact, some energy providers had already started to provide households with smart meters upon request (e.g. Oxxio from 2005).

After the assessment by the Dutch privacy watchdog, the proposal was amended by requiring explicit consumer consent for transferring detailed consumption data to energy suppliers (however, daily usage would be mandatorily forwarded). Also, addition of purpose specification and use limitation, data subjects' right of access, data removal after use, and suitable security measures were required according to the Dutch privacy law (Cuijpers and Koops 2013). The October 2008 report by the Consumer Union concluded that smart meters also put pressure on the right to inviolability of the home, and the right to respect for family life (Hoenkamp *et al.* 2011).

On the basis of this analysis, the Dutch Senate rejected the bills in 2009, and adopted an adapted version, that allowed for conducting pilot projects (Proeftuinen) involving smart meters. Users, providers and operators experimented with smart grids and meters on a voluntary basis, in selected neighbourhoods. The aim was trying out incentivising users to conserve energy, and to participate in balancing the grid under the presence of sustainable energy sources whose production depends on weather conditions (Cuijpers and Koops 2013). In the meantime, to resolve the issues raised by the Senate, a broad stakeholder collaboration came to define the so-called Dutch Smart Meter Requirements (DSMR4), that implement the adapted version of the bill (in particular, it specifies the last point: defining data granularity for each task). The abolition of the detailed readings was considered to take "the largest privacy sting out of the Dutch law" (Cuijpers and Koops 2013).

Interestingly, the pilot projects have, as far as we have found (Dechesne 2013), not been exploited to explicitly experiment with the effectiveness of the new requirements with respect to the security and privacy issues that were raised when the law was rejected. The pilots are mostly focused on testing the functionality of the technology, and learning how to deal with human participation in balancing the grid. Questions about privacy and security, and associated requirements and values, were not asked to the consumers. In October 2014, a new proposal for the broad smart meter roll out in the Netherlands, on a voluntary basis, was approved by parliament.

User participation in the electricity net is a great paradigm shift, both for users and operators. Experience shows that wrong assumptions are easily made about tasks, responsibilities and risks with respect to (cyber) security. For example, operators are used to thinking in top-down controllable components, which made them neglect privacy issues for consumers, while users are not used to be conscious about the electricity flow, *let al.* one to adapt their behavior—they need incentives. The pilots that are conducted provide a good opportunity for both sides to learn in a relatively controlled environment how roles in the system may shift, and what that would mean for the risks and responsibilities with respect to cyber security. The lack of explicit attention to (cyber) security and privacy in the smart grid pilots (Dechesne 2013) leaves room for reflection on how the pilots could have been used to learn about these aspects for smart metering, by consciously designing them as social experiments.

2.3. OV-Chipkaart

The "OV-chipkaart" is a public transport chipcard gradually rolled out in the Netherlands between 2002 and 2012. The initial version of the card used the MIFARE classic chip. The cryptographic algorithm used to protect the card contents was kept secret by the producer. In 2007, German researchers revealed part of the secret al.gorithm by so-called reverse-engineering (Nohl *et al.* 2008).

In 2008, Dutch researchers of Radboud University Nijmegen found two possible attacks on the card. The attacks would enable them to read, clone, or restore cards by retrieving the cryptographic keys. The researchers demonstrated this possibility with several cards used in practice, including the OV-chipkaart. However, travelling with cloned cards would still be detected by the back-end system in place.

The researchers informed the government and the manufacturer, with publication of the results anticipated 7 months later. The idea of this "responsible disclosure" was giving the responsible authorities enough time to address the issue before the knowledge would become public. The manufacturer then asked the court to prohibit publication.[2] The university claimed their actions were reasonable from the point of view of academic freedom, and the court ruled in the university's favour. The results were published eventually (Garcia *et al.* 2008).

After these events, the organisation responsible for the OV-chipkaart set up a scientific advisory board to enable better handling of feedback in the future. The MIFARE classic card was also gradually replaced with a different one, using standard rather than proprietary cryptography.

---

[2]http://www.sos.cs.ru.nl/applications/rfid/pressrelease-courtdecision.en.html.

2.4. Analysis

In the above examples, it is clear that the problematic aspects of these "social experiments" do not lie in threats to health, environment or safety. Although harm may ensue in domains where information systems are connected to critical (physical) infrastructures, the primary concern is the regulation of access, in this case to information and information systems. In an electronic voting system, there has to be an acceptable balance between who can influence the results, who can verify the results, and who can learn what somebody voted. In the smart metering case, there is a similar question on what data is collected, and who can use what data for which purpose. In the OV-chipkaart case, the possibility of (monetary) fraud by means of manipulation of data was the key issue. Privacy and integrity, next to availability, are core concerns.

Several authors have discussed this fundamental premise of information and cyber security (regulating access) in different terms: ontological friction (Floridi 2005), causal insulation and perimeters (Pieters 2011), and order machines (Vuorinen and Tetri 2012). Security controls are aimed at making it easy for some to get access while at the same time making it difficult for others. This is meant to ensure that the opportunities for adversaries/attackers/criminals gaining illegitimate benefit out of the deployed technology are limited. This is also discussed in the field of crime science, where regulating access is operationalised in terms of making crimes less attractive from an adversary point of view: increasing effort, increasing risk (for the adversaries), reducing rewards, reducing provocations, and removing excuses (Dechesne *et al.* 2014; Gradon 2013).

For each attempt to regulate access, the question can be raised how effective it is (Pieters *et al.* 2014b), which will in turn provide some information on expected harm, as well as the potential of unfair gain for the adversaries. Typically, mathematical reasoning, tests, and experiments can help answer this question. Mathematical reasoning can prove properties of cryptographic tools, testing can reveal security flaws in software, and social experiments (in the narrow social science sense) can provide information on human defences, such as compliance with security rules and guidelines. In many cases, the question on the adequacy of the combined digital, physical and social access controls is still a matter of expert judgement, although risk management tools may help.

In the cases, we have seen several instances in which new information on the security quality of the systems was revealed by external parties after deployment, based on their own assessments. These can be seen as instances of "security-by-experiment", in contrast to the "security-by-design" approaches often advocated. This raises the question which techniques are already available for experimenting with security both in the design and in the deployment stage.

In the next section, we will look into experimental practices on the effectiveness of regulating access before deployment. We will then use this as a basis for discussing practices on evaluating and improving regulation of access after deployment, as in the NTaSE paradigm.

3. Adversarial Experiments in Cyber Security

Where the traditional paradigm of new technologies as social experiments focused on safety, the application to security-sensitive technologies requires taking into account other aspects of the notion of "experiment". In this case, the uncertain outcomes are not only due to natural events and human mistakes, but also to strategic behaviour of adversaries, or attackers.

Experimental approaches differ for adversarial and non-adversarial contexts. This does not only hold for new technologies as social experiments, but also for experiments in a narrower sense, i.e. observing the responses of systems to controlled conditions. To understand the consequences for the notion of new technologies as social experiments, in this case cyber security as social experiment, it is therefore helpful to investigate existing experimental approaches, in the narrower sense, in cyber security. Typically, those are controlled studies on the effect of adversarial behaviour on systems, which can involve technical as well as human components.

3.1. Security Testing

One such context occurs in security testing, which is quite different from regular testing of information technology. For functional testing, i.e. testing whether the behaviour of the system conforms to the specification, it is tested whether the right behaviour occurs under the specified conditions. If I want to send a secret from A to B, I can observe whether the secret submitted at A is actually produced at B. What I cannot observe is who else might have learnt the secret in the meantime.

It is widely accepted in the security community that security is not a functional property. Although certain aspects of security can be embedded in functions, such as authentication, in general security requirements say something about what should not happen in a system, and not in particular on the relation between inputs and outputs. An electronic voting system may produce the right result for any sequence of input votes submitted through the intended channel (and therefore be functionally correct), but that does not mean that it is secure (as votes may also be input in non- standard ways).

This also means that security flaws do not show up in regular software testing procedures, unless specific tests for security are executed. For example, a database may return correct results for all "normal" queries, but a specifically crafted "odd" query may allow an attacker to retrieve sensitive information he is not authorised for, or even to delete (parts of) the database (so-called SQL injection). Lack of input validation is one of the important causes of security weaknesses (Tsipenyuk *et al.* 2005), and it requires tests that intentionally provide "odd" inputs.

Even when specific security testing is in place, the sheer number of possible security flaws may cause the testers to overlook important aspects. For example, side-channel attacks make use of measurements on power consumption, timing, or electromagnetic radiation to deduce properties of the information being processed in a system (Standaert *et al.* 2009). This may for example allow an attacker to eavesdrop on voter's choices in an election using electronic voting machines, as we have seen in the electronic voting case (Pieters 2009). Security testing thus requires taking the perspective of the adversary to see what tests should be executed (Rennoch *et al.* 2014). This requires some idea about possible adversary behaviour.

Although components may be tested separately, taking the adversarial perspective happens in particular on a system level, where so-called "penetration tests" can be executed to find paths through which attackers can gain access to critical system assets. In penetration tests, one hires ethical hackers to try and break into the system, reporting the results to the system owner. These hackers play the role of the adversaries in the tests, relieving the system owners from having to play this role themselves.

3.2. Security Awareness Experiments

In the above, we have explained how adversarial testing is different from functional testing, focused on information systems in the technical sense, and illustrating the adversarial aspects. However, humans often play a very important role in the protection of information and connected infrastructures, and this aspect can be subjected to tests or experiments as well. Adversarial roles are therefore also important in experiments on so-called social engineering.

Social engineering (Tetri and Vuorinen 2013) refers to the exploitation of human weaknesses or lack of awareness to gain access. Phishing attacks, in which fake e-mails request user credentials, are the most widely known example. Social engineering may also involve face-to-face interaction or phone calls, in which the victim is asked for specific information that can be used by the attacker. To this end, the attacker may pretend to be someone else, such as a system administrator (impersonation). Social engineering is thus fundamentally tied to the manipulation of people.

Because of this, associated experiments inevitably have elements of manipulation and persuasion, which makes them different from standard experiments. Social science studies on manipulation have a long history, and go back to the notorious Milgram experiments, in which obedience to authority was measured. In the experiment, participants were asked to administer high voltage shocks to a fictive person (Milgram 1974). In security, such "adversarial experiments" are essential to test for the feasibility of executing attacks by manipulating people. Experiments have been conducted with phishing e-mails (Finn and Jakobsson 2007), and handing keys over to strangers (Bulle´e et al. 2015). Another variant is the digital equivalent of lost letter experiments, measuring the pick-up rate of dropped USB keys (Lastdrager et al. 2013), which could contain malware. Penetration testing has also been generalised to include social engineering, allowing interaction between the penetration testers and employees of the organisation being tested

(Dimkov *et al.* 2010).

In adversarial experiments, adversarial roles need to be assigned. This can be penetration testers, people designing phishing e-mails for the experiment, or people asking a favour from others in social engineering setups. In controlled experiments, the adversarial roles are carefully scripted, to make sure the experiment is realistic, respectful, reliable, repeatable, and reportable (Dimkov *et al.* 2010). In penetration tests, the penetration testers typically have more freedom to try out scenarios that they think might work. Although such tests may show that weaknesses exist in complex systems, they are less repeatable (etc.) than controlled experiments. Other participants are typically not aware of the adversarial roles, to make sure they behave as they would in a natural setting.

3.3. Analysis

In various different contexts, social as well as technical, experiments are different when they try to take security into account. This is because one needs to consider the adversaries in the experiments. Contrary to a safety context, a single weakness may be sufficient for low security, because adversaries behave strategically, and direct their efforts towards weak spots in the system. Adversarial behaviour therefore needs to be "designed" into the experiments.

When broadening the experiments to socio-technical systems rather than technical systems, another concern is the moral acceptability of the experiments. Adversarial experiments with human subjects require special care, because (i) they can typically not be informed about the true purpose of the experiment and need to be debriefed afterwards, (ii) this may cause additional negative effects within the experiment, and (iii) the subjects may be punished for their behaviour after the end of the experiment, for example by their company, because they "did the wrong thing". Most people would no longer consider the Milgram experiment acceptable, and the acceptability question should also be asked for adversarial experiments in security.

What does the notion of adversarial roles in experiments imply for the NTaSE paradigm? Most of the experimental techniques (in a narrow sense) outlined above provide general information about the susceptibility of people to social cyber attacks (social engineering experiments), or about a specific technology when it is being developed (security testing). After deployment, the environment is often even less controlled. Still, under the assumptions of the NTaSE paradigm, there will be side- effects that only show up after deployment. In a security context, adversarial roles will be needed to find such effects. We will first investigate existing approaches for responsible deployment in cyber security, and then discuss implications for the NTaSE paradigm.

How does the cyber security community leverage the notion of adversarial roles in security experiments in the deployment context?

4. Techniques for Responsible Deployment

In this section, we take social experiments in cyber security out of the lab and into the real world. After deployment of security-sensitive technologies in cyberspace, what methods are in place to gather security-relevant information, in particular related to security weaknesses, and who play the adversarial roles in such experiments? And which conditions for responsible experimentation (Van de Poel 2009) do these techniques relate to?

4.1. Beta Versions

Many software vendors distribute so-called "beta versions" prior to official release. These versions allow interested users/early adopters to get an early experience with the new features, while at the same time knowing that they may still contain mistakes. Before the software is rolled out for a wider audience, mistakes that are found can be fixed, limiting the potential harm that such mistakes could cause. This is an instance of the condition of consciously scaling up.

For systems involving hardware, such as the cases discussed earlier, it is typically natural to scale up gradually. However, what is problematic is that security often does not play a role in the decision to scale up. For example, we have identified that security was hardly accounted for in Dutch smart grid pilots (Dechesne *et al.* 2014). If security is not taken into account, security issues may show up later, in particular because larger-scale systems are more interesting for (cyber) attackers.

4.2. Open Source

Another way to organise feedback on software is to publish the source code, i.e. the program in a human-readable programming language from which the machine-executable code is derived (Hoepman and Jacobs 2007; Payne 2002). By allowing others to see the source code, they may be able to spot and report mistakes, also those with potential security consequences. A further benefit is that actors independent of the original developer may continue the development and maintenance of the software if the original developer is unable or unwilling to do so.

Open source enables better monitoring and feedback with respect to the deployed technology. The effect of open source is typically dependent upon the willingness of external parties to inspect the code and report issues.

Many newer systems for electronic voting are required to be open source by the responsible governments, in order to improve monitoring, feedback, and continuity. Still, open source does not guarantee that this particular software is actually running on the machines, which is a tricky technical question.

4.3. Bug Bounties

Bug tracking systems are widely used for the reporting and fixing of mistakes found in software. These systems may be confidential, but are typically public in case of open source code.

In bug bounty programs, software companies pay those who find mistakes in their software, which often have security consequences (Bo¨hme 2006; Just *et al.* 2008). The idea is that users, or "white-hat" hackers, have a greater incentive to report the vulnerabilities if there is a reward attached. This becomes increasingly important in the context of black markets, in which the same vulnerabilities could be sold to "real adversaries", in particular when the vulnerability is unknown and there is no patch [zero-day, Bilge and Dumitras (2012)]. There is typically a relation between the severity of the bug and the amount paid.

More generally, guidelines for what is called "responsible disclosure" have emerged (Cavusoglu *et al.* 2005), for example when scientists find issues in widely deployed technologies, as in the OV-chipkaart case. It is considered appropriate to give the vendor a reasonable amount of time to resolve the issue before going public.

Bug bounties and responsible disclosure ensure feedback in systems with adversarial risk, when systems are too complex to test completely before deployment. The monetary aspect (payment) is particularly important when there are other markets for identified weaknesses. The payment may incentivise actors like the pressure group in the e-voting case and the Nijmegen researchers in the OV-chipkaart case, but in particular less benevolent ones, to (a) perform such studies in the first place, and (b) report the results to the problem owners first. Thereby, they make it more likely that security weaknesses are found and resolved before they can be exploited.

4.4. Red-Team–Blue-Team/Gaming

When the goal is to make system operators more aware of security issues, a red team/blue team training exercise may be considered (Mirkovic *et al.* 2008). In this model, half of the participants play attacker roles whereas the other half play defender roles, with respect to a simulated system or the real system. The attackers try to find possible attack vectors, whereas the defenders try to block those. Next to training effects, the proceedings of the red team may also point to previously unknown attack strategies, providing feedback opportunities. This approach has also been applied outside the computing domain (Grayman *et al.* 2006). The gamification aspect makes this approach attractive to participants. Still, even a low-profile "argumentation game" already provides opportunities for adversarial roles in risk assessment (Prakken *et al.* 2013).

This strategy will also increase long-term monitoring capability by raising awareness among personnel.

4.5. Honeypots

Honeypots (Kreibich and Crowcroft 2004) are parts of information systems that look like they have important functions, but in fact are only meant to deflect attackers. In this way, they serve two purposes. Firstly, they make it harder for attackers to find the real targets of their attack. Secondly, they gather information about the behaviour of the deflected attackers. This information may in turn be used to optimise the defences of the real assets.

In this way, honeypots are a means to implement the monitoring condition in computer systems with adversarial risks, and at the same time to contain hazards by increasing efforts for the adversaries. With respect to monitoring, the outcome of evaluations of the experiment is dependent on the behaviour of real adversaries, which may be both an advantage and a disadvantage. On the one hand, one does not need to question whether assigned adversarial roles are representative for the real adversary community. On the other hand, there is no guarantee that any useful information whatsoever will be obtained, as one has no control over the adversaries.

4.6. Socio-Technical Penetration Testing

Penetration tests may focus on technical vulnerabilities, such as software flaws that enable the attacker to have his own code executed, but they may also include physical attack vectors (access to the premises) or social engineering. In this case, it is important to follow ethical guidelines in order to avoid harm (Dimkov *et al.* 2010).

One particular scientific case study with such approaches involved the retrieval of distributed laptops by students, basically "stealing back" the laptops (Dimkov *et al.* 2010). In this particular context, it was both meant as an evaluation of security controls and as a training exercise (red team) for the students. In reality, professional penetration testers assist organisations in such tests.

For many systems now deployed, technical penetration tests (white-hat hacking) will be part of the procedure. However, including also the socio-technical context (physical access and social engineering) in the tests is not obvious, as such tests are generally more difficult because of social and ethical concerns.

4.7. Analysis

The abovementioned approaches show the spectrum of responsible deployment techniques in the cyber security domain. One of the main questions in relation to these different approaches is who plays the adversarial roles (**Table 1**). Such roles can be played by real adversaries (honeypots), professional testers (penetration testing), operators (red team/blue team), or users/crowdsourcing (bug bounties). This has consequences for the type of information that is provided about the system, as well as the learning effects.

Based on the above examples, we can conclude that the notion of adversarial roles is an important feature in responsible deployment in the cyber security domain. As we have argued in our earlier paper (Pieters *et al.* 2014a), the cyber security field does not generally understand these deployment techniques as instances of responsible deployment or NTaSE. For the cyber security community itself, the NTaSE paradigm thus helps to make better decisions on responsible deployment, complementing security-by-design with security-by-experiment.

A specific issue for the cyber domain is the very short update cycle of software. Because of frequent updates, experimental knowledge obtained for one software version may already lose its value when the next version becomes available. These service aspects of software require different perspectives on responsible design than traditional products and architectures (Pieters 2013). Conversely, whereas bug bounties and responsible disclosure work for software, because

Table 1. Responsible deployment methods per adversarial role type.

Adversarial role type	Example method	Example usage
Users	Beta versions	Software
Users	Open source	Software
Users	Bug bounties	Software
Users	Responsible disclosure	Software, systems
Operators	Gaming	Systems, organisations
Professional testers	Penetration testing	Systems, organisations
Real attackers	Honeypots	Computer networks

patches can be distributed quickly, they may be less effective in case of hardware, since it is not possible to change, say, the type of OV-chipkaart on short notice.

Most of the above approaches implement conditions on the design of responsible experiments: monitoring, feedback, conscious scaling up and containment of hazards. It needs to be assumed that preconditions for responsible experimentation have been fulfilled (Van de Poel 2009), specifically absence of alternative testing methods. In particular, offline tests of the systems should have been conducted where possible. While we won't discuss proportionality and controllability here [see Pieters *et al.* (2014a) for a short discussion], we do want to point out that informed consent is another important precondition.

NTaSE could learn from cyber security by showing ways to include adversarial roles in responsible deployment. Generalisations of the cyber security practices discussed in this section can provide techniques to include security and adversarial aspects in the paradigm. The obvious question is to what extent these lessons can indeed be applied to other security-sensitive domains, for example nuclear energy. Some approaches will lend themselves better than others. For example, it may be unrealistic to open source plans of nuclear facilities, but gamification of security risk in such facilities would be possible. The differences between physical, digital and social/institutional aspects of technology deployment are key here, and many questions can be asked in terms of whether responsible deployment technique X works for domain Y. Can we make bug bounties work for institutional arrangements rather than software, such as government policies? Or can we create "social honeypots" by fake employee profiles to deflect social engineering attempts?[3]

Even with such techniques, security aspects inevitably change the experimental settings of the deployment, and therefore require reconsidering the conditions for responsible deployment in NTaSE.

5. Conditions for Responsible Deployment

In our previous paper (Pieters *et al.* 2014a), we reinterpreted the conditions

---

[3]Thanks to Demetris Antoniou for this suggestion.

for responsible deployment of new technologies can be seen to apply in a cyber security setting. For example, we pointed out that controllability now also applies to adversary behaviour—an additional concern for the condition of controllability
and that scaling-up makes a system more attractive for adversaries—an additional concern for the condition of consciously scaling-up. We refer the interested reader to the abovementioned paper.

Here, we are addressing the generalisation of lessons from responsible deployment in the cyber security domain to other technologies. For general adoption of security considerations in the NTaSE framework, we now discuss the conditions for responsible deployment that we think are most interesting in the light of adversarial risks.

5.1. Informed Consent and Debriefing

Typically, openness on the goals and design of an experiment would be needed to make sure participants understand and agree with the setup. This also holds for social experiments with new technologies. However, knowledge of the design of adversarial aspects may influence the behaviour of the participants, thereby reducing the value of the experiment. For example, if I know that penetration tests will be executed, I will be less likely to comply with the assigned adversaries, because the knowledge has made me more aware.

This issue has already been discussed in the context of penetration testing, and debriefing is seen as an important solution when informed consent is not fully possible. However, that is not the whole story, because it assumes professional adversarial roles. In settings where adversarial roles are played by others than professionals, several other issues emerge. What if real attackers enter the scene? What information should participants in pilots receive in order to prevent harm to, say, their personal data? And if users or operators are allowed to play adversarial roles themselves, as in bug bounties or games, what are the limits of acceptable behaviour?

Although specific guidelines exist for penetration testing and bug bounties, the general question of how informed consent can be reconciled with adversarial roles in the experiment still needs to be answered for the case where experiments are real deployments.

5.2. Learning

Learning is an important goal of security-by-experiment.[4] In essence, one tries to induce learning from incidents in pilots (Drupsteen and Guldenmund 2014) before real large-scale incidents occur.

In this context, there is one important aspect for discussion. When we enable learning for cyber security by means of social experiments, it may not only be the defenders who learn, but also the adversaries. These can be external adversaries, or participants in the experiments who have been assigned adversarial roles and then misuse what they learnt later. This is the same issue that we considered when letting students play adversarial roles (Dimkov *et al.* 2011). In other words, aren't we providing adversaries with malicious ideas when experimenting with security-sensitive technologies, designing adversarial aspects into such experiments, and trying to learn from those? One could argue that hiding such information from adversaries would not be possible in the long run anyway, but still the experiment may require an additional condition, stating that learning by potential adversaries should be minimised in the design of the experiment.

In particular, (i) a set of criteria should be established on who is eligible for assigned adversarial roles, (ii) the adversaries should have anonymous means of raising questions and concerns during the experiment, and (iii) the adversaries should participate in an evaluation session afterwards. Finally, as it should not be assumed that the assigned roles are the only adversaries active in the experiment, a discussion on who else might have acted as an adversary and/or obtained information that would enable this in the future (NGOs, but also criminal organisations) should be part of the evaluation.

5.3. Governance

In addition, there is the question whether deployment experiments in the

---

[4]This subsection is largely the same as in the preceding conference paper (Pieters *et al.* 2014a).

cyber world need more high-level guidance. We have seen in the Dutch examples that each deployment seems to be rather ad-hoc, without clear guidance on evaluating security while scaling up. Several people have suggested to us that institutional arrangements, such as the FDA (US Food and Drugs Administration) in case of drugs, might assist in developing knowledge and guidance on deployment processes and associated experimental practices. Maybe national cyber security centres could play a similar role for cyberspace, and there may be similar options for other security-sensitive technologies. Although this is undoubtedly controversial, and it is not our core field of study, we think that the question would be worth discussing from a policy point of view.

In this light, it is interesting to observe that new regulation on cyber security is already on its way, for example in the form of the EU Directive on Network and Information Security. Responsible deployment, in the form of "security-by- experiment" could have a place in the implementation of the Directive by the member states, next to the commonly advocated "security-by-design".

6. Conclusions and Discussion

In our earlier paper (Pieters *et al.* 2014a), we highlighted for the cyber security community how NTaSE could provide a useful perspective on deployment of security sensitive technologies in the real world. The technological infrastructure and socio-political context is simply too complex for security assessments to be able to predict all potential side-effects and challenges the technology will face. So we argued, in line with NTaSE, that deployment of security-sensitive technologies should be regarded as an experiment, and discussed how conditions for responsible experimentation from the paradigm may be used to address issues that arise in the cyber security setting. For example, the condition of feedback points to the observation that feedback may also be available to adversaries. Similarly, the condition of conscious scaling up points to the observation that larger-scale systems become more attractive for adversaries.

In the current contribution, conversely, we have investigated what it would mean for the NTaSE paradigm to take security aspects into account next to safety. We did so based on an overview of experimental and responsible deployment approaches in the domain of cyber security, drawing lessons from the concept of adversarial roles in such approaches (**Table 1**). These approaches come at different costs, and may not be feasible for all types of systems. A careful selection of applicable methods is therefore needed when a deployment plan is made.

One open issue is that with respect to adversarial roles, different adversaries may be interested in different outcomes. For example, individual travellers may try to clone their own OV-chipkaart, criminal organisations may be interested in selling large numbers of cloned cards, marketing firms may be interested in profiling customers based on their electricity consumption, and hostile foreign governments may be interested in influencing the outcome of an election. The question is to what extent adversarial roles in security-by-experiment need to represent those different adversaries, and if so, how this could be achieved in practice. At the very least, we can assume that the goals of the adversarial roles will influence the outcome of the experiment, and we should be aware of this in the evaluation.

With respect to learning, a balance needs to be sought between the adversarial roles that are foreseen in providing feedback on security issues, and the increase in security risk induced by the publication of information that enables such feedback. In cyber security, there is a general tendency towards openness of designs (if not software), based on Kerckhoffs' principle. Following that principle, "security by obscurity" has been criticised as bad practice, although there have been some efforts to rehabilitate obscurity as a sensible security control (Pavlovic *et al.* 2011; Stuttard 2005).

As for the importance of adversarial aspects, these are of course broader than NTaSE only. Security-by-design and privacy-by-design require adversarial perspectives as well, as exemplified by security testing as opposed to regular testing. Even moral principles related to risk may have different interpretations in an adversarial context, for example the precautionary principle (Pieters and Van Cleeff 2009), primarily because at least part of the responsibility for the effects seems to lie with the adversary rather than the designer. Thus, the adversarial question extends to the whole risk management chain (Rios Insua *et al.* 2009), from principles to the evaluation of social experiments.

In a sense, we have also taken a somewhat adversarial role in this paper with respect to the deployment of the NTaSE paradigm, pointing to a vulnerability that may give adversaries an advantage. We hope that our adversarial role was scripted carefully enough to prevent harm, and to enable some learning indeed.

Acknowledgements

The authors wish to thank Neelke Doorn, Sean Peisert, Ibo van de Poel and Shannon Spruit for their support of initial ideas for investigating this topic. The research leading to these results has received funding from the European Union's Seventh Framework Programme (FP7/2007-2013) under grant agreement ICT-318003 (TRESPASS). This publication reflects only the authors' views and the Union is not liable for any use that may be made of the information contained herein.

Open Access

This article is distributed under the terms of the Creative Commons Attribution 4.0 International License (http://creativecommons.org/licenses/by/4.0/), which permits unrestricted use, distribution, and reproduction in any medium, provided you give appropriate credit to the original author(s) and the source, provide a link to the Creative Commons license, and indicate if changes were made.

Source: Pieters W, Hadžiosmanović D, Dechesne F. Security-by-Experiment: Lessons from Responsible Deployment in Cyberspace [J]. Science & Engineering Ethics, 2015: 1–20.

References

[1] Bilge, L., & Dumitras, T. (2012). Before we knew it: An empirical study of zero-day attacks in the real world. In Proceedings of the 2012 ACM conference on computer and communications security (pp. 833–844). New York, NY, USA: ACM. doi:10.1145/2382196.2382284.

[2] Bo¨hme, R. (2006). A comparison of market approaches to software vulnerability disclosure. In G. Mu¨ller (Ed.), Emerging trends in information and communication security (Vol. 3995, pp. 298–311). Berlin: Springer. doi:10.1007/11766155_21.

[3] Bulle´e, J.-W. H., Montoya, L., Pieters, W., Junger, M., & Hartel, P. H. (2015). The

persuasion and security awareness experiment: Reducing the success of social engineering attacks. Journal of Experimental Criminology. doi:10.1007/s11292-014-9222-7.

[4] Cavusoglu, H., Cavusoglu, H., & Raghunathan, S. (2005). Emerging issues in responsible vulnerability disclosure. In Proceedings of the workshop on the economics of information security (WEIS).

[5] Commissie onderzoek elektronisch stemmen in het stemlokaal. (2013). Elke stem telt: Elektronisch stemmen en tellen. http://tinyurl.com/nkg5m2s Ministerie van Binnenlandse Zaken en Koninkrijksrelaties.

[6] Cuijpers, C., & Koops, B.-J. (2013). Smart metering and privacy in europe: Lessons from the dutch case. In S. Gutwirth, R. Leenes, P. de Hert, & Y. Poullet (Eds.), European data protection: Coming of age (pp. 269–293). Netherlands: Springer. doi:10.1007/978-94-007-5170-5_12.

[7] Dechesne, F. (2013). (Cyber)security in smart grid pilots. http://tinyurl.com/pm4a43o TU Delft.

[8] Dechesne, F., Hadžiosmanović, D., & Pieters, W. (2014). Experimenting with incentives: Security in pilots for future grids. IEEE Security & Privacy, 12(6), 59–66.

[9] Dimkov, T., Pieters, W., & Hartel, P. (2010). Effectiveness of physical, social and digital mechanisms against laptop theft in open organizations. In Green computing and communications (GreenCom), 2010 IEEE/ACM Int'l conference on Int'l conference on cyber, physical and social computing (CPSCom) (pp. 727–732). 2010. doi:10.1109/GreenCom-CPSCom.165.

[10] Dimkov, T., Pieters, W., & Hartel, P. (2011). Training students to steal: A practical assignment in computer security education. In Proceedings of the 42nd ACM technical symposium on computer science education (pp. 21–26). New York, NY, USA: ACM. doi:10.1145/1953163.1953175.

[11] Dimkov, T., van Cleeff, A., Pieters, W., & Hartel, P. (2010). Two methodologies for physical penetration testing using social engineering. In Proceedings of the 26th annual computer security applications conference (pp. 399–408). New York, NY, USA: ACM. doi:10.1145/1920261.1920319.

[12] Drupsteen, L., & Guldenmund, F. W. (2014). What is learning? A review of the safety literature to define learning from incidents, accidents and disasters. Journal of Contingencies and Crisis Management, 22(2), 81–96. doi:10.1111/1468-5973.12039.

[13] Election Process Advisory Commission. (2007). Voting with confidence. http://www.kiesraad.nl/nl/Overige_Content/Bestanden/pdf_thema/Voting_with_confidence

[14] Finn, P., & Jakobsson, M. (2007). Designing ethical phishing experiments. Technology and Society Magazine, IEEE, 26(1), 46–58. doi:10.1109/MTAS.2007.335565.

[15] Floridi, L. (2005). The ontological interpretation of informational privacy. Ethics and Information Technology, 7, 185–200.

[16] Garcia, F. D., de Koning Gans, G., Muijrers, R., van Rossum, P., Verdult, R., Wichers

Schreur, R., et al. (2008). Dismantling mifare classic. In S. Jajodia & J. Lopez (Eds.), Computer security—ESORICS 2008 (Vol. 5283, pp. 97–114). Berlin: Springer. doi:10.1007/978-3-540-88313-5_7.

[17] Gonggrijp, R., & Hengeveld, W.-J. (2007). Studying the Nedap/Groenendaal ES3B voting computer: A computer security perspective. In Proceedings of the USENIX workshop on accurate electronic voting technology (pp. 1–1). Berkeley, CA, USA: USENIX Association. http://dl.acm.org/citation.cfm?id=1323111.1323112.

[18] Gradon, K. (2013). Crime science and the internet battlefield: Securing the analog world from digital crime. Security & Privacy, IEEE, 11(5), 93–95. doi:10.1109/MSP.2013.112.

[19] Grayman, W., Ostfeld, A., & Salomons, E. (2006). Locating monitors in water distribution systems: Red team-blue team exercise. Journal of Water Resources Planning and Management, 132(4), 300–304. doi:10.1061/(ASCE)0733-9496(2006)132:4(300).

[20] Hermans, L., & van Twist, M. (2007). Stemmachines: een verweesd dossier. Rapport van de Commissie Besluitvorming Stemmachines. Ministerie van Binnenlandse Zaken en Koninkrijksrelaties. (Available online: http://www.minbzk.nl/contents/pages/86914/rapportstemmachineseenverweesddossier consulted April 19, 2007).

[21] Hoenkamp, R., Huitema, G. B., & de Moor-van Vugt, A. J. C. (2011). The neglected consumer: The case of the smart meter rollout in the Netherlands. Renewable Energy Law and Policy Review, 4, 269–282.

[22] Hoepman, J.-H., & Jacobs, B. (2007). Increased security through open source. Communications of the ACM, 50(1), 79–83. doi:10.1145/1188913.1188921.

[23] Jacobs, B., & Pieters, W. (2009). Electronic voting in the Netherlands: From early adoption to early abolishment. In A. Aldini, G. Barthe, & R. Gorrieri (Eds.), Foundations of security analysis and design V (Vol. 5705, pp. 121–144). Berlin: Springer. doi:10.1007/978-3-642-03829-7_4.

[24] Jacobs, J. F., Van de Poel, I., & Osseweijer, P. (2010). Sunscreens with titanium dioxide (TiO2) nano-particles: A societal experiment. NanoEthics, 4(2), 103–113. doi:10.1007/s11569-010-0090-y.

[25] Just, S., Premraj, R., & Zimmermann, T. (2008). Towards the next generation of bug tracking systems. In Visual languages and Human-Centric computing, 2008. VL/HCC 2008. IEEE symposium on (pp. 82–85). doi:10.1109/VLHCC.2008.4639063.

[26] Kreibich, C., & Crowcroft, J. (2004). Honeycomb: Creating intrusion detection signatures using honeypots. SIGCOMM Computer Communication Review, 34(1), 51–56. doi:10.1145/972374.972384.

[27] Krohn, W., & Weingart, P. (1987). Commentary: Nuclear power as a social experiment-European political "fall out" from the Chernobyl meltdown. Science, Technology, & Human Values, 12(2), pp. 52–58. http://www.jstor.org/stable/689655.

[28] Lastdrager, E., Montoya, L., Hartel, P., & Junger, M. (2013). Applying the lost-letter technique to assess it risk behaviour [conference proceedings]. In Socio-technical aspects in security and trust (STAST), 2013 third workshop on (pp. 2–9). doi:10.1109/

STAST.2013.15.

[29] Lehtveer, M., & Hedenus, F. (2015). Nuclear power as a climate mitigation strategy—technology and proliferation risk. Journal of Risk Research, 18(3), 273–290. doi:10.1080/13669877.2014.889194.

[30] Milgram, S. (1974). Obedience to authority: An experimental view. London: Tavistock Publications.

[31] Mirkovic, J., Reiher, P., Papadopoulos, C., Hussain, A., Shepard, M., Berg, M., et al. (2008). Testing a collaborative DDoS defense in a red team/blue team exercise. Computers, IEEE Transactions on, 57(8), 1098–1112. doi:10.1109/TC.2008.42.

[32] Nohl, K., Evans, D., Starbug, & Plo¨tz, H. (2008). Reverse-engineering a cryptographic RFID tag. In Usenix security symposium (Vol. 28, pp. 185–193).

[33] Pavlovic, D. (2011). Gaming security by obscurity. In Proceedings of the 2011 new security paradigms workshop (pp. 125–140). New York, NY, USA: ACM. doi:10.1145/2073276.2073289.

[34] Payne, C. (2002). On the security of open source software. Information Systems Journal, 12(1), 61–78. doi:10.1046/j.1365-2575.2002.00118.x.

[35] Pieters, W. (2008). La volonté machinale: understanding the electronic voting controversy. Unpublished doctoral dissertation, Radboud University Nijmegen. http://eprints.eemcs.utwente.nl/13896/.

[36] Pieters, W. (2009). Combatting electoral traces: the Dutch tempest discussion and beyond. In P. Ryan & B. Schoenmakers (Eds.), E-Voting and identity: Second international conference, VOTE-ID 2009 (Vol. 5767). Springer.

[37] Pieters, W. (2011). The (social) construction of information security. The Information Society, 27(5), 326–335. doi:10.1080/01972243.2011.607038.

[38] Pieters, W. (2013). On thinging things and serving services: Technological mediation and inseparable goods. Ethics and Information Technology, 15(3), 195–208. doi:10.1007/s10676-013-9317-2.

[39] Pieters, W., Hadžiosmanović, D., & Dechesne, F. (2014a). Cyber security as social experiment. In Proceedings of the 2014 new security paradigms workshop. ACM.

[40] Pieters, W., Probst, C. W., Lukszo, S., & Montoya Morales, A. L. (2014b). Cost-effective- ness of security measures: A model-based framework. In T. Tsiakis, T. Kargidis, & P. Katsaros (Eds.), Approaches and processes for managing the economics of information systems (pp. 139–156). Hershey, PA, USA: IGI Global. doi:10.4018/978-1-4666-4983-5.ch009.

[41] Pieters, W., & Van Cleeff, A. (2009). The precautionary principle in a world of digital dependencies. IEEE Computer, 42(6), 50–56.

[42] Pieters, W., & Van Haren, R. (2007). Temptations of turnout and modernisation: E-voting discourses in the UK and The Netherlands. Journal of Information, Com-

munication and Ethics in Society, 5(4), 276–292.

[43] Prakken, H., Ionita, D., & Wieringa, R. (2013). Risk assessment as an argumentation game. In J. Leite, T. Son, P. Torroni, L. van der Torre, & S. Woltran (Eds.), Computational logic in multi-agent systems (Vol. 8143, pp. 357–373). Berlin Heidelberg: Springer. doi:10.1007/978-3-642-40624-9_22.

[44] Rennoch, A., Schieferdecker, I., & Großmann, J. (2014). Security testing approaches for research, industry and standardization. In Y. Yuan, X. Wu, & Y. Lu (Eds.), Trustworthy computing and services (Vol. 426, pp. 397–406). Berlin Heidelberg: Springer. doi:10.1007/978-3-662-43908-1_49.

[45] Rios Insua, D., Rios, J., & Banks, D. (2009). Adversarial risk analysis. Journal of the American Statistical Association, 104(486), 841–854. doi:10.1198/jasa.2009.0155.

[46] Robaey, Z. (2013). Who owns hazard? The role of ownership in the GM social experiment. In H. Rcklinsberg & P. Sandin (Eds.), The ethics of consumption (pp. 51–53). Wageningen: Wageningen Academic Publishers. doi:10.3920/978-90-8686-784-4_7.

[47] Standaert, F.-X., Malkin, T. G., & Yung, M. (2009). A unified framework for the analysis of side-channel key recovery attacks. In A. Joux (Ed.), Advances in cryptology-EUROCRYPT 2009 (Vol. 5479, pp. 443–461). Berlin: Springer. doi:10.1007/978-3-642-01001-9_26.

[48] Stuttard, D. (2005). Security & obscurity. Network Security, 2005(7), 10–12. http://www.sciencedirect.com/science/article/pii/S1353485805702592. doi:10.1016/S1353-4858(05)70259-2.

[49] Taebi, B., Roeser, S., & van de Poel, I. (2012). The ethics of nuclear power: Social experiments, intergenerational justice, and emotions. Energy Policy, 51(0), 202–206. http://www.sciencedirect.com/science/article/pii/S0301421512007628. (Renewable Energy in China). doi:10.1016/j.enpol. 2012.09.004.

[50] Tetri, P., & Vuorinen, J. (2013). Dissecting social engineering. Behaviour & Information Technology, 32(10), 1014–1023. doi:10.1080/0144929X.2013.763860.

[51] Tsipenyuk, K., Chess, B., & McGraw, G. (2005). Seven pernicious kingdoms: A taxonomy of software security errors. Security & Privacy, IEEE, 3(6), 81–84. doi:10.1109/MSP.2005.159.

[52] Van de Poel, I. (2009). The introduction of nanotechnology as a societal experiment. In S. Arnaldi, A. Lorenzet, & F. Russo (Eds.), Technoscience in progress. Managing the uncertainty of nanotechnology (pp. 129–142). Amsterdam: IOS Press. doi:10.3233/978-1-60750-022-3-129.

[53] Van de Poel, I. (2011). Nuclear energy as a social experiment. Ethics, Policy & Environment, 14(3), 285–290.

[54] Vuorinen, J., & Tetri, P. (2012). The order machine—the ontology of information security. Journal of the Association for Information Systems, 13(9), 695–713.

Chapter 14
Taking Back Control of Privacy: A Novel Framework for Preserving Cloud-Based Firewall Policy Confidentiality

Tytus Kurek, Marcin Niemiec, Artur Lason

AGH University of Science and Technology, Mickiewicza 30, 30-059 Krakow, Poland

Abstract: As the cloud computing paradigm evolves, new types of cloud-based services have become available, including security services. Some of the most important and most commonly adopted security services are firewall services. These cannot be easily deployed in a cloud, however, because of a lack of mechanisms preserving firewall policy confidentiality. Even if they were provided, the customer traffic flowing through the Cloud Service Provider infrastructure would still be exposed to eavesdropping and information gaining by performing analysis. To bypass these issues, the following article introduces a novel framework, known as the Ladon Hybrid Cloud, for preserving cloud-based firewall policy confidentiality. It is shown that in this framework, a high level of privacy is provided thanks to leveraging an anonymized firewall approach and a hybrid cloud model. A number of optimization techniques, which help to further improve the Ladon Hybrid Cloud privacy level, are also introduced. Finally, analysis performed on the framework shows that it is possible to find a trade-off between the

Ladon Hybrid Cloud privacy level, its congestion probability, and efficiency. This argument has been demonstrated through the results of conducted experiments.

Keywords: Firewall, Cloud Computing, Privacy, Bloom Filter

1. Introduction

During the past couple of years, the cloud computing paradigm has evolved from an experimental approach to hosting *Information and Communications Technology* (ICT) services in a distributed systems environment, to a leading trend in the ICT market[1]. Thanks to this, most types of services are available in a cloud today, including security services. The model of hosting security services in a cloud is referred to as *Security as a Service* (SecaaS)[2].

Following the needs of business which keep increasing due to the expansion of the technology, many ICT companies, including leaders such as AT&T with its *Network-Based FireWall Services* (NBFWS)[3], have already begun offering security services in a cloud. These include firewall services, *Intrusion Prevention System* (IPS) services, e-mail filtering, and web filtering. In most cases, including AT&T NBFWS and Cloudera *Enterprise Services Cloud* (ESC)[4], the security services are deployed by leveraging a hybrid cloud model with customers connected to the *Cloud Service Provider* (CSP) via a secure *Virtual Private Network* (VPN) connection. In such a system, most of the customer security services are hosted in a cloud, while the basic security infrastructure, responding to last mile attacks for example, remains on its premises. The on-premises infrastructure can be managed by the CSP or the customer. Alternatively, a hybrid management system can be applied with the CSP being responsible for the on-premises infrastructure installation, its initial configuration, monitoring, etc., and the customer being responsible for the entire security policy management.

One of the core security services adopted by the vast majority of organizations are firewall services. It is hard to imagine an enterprise, government unit, university, or even home business running its network services without being protected by a firewall. Thanks to such technologies as AT&T NBFWS or Virtela ESC, these can be outsourced to the cloud, resulting in significantly reduced management overhead, decreased *Total Cost of Ownership* (TCO), improved business agility, and so on[5]. However, because of a lack of mechanisms preventing the CSP from having an insight into the customer' firewall policy, there are still issues of information confidentiality and privacy[6]–[9].

In addition, another threat is information gaining by traffic eavesdropping and analysis. Since in a hybrid cloud SecaaS model all the traffic flows unencrypted through the CSP infrastructure and there are no mechanisms protecting against eavesdropping, sensitive information such as that regarding allowed *Internet Protocol* (IP) addresses can be easily gained by the CSP based on traffic analysis. This exposes a serious vulnerability of such systems, as according to recent reports, most data harvesting events take place during transit[10][11]. In addition, although the CSP itself is obliged by contract to maintain information confidentiality, according to research shown in[12][13], employees would not hesitate to steal such sensitive information if laid off, for example.

This leads to the following conclusion. Until mechanisms preserving firewall policy confidentiality and preventing information gaining by traffic eavesdropping and analysis are designed, organizations will not be able to run their firewall services in a cloud in a way that is sufficiently confidential to preserve their privacy. This problem seems to be an unresolved security hole, as only one solution has been proposed so far.

Referred to as the Ladon framework by its authors, it attempts to preserve cloud-based firewall policy confidentiality[14]. It is supposed to achieve it by leveraging an anonymized firewall in the public cloud. In such frameworks, the CSP is prevented from having an insight into the original firewall policy. However, the Ladon framework provides no mechanism for preventing the CSP from deducing the original firewall policy by traffic eavesdropping and analysis. As the final decision on network packets is still known to the CSP, it can determine the original firewall policy over time. As such, the privacy of cloud-based firewall policies cannot be preserved using Ladon.

Motivated by the above observations, the following contributions are made in this paper. Firstly, a novel framework for preserving cloud-based firewall policy confidentiality, known as the Ladon Hybrid Cloud, is introduced as an extension and augmentation to the regular Ladon framework. It is shown that by introducing the purposefulness of packet decision uncertainty, the main drawback of Ladon—the risk of firewall deanonymization by packets eavesdropping and analysis—is significantly reduced. Additional optimization techniques which help improve Ladon Hybrid Cloud privacy level based on the type of firewall policy in use are also introduced. It is shown that after deploying the Ladon Hybrid Cloud according to best practices, the risk of information gaining by the CSP does not differ significantly from that of a regular *Internet Service Provider* (ISP). Finally, by performing mathematical framework analysis, the results of which have been confirmed through the results of the experiment performed, the article shows that it is possible to find a trade-off between the Ladon Hybrid Cloud privacy level, its congestion probability, and efficiency.

The rest of the paper proceeds as follows. First of all, related work is reviewed in Sect. 2. Section 3 includes a presentation of the Ladon framework. In Sect. 4, a novel framework for preserving cloud-based firewall policy confidentiality, known as the Ladon Hybrid Cloud, is introduced, along with its optimization techniques. In Sect. 5, all the mathematical framework analyses are shown. Their experimental results follow in Sect. 6. All observed Ladon Hybrid Cloud limitations and directions for future work are noted in Sect. 7. Finally, Sect. 8 contains the conclusions.

2. Related Work

Khakpour and Liu[14] presented the Ladon framework as a first step toward cloud-based firewalling. The Ladon leverages an anonymized firewall based on a set of *Bloom Filter Firewall Decision Diagrams* (BFFDDs) which are compiled from regular *Firewall Decision Diagrams* (FDDs)[15] in which edge sets are replaced by *Bloom Filters* (BFs)[16]. Thanks to the merging of these elements that are explained in detail in the next section, regular *Access Control List* (ACL) rules are transformed into a structure which is still visible to the CSP, although it does not provide it with straightforward information regarding the original ACL structure. In such a framework, the ACL rules of the customer's firewall can neither be directly read by the CSP, nor easily cracked using brute-force techniques. However, as described below, these can be determined by packets eavesdropping and analysis.

Other studies related to the topic of this article are those related to moving target defense. In[17], the authors have studied techniques of substituting different targets for any given request in order to create a dynamic and uncertain attack surface area of a given system. This enabled them to demonstrate that such systems are less vulnerable and more secure. The Ladon Hybrid Cloud framework presented in this article also intentionally introduces uncertainty to the attack surface area; however, it achieves this by using a BF false-positive rate, as explained below. All targets remain unchanged for all given requests over time.

3. From ACL to Ladon

An FDD, presented by Gouda and Liu in[16], is a mathematical structure which is a formal firewall representation. In fact, the FDD transforms a regular firewall policy based on a set of *Access Control Entry* (ACEs) into a tree where packets pass from top to bottom, with particular packet fields being examined at each level. Depending on its particular packet field value, the packet is directed to one of the edges, forming a decision path which finally takes one of the two possible decisions: permit or deny.

This concept is shown in **Figure 1**. Suppose that the firewall takes its final decision based on the source and destination IP addresses alone. The FDD then consists of two levels: one representing the source IP address and the other representing the destination IP address. The edge sets are calculated based on the corresponding ACL. For example, for a packet sourced at 10.10.10.10 and destined for 192.168.192.168, which fits the first ACE in the ACL, its source IP address is examined first on the $F1$ node. 10.10.10.10 fits the 0.0.0.0/1 set, so the packet is passed to the $e11$ edge, where its destination IP address is examined on the $F21$ node. Because 192.168.192.168 fits the 128.0.0.0/1 set, the packet is passed to the $e22$ edge, resulting in a deny decision.

Unlike in a regular firewall, where a packet is examined as a whole by testing it against ACEs from top to bottom until the first match is found, the FDD takes a completely different approach. It splits the packet into fields and examines each field independently on particular tree levels. The resulting path leads to a single, ultimate decision. Sample FDD implementation known as 'Policy Trie' was also presented independently of Gouda and Liu's work by Fulp and Tarsa in[18].

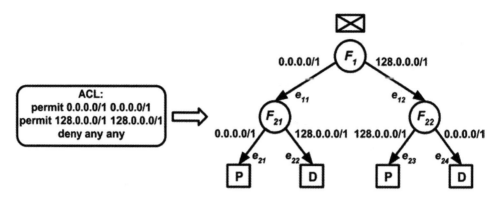

Figure 1. FDD construction.

A BF, presented by Bloom in[16], is a mathematical probabilistic data structure which is used to test whether an object is a member of a set in a time-efficient manner. Mathematically, a BF is a bit array with a size of m which is generated by calculating k-independent hash functions for each of n elements of the set. For each of the results, the corresponding index in BF is set to 1. To check whether an element is a member of the original set, the same hash functions are calculated and corresponding indexes of the BF are checked. If at least one of them is 0, the element is not a member of the original set. If all of them are 1, the element may be a member of the original set. The above indicates that a BF may result in false positives. Moreover, the value of false-positive probability, also known as the BF false-positive rate, can be calculated based on the k, m, and n parameter[19].

This is shown in **Figure 2**. Suppose that a BF with a size of m = 4 using k = 2 hash functions (h_1, h_2) represents a set containing n = 2 elements (s_1, s_2). The BF is generated (case 1) by calculating hash functions for each of the elements ($h_1(s_1)$, $h_2(s_1)$, $h_1(s_2)$, $h_2(s_2)$) and setting up corresponding indexes in the BF (b_0, b_1, b_2 or b_3) to 1. Suppose that the hash function results are as follows: $h_1(s_1) = 0$, $h_2(s_1) = 3$, $h_1(s_2) = 3$, $h_2(s_2) = 2$. The following BF indexes are set to 1 as a result: b_0, b_2, b_3. At this point, the BF can be used to test object presence in the source set (case 2). For a given object (s_x) to be tested, the same hash functions are calculated ($h_1(s_x)$, $h_2(s_x)$) and corresponding BF indexes are examined to see whether they are set to 1. Suppose that the hash function results are as follows: $h_1(s_x) = 0$, $h_2(s_x) = 1$. It is then clear that the object is not an element of the source set—b_1 is not set to 1. On the other hand, if the hash function results for some other object (s_y) to be tested

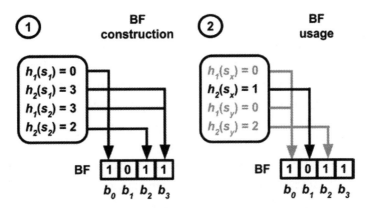

Figure 2. BF construction and usage.

are as follows: $h_1(s_y) = 0$, $h_2(s_y) = 2$, the object is considered to be a member of the source set with some permissible misclassification probability, known as the BF false-positive rate. While the object s_y is considered to be a member of the source set, it actually is not.

So far, brick level structures which build up BFFDD have been covered. Based on them, the BFFDD definition can be explained as follows. According to[14], the BFFDD is a data structure formed from regular FDD where, for a given edge, the edge set is represented by a BF. Because it is in the nature of the BF that it may result in false positives, ambiguities may occur in BFFDD, leading to multiple decision paths and as a result to multiple decisions. To eliminate such ambiguities N, independent BFFDDs are implemented and executed simultaneously. The resulting decision paths are then compared looking for a single, common path which leads to a common, final decision.

The concept of BFFDD and its construction algorithm is shown in **Figure 3**. Suppose that the firewall takes its final decision based on the source IP address only. The original ACL is then transformed into an FDD with one level only. Next, the edge sets e_1(0.0.0.0/1) and e_2(128.0.0.0/1) are transformed into BF1 and BF2 correspondingly. The process of BF2 construction is shown within the gray round rectangle. However, the BF shown in this example has a size of $m = 4$ and uses $k = 2$ hash functions; these are obviously much greater in a real scenario.

Section 1 shows a standard business model for hosting firewall services in the cloud. Outsourced firewall services are hosted in the public cloud located in the data center owned and managed by the CSP. All traffic destined to the customer first enters the public cloud, which is connected with customer premises via a secure VPN connection. The technology used to deliver firewall services is not visible to the customer. Assume that it is based on a set of independent BFFDDs as described above. A framework of cloud-based firewall services based on BFFDDs is shown in **Figure 4**.

Packets permitted on customer premises, referred to as 'good packets' in the rest of the article, are represented there by plain envelopes. In turn, packets denied on customer premises, referred to as "bad packets" in the rest of the article, are represented there by striped envelopes. All packets enter the public cloud first (step 1) where bad packets are discarded (step 2). Next, good packets are sent to customer LAN (step 3). Such framework was referred to as the Ladon framework by its authors in[14].

By implementing and testing Ladon in a live environment, Khakpour and Liu demonstrated that it is an effective framework for the outsourcing of the firewall services. It was also shown that any attempts to deanonymize the BFFDD can be extremely time-consuming.

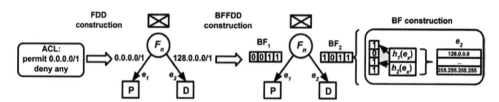

Figure 3. BFFDD construction.

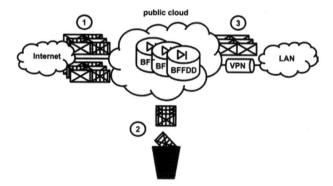

Figure 4. Ladon framework.

4. Framework Design

Assume there is Ladon framework implemented with the firewall services hosted by an honest-but-curious CSP wishing to get an insight into the customer's firewall security policy. The CSP has an insight into the BFFDDs delivering the firewall services, but as these are just binary structures representing the BFs on the edges of the original FDDs used to generate the BFFDDs, it gives it no information on the original firewall security policy. However, the CSP has an insight into whether a packet flowing via the Ladon framework is permitted on the customer side. This is because only packets permitted on the customer side are flowing via the VPN connection between the CSP and the customer. Packets denied on the customer side are discarded in the public cloud. Although the VPN connection is encrypted, the CSP, as one of its initiators, can intercept packets entering the VPN.

Therefore, although the regular Ladon framework resolves the issue of firewall policy confidentiality in such a way that it cannot be directly read, the final decision for a packet is still known by the CSP. This means that after an appropriately long period of time, the CSP can build up an almost full knowledge base regarding packets which are permitted on the customer side. While the underlying ACL structure is protected, traffic flowing between the CSP and the customer can still be easily eavesdropped and analyzed to obtain information about good packets.

Such a framework allows the CSP to bypass the reconnaissance phase, and most of the scanning phase of an attack, which according to[20], can significantly reduce the time required to perform the attack. To address this issue, a novel framework, called the Ladon Hybrid Cloud, which introduces the purposefulness of packet decision uncertainty based on BFFDD and a hybrid cloud model, was designed and is presented in the following section of the article.

4.1. Introducing the Ladon Hybrid Cloud

In a regular Ladon framework, based on a set of BFFDDs, the final decision for a packet is always certain. However, in a single BFFDD, ambiguities may occur as a result of BF false positives. This is because a BF false positive leads to a situation where the packet field matches both edges on a particular BFFDD level, resulting in multiple decision paths, and as a result in multiple decisions for the packet.

Instead of eliminating such ambiguities by using a set of BFFDDs and searching for a single, common path with a single, common decision, the Ladon Hybrid Cloud takes a completely different approach. It leverages a single BFFDD instead and intentionally allows some of the packets to result in multiple decisions. This approach leverages a hybrid cloud model, as it is a leading trend in the SecaaS market[3][4], with BFFDD in a public cloud and a regular firewall in a private cloud. In order to simplify the management of the private cloud, hybrid management can be applied.

In this novel framework, after passing through BFFDD in a public cloud on the CSP side, packets resulting in certain deny decisions are directly discarded, while those resulting in certain permit decisions are sent directly to the customer over a trusted network. Additionally, packets resulting in multiple decisions are sent to the private cloud on customer premises over an untrusted network for additional filtering. Segregation of packets resulting in certain permit decisions from those resulting in multiple decisions can be organized based on the *Virtual Local Area Network* (VLAN) logic.

The Ladon Hybrid Cloud concept is shown in **Figure 5** inside the gray round rectangle. Good packets are represented there by plain envelopes, while bad packets are represented by striped envelopes. All packets enter the public cloud first (step 1) where those resulting in certain deny decisions are directly discarded (step 2). Those resulting in certain permit decisions are sent directly to the customer *Local Area Network* (LAN) over a trusted VLAN, represented by a continuous line (steps 3 and 7). Finally, packets resulting in multiple decisions are sent to the private cloud for additional filtering over an untrusted VLAN, represented by the dotted line (step 4). The private cloud performs additional filtering by discarding of the rest of the denied packets (step 5) and sends permitted packets into the customer LAN (step 6).

Although customer traffic is still exposed to eavesdropping and analysis by the CSP in the Ladon Hybrid Cloud, the amount of information carried by particular packets is reduced compared to a regular Ladon framework. This is because

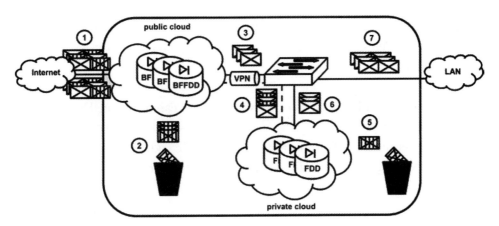

Figure 5. Ladon hybrid cloud.

some packets result in multiple decisions after passing through BFFDD which causes the knowledge base built by the hostile CSP not to be 100% accurate. Moreover, assuming that over time the number of bad packets with different packet headers is growing, packets diversity causes the knowledge base to grow too. The above assumption takes into account *Distributed Denial of Service* (DDoS) attacks, for example, during which most of the source IP addresses are new[21].

An inaccurate and constantly growing knowledge base regarding packets perceived by the CSP as good packets causes the CSP not to fully trust it and forces it to perform additional reconnaissance and scanning in order to obtain accurate information. Moreover, over time, demand for resources required to store and analyze the data can significantly increase. It is therefore clear that the Ladon Hybrid Cloud eliminates the main drawback of a regular Ladon framework: the risk of firewall deanonymization by packets eavesdropping and analysis. This is achieved by introducing the purposefulness of packet decision uncertainty. The following sections cover additional framework optimization mechanisms which help increase this uncertainty even further and as a result provide a high level of privacy.

4.2. Ladon Hybrid Cloud Optimization

Although the BFFDD may result in multiple decisions for some of the packets, for the others, a final decision remains certain. Part of the knowledge base maintained by the hostile CSP will therefore always be accurate. It is possible to eliminate this vulnerability, however, by redesigning the BFFDD in such a way that it always results in multiple decisions for either good or bad packets, based on the adopted firewall policy type.

In the real world, two types of firewall policies can be adopted based on organization requirements:
- Closed: Permitting only a specific subset of traffic and denying the rest,
- Open: Denying only a specific subset of traffic and permitting the rest.

For inbound traffic flow, considered in this article, most organizations apply the closed firewall policy rather than the open one, because it minimizes the risk of malicious traffic passing through. In such a case, the BFFDD is redesigned in such a way that it always results in multiple decisions for good packets. As closed firewall policy is a leading trend in most of the organizations today, it will be used as an example in further arguments in this article. Likewise, in an organization applying open firewall policy, the BFFDD can be redesigned so it always results in multiple decisions for bad packets accordingly.

As has been mentioned, in the case of a closed firewall policy type, the BFFDD is updated to always result in multiple decisions for good packets. The only packets that may still result in certain decisions are therefore bad packets. The framework is designed in this way, because when adopting closed firewall policy type, good packets carry significantly more information for the CSP regarding the original ACL structure compared to bad packets. This is because a characteristic of closed firewall policy type is that the subset of traffic which is permitted is much smaller than the subset of traffic which is denied.

Figure 6 represents a BFFDD with one level and all the cases that it can result in:
- Case 1: Certain permit decision for good packets,
- Case 2: Multiple decisions for good packets,
- Case 3: Certain deny decision for bad packets,
- Case 4: Multiple decisions for bad packets.

As mentioned above, case 1 should be fully eliminated. To achieve this, a regular BFFDD is compiled first and then tested against all good packets. For

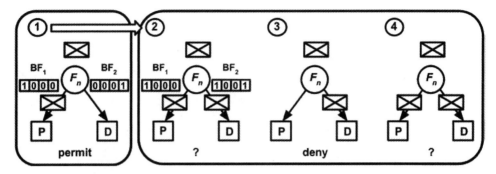

Figure 6. BFFDDCF construction.

those resulting in certain permit decisions (case 1), the BF representing the set of the edge that leads to a deny decision (BF2 in this case) is updated so that it results in a forced false positive. As a consequence, multiple decision paths are applied to all good packets (case 2) leading to multiple decisions applied to all of them. In other words, the redesigned BFFDD eliminates case 1 by transforming it into case 2, resulting in three possible cases (2, 3, and 4) shown inside the gray round rectangle.

The above transformation can be performed on any BFFDD level; however, for analysis and implementation simplicity, it is assumed that it is performed on the last level representing the last examined packet field. Such a redesigned BFFDD will be referred to as BFFDDCF (*Bloom Filter Firewall Decision Diagram for Closed Firewalls*). Likewise, such a redesigned Ladon Hybrid Cloud which leverages a BFFDDCF in a public cloud will be referred to as *Ladon Hybrid Cloud for Closed Firewalls* (LHCCF).

As a consequence, the traffic flowing between public and private clouds consists of all good packets and some bad packets, while multiple decisions are applied to all of them by the BFFDDCF. This leads to a situation where all packets flowing between the CSP and the customer go via an untrusted VLAN represented by the dotted line, so the traffic segregation engine can be fully eliminated from the LHCCF, as shown in **Figure 7**. As all good packets require additional filtering in the private cloud, there is no traffic flowing between public and private clouds over the trusted VLAN represented by the continuous line. It is clear that in this case, the private cloud needs to process more packets.

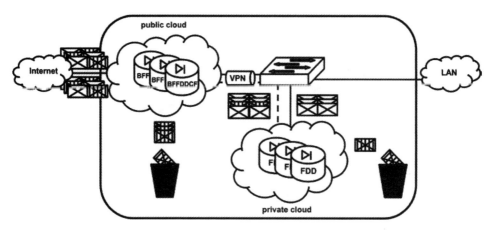

Figure 7. Ladon hybrid cloud for closed firewalls.

In such a framework, the CSP cannot draw any additional information from the traffic, except of a fact that part of it is permitted on customer premises. However, the CSP can still maintain a certain knowledge base regarding packets which are explicitly denied in the BFFDDCF, and this gives it a very limited amount of information, as has been stated before. At this point, the amount of information which the CSP can extract by performing traffic eavesdropping and analysis does not differ greatly from that of the regular ISP which the customer is connected to. The original firewall policy cannot be directly read by the CSP or be assumed by performing traffic eavesdropping and analysis. While the first of these two features is provided by the regular Ladon framework, the second is provided by the Ladon Hybrid Cloud only.

5. Framework Analysis

Because of the uncertainty of packet decision making in the LHCCF, the rate of traffic flowing between public and private clouds is increased compared to a regular Ladon framework. This is because the traffic consists not only of good packets, but also of some bad packets which result in multiple decisions after passing through the BFFDDFC. The factor of bad packets which are transmitted is determined by the probability that the BFFDDCF results in multiple decisions, referred to as "BFFDDCF multiple decision probability" in the rest of this article. Furthermore, the BFFDDCF multiple decision probability is a result of the false-positive rates of the particular BFs that make it up. The following section shows how the parameters of the particular BFs in the BFFDDCF can be used to control the rate of bad packets flowing between public and private parts of the LHCCF. As a result, an overall rate of traffic flowing between the CSP and the customer can be controlled as well and a trade-off can be found between the LHCCF privacy level, its congestion probability, and efficiency.

5.1. Controlling Traffic Rate

So far, it has been shown that the greater the value of the BFFDDCF multiple decision probability is, the higher the privacy level is provided by the LHCCF. This is because increasing the BFFDDCF multiple decision probability increases the number of bad packets that result in multiple decisions and, as a result, decreases the amount of information carried by particular packets.

But can the BFFDDCF multiple decision probability be increased without limits? What is its effect on LHCCF congestion probability and efficiency? The following analysis attempts to answer these questions. Let us define:

- p—BFFDDCF multiple decision probability
- r—Rate of traffic at the public cloud entrance (packets/s)
- s—Rate of traffic at the public cloud exit (packets/s)
- g—Rate of traffic consisting of good packets (packets/s)
- u—Good packet ratio
- $SMAX$—Throughput of the private cloud
- pc—LHCCF congestion probability
- e—LHCCF efficiency

These are shown in **Figure 8**.

Based on the above, the rate of traffic at the public cloud exit is:

$$s = g + p(r - g) \quad (1)$$

as it consists of all good packets and those bad packets that result in multiple decisions after passing through the BFFDDCF. It is then possible to control the rate of

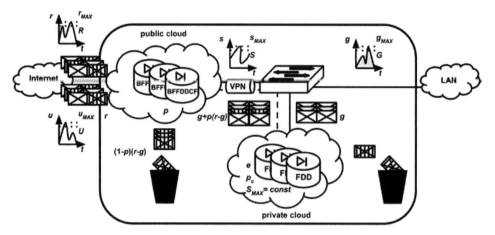

Figure 8. Traffic flow in the LHCCF.

traffic at the public cloud exit while knowing the value of the BFFDDCF multiple decision probability, the rate of traffic at the public cloud entrance, and the rate of traffic consisting of good packets.

By expressing the ratio, referred to as "good packet ratio" in the rest of the article, of the rate of traffic consisting of good packets and the rate of traffic at the public cloud entrance as:

$$u = \frac{g}{r} \tag{2}$$

Formula 1 can be transformed to a function of good packet ratio and is:

$$s = ur + p(r - ur) \tag{3}$$

Again, it is then possible to control the rate of traffic at the public cloud exit knowing the value of the BFFDDCF multiple decision probability, the rate of traffic at the public cloud entrance, and the good packet ratio.

The rate of traffic at the public cloud exit is also the rate of traffic at the private cloud entrance. It is therefore important to control the rate of traffic at the private cloud entrance to ensure that it never exceeds its throughput. This is because a higher traffic rate results in private cloud congestion and, as a result, in congestion of the whole LHCCF. The equation from Formula 3 can be then replaced with the inequality:

$$S_{MAX} \geq ur + p(r - ur) \tag{4}$$

Therefore, by knowing the values of the rate of traffic at the public cloud entrance and good packet ratio, it is then possible to adjust the BFFDDCF multiple decision probability value so that the rate of traffic at the private cloud entrance never exceeds its throughput.

The values of the rate of traffic at the public cloud entrance and good packet ratio change over time; however, the BFFDDCF multiple decision probability cannot change over time, because of computational limitations as discussed below. Therefore, while designing BFFDDCF, some constant values of the rate of traffic at the public cloud entrance (R) and the good packet ratio (U) need to be assumed instead.

Those could be taken based on maximum values of particular variables (r_{MAX}, u_{MAX}) observed over time to ensure LHCCF congestion probability on the lowest possible level. However, as shown later in this section, high values of R and U enforce a lower BFFDDCF multiple decision probability and then, as a result, a lower LHCCF privacy level. As such, instead of selecting maximum values, it is better to relax them somewhat (e.g., by selecting average values). On the other hand, lower R and U values result in a higher privacy level while allowing LHCCF congestion during r and u peak periods. This is because instantaneous r and u values exceeding R and U constants result in the rate of traffic at the private cloud entrance exceeding its throughput.

Another parameter to consider when designing the LHCCF is a time period for which constant values of R and U are selected. As traffic statistics tend to be similar over recurrent time periods, it may be worth running different BFFDDCFs during these periods (e.g., one BFFDDCF during the day and another during the night). Thus, constant values of R and U should be selected, and the following analysis should be performed for each recurrent time period independently when designing the LHCCF.

Assuming that statistics of the rate of traffic at the public cloud entrance are known, it is then possible to select a constant value:

$$R \approx r_{MAX} \qquad (5)$$

while ensuring that the instantaneous value of r variable exceeds R constant with a probability of:

$$p = P(r > R) \qquad (6)$$

Likewise, assuming that good packet ratio statistics are also known, it is then possible to select a constant value:

$$U \approx u_{MAX} \qquad (7)$$

while ensuring that the instantaneous value of the u variable exceeds the U constant with a probability of:

$$pu = P(u > U) \qquad (8)$$

Based on approximations made in Formulas 6 and 8, the LHCCF congestion probability is as follows:

$$\begin{aligned} pc &= P(r > R \cup u > U) \\ &= pr + pu - P(r > R \cap u > U) \end{aligned} \qquad (9)$$

It can also be expressed as follows:

$$pc = P(s > S_{MAX}) = P\big((ur + p(r - ur)) \geq S_{MAX}\big) \qquad (10)$$

Assuming that both fragmentary probabilities and the probability of a conjunction in Formula 9 are known based on the traffic analysis, it is then possible to control the LHCCF congestion probability.

Moreover, it can be seen that the higher the value of the BFFDDCF multiple decision probability, the higher the LHCCF congestion probability. On the other hand, it has already been shown that the higher the value of the BFFDDCF multiple decision probability, the higher the LHCCF privacy level. Thus, finding a trade-off between the LHCCF privacy level and its congestion probability is a matter of selecting such a value of the BFFDDCF multiple decision probability that satisfies both.

Let us also define the LHCCF efficiency as a ratio of the rate of the traffic at the public cloud entrance and the rate of the traffic at the public cloud exit. It is then:

$$e = \frac{r}{s} = \frac{1}{u + p(1-u)} \qquad (11)$$

However, as the R constant value has been assumed instead, the LHCCF efficiency is constant too and is:

$$E = \frac{1}{U + p(1-U)} \qquad (12)$$

It is then possible to control the LHCCF efficiency while knowing the value of the BFFDDCF multiple decision probability.

Moreover, it can be seen that the higher the value of the BFFDDCF multiple decision probability, the lower the LHCCF efficiency. Thus, finding a trade-off between the LHCCF privacy level and its efficiency is a matter of selecting such a value of the BFFDDCF multiple decision probability that satisfies both. Finally, finding a trade-off between the LHCCF privacy level, its congestion probability and efficiency is a matter of selecting such a value of the BFFDDCF multiple decision probability that satisfies all three parameters.

By replacing the r and g variables in Formula 4 by the R and U constants from Formulas 5 and 7, it is transformed thus:

$$S_{MAX} \geq UR + p(R - UR) \qquad (13)$$

therefore, the BFFDDCF multiple decision probability is:

$$p \leq \frac{S_{MAX} - UR}{R(1-U)} \qquad (14)$$

It is then possible to control the rate of traffic at the private cloud entrance by adjusting the BFFDDCF multiple decision probability value while knowing private cloud throughput and base statistical traffic parameters. These include the rate of traffic at the private cloud entrance and the good packet ratio, both on an agreed level.

As shown in the next subsection, the BFFDDCF multiple decision probability can take any value between U and 1. By taking advantage of this property $(p \in (U,1])$ and by substituting edge values into Formula 10, it can be seen that it applies to the $R \in \left[S_{MAX}; \frac{S_{MAX}}{U(2-U)} \right)$ interval only. This is because for the rate of traffic at the public cloud entrance being lower than the private cloud throughput, the BFFDDCF multiple decision probability can take any value from the $(U; 1]$ interval. The generalized Formula 14 is then:

$$P \leq \begin{cases} 1 & \text{for } R \in [0; S_{MAX}) \\ \dfrac{S_{MAX} - UR}{R(1-U)} & \text{for } R \in \left[S_{MAX}; \dfrac{S_{MAX}}{U(2-U)} \right) \end{cases} \qquad (15)$$

A plot of the maximum allowable BFFDDCF multiple decision probability of the R argument including the U parameter is shown in **Figure 9**.

5.2. BF False-Positive Rate: The Core Control Engine

So far, it has been shown that the BFFDDCF multiple decision probability can be used to control the rate of traffic flowing between the CSP and the customer. The following section takes a closer look at the BFFDDCF multiple decision probability itself, which is the result of false-positive rates of particular BFs which make it up.

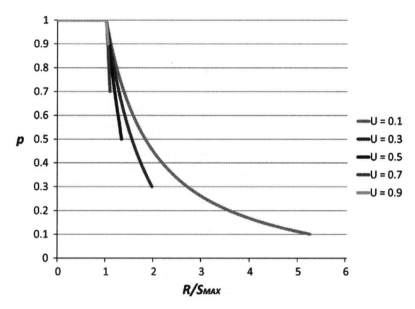

Figure 9. Maximum allowable BFFDDCF multiple decision probability.

Based on studies conducted in[18], a classic FDD consists of five fields which include protocol, source IP, source port, destination IP, and destination port. Suppose that the BFFDDCF presented in the following example consists of the same five fields and has $j = 5$ levels as shown in **Figure 10**. Let us also define:

- p_g—BFFDDCF multiple decision probability for good packets,
- p_b—BFFDDCF multiple decision probability for bad packets,
- p_j—Comprehensive multiple decision probability on the j-th BFFDDCF level,
- f_{ij}—False-positive rate of the i-th BF on the j-th BFFDDCF level.

These are represented in **Figure 10** as values associated with particular edges and values associated with particular summary buckles, respectively.

The BFFDDCF multiple decision probability is a sum of the BFFDDCF multiple decision probability for good packets multiplied by the good packet ratio and the BFFDDCF multiple decision probability for bad packets multiplied by the bad packet ratio; it is then:

$$p = Up_g + (1-U)p_b = U + (1-U)p_b \qquad (16)$$

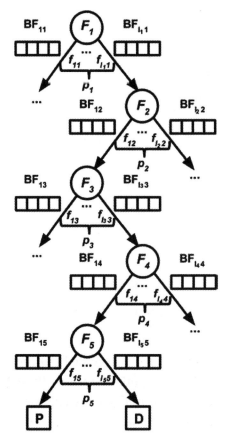

Figure 10. BFFDDCF internal structure.

This is because good packets, which make up $U * 100\%$ of the total traffic, always result in multiple decisions.

The BFFDDCF multiple decision probability for bad packets is a comprehensive multiple decision probability on at least one of its j levels and is then:

$$p_b = 1-(1-p_1)(1-p_2)(1-p_3)(1-p_4)(1-p_5) \qquad (17)$$

In other words, it is a probability of an opposite event that ambiguity occurring on none of j levels. No ambiguities result in comprehensive multiple decision probabilities on each BFFDDCF level equal to 0 and in the BFFDDCF multiple decision probability equal to 0 as well. On the other hand, even a single ambiguity on any of j levels leads to multiple decision paths and to multiple decisions. The above is an internal FDD characteristic[15].

On each of the five levels, multiple decisions for bad packets occur when at least one of the BFs on this level results in a false positive. Furthermore, all BFs on the same level cannot result in a false positive at the same time—at least one of them must result in a true positive. Because of this, the comprehensive multiple decision probability for bad packets on the j-th level is as follows:

$$p_j = 1 - \prod_{i=1}^{Ij} f_{ij} - \prod_{i=1}^{Ij}(1 - f_{ij}) \tag{18}$$

In other words, it is a probability of an opposite event that none or all of the BFs on the j-th level result in a false positive.

Formula 18 applies to BFFDDCFs without edges with a set containing all possible values of a particular packet field (known as 'default edges'[14]). This is because in the BFFDDCF with default edges on a particular level, ambiguities cannot occur on that level, as the packet field always matches the default edge while it never matches an opposite edge. As a result, comprehensive multiple decision probability on that level is equal to 0, which can then result in the BFFDDCF multiple decision probability being equal to 0 as well. To ensure an absence of default edges in BFFDDCF when designing the ACL used to compile it, it is important to avoid general statements, but to be as strict as possible instead. However, if the above cannot be completed because of the internal ACL structure (e.g., a decision is taken regardless of the destination IP address), some redundant statements can be added which eliminate default edges while not breaking ACL structure. This can be completed by modifying the ACL to work differently for destination IP addresses which do not actually belong to the customer.

By putting Formulas 17 and 18 into Formula 16, the BFFDDCF multiple decision probability is then:

$$p = U + (1-U)\left(1 - \prod_{j=1}^{5}\left(\prod_{i=1}^{Ij} f_{ij} + \prod_{i=1}^{Ij}(1 - f_{ij})\right)\right) \tag{19}$$

Now suppose that the presented BFFDDCF consists of J levels instead of five. By performing analysis as shown above, it is possible to find a generalized version of Formula 19 for the *J*-level BFFDDCF multiple decision probability which is:

$$p = U + (1-U)\left(1 - \prod_{j=1}^{J}\left(\prod_{i=1}^{Ij} f_{ij} + \prod_{i=1}^{Ij}(1-f_{ij})\right)\right) \quad \bullet \quad (20)$$

The above equation cannot be solved by analytical methods. However, knowing that the BF false-positive rate is[16][19]:

$$f \approx \left(1 - \left(1 - \frac{1}{m}\right)^{kn}\right)^k \quad (21)$$

it is possible to design BFFDDCF so that its maximum allowable multiple decision probability is as close to the value from Formula 12 as possible by selecting appropriate values of the BFs that build it up. As n, which represents the edge set size, is known, these parameters include the number of hash functions (k) and BF size (m).

Consequently, it is possible to design BFFDDCF by selecting appropriate values of BF parameters so that the rate of traffic at the private cloud entrance can be controlled while knowing the private cloud throughput, base statistical traffic parameters, and the particular edge set sizes. Selecting appropriate BF parameters makes it possible to find a trade-off between Ladon Hybrid Cloud privacy level, its congestion probability, and efficiency.

5.3. Summary

The BFFDDCF multiple decision probability affects the rate of traffic at the public cloud exit, both with the rate of traffic at the public cloud entrance and good packet ratio. The rate of traffic at the public cloud exit is also the rate of traffic at the private cloud entrance and is limited by the private cloud throughput. By replacing the rate of traffic at the public cloud entrance and the good packet ratio variables with some constant values, the rate of traffic at the private cloud entrance depends on the BFFDDCF multiple decision probability only. These constant values are calculated based on the traffic parameters while ensuring LHCCF congestion probability and its efficiency at a desired level.

Moreover, the BFFDDCF multiple decision probability depends on false-positive rates of the particular BFs that make it up. BFs can then be designed in such a way that their false-positive rates result in a desired value of the BFFDDCF multiple decision probability. Finally, as the BF false-positive rate depends on particular BF parameters, these can be used as a core engine to control the rate of traffic flowing between the CSP and the customer. Adjusting these parameters and then their derivatives makes it possible to find a trade-off between LHCCF privacy level, its congestion probability, and efficiency.

6. Experimental Results

In order to confirm the validity of the analysis performed and to demonstrate the veracity of its key findings, the following experiments were conducted. First, a simulation of LHCCF was performed based on traffic statistics from two real firewalls. The experiment demonstrated in practice that it is possible to find a trade-off between the LHCCF privacy level, its congestion probability, and efficiency by selecting an appropriate value of the BFFDDCF multiple decision probability. Next, software for generating BFFDDCF and testing purposes was implemented. It was demonstrated in practice that it is possible to design BFFDDCF with the desired value of multiple decision probability by selecting appropriate parameters of BFs that build it up.

6.1. BFFDDCF Multiple Decision Probability Selection

Section 5 shows that the core parameters of the LHCCF which affect the LHCCF privacy level, its congestion probability, and efficiency, are the rate of traffic at the public cloud entrance, good packet ratio, and throughput of the private cloud. It has been shown that while throughput of the private cloud is constant, the remaining parameters change over time. As such, the analysis has been based on the constant values of the rate of the traffic at the public cloud entrance and the good packet ratio and selected for each recurring period of time. Contrary to the analysis from Sect. 5, the following simulations demonstrated in practice how the instantaneous values of the LHCCF congestion probability and efficiency change over time depending on real-time statistics of the rate of traffic at the public cloud entrance and the good packet ratio.

The real-time statistics of the rate of the traffic at the public cloud entrance were gathered from two real firewalls, referred to as "Firewall 1" and "Firewall 2" in the rest of the section, observed during a one-day period, and sampled every five minutes. These are shown in **Figure 11(a)**, **Figure 11(b)**, respectively. It was assumed that Firewall 1 and Firewall 2 represent the BFFDDCF service hosted in the public cloud of the LHCCF. In turn, the real-time statistics of the good packet ratio were generated separately for Firewall 1 and Firewall 2 using the Chi-squared distribution with an average of 0.7 and 0.3, respectively. The statistics were generated for a one-day period with particular points representing samples taken every five minutes. These are shown in **Figure 11(c)**, **Figure 11(d)**, respectively. A 10 Mb/s value of the throughput of the private cloud was assumed for both firewalls.

Next, the following simulations were performed for each firewall. First, the BFFDDCF multiple decision probability with a value of 1 was assumed, and instantaneous values of the LHCCF congestion probability and its efficiency were computed for each sample. Average values of the above parameters were computed later. These are referred to as 'reference values' in the rest of this section. The same method was used to calculate average values of the LHCCF congestion probability and its efficiency for 11 values of the BFFDDCF multiple decision probability, varying from its minimum value of 0.7 and 0.3, respectively, to its maximum value of 1. Finally, rates of the average values and reference values were computed. These are shown in **Figure 11(e)-(h)**, and they are referred to as "LHCCF congestion probability ratio" ($p!$) and "LHCCF efficiency ratio" (e'), respectively.

It can be seen that the experimental results confirm the validity of conclusions drawn in Sect. 5: the higher the value of the BFFDDCF multiple decision probability, the higher the LHCCF congestion probability and the lower its efficiency. There is no easy way of simulating the LHCCF privacy level, but its relation to the BFFDDCF multiple decision probability is intuitive: the higher the value of the BFFDDCF multiple decision probability, the higher the LHCCF privacy

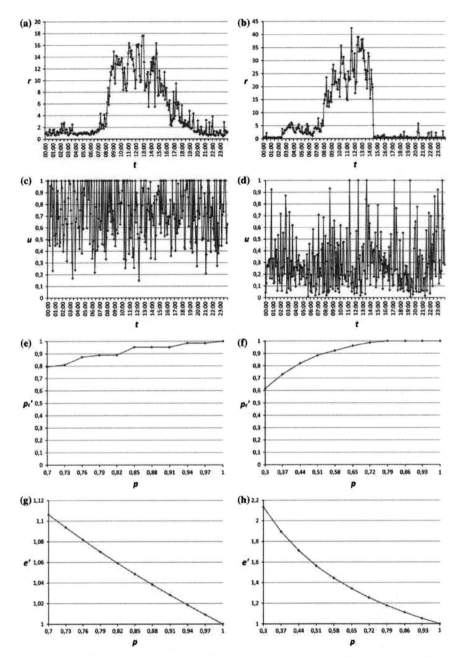

Figure 11. Experimental results—BFFDDCF multiple decision probability selection. (a) Firewall 1—rate of the traffic at the public cloud entrance, (b) Firewall 2—rate of the traffic at the public cloud entrance, (c) Firewall 1—good packet ratio, (d) Firewall 2—good packet ratio, (e) Firewall 1—LHCCF congestion probability ratio, (f) Firewall 2—LHCCF congestion probability ratio, (g) Firewall 1—LHCCF efficiency ratio, (h) Firewall 2—LHCCF efficiency ratio.

level. Therefore, each of these parameters can be controlled by selecting an appropriate value of the BFFDDCF multiple decision probability. Thus, finding a trade- off between them is possible and is a matter of selecting such a value of the BFFDDCF multiple decision probability that satisfies business requirements. Although there is no single optimal value, a satisfactory range can be found by knowing basic traffic statistics of the designed LHCCF and by defining minimum values of its privacy level and efficiency, and the maximum value of its congestion probability.

6.2. BF Parameters Selection

The second part of the experiment aimed to demonstrate that it is possible to design BFFDDCF with the desired value of multiple decision probability. Thus, a program which generates BFFDDCF on the basis of the set of ACL rules and statistical traffic parameters was implemented and used. This software is referred to as "generator" in the rest of the article. Another program which computes the BFFDDCF multiple decision probability based on the set of ACL rules and statistical traffic parameters was implemented and used by the authors too. This second piece of software is referred to as "tester" in the rest of the article. The experiment aimed to use software to compare two values of BFFDDCF multiple decision probabilities: one determined by the generator and the other computed by the tester.

The basic operation of the generator is described below. First, the generator computes FDD based on a given set of ACL rules. Then, it determines the value of the BFFDDCF multiple decision probability based on given statistical traffic parameters, including the throughput of the private cloud, good packet ratio, and rate of traffic at the public cloud entrance, and analysis performed in Sect. 5.1. Knowing the FDD structure and the desired value of the BFFDDCF multiple decision probability, the generator determines BF parameters for each of the BFs in BFFDDCF based on analysis performed in Sect. 5.2. Knowing these parameters, it then transforms FDD to BFFDD by computing the BFs for each of the FDD edges. Finally, the generator transforms BFFDD to BFFDDCF by updating the BFs in such way that good packets always result in multiple decisions.

The multiple decision probability of the BFFDDCF generated by the generator is later computed by the tester. It does so by testing each of the packets from a sample in the BFFDDCF and checking whether it results in a certain decision or multiple decisions. Based on the results for all the packets in the sample, it then computes the BFFDDCF multiple decision probability.

The experiment was conducted based on security policies from two different firewalls, the first consisting of a small number of ACEs and the second consisting of a large number of ACEs, referred to as "Firewall 1" and "Firewall 2", respectively. Moreover, a few assumptions regarding statistical traffic parameters were made. A good packet ratio of 0.1 was assumed in all cases, and six different values of the rate of traffic at the public cloud entrance were considered. Based on the above assumptions, the experiment was conducted on automatically generated samples consisting of 3,932,160 and 62,914,560 network packets, respectively. The results are given in **Table 1**. The cells contain both the BFFDDCF multiple decision probability determined by the generator and the BFFDDCF multiple decision probability computed by the tester for each of the two firewalls and for each of the six values of the rate of traffic at the public cloud entrance. To better illustrate data from **Table 1**, these are shown as a plot in **Figure 12**.

Based on the experimental results from **Table 1** and **Figure 12**, it can be seen that in each case the value of the BFFDDCF multiple decision probability computed by the tester is close to the value of the BFFDDCF multiple decision probability determined by the generator. Also, in most of the cases, the value computed by the tester is lower than the value determined by the generator. A few cases in which the value computed by the tester is higher than the value determined by the generator may be the result of a library used to implement the BF[22],

Table 1. Experimental results.

R	p (Determined)	p (Firewall 1)	p (Firewall 2)
$0.1 S_{MAX}$	1	1	1
S_{MAX}	1	1	1
$2 S_{MAX}$	0.444	0.337	0.393
$3 S_{MAX}$	0.259	0.236	0.276
$4 S_{MAX}$	0.167	0.147	0.147
$5 S_{MAX}$	0.111	0.163	0.128

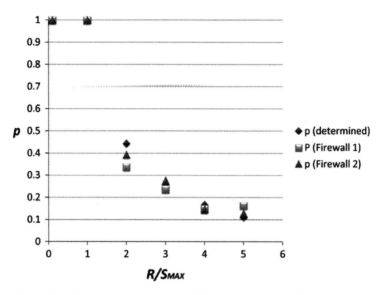

Figure 12. Experimental results—BF parameters selection.

a computational error, or the internal structure of the security policy. In most cases, the experimental results coincide with the theoretical values determined based on Formula 15 and shown in **Figure 9**.

All operations were performed on a virtual machine running the CentOS 6.5 OS with a 2.6 GHz processor and 4 GB of *Random-Access Memory* (RAM). In the course of the experiment, it turned out that the operation which is the most computationally expensive in the BFFDDCF generation process is the operation which transforms the BFFDD to the BFFDDCF. It was noticed that a relation of the BFFDDCF generation time to the number of combinations of good packets is linear and is around 1 s per 100,000. The experiment demonstrated that it is possible to design BFFDDCF in which its multiple decision probability is as close as possible to the desired value by selecting appropriate parameters of the BFs that build it up.

7. Limitations and Future Work

While it has been shown that the Ladon Hybrid Cloud is an effective framework for preserving cloud-based firewall policy confidentiality, it is important to mention its limitations and suggest directions for future work.

The first is a concern regarding the return traffic. In Sect. 3, it was stated that the main drawback of the Ladon framework—the risk of firewall deanonymization by packet eavesdropping and analysis—can be eliminated by introducing the purposefulness of packet decision uncertainty in the public cloud. Although this is true for the forward traffic, such information can still be gained by packets eavesdropping and performing analysis of the return traffic. This limitation applies to the *Transmission Control Protocol* (TCP) only. This is because the TCP requires all packets to be acknowledged so the packets permitted on the customer side will result in an acknowledgment in the return traffic. Despite the fact that this limitation raises a concern, a target solution is left for future consideration. However, three potential solutions are proposed below.

The first solution is to route the return traffic via another link which does not pass through the public cloud on the CSP side. As a result, the CSP is simply deprived of information regarding the return traffic. Another solution is to encrypt the return traffic. Although the forward traffic needs to pass unencrypted via the BFFDD on the CSP side, the return traffic may be encrypted to ensure that no information, including acknowledgments, can be retrieved. The third potential solution is to extend the return traffic with fake packets including fake acknowledgments. As a result, the knowledge base regarding the acknowledged packets built by the hostile CSP becomes inaccurate and expands over time.

Another limitation is the dependence of the Ladon Hybrid Cloud on statistical traffic parameters. In Sect. 4, it was mentioned that both the rate of traffic at the public cloud entrance and the good packet ratio values change over time. To address this, some constant values have been used instead which can be calculated based on the statistical traffic parameters. However, assuming that computational power will continue to increase, it is likely that in the future it will be possible to adjust the BFFDDCF multiple decision probability continuously based on real-time traffic analysis. This is because BFFDDCF computation is time-consuming even for modern, powerful computers. The above may also be a limitation for organizations which change their firewall policy frequently.

An interesting direction for future development of the Ladon Hybrid Cloud is analysis of the Ladon Hybrid Cloud security level in architectures with a federation of multiple clouds. In this article, just one public and one private cloud were assumed. However, as cloud federation technologies are becoming increasingly popular, it is possible to implement the Ladon Hybrid Cloud in one of the following architectures: many public clouds and one private cloud, one public cloud and many private clouds, and many public clouds and many private clouds. With fewer packets flowing between the public and private clouds, the CSP is able to obtain less information regarding the original firewall security policy; therefore, the expected Ladon Hybrid Cloud security level is higher in architectures with a cloud federation. Detailed analysis of such frameworks, required in order to demonstrate the validity of the assumptions, should further be performed.

8. Conclusions

The number of cloud-based services increases every year. The maturity of cloud computing technology encourages organizations to move subsequent types of services, previously impossible to outsource, into the cloud. This includes security services which include firewall services. However, those which have already begun to be widely adopted continue to suffer from information confidentiality and privacy issues as a result of firewall policy outsourcing.

While a framework, referred to as Ladon by its authors, preserving the confidentiality of the original firewall policy by introducing BFFDD has been proposed, it has a drawback: There is a risk of firewall deanonymization by traffic eavesdropping and analysis. To bypass this issue and limit the amount of information regarding the original firewall structure carried in packet headers, a novel framework introducing the purposefulness of packet decision uncertainty has been proposed in this article as an extension to Ladon. This extension known as the Ladon Hybrid Cloud leverages a hybrid cloud model and performs additional filtering of packets resulting in multiple decisions after passing through BFFDD in a private cloud on customer premises. Additional optimization techniques which help minimize the amount of information carried by particular packets based on the firewall policy type in use have also been proposed.

As computational resources of the private cloud are usually limited, an analysis of the Ladon Hybrid Cloud has been performed to check how the framework deals with this. It has been shown in the results of the analysis and confirmed in the results of the experiment that it is possible to control the rate of traffic at the private cloud entrance by selecting appropriate values of BF parameters while knowing basic traffic statistics. It has also been demonstrated that it is possible to find a trade-off between the Ladon Hybrid Cloud privacy level, its congestion probability, and efficiency.

The Ladon Hybrid Cloud allows organizations to take back control of privacy by helping them preserve their firewall policy confidentiality when outsourcing firewall services into the cloud. It extends the regular Ladon framework by eliminating its main drawback—the risk of firewall deanonymization by packets eavesdropping and analysis. The Ladon Hybrid Cloud is the final missing part of the puzzle which resolves the key issue of cloud-based firewall services: information confidentiality and privacy.

Open Access

This article is distributed under the terms of the Creative Commons Attribution 4.0 International License (http://creativecommons.org/licenses/by/4.0/), which permits unrestricted use, distribution, and reproduction in any medium, provided you give appropriate credit to the original author(s) and the source, provide a link to the Creative Commons license, and indicate if changes were made.

Source: Kurek T, Niemiec M, Lason A. Taking back control of privacy: a novel framework for preserving cloud-based firewall policy confidentiality [J]. International Journal of Information Security, 2015: 1–16.

References

[1] Furth, B.: Cloud computing fundamentals. In: Furth, B., Escalante, A (eds.) Handbook of Cloud Computing, pp. 3–20. Springer, New York (2010).

[2] SecaaS Working Group: Defined Categories of Service (2011). Cloud Security Aliance (CSA). https://cloudsecurityalliance.org/wp-content/uploads/2011/09/SecaaS_V1_0. Accessed 1 Aug 2014.

[3] AT&T Intellectual Property: Managed Firewall Service Network-Based. AT&T, Inc. http://www.business.att.com/content/productbrochures/Network-Based-Firewall. Accessed 1 Aug 2014.

[4] Virtela Inc: Virtela Enterprise Services Cloud (ESC). Virtela, Inc. http://www.Virtela.net/services/virtela-esc/. Accessed 1 Oct 2013.

[5] VMware Inc: Business and Financial Benefits of Virtualization. VM ware, Inc. http://www.vmware.com/files/pdf/cloud-journey/VMware-Business-Financial-Benefits-Virtualization-Whitepaper. Accessed 1 Aug (2014).

[6] Websence Inc: Seven Criteria for Evaluating Security-as-a-Service (SaaS) Solutions. Websence, Inc. http://www.websense.com/assets/white-papers/whitepaper-seven-criteria-for-evaluation- security-as-a-service-solutions-en. Accessed 1 Aug 2014.

[7] Zhou, M., Zhang, R., Xie, W., Qian, W. and Zhou, A.: Security and privacy in cloud computing: a survey. In: 6th International Conference on Semantics, Knowledge and Grids, pp. 105–112 (2010).

[8] Chen, D., Zhao, H.: Data security and privacy protection issues in cloud computing. In: International Conference on Computer Science and Electronics Engineering, pp. 647–651 (2012).

[9] Kurek, T., Lason, A., Niemiec, M.: First step towards preserving the privacy of cloud-based IDS security policies. Secur. Comm. Netw. 9999 (2015). doi:10.1002/sec.1272.

[10] Trustwave SpiderLabs: 2012 Global Security Report. Trustwave Holdings, Inc. http://www2.trustwave.com/rs/trustwave/images/2012-Global-Security-Report. Accessed 1 Oct 2013.

[11] Trustwave SpiderLabs: 2013 Global Security Report. Trustwave Holdings, Inc. http://www2.trustwave.com/rs/trustwave/images/2013-Global-Security-Report. Accessed 1 Aug 2014.

[12] Scoop Media: Survey: 88 % of ICT Employees Would Steal. Scoop Top Stories. http://www.scoop.co.nz/stories/BU0811/S00203.htm. Accessed 1 Aug 2014.

[13] McAfee, R.B., Champagne, P.J.: Effectively Managing Troublesome Employees, Westport (1994).

[14] Khakpour, A.R., Liu, A.X.: First step toward cloud-based firewalling. In: 31st International Symposium on Reliable Distributed Systems, pp. 41–50 (2012).

[15] Gouda, M.G., Liu, A.X.: Structured firewall design. Comput. Netw. J. 51(4), 1106–1120 (2007).

[16] Bloom, B.: Space/time trade-offs in hash coding with allowable errors. Commun. ACM. 13(7), 442–426 (1970).

[17] Jajodia, S., Ghost, A.K., Swarup, V., Wang, C., Sean Wang, X.: Moving Target Defense. Springer, New York (2011).

[18] Fulp, E.W., Tarsa, S.J.: Trie-based policy representations for network firewalls. In: 10th IEEE Symposium on Computers and Communications, pp. 434–441 (2005).

[19] Bose, P., Guo, H., Kranakis, E., Mahashwari, A., Morin, P., Morrison, J., Smid, M., Tang, Y.: On the False Positive Rate of Bloom Filters. Report, School of Computer

Science. doi:10.1016/j.ipl.2008.05.018.

[20] EC-Council: The 5 phases every hacker must follow. EC-Council. http://www.clrgroup.com/Site/pdf/EthicalHacking. Accessed 1 Oct 2014.

[21] Jung, J., Krishnamurthy, B., Rabinovich, M.: Flash Crowds and Denial of service attacks: characterization and implications for CDNs and websites. In: Proceeding of 11th World Wide Web conference, pp. 293–304 (2002).

[22] Burstein, M.: Creating a simple Bloom filter. Max Burstein Blog. http://maxburstein.com/blog/creating-a-simple-bloom-filter/. Accessed 1 Aug 2014.

Chapter 15

Terrorist Networks and the Lethality of Attacks: An Illustrative Agent Based Model on Evolutionary Principles

Paul Ormerod

Volterra Partners LLP, London, UK

Abstract: A data base developed from the Memorial Institute for the Prevention of Terrorism's (MIPT) Terrorism Knowledge Base for the years 1998–2005 was provided to participants in the workshop. The distribution of fatalities in terrorist attacks is, like many outcomes of human social and economic processes, heavily right-skewed. We propose an agent based model to analyse this, and to enable generalisations to be made from the historical data set. The model is inspired by modelling developments in cultural evolutionary theory. We argue that a more appropriate "null" model of behaviour in the social sciences is on based upon the principle of copying, rather than the economic assumption of rationality in the standard social science model.

Keywords: Agent Based Model, Terrorist Fatalities, Cultural Evolution, Mutations

Terrorist Networks and the Lethality of Attacks: An Illustrative Agent Based Model on Evolutionary Principles

1. Introduction

Asal and Rethemeyer[1] report an econometric analysis of a data base developed from the Memorial Institute for the Prevention of Terrorism's (MIPT) Terrorism Knowledge Base. The dataset is complete for the years 1998–2005. Of the 395 clearly identified terrorist organizations operating throughout the world over this period, only 68 killed 10 or more people during that period. Indeed, only 28 killed more than 100 people. The econometric study analyses the factors which can account for this dramatic difference in organizational lethality.

They conclude that "1) large organizations, 2) organizations that address supernatural audiences through religious ideologies, 3) organizations with religious-ethnonationalist ideologies—ideologies that define another and play to the supernatural, 4) organizations that build and maintain extensive alliance connections with peers, and 5) organizations that maintain control over territory are the primary actors in this story. Though much of the organizational and social movements literature suggest that new organizations are less effective and able, our data was unable to find evidence that newness matters. Some widely held theories about the correlates of lethality—including the belief that state sponsorship and "homebase" regime-type would affect organizational lethality—could not be substantiated with our data. In fact, there is equivocal evidence that state sponsorship tend to restrain killing by client organizations. Size coupled with religious and ethnonationalist ideology generates the capability needed to pursue deadly ends".

This paper considers a potential generalisation of the econometric approach using the methodology of agent based modelling (ABM). Section 2 briefly considers methodological aspects of the issue, and section 3 discusses some principles of agent behaviour. Section 4 sets out the model and section 5 discusses some illustrative results.

1.1. Some Methodological Reflections

The Asal and Rethemeyer paper is detailed and thorough, and takes proper account of the heavily right-skewed nature of the dependent variable, the number of people killed in each incident. The dependent variable ranges from 0 to 3505 with a median of 0, a mean of and a standard deviation of 202.04. Of the 395 organizations for which there are data, 240 of those organizations perpetrated one or more incidents that resulted in no fatalities.

However, econometric analysis of data, no matter how sophisticated, essentially involves fitting a plane through the n dimensions of the explanatory data. Even when the regression technique is based upon the principle of maximum likelihood, the result can still be given this geometric interpretation. Helbing[2] offers a more detailed description of the potential restrictions of this approach, but one of the essential problems can be summarised as follows.

A restriction on the ability to generalise from econometric results is that a small number of data points may exercise a strong influence on the fit of the n-dimensional plane. Consider, for example, the standard linear regression model $y = X\beta + \varepsilon$, where the vector of estimated parameters is $b = (X^T X)^{-1} X^T y$ and the fitted values are $y^* = Xb = X(X^T X)^{-1} X^T y$.

The hat matrix, H, maps the vector of fitted values to the vector of observed values, and describes the influence each observed value has on each fitted value[3], where $H = X(X^T X)^{-1} X^T$, It is the orthogonal projection onto the column space of the matrix of explanatory factors, X.

This matrix can be used to identify observations which have a large influence on the results of a regression. If such observations exist, ideally we would like to have more data from this part of the observation space. So if we have a small number of observations in the tail of the distributions of both an explanatory and the dependent variable and these are correlated, such observations will inevitably exercise a strong influence on the results.

Perhaps not surprisingly, therefore, large terrorist organisations, and particularly those which maintain and build extensive alliances, are identified in econometric analysis as being two of the key factors which are linked to effective attacks leading to high levels of fatality.

However, more generally, we are interested not so much in projecting the single path of history which we actually observe into the future, but in considering the evolutionary potential of the groups. The policy message from the econometrics is to focus on large, well connected groups in the prevention of attacks. But, for example, what is the potential for small, less connected groups to acquire the ability to develop the capacity to carry out highly effective attacks?

Agent based modelling is a way of trying to examine the evolutionary potential of any given system. Helbing and Barietti[4], for example, state that "computer simulation can be seen as experimental technique for hypothesis testing and scenario analysis". They go on to argue that "agent-based simulations are suited for detailed hypothesis-testing, i.e. for the study of the consequences of ex-ante hypotheses regarding the interactions of agents. One could say that they can serve as a sort of magnifying glass or telescope ("socioscope"), which may be used to understand our reality better. by modeling the relationships on the level of individuals in a rule-based way, agent-based simulations allow one to produce characteristic features of the system as emergent phenomena without having to make a priori assumptions regarding the aggregate ("macroscopic") system properties".

1.2. Approaches to Agent Behaviour

The standard socio-economic science model, SSSM[5], postulates a high level of cognitive ability on the part of agents. Agents are assumed to be able to gather all relevant information, and to process it in such a way as to arrive at an optimal decision, given their (fixed) tastes and preferences. Even with the relaxation of the assumption of complete information[6], agents are still presumed to have formidable cognitive powers. This, the core model of economics, has a certain amount of explanatory power, though as[5] goes on to state; "Within economics there is essentially only one model to be adapted to every application: optimization subject to constraints due to resource limitations, institutional rules and/or the behavior of others, as in Cournot-Nash equilibria. The economic literature is not the best place to find new inspiration beyond these traditional technical methods of modeling".

Perhaps the most important challenge to this approach comes when decisions do not depend not on omniscient cost-benefit analysis of isolated agents with fixed tastes and preferences, but when the decision of any given agent depends in part directly on what other actors are doing. In such situations, which are probably the norm rather than the exception in social settings, not only do choices involve many options for which costs and benefits would be impossible to calculate (e.g., what friends to keep, what job to pursue, what game to play, etc.), but the preferences of agents themselves evolve over time in the light of what others do.

Complex choices can be fundamentally different from simple two-choice scenarios, such that the problem becomes very difficult to predict, as has been demonstrated in ecological[7] and human settings[8]. Such scenarios are where so-called zero-intelligence models[9] do better at understanding emergent patterns in collective behaviour. However, despite their empirical success e.g.[10]–[12], they have met with resistance amongst social scientists.

The zero intelligence model is based upon the particle model of physics. Indeed, one of the fastest-rising keywords in the physics literature is "social". Literally thousands of papers in physics are devoted to modeling social systems, and indeed regular sections of leading journals such as Physica A and Physical Review E are devoted to this topic. The analogy between people and particles has been so consistent that a recent popular review was appropriately titled The Social Atom[11].

This approach has provided significant insights into modeling collective interactions in social systems, from Internet communities to pedestrian and vehicular traffic, economic markets and even prehistoric human migrations e.g.[13]–[15].

The idea that there are serious limits to human cognitive powers in complex systems is one which has strong empirical support. Kahneman, for example[16] argues that "humans reason poorly and act intuitively". Decisions may often be taken in circumstances in which the assumption that individuals have no knowledge of the situation is a better approximation to reality than is the assumption that they possess complete information and have the capacity to process this information to make optimal decisions.

An illustration of the limits to human awareness and social calculation is the well known Prisoner's Dilemma game, invented by Drescher and Flood in 1950. The optimal strategy or "Nash equilibrium" for the one period game was discovered very quickly. However, as documented in detail in[17]. Flood recruited distinguished RAND analysts John Williams and Armen Alchian, a mathematician and economist respectively, to play 100 repetitions of the game. The Nash equilibrium strategy ought to have been played by completely rational individuals 100 times. It might, of course, have taken a few plays for these high-powered academics to learn the strategy. But Alchian chose co-operation rather than the Nash strategy of defection 68 times, and Williams no fewer than 78 times. Their recorded comments are fascinating in themselves. Williams, the mathematician, began by expecting both players to co-operate, whereas Alchian the economist expected defection, but as the game progressed, co-operation became the dominant choice of both players.

Even now, after almost 60 years of analysis and literally thousands of scientific papers on the subject, when sufficient uncertainty is introduced into the game, the optimal strategy remains unknown. Certainly, some strategies do better than most in many circumstances, but no one has yet discovered the optimal strategy even for a game that is as simple to describe as the Prisoner's Dilemma.

Assumptions of optimality and rationality can be certainly be useful when payoffs are predictable from one event to the next—hunting and gathering in a consistent environment, for example, or even modern situations where the complexity of choices is low[18]. But more generally, the zero intelligence model may be more useful as the "null".

However, as is argued in[19], human agents are fundamentally different from particles in physics in that, however imperfectly, they can act with purpose and intent. So the basic null model of zero intelligence needs to be modified to incorporate some aspect of this functionality[20].

Simon in his seminal paper on behavioural economics[21] argued that the fundamental issue which all sentient beings have to take into account when taking decisions is to reduce the massive dimensionality of the choice set which they face: "Broadly stated, the task is to replace the global rationality of economic man with a kind of rational behavior that is compatible with the access to information, and the computational capacities that are actually possessed by organisms, including man, in the kinds of environments in which such organisms exist".

An important way of coping with an evolving, complex environment is to follow a strategy of copying, or social learning as it is described in cultural evolution. In social and economic systems, decision makers often pay attention to each other either because they have limited information about the problem itself or limited ability to process even the information that is available[22].

A striking example of the power of simple copying strategies in an evolving environment is provided by the computer tournament described by[23]). As the abstract states: 'Social learning (learning through observation or interaction with other individuals) is widespread in nature and is central to the remarkable success of humanity, yet it remains unclear why copying is profitable and how to copy most effectively. To address these questions, we organized a computer tournament in which entrants submitted strategies specifying how to use social learning and its asocial alternative (for example, trial-and-error learning) to acquire adaptive behavior in a complex environment. Most current theory predicts the emergence of mixed strategies that rely on some combination of the two types of learning. In the tournament, however, strategies that relied heavily on social learning were found to be remarkably successful, even when asocial information was no more costly than social information. Social learning proved advantageous because individuals frequently demonstrated the highest-payoff behavior in their repertoire, inadvertently filtering information for copiers. The winning strategy relied nearly exclusively on social learning and weighted information according to the time since acquisition'.

In other words, in a complex environment in which the pay-off to various strategies was, by design, constantly evolving, simple copying proved a very effective strategy.

Copying, of course, is the essence of the principle of preferential attachment, initially formulated in a general way by Simon[24] and rediscovered by Barabasi and Albert[25]. In its more recent incarnation, the model has been hugely influential because it resulted in a power law (or at least long-tailed) degree distribution (connections per network node)—a kind of distribution so intriguing to many that the editor of Wired magazine wrote an entire book, ten years later, about its significance to modern online economies[26].

However, a key drawback of the approach is that it is ultimately static. The rankings in the distribution, in other words, gradually become fixed, and attempts to modify the basic model are rather artificial[27]. Turnover in rankings is not just a feature of modern markets of popular culture, but operates on the time scale required for cities to evolve[28].

Preferential attachment is a special case of a model developed in cultural evolutionary theory e.g.,[29][30], in which it remains the basic principle of decision making, but an agent is also able (with a small probability) to make an innovative choice. The model is developed from the concept of genetic evolution, which is based on the principles of copying and mutation (innovation).

In its most recent and most general formulation[31] the model of social learning in cultural evolution is, with just two parameters, capable of replicating any right skewed distribution and of accounting for the turnover which is observed in relative standings in any evolutionary environment. One parameter describes the relative frequencies with which preferential attachment (copying) and innovation are used to make decisions, and the other describes the span of historical time used to observe the decisions which other agents have made. Clearly, this latter is different for a firm choosing its location to a teenager choosing which video to download from YouTube.

1.3. The Model

In order to try to generalise from the data set used by Asal and Rethemeyer, we retain copying as one of the basic principles used by terrorist organisations. They can observe and copy, exactly as in the evolutionary learning computer tournament, tactics and strategies used by other organisations. In a situation in which the environment in which they operate evolves, the pay off (*i.e.* number of fatalities inflicted) may of course differ from previously enacted versions of the same tactic. But copying is one of the building blocks of the model.

We augment the model with further behavioural principles which seem appropriate in this context. By its very nature, the terrorist world is clandestine, and organisations will differ in their propensity to share information about tactical capabilities with other such organisations. In addition, organisations will differ in their willingness or ability to absorb and execute a tactic they have not previously used, even if it is explained to them by another organisation.

We consider these factors in the context of the ability of the organisations to acquire the capacity to carry out an innovation, or tactical mutation, developed by a terrorist group. An overview of the model is shown in **Figure 1**, which is similar to the model developed to account for the diffusion of technological innovations across companies in four industries in the Greater Manchester region of the UK[32]: The model takes an initial mutation to be exogenous and it is taken up by one agent/organisation at the outset. The characteristics of the agents are governed by their willingness to innovate, their desire to keep innovation to themselves, and their willingness to communicate with others. The innovating agent will be connected to other agents via the network structure, and at the next step of the model the innovation will be passed on according to the extent agents discover the mutation and their own willingness to take it up. At further steps of the model further agents may be able to discover and take up the mutation, until eventually no further take up occurs.

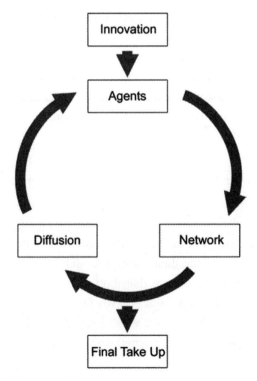

Figure 1. Model overview.

We define two different methods by which tactical mutations may be passed on via the network linkages. The first is a direct relationship between two partners, while the second is a group relationship.

First, an organisation with an innovation will provide it to another only if its level of secrecy, or the propensity of a group to try to retain the benefits of its innovations, is less than the absorptive capacity, or the degree to which a terrorist actively engages in activities which enable it to identify and adopt new innovations, of the organisations it is linked with. This method of adopting an mutation represents a mutual relationship or exchange between terrorist organisations and implies a degree of trust or collaboration.

The second method for spreading a mutation is based on the principle of copying. Here if a group looks at the spectrum of organisations to which it is linked and finds that the proportion that have adopted an innovation is higher than their own personal threshold, they will mimic their behaviour and adopt the mutation. In some circumstances this threshold may be very high and only when all or nearly all of the other groups an organisation has relationships with have taken up an innovation will they be persuaded to do the same. For other organisations relatively few outfits may have to have the same innovation before they adopt it. This mechanism represents a copying behaviour. This may occur even when a terrorist group may not fully understand the reasons and benefits of a mutation but relies on observing that others have adopted it. This behaviour is more likely to be a response to competitor behaviour.

Figure 2 plots the distribution of the numbers of connections between organisations in the data set.

In each separate solution of the model, a network is generated which has the same distribution as that of the data. More precisely, the null hypothesis that the model network has the same degree distribution as the data is not rejected on a Kolmogorov-Smirnov test at a p-value < 0.01, far below the conventional level of statistical significance. In other words, we can regard the model networks as being identical in degree distribution to the actual data.

In this illustrative application of the model:

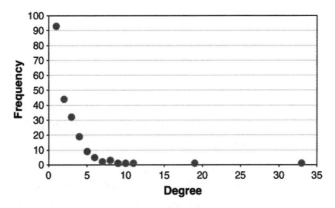

Figure 2. Degree distribution of connections between terrorist organisations.

- Each agent is allocated at random a willingness to seek and adopt an innovation/mutation, α_i, drawn from a uniform on [0,1].

- Each agent is allocated at random a willingness to share an innovation, σ_j, drawn from a uniform on [0,1].

- An agent adopts the innovation of another if $\alpha_i > \sigma_j$.

- Each agent has a threshold for imitating the innovations of its neighbours, τ_i drawn from uniform on [0.5,1].

This latter rule is the principle of binary choice (adopt or not adopt) with externalities[21][33].

The model proceeds in a series of steps, in each of which the following procedure is operated. First, all agents not in possession of the mutation who are not connected to any other organisation which does have the innovation are identified. These agents play no further part in this particular step of the model. Of course, in subsequent steps, agents to which they are connected may by then have acquired the mutation. So whilst they may not acquire it in this period, they are not precluded from so doing in future steps within the same solution of the model.

If the absorptive capacity, α_i of the agent without the innovation is greater than the willingness to share (or secrecy), σ_j, of the agent with the innovation, the agent is assumed to adopt the innovation. If not, the copying rule is then invoked, and the agent adopts it if its threshold is lower than the proportion of agents to which it is connected which has the innovation.

Any particular solution of the model ends when no agent adopts the mutation in a given step of the model. We measure and record the proportion of agents which have adopted the innovation.

This process is the basic building block of the illustrative approach. We can readily imagine, however, that certain capabilities are inherently harder to acquire than others. Shooting a civilian (e.g. a member of the IRA shooting a Protestant at random in Northern Ireland) requires much lower levels of skill and expertise than a coordinated bombing. We illustrate the effects of introducing this into the model in the following heuristic way.

We populate the model with 1000 agents and solve it 1000 times. At the end of each solution, the organisations which have acquired the innovation are deemed capable of carrying out a relatively low level attack, which we describe as Level 1 capability. The model is reinitialised, a new agent is selected at random to acquire an innovation, and the process is repeated. At the end of this, more agents will have acquired Level capability. And those which acquired it in the first solution and in addition acquire it in this new one are deemed to have acquired Level 2 capability. In other words, they have acquired sufficient technical skill to mount more serious attacks, with presumed higher levels of fatalities. Finally, we repeat the process again, and those agents which acquire the innovation on all three occasions are deemed capable of Level 3 attacks.

2. Discussion and Results

Some properties of the model are illustrated in the three graphs below. These are the results averaged across 1000 separate solutions of the model, and show the total percentage of organisations with a given degree (number) of links which acquire the capabilities to acquire the ability to carry out Levels 1, 2 and 3-type attacks.

First, **Figure 3** shows the results for Level 1 capabilities.

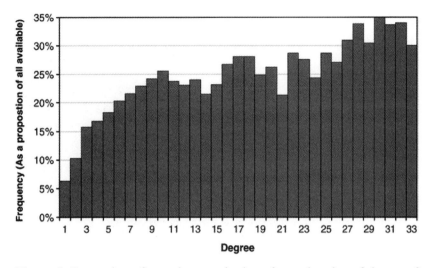

Figure 3. Proportion of terrorist organisations for each value of degree of connections which acquire the capability to carry out Level 1 attacks, average across 1000 solutions.

Quite rapidly, the connection between the acquisition of this level of capability and the degree of an organisation's links with other terrorist groups falls away. Only a very small percentage of groups with a very low number of connections are able to acquire even this level of capability. But the proportion of organisations which acquire it and which have, say, just 6 links is not very different from the proportion of those with 26 links which acquire. Only for those with a very high number of connections does the proportion rise.

Figure 4 and **Figure 5** show the results for the proportions which acquire Levels 2 and 3 capabilities.

The precise topology of the network and the parameters allocated to each agent differ across each of the 1000 solutions of the model, but the same network and parameter values are retained within each solution for each of the steps of acquisition capability. So, for example, the structure may be such that an agent is connected to another which is very likely to acquire the mutation, and the parameters are such (in particular the absorption and secrecy parameters) that it, too, is therefore likely to acquire it. So the proportion of agents of any given degree acquiring Level 3 capability is not simply (as an approximation) one third the value of those acquiring Level 1, but is much higher.

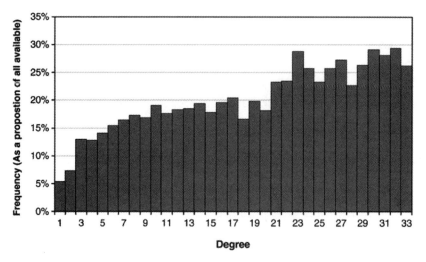

Figure 4. Proportion of terrorist organisations for each value of degree of connections which acquire the capability to carry out Level 2 attacks, average across 1000 solutions.

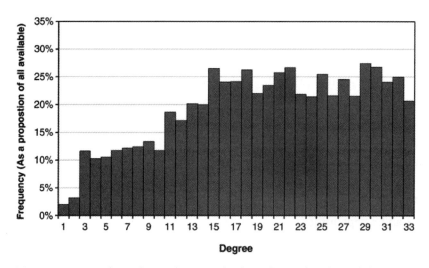

Figure 5. Proportion of terrorist organisations for each value of degree of connections which acquire the capability to carry out Level 3 attacks, average across 1000 solutions.

These results are illustrative of a principle of how to try to generalize from the data set. Indeed, we see that the proportion of organizations with only weak connections to others which acquire level 3 capabilities is, apart from those with virtually none, non-trivial. So although in the one actual history we are able to observe, organizations with low levels of connections tended not to carry out attacks involving high levels of fatalities, this does not mean that in future they will be unable to acquire such characteristics.

We are not claiming that this is a definite model with which to inform future policy, though it does offer some potential guidelines. Rather, it illustrates how an agent based model constructed on evolutionary principles in a complex environment can be used to extract more information from a data set than is possible using conventional analytical methods such as econometrics. One potential extensions is to endogenise the network, such that agents have incentives to develop new links, but also introducing extinctions of agents, which may be enhanced (made easier for the authorities) as the number of connections of an organization increase ([34]offers a general model of extinctions in an evolving network). Another is to endogenise the parameters, so that agent behavior becomes reinforced both by reference to their own previous decisions and by reference to the properties of their neighbours.

Competing Interests

The author declares that he has no competing interests.

Paper presented at the Department of Homeland Security Workshop on Biologically-Inspired Approaches to Understanding and Predicting Social Dynamics, Washington DC, August 2009.

Source: Ormerod P. Terrorist networks and the lethality of attacks: an illustrative agent based model on evolutionary principles [J]. Security Informatics, 2012, 1(1): 1–8.

References

[1] V Asal, KR Rethemeyer, The nature of the beast: organizational structure and the lethality of terrorist attacks. J. Polit. 70, 437–449 (2008).

[2] D Helbing, Pluralistic Modeling of Complex Systems (2010). arXiv:1007.2818v1. http://arxiv.org/CornellUniversityLibrary.

[3] DC Hoaglin, RE Welsch, The hat matrix in regression and ANOVA. Am. Stat. 32,

17–22 (1978).

[4] D Helbing, S Balietti, Agent Based Modeling,' FuturICT meeting, Zurich, June 2011 (2011). http://dl.dropbox.com/u/6002187/16June_AgentBasedModelling.pdf.

[5] V Smith, Constructivist and ecological rationality in economics. Am. Econ. Rev. 93, 465–508 (2003).

[6] GA Akerlof, The market for lemons: quality uncertainty and the market mechanism. Q. J. Econ. 84, 488–500 (1970).

[7] BA Melbourne, A Hastings, Highly variable spread rates in replicated biological invasions: fundamental limits to predictability. Science 325, 1536–1539 (2009).

[8] MJ Salganik, PS Dodds, DJ Watts, Experimental study of inequality and unpredictability in an artificial cultural market. Science 311, 854–856 (2006).

[9] JD Farmer, P Patelli, I Zovko, The predictive power of zero intelligence in financial markets. Proc. Natl. Acad. Sci. 102, 2254–2259 (2005).

[10] P Ball, Critical Mass: How One Thing Leads to Another (Heinemann, London, 2004).

[11] M Buchanan, The Social Atom (Bloomsbury, London, 2007).

[12] MJ Newman, A-L Barabási, DJ Watts, The Structure and Dynamics of Networks (Princeton University Press, Princeton, 2006).

[13] GJ Ackland, M Signitzer, K Stratford, MH Cohen, Cultural hitchhiking on the wave of advance of beneficial technologies. Proc. Natl. Acad. Sci. 104, 8714–8719 (2007).

[14] I Farkas, D Helbing, T Vicsek, Mexican waves in an excitable medium. Nature 419, 131–132 (2002).

[15] X Gabaix, P Gopikrishnan, V Plerou, HE Stanley, Institutional investors and stock market volatility. Q. J. Econ. 121, 461–504 (2006).

[16] D Kahneman, Maps of bounded rationality: psychology for behavioral economics. Am. Econ. Rev. 93, 1449–1475 (2003).

[17] P Mirowski, Machine Dreams: Economics Becomes a Cyborg Science (CUP, Cambridge UK, 2002).

[18] B Winterhalder, EA Smith, Analyzing adaptive trategies: human behavioral ecology at twenty-five. Evol. Anthropology 9, 51–72 (2000).

[19] RA Bentley, P Ormerod, Agents, Intelligence and Social Atoms, in Integrating Science and the Humanities, ed. by M. Collard, E. Slingerland (Oxford University Press, Oxford, 2011).

[20] P Ormerod, M Trabbati, K Glass, R Colbaugh, Explaining Social and Economic Phenomena by Models with Low or Zero Cognition Agents, in Complexity Hints for Economic Policy: New Economic Windows, Part IV, ed. by (Springer, Milan, 2007), pp. 201–210. doi:10.1007/978-88-470-0534-1_10.

[21] HA Simon, A behavioral model of rational choice. Q J Econ 69, 99–118 (1955).

[22] TC Schelling, Hockey helmets, concealed weapons, and daylight saving: a study of binary choices with externalities. J Confl. Resolut. 17, 381–428 (1973).

[23] L Rendell, R Boyd, D Cownden, M Enquist, K Eriksson, MW Feldman, L Fogarty, S Ghirlanda, T Lillicrap, KN Laland, Why copy others? Insights from the social learning strategies tournament. Science 328, 208–213 (2010).

[24] HA Simon, On a class of skew distribution functions. Biometrika 42, 425–440 (1955).

[25] R Albert, A-L Barabási, Statistical mechanics of complex networks. Rev. Mod. Phys. 74, 47–97 (2002).

[26] C Anderson, The Long Tail: Why the Future of Business Is Selling Less of More (Hyperion, New York, 2006).

[27] SN Dorogovtsev, JFF Mendes, Evolution of networks with ageing of sites. Physical Review E. 62, 1842 (2000).

[28] M Batty, Rank Clocks. Nature 444, 592–596 (2006).

[29] SJ Shennan, JR Wilkinson, Ceramic style change and neutral evolution: a case study from Neolithic Europe. Am. Antiq. 66, 577–594 (2001).

[30] MW Hahn, RA Bentley, Drift as a mechanism for cultural change: an example from baby names. Proc. R. Soc. B 270, S1–S4 (2003).

[31] RA Bentley, P Ormerod, M Batty, Evolving social influence in large populations. Behav. Ecol. Sociobiol. 65, 537–546 (2011).

[32] P Ormerod, B Rosewell, G Wiltshire, Network Models of Innovation Processes and the Policy Implications, in Handbook on the Economic Complexity of Technological Change, ed. by C. Antonelli (Edward Elgar, Cheltenham UK, 2010).

[33] DJ Watts, A simple model of global cascades on random networks. Proc. Natl. Acad. 99, 5766–5771 (2002).

[34] P Ormerod, R Colbaugh, Cascades of failure and extinction in evolving social networks. J. Artif. Soc. and Soc. Simul. 9, 4 (2006).

Chapter 16

Threat Driven Modeling Framework Using Petri Nets for e-Learning System

Aditya Khamparia[1], Babita Pandey[2]

[1]Department of Computer Science and Engineering, Lovely Professional University, Phagwara, Punjab, India
[2]Department of Computer Applications, Lovely Professional University, Phagwara, India

Abstract: Vulnerabilities at various levels are main cause of security risks in e-learning system. This paper presents a modified threat driven modeling framework, to identify the threats after risk assessment which requires mitigation and how to mitigate those threats. To model those threat mitigations aspects oriented stochastic petri nets are used. This paper included security metrics based on vulnerabilities present in e-learning system. The Common Vulnerability Scoring System designed to provide a normalized method for rating vulnerabilities which will be used as basis in metric definitions and calculations. A case study has been also proposed which shows the need and feasibility of using aspect oriented stochastic petri net models for threat modeling which improves reliability, consistency and robustness of the e-learning system.

Keywords: Threat Modeling, AOSPNs, Security Metrics, Petri Nets, Aspects, e-Learning

1. Background

Due to enhancement in security problems for e-learning systems (Hecker 2008), it is essential that security concerns to be addressed in early stages of system development cycle (Jalal *et al.* 2008). Various e-learning systems are designed which are based on formal techniques and provides threat modeling only in requirement phase but not in design and analysis phase of existing system. Due to this there will be no guarantee that design vulnerabilities of system can be removed easily.

A petri net is one of mathematical modeling language or tool used for description of discrete distributed systems. It is a directed bipartite graph in which the nodes represent transitions, places and directed arcs (Murata 1989). It is a graphical based model used for stepwise processes which includes choice, iterations, and executions. Petri nets performed process analysis by using theory based on mathematical cases. Various types of petri nets are used to model behavior of system like colored petri nets (Houmb and Sallhammar 2012), timed petri nets and stochastic petri nets. Stochastic Petri Nets (SPNs) models distributed computing architectures and other software (Peterson 1977).

The proposed paper uses threat modeling for threat identification in system and categorizes those threats according to their categories (STRIDE) like spoofing identify, tampering data, repudiation, information disclosure, denial of service and elevation of privilege (Howard 2003). In proposed framework, new phases of threat modeling were added to fit with aspects and SPNs. Threat modeling offers various benefits such as 1) easier for team members to understand their application in better way; 2) easier to identify faults in system; 3) complex design faults can be identified easily which was not able to retrieve earlier in easy way.

The main system functions are modeled using SPNs whereas the threat mitigations are modeled using aspect oriented stochastic petri nets (AOSPN) which we have developed in this proposed research. Our modified threat driven framework measures the correctness, soundness and completeness of the SPN and AOSPN models (Dehlinger and Nalin 2006). Threat analysis (risk assessment), disintegration correction assessment, mitigation (attenuation) correction assessment and mitigation (attenuation) assessment are introduced phases that were added to threat modeling framework. In risk analysis phase risk of threat is measured by assigning the likelihood of occurrence and impact to system. Correctness assessment is measured using three main behavioral criteria of petri nets which are reachability, boundness and liveness. Mitigation (attenuation) assessment is calculated using a security metric that was adapted in this proposed work. Augmented metric is based on CVSS (Mell and Romanosky 2007) and proposed methods of Wang et al. (Wang et al. 2009). A modification in weight metric score was given to compute a quantitative score after applying the mitigations. On the basis of security metrics calculation (Payne 2006), we are able to observe how effective the mitigations were which enable e-learning researchers to compare mitigation effectiveness.

The rest of paper is organized as follows. "Literature review" section describes related work. "Aspect oriented SPN model (AOSPN)" section deals with aspect oriented SPNs. "Modified threat driven modeling framework" section describes the modified proposed threat framework model. "Proposed security metric" section shows the extended security metric and its calculations. "Case study" section deals with systematic case study, applying the threat driven framework and security metrics to specific question answer system and shows performance evaluation with respect to other frameworks. "Conclusion" section concludes the paper.

2. Literature Review

The proposed e-learning system mitigates the threats by performing threat modeling, aspect oriented development, usage of stochastic petri nets and security metric computation. The threat modeling is used to identify the threats which require mitigation and how to mitigate them. The process starts by disintegrating the applications, then determining and rank threats. Adapt methodology to respond to threats, choose best possible way to mitigate the threats and finally choose the appropriate technologies for the identified techniques.

Dehlinger and Nalin (2006) developed an aspect oriented model which provides UML based security and includes security policies as an aspect while designing a secure system. They have reviewed a security framework whose purpose is to provide the authors lessons derived from its design and use. They have verified the security of software using aspect oriented nets (Xu and Nygard 2006a, b). Their approach distinguished the software modeling and threat mitigations which are modeled by petri nets and aspect oriented nets simultaneously.

Sometimes the behavior of model not only depends on its structure but also on the timing. There is a requirement of stochastic petri nets (SPNs) which adds non deterministic time through adjustable randomness of the transitions (Haas 2002). These nets are modeled on basis of exponential random distributions and their performance analysis is based upon Markov theory (Balogh and Turcáni 2011). SPNs offers numerous advantages over original petri nets like ease of functional behavior analysis and testing with aid of graphical format, describe concurrency, synchronizations and show correlation among activities which describes the qualitative and quantitative properties of specified system like number of tokens firing from one place, how many tokens are expected to reach from one state to another at given time duration etc.

Over the last few years, developing methods to measure security loop holes is biggest challenge and concern among researchers. The NIST provided a paper as an overview of the security metrics area and looks at the possible possibilities of research that could be followed to advance the state of art (Jansen 2009). Some researchers distinguished between low level metrics and high level metrics for performing various estimations related to security. (Jensen 2008) created a tool SODAWeb which adapts and filter security techniques by using various applications supported by tools. (Heyman *et al.* 2008) have presented method of using security patterns to combine security metrics.

In our proposed security model, we have considered Common Vulnerability Scoring System (CVSS) (Mell and Romanosky 2007) which consists of three groups: Base, Temporal and Environmental. A numeric score has been produced by individual groups ranges from 0 to 10. A new approach was proposed by (Wang *et al.* 2009) to define software security metrics based on vulnerabilities included in software systems and their impacts on quality of software. We have utilized the approach in e-learning based systems. It uses the Common Vulnerabilities and Exposures (CVE) and CVSS in their metric definition and calculation. A complete comparative view of similarity and differences of proposed method with existing methods are given in **Table 1**.

Table 1. Similarity and differences of proposed method with existing models

Author	Proposed method	Similarity	Differences
Dehlinger and Nalin (2006)	Developed aspect oriented model to provide UML based security feature	They have used Aspect oriented net	The approach distinguished the software modeling and threat mitigations but without consideration of CVSS features
Omrani et al. (2011)	Proposed an adaptive e-learning system based on high level petri nets by considering learners learning style, score and knowledge level	They also evaluated the performance of e-learning system	They considered high level petri nets (HLPN) for performance evaluation without considering threats in system but in our security based model we have used stochastic petri nets (SPN) to improve robustness of system by using before and after mitigation strategy which improves learning performance
Balogh et al. (2012a, b)	Designed petri net based LMS system to regulate the communication according to student knowledge and ability and deliver learning material according to their needs	NA	They have focused on personalization using petri nets without consideration of security metrics which improves consistency and reliability of e-learning system
Hammami and Mathkour (2013)	Develop an e-learning system architecture which includes multi agent system and adaptive e-learning	NA	They have used object petri nets to build multi-agent architecture which adapts learner according to the learning preference and controls the communication and interaction among different agents. But in our proposed model we have used aspect oriented and stochastic petri nets with consideration of security, threat and risk assessment which has not been considered by Hammami and Mathkour (2013)

3. Aspect Oriented SPN Model (AOSPN)

It incorporates the fundamental features of aspect oriented development. Aspects are the units that modularize the cross cutting concerns (cross cut the boundaries of traditional programming constructs). An aspect oriented program consists of a number of base modules and aspects that can be merged into an executable whole. AOSPN includes the basic concepts like join points, advices, pointcuts and introduction (Schauerhuber *et al.* 2006). An advice is contained by an aspect and is a piece of code that is inserted at one or more specific points of core concern. A join point is point in the execution where an advice is inserted. Join points may be transitions, predicates, and arcs in the SPN. A point cut is a language construct that designates a join point. Point cut defines whether a given join point matches according to defined criteria. An introduction net introduces new members to base modules. It allows aspects to modify the static structure of program.

In AOSPNs there are three types of pointcuts as described by: transition, predicate and arc. A stochastic petri net-based aspect A is a structure <P, D, I> where P is set of pointcuts, D is a set of advice nets and I is a set of introduction nets. Processing timed transition pointcuts remove all the transitions selected by each transition pointcuts and replace it with the corresponding introduction nets according to the advice specifications.

Suppose there is a threat in the timed transition T1 in the stochastic petri net N1 in **Figure 1**. We define the aspect as shown in **Figure 2** where the pointcut specifies the place of the threat, advice net described how the mitigation will be weaved and introduction net illustrates the mitigation. For clarity, the weaving mechanism assumes that a base net does not share names with SPNs in aspects. Aspects weaving with the base net results in a new stochastic petri net. It can further be weaved with other aspects that involve the original base net. The order in which aspects are applied to a base net is not significant.

AOSPN model alone cannot tackle the increasing challenge of lack of data, how a system may react to certain security attacks although the chances of future security attacks are still unknown. There is little information known about the motivation and behaviour of attackers at this stage. To identify the attack trends report

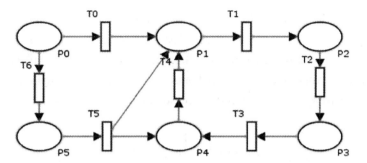

Figure 1. Stochastic petri net N1.

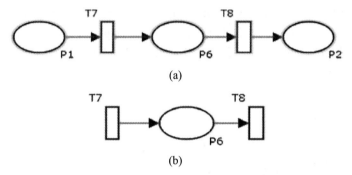

Figure 2. An aspect model with advice and introduction net. (a) Advice tcut. (b) Introduction net.

and vulnerabilities bulletin information in terms of CVSS is known. The benefit of CVSS is that it addresses the vulnerabilities directly and in collaboration with the vendors of the affected products. That is, CVSS tries to be specific and do not attempt to categorize attacks on a general basis nor does it provide a general model for estimating risk level. CVSS purely provides information about vulnerabilities on an operational level and leaves it to the vendors to add the information specific for their products and to the customers to interpret the information in the perspective of a particular Target of Evaluation. It is always better to use environmental metrics along with base and temporal metrics which we incorporated in our approach as given in CVSS to evaluate integrity, availability and confidentiality rather than productivity, reputation and privacy (Houmb and Franqueira 2009).

4. Modified Threat Driven Modeling Framework

The threat driven framework has been illustrated in **Figure 3** which is used to provide security in software based e-learning systems. This proposed framework comprises of six steps which are Disintegrate application (decomposition), Disintegration correction assessment, Threat analysis (Threat identification, Identify Application vulnerability, and Risk assessment), Threat mitigation, Mitigation correction assessment and Mitigation assessment.

Our framework has proposed some modifications over the threat driven framework proposed by (Shrief *et al.* 2010) and traditional framework. Out of these steps Disintegrate application, Threat identification and Threat mitigation are taken from traditional framework and framework proposed by (Shrief *et al.* 2010) as shown in (Howard 2003), while remaining steps were customized according to their usage with SPNs (Murata 1989; Peterson 1977; Haas 2002; Wang *et al.* 2009).

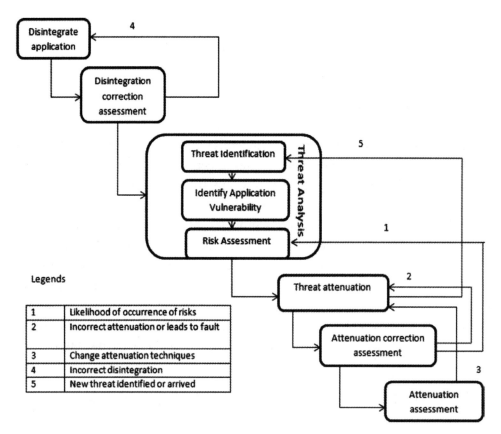

Figure 3. Threat driven modeling framework.

1) Disintegrate application: In this phase based on systems requirement the main module will be modeled using SPNs. The existing models using UML can be easily transferred to SPNs. Further SPNs utilized for system functions as deliverables.

2) Disintegration correction assessment: In this phase the behavioral properties for SPNs will be tested using basic properties of nets like Reachability, boundness, liveness and safeness. Due to changes in behavioral properties if the SPN leads to deadlock or starvation then changes can be made by reverting back to previous phase.

3) Threat analysis: This phase is carried out in three steps: Threat identification, Identification of application vulnerability and risk assessment. After disintegration phase threat has been identified and modeled through SPNs. Threat has been categorized using STRIDE in which identified threats marked on the SPNs as pointcuts. In next step security vulnerabilities for individual applications in e-learning were identified. Some of vulnerabilities are (authentication, authorization, input and data validation, configuration management, session management, auditing and logging etc.) for applications like virtual learning environment, student administration, mobile learning, virtual learning, certification etc. (Hayaati and Fan 2010). In last phase, the effect of threat can be identified on e-learning system using risk assessment. Threat matrix has been generated along with threats corresponding to vulnerabilities and prioritizes them on the basis of their likelihood of occurrence.

4) Threat mitigation: In this phase, the techniques to attenuate the threats are chosen. The deliverable of this phase is set of aspects describing the mitigations (introduction net) and how they will be inserted to original system with help of (advice nets) in specified pointcuts. If the new or unidentified threats occur after applying mitigation then they can be identified back in the previous phase i.e. threat analysis and then attenuated.

5) Mitigation correction assessment: In this phase if applying the mitigation leads to fault in the behavioral properties or due to incorrect mitigation then changes can be made by going back to previous phase and redesign the mitigations. After applying mitigation, if there is chance of likelihood of occurrence of risks then it can be redirected back to risk assessments phase in threat analysis to mi-

nimize threat affect in nets.

6) Mitigation assessment: This is a recurrent phase which will be repeated before and after mitigation in which various security metrics will be applied to determine the potency of selected threat mitigations. The system threat level will be indicated by numeric values. If there is no changes in results obtained by numeric values in decreasing order then that appropriate mitigation were not chosen. So better mitigation techniques will be selected further by returning back to previous phases.

5. Proposed Security Metric

The proposed security metric is based on the CVSS (Common Vulnerability Scoring System) which is customization and modification of work done by Wang *et al.* The security steps were customized so that they can be used with SPN models and they relied on the weakness of the e-learning system software. The proposed security metric process has been carried out in eight steps as follows:

1) Identification of weaknesses and vulnerabilities in applications.
2) Calculate severity for individual vulnerabilities.
3) Calculate the probability of vulnerability occurrence.
4) Calculate the probability of threat occurrence and risk assessment.
5) Calculate the percentage of each weakness.
6) Calculate the security metric.
7) Again calculate threats severity after mitigation.
8) Recalculate the security metric.

Various equations have been used for depiction of security metrics. The security metric (SM(s)) is calculated by product of severity of weakness (W_n) and risk of corresponding weakness (P_n) as shown in Equation (1). Here $n = 1, 2, 3, \ldots, m$.

$$SM(s) = \sum_{n=1}^{m} (P_n \times W_n) \tag{1}$$

Now, W_n is defined as average base score of its k vulnerabilities, as shown in Equation (2).

$$W_n = \sum_{i=1}^{k} \frac{V_i}{K} \qquad (2)$$

The percentage each representative weakness occurs in the overall weakness occurrences is used to calculate P_n as shown in Equation (3).

$$P_n = \frac{R_n}{\sum_{i=1}^{m} R_i} \qquad (3)$$

where R_n is the frequency of occurrences for each representative weakness in the SPN as shown in Equation (4), where K is the number of weaknesses and A is the sum of affected nodes in SPNs.

$$R_n = \frac{K}{\sum_{i=1}^{m} A} \qquad (4)$$

To make the value of SM(s) value to range from 0 to 10 is required to hold for P_n.

$$\sum_{n=1}^{n} P_n = 1 \qquad (5)$$

The severity of each weakness in e-learning systems after mitigation is re-calculated as shown in Equation (6). Here E denotes Exploitability, RL denotes Remediation Level and RC for Report Confidence which are important temporal metrics of CVSS. CR denotes Confidentiality Requirement, IR denotes Integrity Requirement and AR denote Availability Requirement which are environmental metrics of CVSS.

$$W_{n_{new}} = \sum_{i=1}^{k} \frac{V_i \times E \times RL \times RC}{K \times CR \times IR \times AR} \qquad (6)$$

For each mitigation weakness, if there exists certain vulnerabilities that still occurred are identified by recalculating Equation (4). If the number of affected nodes become same compared to the results obtained after applying mitigations then security metric has to be recalculated with help of Equation (1). The proposed system intended to identify threats and their analysis in design and analysis phase, therefore the number of nodes affected in the SPN will be compromised due to threat occurrence is used. To re compute the threat's severity after applying the mitigation CVSS based Equation (6) is added for solving computations.

6. Case Study

The proposed framework modules have been applied to case study on modeling of Question-Answer system for Udutu based e-learning system. (Shrief *et al.* 2010) developed their threat framework and applied their framework on AI specific question answering system which is different from our question answering system.

6.1. Decompose Application

The Question-Answer application in Udutu based system allows users to ask questions on Java Programming and system processes those questions (objective or subjective) and produces an answer. After user get authorized and authenticated by the system, he/she could enter question on Java modules. The system responsibility is to check whether there exists direct answer to that question or not. If direct answer exists, then it can be retrieved from knowledge base and displayed. Otherwise, system process the same question by searching the possible collective keywords to the nearest possible answers stored in knowledge base. From the available data, all possible answers have been created and from these answers select the best answer specified by user and finally display the appropriate answer.

The SPNs are best designed and modeled by Petri nets model which contain random events and perform processing of input data. The Petri net modeling is shown in **Figure 4** in which initial marking starts by one token in P0 that carries out different values throughout the transition firings from one place to another.

For better understanding of above depicted model, meanings of places and transitions are shown in **Table 2** respectively.

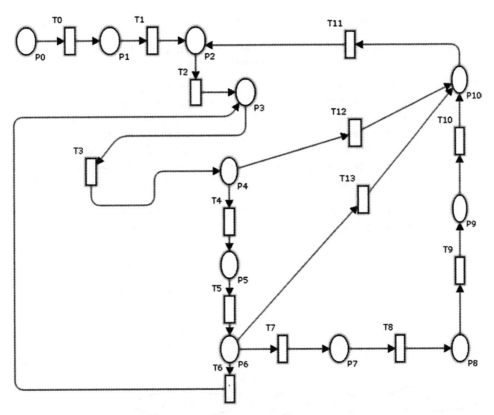

Figure 4. Petri net model for question answering system.

6.2. Decomposition Correction Assessment

To check correctness assessment for e-learning based system three main behavioral properties are required: reachability, boundness and liveness. Reachability determines whether a state can be reachable from one to other (Haas 2002). SPNs are k bounded if they doesn't contain more than k tokens in all reachable markings, including initial marking. Liveness determines that any state which is reachable can be fired without coming into deadlock situation. The reachability graph shows in **Figure 5** shows the different markings and various states of SPNs that can be reached. This case study on e-learning based system is 1-bounded, live and also known as safe SPNs. The nodes in **Figure 5** show different markings while arcs are labeled with transition names to show that marking is reached by firing of certain transitions.

Table 2. Place and transitions for question answer system.

Place number	Place description	Transition number	Transition description
P0	User login	T0	Authenticate
P1	Authenticated/authorized	T1	Go to main page
P2	System ready state	T2	Enter an objective question
P3	Objective question asked	T3	Search if a direct objective answer exists
P4	Direct answer yes/no	T4	Enter a subjective question
P5	Subjective question asked	T5	Search if a direct subjective answer exists
P6	Direct answer yes/no	T6	Searching the data in objective question knowledge base
P7	Data found	T7	Create the answers
P8	Answer formed	T8	Select from the answers
P9	Answer selected	T9	Display an answer
P10	Response displayed	T10	Getting response from answers
		T11	Exit
		T12	Decrypt answer formation decision and retrieve direct response for objective
		T13	Decrypt answer formation decision and retrieve direct response for subjective

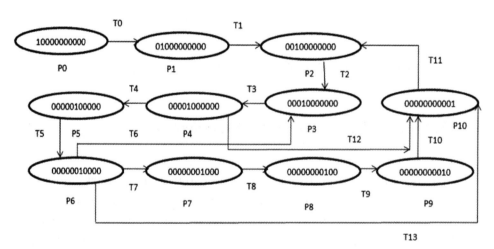

Figure 5. Reachability graph for SPNs of question answer system.

6.3. Threat Analysis

This module is divided into three phases as threat identification, application vulnerability and risk assessment.

6.4. Identify Threats

Various kinds of threats are used to be mitigated in present case study on e-learning system. First, when the user log into the system due to possibility of threats like network eavesdropping, password guessing, cookie reply there is chance to lead authentication vulnerability. Second, when system starts searching to check whether direct answer exists or not, an attacker can tamper the data and change the response formation mode. Third, an elevation of privilege can occur if an unauthorized user tries to decrypt the answer formation decision and search for data or display direct response. Finally, while creating possible answers from data gathered, an attacker can tamper the data and influence answer creation.

After identification of threats the security vulnerabilities for different applications in e-learning were identified then finally the threat is being analyzed by risk assessment matrix (Hayaati and Fan 2010). Risk of each threat is measured by assigning the likelihood of occurrence and impact to system. The risk evaluation done by using the risk evaluation grid proposed by (Barbeau 2005) Risks derived from threat analysis were classified in three main groups minor, major and critical which are decided by expert. The threat analysis result has been converted to the e-learning threats risk matrix as shown in **Table 3**.

Table 3. Threat matrix.

			E-learning Applications									
	Legend		VLE									
	CR	Critical	Online course admin	Course management	Communication tools							
	MA	Major										
	MI	Minor										
		Not relevant										
No	Vulnerability Category	Threats	Course learning	Grading center	Deliver learning content	Online session	Email	Personal Portfolio	File storage	Assessment tool	Mobile learning	Virtual library
1	Input & data validation	Buffer overflow	MI	MA	MA	MI	MA	MI	MA	CR	MA	MI
2	Authentication	Network Eavesdropping	MI	CR	MI	MI	MA	MA	MA	CR	MA	MI
3	Authorization	Elevation of privilege	MA	CR	MA							
4	Encrypt response decision	Change response formation mode	MA	MA	MA	MI	MI	MA	CR	CR	MI	MI
5	Encrypting system call arguments	Influence answer creation	MI	MA	MI	MI	MA	CR		CR	MI	
6	Auditing and logging	Malicious code	MI	MI	MA							MI

CR critical, *MA* major, *MI* minor, - not relevant.

6.5. Mitigate Threats

Various types of threats have been identified and determined from categories of STRIDE. Here, as a sample the aspect for threat mitigation of tampering with data threat is provided. Some other threats can also be mitigated as shown as encircled in **Figure 6**. The tampering with data threat is mitigated by encryption to prevent a code injection attack to influence the answer creation (Wang *et al.* 2009). As shown in aspect of threat in **Figure 6**, T19 is the transition which represents the encrypting system call arguments (Oyama 2006); P15 represents the state of system where all calls are encrypted so that no attack can happen and tamper the data; P14 is process where arguments are decrypted after processing and then encrypted

Advice net:

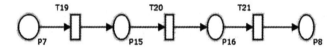

Introduction net:

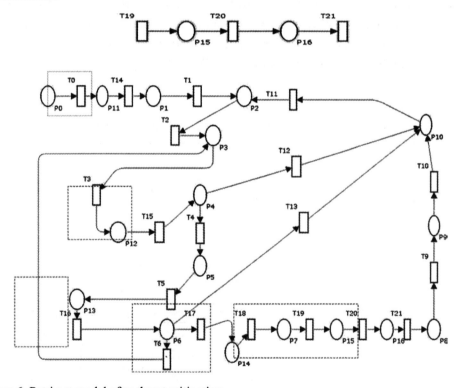

Figure 6. Petri net model after threat mitigation.

Chapter 16

again; P16 is the system state where answer formation gets completed; and T21 is the process of decrypting the system call arguments. Other threats like elevation of privilege threat will be mitigated by authorization; similarly networking eavesdropping will be mitigated by authentication. For given aspect of data tampering threat the pointcut is T8; the advice net and introduction net will be.

6.6. Security Metric Calculation (Mitigation Correction Assessment)

6.6.1. Identify Weakness and Vulnerabilities

Some of the important weaknesses in e-learning system were identified like network eavesdropping; change response formation mode; influence answer creation and elevation of privilege.

6.6.2. Calculate Severity for Each Vulnerability

The CVSS for each vulnerability should be calculated by assigning values to each of the six base metrics and creating the base vector as follows:

1) Network eavesdropping: The base vector will be AV:[A]/AC:[H]/Au:[S]/C:[N]/I:[N]/A:[P] = 1.5

2) Change response formation mode: The base vector will be AV:[A]/AC:[H]/Au:[S]/C:[N]/I:[C]/A:[N] = 4.3.

3) Influence answer creation: The base vector will be AV:[A]/AC:[H]/Au:[S]/C:[C]/ I:[C]/A:[P] = 6.2.

4) Elevation of privilege: The base vector will be AV:[A]/AC:[H]/Au:[S]/C:[C]/I:[N]/ A:[P] = 5.

6.6.3. Calculate the Probability of Vulnerability Occurrence

The probability of vulnerability occurrence can be calculated by identifying the weakness and vulnerabilities occurrence in the software. These calculations are computed or obtained from Equation (4) (R_n). 1) Network eavesdropping: $R_1 = 1/20$, 2) Change response formation mode: $R_2 = 1/20$, 3) Influence answer creation: $R_3 = 1/20 + 1/20 = 1/10$, 4) Elevation of privilege: $R_4 = 1/20 + 1/20 + 1/20 = 3/20$.

6.6.4. Calculate the Percentage of Each Weakness

The percentage of each weakness in the software is calculated from Equation (2) (Wn) and Equation (3) (Pn). 1) Network eavesdropping: $P_1 = R_1/(R_1 + R_2 + R_3 + R_4) = 0.15$, 2) Change response formation mode: $P_2 = R_2/(R_1 + R_2 + R_3 + R_4) = 0.15$, 3) Influence answer creation: $P_3 = R_3/(R_1 + R_2 + R_3 + R_4) = 0.28$, 4) Elevation of privilege: $P_4 = R_4/(R_1 + R_2 + R_3 + R_4) = 0.42$.

6.6.5. Calculate Security Metric

The outputs of Equations (2) and (3) are require to substituted in Equation (1) to obtain the security metric value. The security metric score is calculated based on Equation (1):

$$\begin{aligned} SM(s) &= W_1 \times P_1 + W_2 \times P_2 + W_3 \times P_3 + W_4 \times P_4 \\ &= (1.5 \times 0.15 + 4.3 \times 0.15 + 6.2 \times 0.28 + 5 \times 0.42) = 4.7. \end{aligned}$$

6.6.6. Recalculation of Severity of Threats after Mitigation

For obtaining a comparative analysis between the state before and after mitigation the security metric SM(s) should be recomputed again. The resulting value obtained after computation should be less than the one computed before mitigations.

The CVSS temporal score should be calculated for each mitigated threat by assigning values to each of temporal metrics and created the temporal vector. The temporal score for the mitigations of four identified threats are:

1) Authentication: The temporal vector will be E:[F]/RL:[W]/RC:[C] = 1.35.

2) Encrypt response decision: The temporal vector will be E:[POC]/RL:[W]/RC:[UR] = 3.5.

3) Encrypting system call arguments: The temporal vector will be E:[H]/RL:[W]/RC:[C] = 5.9.

4) Authorization: The temporal vector will be E:[F]/RL:[W]/RC:[C] = 4.5.

We have considered only confidentiality requirement (CR), integrity requirement (IR) and availability requirement (AR) (Heyman *et al.* 2008) metrics for calculation of Wnnew.

The environment metrics for identified threats are:

1) Authentication: The required environmental vector will be CR:[M]/IR:[H]/AR:[H].
Where M is 1.0, H is 1.51.

2) Encrypt response decision: The required environmental vector will be CR:[H]/IR:[L]/AR:[M]
Here M is 1.0, H is 1.51 and L is 0.5.

3) Encrypting system call arguments: The required environmental vector will be C:[H/IR:[H]/AR:[M].

4) Authorization: The required vector will be CR:[M]/IR:[H]/AR:[H].

From Equation (6) the new obtained value for $W_{n_{new}}$ need to be calculated which gives new value for severity of weakness after applying mitigations as:

$$W_{n_{new}} = \frac{1.35}{(1\times1.51)} + \frac{3.5}{(1\times1.51\times0.5)} + \frac{5.9}{(1.51\times1.51\times1)} + \frac{4.5}{(1\times1.51\times1.51)}$$

Recalculate the security metric:

The security metric score SM(s) could be computed based on Equation (1) after substituting $W_{n_{new}}$.

$$SM(s) = P1 \times \frac{1.35}{(1 \times 1.51)} + P2 \times \frac{3.5}{(1 \times 1.51 \times 0.5)} + P3 \times \frac{5.9}{(0.5 \times 1.51 \times 1)}$$

$$+ P4 \times \frac{4.5}{(1 \times 1.51 \times 1.51)}$$

$$= 0.15 \times 0.59 + 0.15 \times 4.63 + 0.28 \times 2.58 + 0.42 \times 1.97 = 2.33.$$

After evaluation of complete case study it was observed that before applying mitigations the threats determined in system and metric value was 4.7, whereas after applying the mitigations the threat mitigations the security metric was re-computed to check the effectiveness of the applied mitigations and scored metric value obtained was 2.33. It indicates that the mitigations were very effective in places where applied in system on basis of their occurrences. These security metric values indicate the effectiveness of applied mitigations and provide comparative analysis between different mitigations.

7. Performance Evaluation

We have compared our framework with two existing threat frameworks 1) traditional framework (Howard 2003) and framework proposed by (Shrief *et al.* 2010). The traditional framework only considered base metrics whereas the (Shrief *et al.* 2010) considered base and temporal metrics for the measurement of severity of threat. Our proposed framework is based on base, temporal and environment metrics therefore it gives better results as compared to the two existing frameworks. The comparative view of threat driven frameworks are shown in **Table 4**.

8. Conclusion

This paper has shown an effective security threat driven modeling framework, modified security metric with usage of CVSS and AOSPN models. In threat modeling framework correction assessment has been involved, mitigation correctness to measure the behavioral properties of SPNs and AOSPNs, and mitigation assessment to measure the mitigations effectiveness. These SPNs model weaved a

Table 4. Comparative view of various threat driven frameworks.

Author	Modeling	Framework	Equation
Howard (2003)	NA	Framework consists of 3 modules: 1. Decompose application 2. Identify threats 3. Mitigate threats	After mitigation the severity is calculated only on basis of base metrics i.e. $W_{n_{new}} = \frac{V_i}{K}$
Shrief et al. (2010)	Stochastic petri net	Framework consists of 6 modules: 1. Decompose application 2. Decomposition correction assessment 3. Identify threats 4. Mitigate threats 5. Mitigation correction assessment 6. Mitigation assessment	After mitigation the severity is calculated only in terms of base and temporal metrics i.e. $W_{n_{new}} = \frac{V_j \times E \times RL \times RC}{K}$
Our proposed approach	Aspect oriented stochastic petri nets	Framework consists of 6 modules and threat identification is divided into sub modules. 1. Disintegrate application 2. Disintegration correction assessment 3.1. Threat identification 3.2. Identify application vulnerability 3.3. Risk assessment matrix 4. Mitigate (Attenuate) threats 5. Mitigation (Attenuation) correction assessment 6. Mitigation (Attenuation) assessment	After mitigation the severity is calculated only in terms of base, temporal and environmental metrics i.e. $W_{n_{new}} = \frac{V_j \times E \times RL \times RC}{K \times CR \times IR \times AR}$

point cut, advice nets and introduction nets into existing petri net system. Finally, security metric calculations were computed for SPNs with usage of CVSS and a new modified equation introduced by using base, temporal and environmental metrics to calculate the metric after mitigations to perform comparison among them.

Authors' Contributions

AK developed proposed modified threat modeling framework and calculate the security metrics before and after mitigations. BP developed threat matrix and calculated the probabilities of vulnerabilities occurrences. Both authors read and approved the final manuscript.

Acknowledgements

The authors thank infotech services for their valuable contributions in collection of vulnerabilities data identification in e-learning. The authors thank Dr. Shrikant for his advices concerning the threat analysis design.

Competing Interests

The authors declare that they have no competing interests.

Source: Khamparia A, Pandey B. Threat driven modeling framework using petri nets for e-learning system [J]. Springerplus, 2016, 5(1): 1–16.

References

[1] Balogh Z, Turcáni M (2011) Possibilities of modelling web-based education using IF-THEN rules and fuzzy petri nets in LMS. Commun Comput Inf Sci 251(1):93–106.

[2] Balogh Z, Turcáni M, Magdin M, Burianová M (2012a) Creating model educational processes using petri nets implemented in the LMS. In: Efficiency and responsibility in education 2012: 9th international conference, FEM CULS, Prague, pp 7–16.

[3] Balogh Z, Magdin M, Turčáni M (2012b) Development of a universal model of e-course using petri nets. Development 50:231–248.

[4] Barbeau M (2005) Wimax/802.16 threat analysis, pp 8–15.

[5] Dehlinger JS, Nalin V (2006) Architecting secure software systems using an aspect oriented approach: a survey of current research. IOWA State University, Computer Science.

[6] Haas PJ (2002) Stochastic petri nets: modeling, stability simulation. Springer, New York.

[7] Hammami S, Mathkour H (2013) Adaptive e-learning system based on agents and object petri nets. Comput Appl Eng Educ 23(2):170–190.

[8] Hayaati AN, Fan I (2010) Information security threat analysis for e-learning. Tech education, CCIS 73. Springer, Berlin, pp 285–291.

[9] Hecker A (2008) On system security metrics and definition approaches in the second international conference on emerging security information systems and technologies.

[10] Heyman T, Huygens C, Joosen W (2008) Using security patterns to combine security metrics in proceedings of third international conference on availability, reliability and security.

[11] Houmb SH, Franqueira VNL (2009) IEEE proceedings on forum of incident response

and security terms, pp 1–9.

[12] Houmb SH, Sallhammar K (2012) Modeling system integrity of a security critical system using colored petri nets. Springer, Berlin.

[13] Howard M (2003) Writing secure code, 2nd edn. Microsoft Press, New York.

[14] Jalal A, Zeb MA, Peshawar P (2008) Security enhancements for e-learning portal. Int J Comput Sci Netw Secur 2:236 Jansen W (2009) Directions in security metrics research, computer security division. National Institute of Standards and Technology, Gaithersburg.

[15] Jensen PHM (2008) Secure software design in practice in the third international conference on availability, reliability and security.

[16] Mell P, Romanosky S (2007) A complete guide to the common vulnerability scoring system. National Institute of Standards and Technology, Carnegie Mellon University, Gaithersburg.

[17] Murata T (1989) Petri nets: properties, analysis and applications. In: Proceedings of IEEE, pp 541–580.

[18] Omrani F, Harounabadi A, Rafe V (2011) An adaptive method based on high level petri nets for e-learning. J Softw Eng Appl Issue 4:559–570.

[19] Oyama Y (2006) Prevention of code injection attacks by encrypting system call arguments, Technical report TR0601 Payne SC (2006) A guide to security metrics, SANS Institute Reading Room.

[20] Peterson JL (1977) Petri nets. ACM Comput Surv 9(3):223–252.

[21] Schauerhuber A, Kapsammer E, Retschitzegger W, Wimmer M (2006) Towards a common reference architecture for aspect oriented modeling. In: Proceedings of the 8th international workshop on aspect oriented modeling (AOM), Germany, pp 876–879.

[22] Shrief NH, Hamid A, Mahar KM (2010) Threat driven modeling framework for secure software using aspect oriented stochastic petri nets. In: IEEE proceedings of 7th international conference on informatics and systems, Cairo, pp 238–246.

[23] Wang HW, Guo M, Xia M (2009) Security metrics for software systems. In Proceedings of 47th annual southeast regional conference, Paris, pp 345–357.

[24] Xu D, Nygard KE (2006) Threat driven modeling and verification of secure software using aspect oriented petri nets. IEEE Trans Softw Eng 32(4):212–224.

[25] Xu D, Nygard K (2006) An aspect oriented approach to security requirements analysis in 30th annual international conference computer software and applications conference.

Chapter 17
Secure Data Networks for Electrical Distribution Applications

David M. Laverty, John B. O'Raw, Kang Li, D. John Morrow

School of EEECS, Queen's University Belfast, 125 Stransmillis Road, Belfast BT9 5AH, UK

Abstract: Smart Grids are characterized by the application of information communication technology (ICT) to solve electrical energy challenges. Electric power networks span large geographical areas, thus a necessary component of many Smart Grid applications is a wide area network (WAN). For the Smart Grid to be successful, utilities must be confident that the communications infrastructure is secure. This paper describes how a WAN can be deployed using WiMAX radio technology to provide high bandwidth communications to areas not commonly served by utility communications, such as generators embedded in the distribution network. A planning exercise is described, using Northern Ireland as a case study. The suitability of the technology for real-time applications is assessed using experimentally obtained latency data.

Keywords: Telecoms, Latency, WiMAX, Security

1. Introduction

Security of telecommunications is an important concern in the development of Smart Grid applications, especially those involved with real-time protection and control of grid infrastructure. Since the electrical grid spans a large geographical area, it is necessary to use a wide area network (WAN) for communication. Typically, the applications use packet switched protocols, usually internet protocol (IP). In the last number of years, there have been several high profile cyber security attacks on notable large corporations, and significant vulnerabilities have been found in software technologies which underpin Internet communications[1]–[3]. The security technologies developed for use in enterprises (banks, businesses) may not be sufficient for security of industrial control systems. Although it is desirable to avoid connection with public networks, for some applications this may prove unavoidable. The authors propose the use of WiMAX radio technology as an affordable solution for an electrical utility to create an independent WAN, isolated from public Internet infrastructure. This would have application where secure, high bandwidth communication is required to enable real-time applications. Many such applications are proposed in[4], including control of distribution applications such as embedded generation and electric vehicles[5][6].

This paper presents a technical summary of WiMAX, with range and throughput calculations being developed. Based on these figures, a radio planning study is conducted using professional software to ITU standards. Northern Ireland is used as a case study in this exercise, considering connectivity at wind farm generation sites. The technical suitability of WiMAX for real-time power systems applications is considered in light of experimentally obtained performance data. It is shown that throughput is greatly in excess of the requirement for synchrophasor streaming[7], while the latency easily meets the relevant IEEE standard[8].

The authors recommend that WiMAX is an available solution that may be rapidly deployed to service remote areas of the utility with data telecoms. WiMAX is of low cost, meets immediate needs, and as the utility transitions to fiber for telecoms WiMAX will continue to provide telecoms diversity. WiMAX technology offers a cost effective solution which meets the needs of current Smart Grid applications.

2. Telecoms Network Modernisation in Electric Utilities

Electrical utilities operate many essential telemetry and telecontrol functions over telecoms delivery technologies with many decades of track record in terms of reliability. Modern applications, such as phasor measurement units[4][9], and optimizing control of embedded generation[5][6][10], are demanding bandwidths beyond what established telecoms delivery technologies were designed for[4][11]. The de-facto solution for high bandwidth telecoms needs is to install a data network, typically operating on IP. This can be achieved in a number of ways, including extending the company enterprise network, or provisioning an internet service at the remote substation through a local provider.

Understandably, there is reluctance amongst utility personnel to abandon proven systems for new telecoms solutions, especially those operating IP. Recent press coverage highlights intentional sabotage of high profile networks by 'hacktivist' groups[12], the 'Stuxnet' worm tailored to attack industrial control systems[1], the 'Heartbleed' vulnerability in OpenSSL[2], and the 'Shellshock' vulnerability in Bash[3]. All of these scenarios give rational cause for alarm.

A solution to provide high bandwidth telecoms to remote endpoints that does not traverse public internet would go some way to alleviating many of the intrinsic security concerns.

2.1. Traditional Utility Substation Telecoms

The traditional substation has long operated on the principle of fixed wiring between instrumentation, relays and intelligent electronic devices (IEDs) alongside growing use of low speed serial links[13][14]. Early substation-to-substation communication, syguupervisor control and data acquisition (SCADA), was achieved through use of private or leased telephone lines, microwave, UHF radio or power-line-carrier (PLC) carrying either analogue signals or delivering low data-rates by modem (<56kbps). In Northern Ireland, UHF scanning telemetry operates at 9600baud.

Subsequently, utilities began to install fiber-optic cables between substations, operating E1 and/or synchronous digital hierarchy (SDH) multiplexers. E1 and SDH technologies are based on circuit-switched time division multiplexing (TDM), and are connection orientated. A dedicated circuit is created and used for a specific purpose and available to the application at all times. Bandwidth is constant and traffic transmitted at the speed allowed by the physical media, therefore the latency is very low. Available is of the order of 99.999%[13]. E1 circuits operate at 2.048Mbps in Europe, or the nearest equivalent T1 at 1.544Mbps in USA. A standard application circuit, E0, is 64kbps. This allows 32 circuits on an E1 carrier in Europe, or 24 in USA. Given their excellent latency and constant bandwidth, utility telecoms engineers highly regard the performance of TDM communications.

Whilst highly suitable for many serial telecoms applications including protecting signaling, 64kbps TDM circuits are inadequate in terms of bandwidth for many smart grid applications. While circuits may be multiplexed for additional speed, this adds complexity and is not a long term solution to growing bandwidth needs.

2.2. Transition to Packet Switched Networks

Current practices by IED manufactures are pushing the utility telecoms towards standardization on common communications architecture. Protocols such as DNP3[15], IEC 61850[16] and IEEE C37.118.2[7] have Ethernet as the underlying technology, leading to a growth in use of Ethernet/IP for data transport within and between substations[17].

Ethernet is a packet-switched technology. Messages are separated into segments and transmitted individually across dynamically created connections, which results in a more efficient use of network bandwidth[18]. Ethernet networks are less costly[17] and provide better scalability than TDM networks. The downside is that latency is non-deterministic in the packet network, such that data may arrive early, late or fail to arrive at all. Protocols such as TCP/IP build upon IP to provide correct sequencing of data and guarantee of delivery. Protocols and products exist which attempt to emulate E0/E1 connections over IP networks[14].

As it enters into legacy support and availability, the cost of TDM technology is prohibitive to development of a telecoms network suitable for smart grid. Indeed, given the diversity of applications, it is apparent that no single delivery technology can yield a comprehensive solution to meet all utility telecom needs. However, the use of standardized protocols and addressing schemes will allow messages to traverse many types of media. The authors consider the trend towards a fully IP smart grid is rational in terms of cost and bandwidth, and will show that such networks demonstrate adequate latency, security and reliability. "There is no doubt that Ethernet/IP-based networks are the emerging trend in modern substation communications[17]."

3. WiMAX Technology Overview

Following the success and almost universal adoption of Wi-Fi (802.11 a/b/g/n) on the local area network (LAN), it became clear to equipment vendors and standards groups that demand existed for similarly flexible and utilitarian standards over greater distances.

The IEEE standards board established the IEEE 802.16 working group in August 1998 to develop standards appropriate to fixed wireless access (FWA) broadband. The first standard (802.16-2001) was released in April 2002, dealing with point-to-multipoint line-of-sight propagation in the 10 GHz–66GHz band using single carrier modulation.

To facilitate non line-of-sight, propagation 802.16a was ratified in 2003. This amendment allowed for operation at lower frequencies, 2GHz–11GHz, and the physical standard was extended to use orthogonal frequency division multiplex (OFDM) modulation, allowing increased throughput for the same radio bandwidth. IEEE 802.16d[19] was released in 2004, harmonizing some aspects of the 802.16 standard with the European HIPERMAN[20] and consolidating previous revisions. For the current generation of licensed "last-mile" broadband installations, this is the defacto standard. Alongside the standardized developments in WiMAX, a range of proprietary FWA solutions have emerged, many using adaptations of the original 802.11a protocols. Subsequently IEEE 802.16e-2005, referred to as "mobile WiMAX", was developed. However, in the mobile communications sector, mobile WiMAX has been largely sidelined by greater operator adoption of long-term-evolution (LTE). In this work, the fixed version of WiMAX, IEEE 802.16d, and proprietary FWA solutions are used.

Comparison between WiMAX and Wi-Fi requires caution since the two technologies have different applications. Wi-Fi offers insufficient range for wide area networks such as required in utility applications. **Table 1** lists the key differences, note that Wi-Fi is intended for short range/indoor applications and WiMAX, in its fixed form, is primarily for long range infrastructure applications, i.e. WAN. The quoted maximum speed of typical Wi-Fi and WiMAX products[21][22] are included in **Table 1** for reference but, as will be developed later in this paper, the actual throughput is subject to a complex range of interrelated factors. The quoted typical speeds are based on practical experience and offer a more realistic comparison, but regard that the distance covered by WiMAX is two orders of magnitude greater than for Wi-Fi.

The quality-of-service (QoS) mechanisms of WiMAX and Wi-Fi are different. WiMAX provides quality of service through a connection orientated scheme operated at the base station. This delivers packets to subscriber units based on priority, and similarly schedules subscriber units to transmit based on the priority of their data. Wi-Fi uses contention based quality of service, where subscribers try to access the access point based on random time intervals. This is a poor mechanism for time sensitive data.

Wi-Fi operates in the license exempt industry, science and medical (ISM) bands at 2.4GHz and 5GHz. WiMAX can also operate in these bands, or a license may be applied for and the network operated in a licensed part of the spectrum. This is useful for utilities as there is no mechanism for recourse for interference in a license exempt band. Wi-Fi operates acceptably indoors at short ranges, while WiMAX usually requires line-of-sight when used over long distances.

Table 1. WiMAX and Wi-Fi comparison.

	Wi-Fi (802.11n)	WiMAX (802.16d)
Range	300 m	30 km
Speed (max)	54–300 Mbps	300 Mbps
Speed (observed)	50 Mbps (at 10 m)	30 Mbps (at 15 km)
Quality-of-service	Poor	Good
Frequency	2.4, 5 GHz	900 MHz to 66 GHz
Spectrum	Unlicensed	Licensed/Unlicensed

4. WiMAX Network Topology

The network topology used to service wind farms with high speed data connectivity is that of fixed base stations serving numerous fixed subscriber units. Before radio planning can commence, it is necessary to calculate the distance subscriber units can be from their base station so that base stations can be appropriately placed.

The link between the base station and the subscriber unit, in this work, is considered to be the 'last-mile'. The radio frequency band used is 5.8GHz, a license exempt/light licensed band. In other applications, it may be more appropriate to operate under license in the 3.5GHz band, but the same methodology will apply. Although microwave links can be established without line-of-sight, experience suggests that at 5.8GHz adequate performance in range and throughput can only be achieved with line-of- sight to the base station.

The data throughput possible on a radio link is determined by the link budget, a complex mix of interacting factors typically modeled by standards such as ITU-R P.530[23]. However, in rural areas with predicable and well understood topology and with clear line of sight links (LOS) only, the service area from any base site can be related to the received signal level (RSL) at the receiver. For this exercise, factors such as clutter, noise floor and urban topology are ignored or included in the fade margin. For any given RSL, there are three main parameters which determine the throughput; namely the modulation, coding (e.g. QPSK/ 16-QAM/64-QAM) and spatial streams. Modern equipment dynamically adjusts these parameters to achieve the maximum throughput given the actual RSL. With the test equipment used, the minimum data rate, when modulation and coding are adjusted for poor RSL, is 6.5Mbps. Adjusted for excellent RSL, the equipment can theoretically achieve 300Mbps of throughput. Experience suggests that actual throughput will be of the order of 60% of what the modulation and coding scheme rating allows. In this work, the RSL is engineered so as to achieve an actual throughput of 30 Mbps, indicating a theoretical throughput of 50Mbps is required.

A full derivation of how RSL is determined is beyond the scope of this article. Briefly, the limiting factor is the equivalent isotropically radiated power (EIRP). The maximum allowable EIRP in the UK is 36dBm (4W). EIRP is a function of the output power of the transmitter (Pt) and the gain of the transmit antenna (G_t). When higher gain antennas are used, the transmitter must limit its power so as not to exceed the legal EIRP. That is, $(P_t + G_t) = 36dBm$.

The RSL can be determined using the logarithmic version of the Friis equation[24], where P_r is the power at the receiver (RSL), P_t the power output at the transmitter, G_t and G_r are the transmitter and receiver antenna gains, k is the wavelength of the signal, and d is the distance between the transmitter and receiver. Rearranging and substituting k for f in MHz yields (2) which allows range D to be calculated in kilometers.

$$P_r = (P_t + G_t) + G_r + 20\log_{10}\left(\frac{\lambda}{4\pi d}\right)$$

$$D = \frac{10^{(P_t+G_t+G_r-P_r)/20}}{41.88 \times f}$$

The WiMAX base station is configured with three 1200 sector antennae, or four 900 sector antennae, which allow for communication with subscriber units close to the base station. This mode of operation is described as point-to-multipoint (PtMP), since one base station antenna serves many subscribers. Subscriber units further away will require a dedicated high gain antenna on the base station. This mode of operation is described as point-to-point (PtP). These modes are illustrated in **Figure 1**. Note that subscriber units (SU) close to the base station are served by four 900 sector antennae in PtMP mode, while distant units are served by dedicated high gain antennae in PtP mode.

Since the link is bidirectional, each antenna both transmits and receives. As EIRP is limited to the same value at both the subscriber unit and the base station, the range will be limited by the lower of the two antenna gains. Directional antennae may be used at the subscriber unit, so in practice the base station antennae will limit range.

From the equipment datasheet[21], an RSL of –86dBm is required to achieve 50Mbps throughput. It is normal practice to consider a fade margin to allow for variations in RSL due to changes in environmental conditions. Allowing a 6dB fade margin yields $P_r = -80dBm$. The equipment will pass data with an RSL of –96dBm, thus the functional fade margin is 16dB.

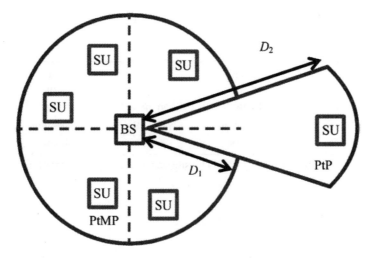

Figure 1. Modes of operation of the fixed WiMAX base station (BS).

Consider transmission from the subscriber unit to the PtMP array on the base station. The subscriber unit has gain $G_t = 27$dB, thus $P_t = 9$dBm (8mW). Frequency f is 5745MHz. If three 120° sector antennae ($G_r = 16$dB) are used at the base station and ($P_t + G_t = 36$dB), from (2) the range D_1 is found to be 16.5km. Changing configuration to four 90° antennae ($G_r = 20$dB), the range D_1 is increased to 26.2km. Repeating for PtP mode, a subscriber unit type antenna at the base station ($G_t = 27$dB) yields $D_2 = 58.7$km. These parameters and results are summarized in **Table 2**.

Since PtP links require that their own frequency and bands are limited at 5GHz, PtP operation is discouraged. OFCOM regulations[25] currently allow for only 4 channels in the 5GHz Band C, the preferred band for FWA. Although 11 channels exist in Band B, these may only be used at much lower power (1W or 30dBm) and are only suited for short range FWA. Thus, it seems prudent to design the network for 4-sector antennae. This will increase the range for PtMP links and reduce the number of base stations required for a small additional equipment cost. Thus, subscriber units closer than 26km from the base station operate in PtMP mode via sector antennae, while more distant units operate using PtP mode up to 58 km. This is in line with the commercial WiMAX operator's practice and experience. Note that subscribers closer to the base station than D_1 or D_2 would expect better than designed performance for their mode. The above discussion is a very simplified summary of the process involved in PtMP FWA planning.

Table 2. WiMAX base station range parameters.

Mode	3-sector	4-sector	PtP
P_t	9 dBm	9 dBm	9 dBm
G_t	27 dB	27 dB	27 dB
G_r	16 dB	20 dB	27 dB
P_r	−80 dBm	−80 dBm	−80 dBm
D	16.5 km	26.2 km	58.7 km

Note: Subscriber unit (P_t, G_t) to base station (P_r, G_r), $f = 5745$ MHz

5. Wireless Network Planning

Using the ranges calculated in the previous section, and assuming line-of-sight operation is required, it is possible to simulate radio propagation and determine appropriate placement of base stations to serve known subscriber sites. This section will consider a case study in which it is desired to provide high speed connectivity to wind farms in Northern Ireland. In this study, there are 51 wind farms that require a connection, **Figure 2**.

The radio propagation is modeled in the ICS software package from ADTI Ltd[26] and uses ITU-R P.525[27] for the calculation of free space attenuation. This software is used professionally for such studies and allows the use of many data sets to assist radio planning. In this study, where line-of-sight operation is required, geographical terrain data is obtained from the NGA/NASA "Shuttle Radar Topography Mission"[28].

5.1. Existing 452 MHz SCADA Towers

It is first worth considering if broadband communication is possible from utility radio towers that presently operate wireless SCADA at 452MHz. At this frequency, line-of-sight is not a requirement. **Figure 3** shows the radio propagation from the 21 SCADA sites available. Since the SCADA sites are low lying, range is limited and many of the SCADA sites are located in areas far from the wind farms. Using these sites, 36 of the 51 wind farms are serviced from 9 SCADA sites. Although this seems favorable, the propagation path is not ideal and individual site surveys would be necessary.

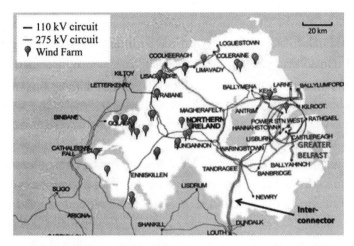

Figure 2. Map of Northern Ireland transmission system and wind farms.

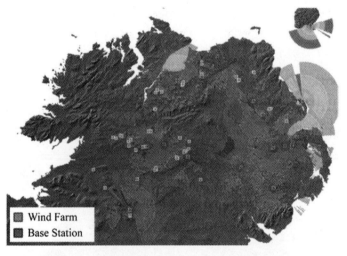

Figure 3. Range from 452MHz SCADA sites if operated as WiMAX base stations (with very limited coverage of wind farms).

5.2. Using Preferred "High-Sites"

The objective of any commercial FWA provider is to provide service to the maximum number of subscribers with the minimum cost in infrastructure. The ideal sites will be those situated on mountain tops with a clear view of the surrounding land and other mountain tops. These are typically already identified on maps as 'trig points' or triangulation pillars, by the government mapping department. It is sometimes possible to acquire a radio site adjacent to such points, or these may be used as a starting point to determine an optimum position. Factors such as local planning considerations and the availability of land for rent/purchase are limiting factors. Additional concerns include vehicular accessibility and the potential for electrical power supply.

Suitable base station locations have been assessed using the method above by an incumbent WiMAX operator in the region. This operator has 103 base station locations, constructed of wooden monopole in a concrete base. Many of these sites provide fill-in for areas where other base stations cannot reach (e.g. valleys). Since wind farms are usually located on high sites, **Figure 4** shows that only a fraction of the available sites are necessary to provide coverage of wind farms. All 51 wind farms can be served using 16 base stations, however optimizing for distance and performance using the ATDI tool, 21 sites would be preferred.

5.3. Costs

The cost of deploying the WiMAX network is subject to a number of factors, including availability and suitability of sites, availability of services (electricity,

Figure 4. Complete coverage of all 51 wind farms using 21 base stations from WiMAX operator's available sites (Available but unused base station sites are indicated in black).

backhaul) and accessibility (roads, rights of way). The figures stated in this paper are indicative of experience and should serve as a guide.

The capital cost of a base station assumes that electrical supply is available nearby and that rights-of-way can be negotiated. The most economical base station structure, a 10m monopole mast, will cost circa US$30k including civil engineering works. For more complex sites using a steel tower, costs will be a multiple of this. Electronic equipment to serve per sector (antennas, transceivers, power supply) will cost of the order $8k, while licensed microwave/MPLS equipment to provide backhaul between base stations is of the order of $80k Ongoing costs will include electricity consumption ($150/year) and site rental (varies between urban/rural sites). A typical 4-sector base station can be considered to cost $120k to install, turnkey and complete.

The cost of subscriber units varies by the technology used. An unlicensed band PtMP subscriber unit typically costs $150, or circa $800 for a licensed band unit. PtP equipment costs circa $3000 for an unlicensed band, or $6000 for a licensed band unit.

6. Technical Performance

The technical performance of WiMAX has been assessed experimentally by using a commercial WiMAX network in operation in Northern Ireland, owned by Northwest Electronics (NWE). Network performance has been assessed using data from two computers connected at wind farms for the purpose of synchrophasor monitoring. The topology of the network is shown schematically in **Figure 5**. Each subscriber unit is connected via unlicensed WiMAX at 5.8GHz, while the core network operates licensed microwave. The two wind farm computers communicate via a total of 9 core network sites. The link between the SUs and their nearest base stations is PtMP. These base stations link to the core network using 5.8GHz PtP links. The core network, comprising 6 of the links, operates licensed, MPLS links which give near fiber like performance. Through agreement with the service provider, the authors' equipment is operated in the same manner as it would be on a private installation. Thus the performance data shown is reflective of what a private utility installation would achieve.

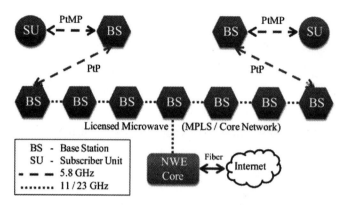

Figure 5. Schematic of system used in testing performance of the WiMAX network.

6.1. Throughput

The throughput has been monitored by performing a speed test using an Internet based tool at hourly intervals. The results, presented in **Figure 6** show that speeds of greater than 7Mbps are consistently available on the PtMP sector of Wind Farm 1 (Tappaghan, Co. Fermanagh), while 15Mbps is the lowest speed seen on Wind Farm 2 (Elliot's Hill, Co. Antrim). Throughput at both wind farms are capped by the ISP at 10Mbps and 20Mbps respectively. These speeds are far in excess of the 75kbps requirement for synchrophasor streaming.

6.2. Latency

Latency requirements for various power systems applications have been defined in IEEE Std. 1646 and are summarized in **Table 3**. The performance of the WiMAX network has been assessed by performing a "ping" test between two wind farm sites connected to the network. A ping sends a packet from one computer to a remote computer and requests an echo. The time until the echo is received is recorded by the sending computer. The ping test operated over the course of one month is summarized as a cumulative probability distribution in **Figure 6**. Note that each wind farm is on a PtMP sector.

Figure 7 shows the round trip time, the time taken by a ping packet to across the WiMAX network both to the remote computer and back to the sending

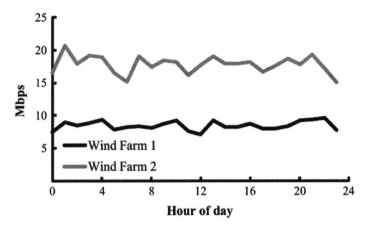

Figure 6. Throughput measured on WiMAX connections at wind farms at hourly intervals.

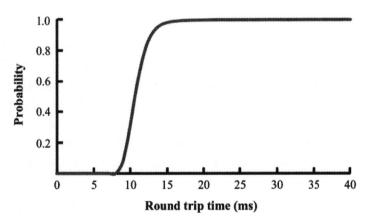

Figure 7. Round trip time (latency) to traverse WiMAX network from one wind farm to another.

Table 3. Latency requirements of IEEE Std 1646.

Speed required	Typical application	One-way latency
Very high	Streaming VT and CT samples and protection signals to switchgear	<2 ms
High	Event notification for protection	2~10 ms
Medium	Exchange non-critical information between protection units, exchange information between control functions, synchrophasor measurements	10~100 ms
Low	Message delivery external to the substation (control centre) or the substation computer or IEDs	>100 ms

computer. After 15ms, 98.0% of ping requests are returned, and by 29ms, 99.9% of pings are returned. The latency in one direction is found by halving the round trip time.

Data can reliably traverse the WiMAX network from one site to another in better than 15ms. This places the performance, according to IEEE 1646, in the "High" category. Further analysis using the "trace route" reveals that most of the latency experienced arises from the base station to subscriber unit link at each end, approximately 3ms in each case. The latency across the core part of the network, which uses licensed microwave/MPLS, is approximately 2ms.

6.3. Reliability

Standard models exist for estimation of link availability and reliability, however a detailed analysis is beyond the scope of this paper[27]. Any detailed analysis will consider the climate, terrain and link parameters to arrive at a fade margin and an availability figure, normally expressed as a percentage. The equipment under test uses 2 × 2 multi-input multi-output (MIMO), there are two spatial streams where both horizontal and vertically polarized signals are utilized. Fading is generally uncorrelated for horizontal and vertically polarized signals and the probability that both signals will fade below their margin is the product each individual fade probability. The overall availability in this case will be better than 99.99% using the carrier's heuristic fade margin of 16dB. No complete losses in connectivity were seen during the duration of the network testing, thus the reliability of the network has been considered on the basis of meeting the 'medium' category of IEEE Std. 1646, which specifies synchrophasor measurements as one of the functions which should meet this standard. Over the course of testing, 99.99% of packets traversed the network in less than 100 ms. To put this into perspective, a PMU reporting at 50 frames per second would experience delays on 464 of the 4.32 million frames reported during each day. Note that no frames would be lost during the time the connection was under observation.

7. Security

From a network engineering perspective, WiMax provides an alternative to a wired Ethernet network. In ISO terms[29], it is the layer 1 physical transport for layer 2 frames. Of the devices used and tested, all supported operation as a layer 2 device (a switch) or as a layer 3 device (a routed connection).

Configuring the WiMax base station unit (BSU) as a layer 2 device in PtMP results in receive stations units (RSU) appearing to be in the same layer 2 network segment. End nodes are accessible to each other and a range of attacks are feasible. Address resolution protocol (ARP) attacks such as impersonation, redirection and man-in-the-middle (MitM) have been demonstrated. Where the RSU connects directly to a remote network segment, local "housekeeping" traffic may be propagated to other RSUs in the same multipoint network, providing opportunities for an external attacker to reconnoitre the remote network segment. These are all clearly undesirable characteristics.

Configuring each link from a BSU as a layer 3 link allows each RSU to appear as a router on the network. Although this makes network planning more complex, it allows for separation of RSUs from a security perspective and depending on the BSU equipment, may allow for the implementation of access control lists (ACLs) and other measures to restrict inter-RSU traffic and unauthorised access to the remote network. It does however restrict layer 2 only protocols to the local network. Such protocols are often found in sub-station equipment, for example DNP3 specifies a layer 2 protocol. In pilot PtMP sites, RSUs were configured as layer 3 devices with a class C address assigned to each BSU. Dividing address space into/30 subnets gave 62 usable RSU connections per BSU.

PtP links were also evaluated. These are expensive in BSU hardware but the main limitation in using these links is that a dedicated frequency and bandwidth required, restricting scalability.

At layer 2 and above, all the normal security concerns exist and all the normal mitigation strategies may be applied. Typical applications tested by the authors use conventional VPN technologies for secure and encrypted communications from the sub-station/RSU to the control centre. Conventional routing protocols operate transparently on this network allowing the underlying secure infrastructure to be completely transparent to the applications running on it and to dynamically respond to changes in topology. This approach is also transparent to the sub-station applications, the underlying network appearing as a two hop routed network.

8. Conclusions

This paper has described the technical background of WiMAX/FWA and discussed the design procedure for using this technology as a delivery mechanism for data in Smart Grid/electric vehicle applications. The performance of WiMAX has been evaluated using empirical data obtained from a commercial WiMAX network in Northern Ireland. The WiMAX system is shown to operate satisfactorily in terms of throughput, latency and reliability.

A planning study has been conducted to determine the suitability of WiMAX for use as a telecoms delivery mechanism for wind farms. The study determined that while existing utility SCADA towers may be of some use, to achieve the necessary line of sight to the wind farms, new base stations are required. To service the 51 wind farms of Northern Ireland, 21 base station locations are preferred. Costs for base stations and subscriber units are indicated.

WiMAX is proposed as an immediately available solution to address the data connectivity needs demanded by synchrophasor applications. WiMAX technology may be rapidly deployed at reasonably low costs to areas where limited data alternatives exist. In utilities with long term objectives to operate ubiquitous fiber optic telecoms, WiMAX can serve as a diverse solution, providing backup connectivity in the event of failure of primary fiber connectivity.

With industry trends moving towards Ethernet/IP communication as standard, the availability of security solutions including licensed bands, MPLS segregation and VPN encryption makes WiMAX an attractive solution for providing data connectivity for remote smart grid applications.

Acknowledgements

This work was supported by the EPSRC/NFSC co-funded Intelligent Grid Interfaced Vehicle Eco-charging (iGIVE) Project (EP/L001063/1). The authors would like to thank Letterkenny Institute of Technology, Ireland, for hosting equipment used during this study.

Open Access

This article is distributed under the terms of the Creative Commons Attribution 4.0 International License (http://creativecommons.org/licenses/by/4.0/), which permits unrestricted use, distribution, and reproduction in any medium, provided you give appropriate credit to the original author(s) and the source, provide a link to the Creative Commons license, and indicate if changes were made.

Source: Laverty D M, O'Raw J B, Kang L I, *et al.* Secure data networks for electrical distribution applications [J]. Journal of Modern Power Systems & Clean Energy, 2015, 3(3): 447–455.

References

[1] Chen TM, Abu-Nime S (2011) Lessons from Stuxnet. Computer 44(4):91–93.

[2] The heartbleed bug. http://heartbleed.com/ Accessed 28 Sept 2014.

[3] What is #Shellshock? https://shellshocker.net/ Accessed 4 Feb 2015.

[4] Patel M, Aivaliotis S, Allen E *et al.* (2010) Real-time application of synchrophasors for improving reliability. North American Electric Reliability Corporation (NERC), Princeton.

[5] Best RJ, Morrow DJ, Laverty DM *et al.* (2010) Synchrophasor broadcast over internet protocol for distributed generator synchronization. IEEE Trans Power Deliver 25(4):2835–2841.

[6] Laverty DM, Best RJ, Morrow DJ (2014) Loss-of-mains protection system by application of phasor measurement unit technology with experimentally assessed threshold settings. IET Gen Transm Distrib 9(2):146–153.

[7] IEEE Std C37.118.2-2011 (2011) IEEE standard for synchrophasor data transfer for power systems. http://ieeexplore.ieee. org/xpl/articleDetails.jsp?arnumber=6111222.

[8] IEEE Std 1646-2004 (2004) IEEE standard communication delivery time performance requirements for electric power substation automation. http://ieeexplore.ieee.org/xpl/mostRecentIssue. jsp?punumber=9645.

[9] Zuo J, Carroll R, Trachian P *et al.* (2008) Development of TVA SuperPDC: phasor

applications, tools, and event replay. In: Proceedings of the IEEE Power and Energy Society general meeting: conversion and delivery of electrical energy in the 21st century, Pittsburgh, PA, 20–24 Jul 2008, 8 pp.

[10] Schweitzer Engineering Laboratories (SEL) (2010) Smart antiislanding using synchrophasor measurements. Schweitzer Engineering Laboratories (SEL), Pullman.

[11] Zweigle G (2009) Archive wide-area information with synchrophasors. AN2009-24, Schweitzer Engineering Laboratories (SEL), Pullman.

[12] BBC News (2011) "Anonymous" defends the use of Web attacks. BBC News, 28 Jan 2011. http://www.bbc.co.uk/news/ technology-12307802.

[13] Duvelson E (2010) Bridging the gap between legacy and modern substation communications. UTC J (2nd Quarter): 37–44. https://www.cavs.msstate.edu/iPCGR-ID_Registration/presentations/2011/Duvelson%20i-PCGRID%202001%20Bridging%20the%20Gap.pdf.

[14] Ward S, Duvelson E (2010) Integrating legacy communications on the smart grid highway. In: Proceedings of the 2010 IEEE PES transmission and distribution conference and exposition, New Orleans, LA, 19–22 Apr 2010, 4 pp.

[15] DNP3 Users Group (2008) DNP3 secure authentication specification version 2.0. DNP3 Users Group Technical Committee. https://www.dnp.org/Lists/Announcements/Attachments/7/Secure%20Authentication%20v5%202011-11-08.pdf.

[16] IEC 61850 (2007) Communication networks and systems in substations. http://www.iec.ch/smartgrid/standards/.

[17] Duvelson E (2011) Global migration toward ethernet/IP networks in substation communications. UTC J (2nd Quarter): 33–41. http://www.bluetoad.com/publication/?i=70749&p=33.

[18] Baran P (1964) On distributed communications networks. IEEE Trans Commun Syst 12(1):1–9.

[19] IEEE Std 802.16-2001 (2001) IEEE standard for local and metropolitan area networks, part 16: air interface for fixed broadband wireless access systems.

[20] Abichar Z, Peng YL, Chang JM (2006) WiMax: the emergence of wireless broadband. IT Prof 8(4):44–48.

[21] NanoStation product datasheet. Ubiquiti Networks. http://www.ubnt.com. Accessed Jun 2012.

[22] Linksys router datasheet. Cisco networking. http://home.cisco. com/en-us/wireless/ Accessed 14 Sept 2014.

[23] ITU-R P.530-14 (2012) Propagation data and prediction methods required for the design of terrestrial line-of-sight systems. https://www.itu.int/dms_pubrec/itu-r/rec/p/R-REC-P.530-14-201202-S!! PDF-E.pdf.

[24] Friis HT (1946) A note on a simple transmission formula. P IRE 34(5):254–256.

[25] Fixed wireless access. OFCOM. http://licensing.ofcom.org.uk/radiocommunication-licences/fixed-wireless-access/?a=0. Accessed 14 Sept 2014.

[26] ICS Telecom. ATDI. http://www.atdi.co.uk/software/software-applications/ics-telecom/. Accessed 14 Sept 2014.

[27] ITU-R P.525-2 (1994) Calculation of free-space attenuation. https://www.itu.int/dms_pubrec/itu-r/rec/p/R-REC-P.525-2-1994 08-I!!PDF-E.pdf.

[28] NASA's Jet Propulsion Laboratory (JPL) (2000) Shuttle radar topography mission. NASA's Jet Propulsion Laboratory (JPL), Pasadena.

[29] ISO/IEC 7498-1:1996 (1996) Information technology—open systems interconnection—basic reference mode. http://webstore.iec.ch/preview/info_isoiec7498-1%7Bed2.0%7Den.pdf.

Chapter 18
Trust Mechanisms for Cloud Computing

Jingwei Huang, David M. Nicol

Information Trust Institute, University of Illinois at Urbana-Champaign 1308 West Main Street, Urbana, Illinois 61801, USA

Abstract: Trust is a critical factor in cloud computing; in present practice it depends largely on perception of reputation, and self assessment by providers of cloud services. We begin this paper with a survey of existing mechanisms for establishing trust, and comment on their limitations. We then address those limitations by proposing more rigorous mechanisms based on evidence, attribute certification, and validation, and conclude by suggesting a framework for integrating various trust mechanisms together to reveal chains of trust in the cloud.

Keywords: Trust, Cloud Computing, Trust Mechanisms, Reputation, QoS, SLA, Transparency-Based Trust, Formal Accreditation, Cloud Audit, Policy-Based Trust, Evidence-Based Trust, Attribute Certification

1. Introduction

Cloud computing has become a prominent paradigm of computing and IT service delivery. However, for any potential user of cloud services, they will ask "can I trust this cloud service?" Furthermore, what exactly does "trust" mean in the context of cloud computing? What is the basis of that trust? If the attributes of a cloud service (or a service provider) are used as evidence for trust judgment on the service (or provider respectively), on what basis should users believe the attributes claimed by cloud providers? Who are authorities to monitor, measure, assess, or validate cloud attributes? The answers to those questions are essential for wide adoption of cloud computing and for cloud computing to evolve into a trustworthy computing paradigm. As addressed in[1], "the growing importance of cloud computing makes it increasingly imperative that we grapple with the meaning of trust in the cloud and how the customer, provider, and society in general establish that trust."

The issues and challenges of trust in cloud computing have been widely discussed from different perspectives[2]–[10]. A number of models and tools have been proposed[11]–[13]. Each contributes a partial view of cloud trust, but lacking still is a complete picture illustrating how cloud entities work together to form a "societal" system, with a solid grounding in trust, serving to facilitate trusted paths to trusted cloud services. The NIST Cloud Computing Reference Architecture[14] identified cloud brokers and cloud auditors as entities who conduct assessment of cloud services; however, there are few studies on trust relation analysis and the chains of trust from cloud users to cloud services (or providers) through those intermediary cloud entities. In this paper, we investigate trust mechanisms for the cloud, present our vision of the "societal systems mechanisms" of trust and a framework for analyzing trust relations in the cloud, and suggest trust mechanisms which combine attribute certification, evidence-based trust and policy-based trust.

Because of the criticality of many computing services and tasks, some cloud clients cannot make decisions about employing a cloud service based solely on informal trust mechanisms (e.g. web-based reputation scores); these decisions need to be based on formal trust mechanisms, which are more certain, more accountable, and more dependable. Here, the word "formal" is meant to carry the sense of "official" assessment in a society. In our suggested cloud trust mechanisms, the attributes of a cloud service (or its provider) are used as evidence for the user's trust judgment on the service (or provider), and the belief in those attributes is based on "formal" certification and chains of trust for validation.

In this paper, we focus somewhat informally on the conceptual basis for analysis of trust in the cloud; we do not at this time address mathematical modeling, which would involve many more precise details, formal languages, and specific use cases. With respect to terminology, an "entity" is an autonomous agent; a "cloud entity" refers to an entity in the cloud, such as a cloud provider, a cloud user, a cloud broker, and a cloud auditor; "semantics of trust" refers to precisely defined meaning of trust, including the relations among the components of trust.

This paper is organized as the following sections: 1) we define the *semantics of trust*; 2) we review the *state-of-the-art trust mechanisms* for cloud computing; 3) we discuss *policy-based trust judgment*, which is a real formal trust mechanism used in Public Key Infrastructure (PKI) practice. By policy-based trust, a cloud service or service provider can be trusted if it conforms to a trusted policy; 4) we present a general structure of *evidence-based trust*, by which particular attributes of a cloud service or attributes of a service provider are used as evidence for trust judgment; 5) we discuss *attribute assessment and attribute certification*, by which some attributes of a cloud service (or service provider) are formally certified, and the belief in those attributes is based on formal certification and chains of trust for validation; 6) we present an integrated view of the trust mechanisms for cloud computing, and analyze the trust chains connecting cloud entities; 7) finally, we give a summary and identify further research.

2. Semantics of Trust

The term "trust" is often loosely used in the literature on cloud trust, frequently as a general term for "security" and "privacy", such as[4]. What exactly does "trust" mean?

Trust is a complex social phenomenon. Based on the concepts of trust developed in social sciences[15][16], we use the following definition[17]:

> *Trust is a mental state comprising*: 1) **expectancy**—*the trustor expects a specific behavior from the trustee* (*such as providing valid information or effectively performing cooperative actions*); 2) **belief**—*the trustor believes that the expected behavior occurs, based on the evidence of the trustee's competence, integrity, and goodwill*; 3) **willingness to take risk**—*the trustor is willing to take risk for that belief.*

It is important to understand that the expected behavior of trustee is beyond the trustor's control; the trustor's belief in that expected behavior of trustee is based on the trustee's capability, goodwill (including intension or motivation), and integrity. The integrity of the trustee gives the trustor confidence about the predictability of the trustee's behavior.

We identify two types of trust, based on the trustor's expectancy: *trust in performance* is trust about what the trustee performs, whereas *trust in belief* is trust about what the trustee believes. The trustee's performance could be the truth of what the trustee says or the successfulness of what the trustee does. For simplicity, we represent both as a statement, denoted as a Boolean-type term, x, called a reified proposition[18]. For the first case, x is what the trustee says; for the second, x represents a successful performance, which is regarded as a statement that the trustee made, describing his or her performance. A *trust in performance* relationship, $trust\_p(d, e, x, k)$, represents that trustor d trusts trustee e regarding e's performance x in context k. This relationship means that if x is made by e in context k, then d believes x in that context. In first-order logic (FOL),

$$trust\_p(d,e,x,k) \equiv madeBy(x,e,k) \supset believe(d, k \supset x) \qquad (1)$$

where $\supset$ is an operator used for reified propositions to mimic the logical operator for implication, $\supset$. A *trust in belief* relationship, $trust\_b(d, e, x, k)$, represents that trustor d trusts trustee e regarding e's belief (x) in context k. This trust relationship means that if e believes x in context k, then d also believes x in that context:

$$trust\_b(d,e,x,k) \equiv believe(e, k \supset x) \supset believe(d, k \supset x). \qquad (2)$$

Trust in belief is transitive; *trust in performance* is not; however, *trust in performance* can propagate through *trust in belief*. A more detailed account can be found in[17][19].

From the definition above, the trustor's mental state of belief in his expectancy on the trustee is dependent on the evidence about the trustee's competency, integrity, and goodwill. This leads to logical structures of reasoning from belief in evidence to belief in expectancy. We will discuss this later in §

'Evidence-based trust'.

The semantics of trust in the context of cloud computing has the same semantic structure as stated above; what still needed are the specific expectancy and the specific characteristics of cloud entities's competency, integrity, and goodwill in the context of cloud computing. We will discuss further in § 'Evidence-based trust'.

3. State-of-the-Art Trust Mechanisms in Clouds

In this section, we discuss existing trust mechanisms in the cloud. From the discussion, we will see that each of the mechanisms addresses one aspect of trust but not others.

3.1. Reputation Based Trust

Trust and reputation are related, but different. Basically, trust is between two entities; but the reputation of an entity is the aggregated opinion of a community towards that entity. Usually, an entity that has high reputation is trusted by many entities in that community; an entity, who needs to make trust judgment on an trustee, may use the reputation to calculate or estimate the trust level of that trustee.

Reputation systems are widely used in e-commerce and P2P networks. The reputation of cloud services or cloud service providers will undoubtably impact cloud users' choice of cloud services; consequently, cloud providers try to build and maintain higher reputation. Naturally, reputation-based trust enters into the vision of making trust judgment in cloud computing[11][13]20].

Reputation is typically represented by a comprehensive score reflecting the overall opinion, or a small number of scores on several major aspects of performance. It is unrealistic to ask a large number of cloud users to rate a cloud service or service provider against a large set of complex and fine-grained criteria. The reputation of a cloud service provider reflects the overall view of a community towards that provider, therefore it is more useful for the cloud users (mostly individual users) in choosing a cloud service from many options without particular requirements. Reputation may be helpful when initially choosing a service, but is inadequate afterwards. In particular, as a user gains experience with the service, the trust placed on that service meeting performance or reliability requirements will evolve based on that experience.

3.2. SLA Verification Based Trust

"Trust, but verify" is a good advice for dealing with the relationships between cloud users and cloud service providers. After establishing the initial trust and employing a cloud service, the cloud user needs to verify and reevaluate the trust. A service level agreement (SLA) is a legal contract between a cloud user and a cloud service provider. Therefore, quality of service (QoS) monitoring and SLA verification is an important basis of trust management for cloud computing. A number of models that derive trust from SLA verification have been proposed[12][13].

A major issue is that SLA focuses on the "visible" elements of cloud service performance, and does not address "invisible" elements such as security and privacy. Another issue is that many cloud users lack the capability to do fine grained QoS monitoring and SLA verification on their own; a professional third party is needed to provide these services. In a private cloud, there may be a cloud broker or a trust authority (e.g. RSA's CTA, to be discussed later in §'Cloud transparency mechanisms'), whom is trusted in the trust domain of the private cloud; so the trusted broker or trust authority can provide the users in the private cloud the services of QoS monitoring and SLA verification. In a hybrid cloud or interclouds, a user within a private cloud might still rely on the private cloud trust authority to conduct QoS monitoring and SLA verification; however, in a public cloud, individual users and some small organizations without technical capability may use a commercial professional cloud entity as trust broker. We discuss this in § 'Trust as a service'.

3.3. Cloud Transparency Mechanisms

Transparency and accountability are a recognized basis for gaining trust on cloud providers. To increase transparency of the cloud, the Cloud Security Alliance (CSA) launched the "Security, Trust & Assurance Registry (STAR)" program[21], a free publicly accessible registry which allows cloud service providers to publish self-assessment of their security controls, in either a "Consensus Assessments Initiative Questionnaire (CAIQ)" or a "Cloud Controls Matrix (CCM)", which embody CSA published best practices. CAIQ contains over 140 questions which cloud users or auditors may ask; CCM is a framework describing how a cloud provider aligns with the CSA security guide[22]. Examples of cloud providers' self assessments can be found at the CSA STAR website[23]. STAR is a useful source for users seeking cloud services. However, the information offered is a cloud provider's *sel*-assessment; cloud users may want assessments performed by some independent third-party professional organizations.

Different from STAR, CSC.com proposed[24] and CSA adopted the Cloud Trust Protocol (CTP)[25], a request-response mechanism for a cloud user to obtain specific information about the "elements of transparency" applied to a specific cloud service provider; the elements of transparency cover aspects of configuration, vulnerability, audit log, service management, service statistics, and so forth. "The primary purpose of the CTP and the elements of transparency is to generate evidence-based confidence that everything that is claimed to be happening in the cloud is indeed happening as described, ..., and nothing else"[26]. CTP provides an interesting channel between cloud users and cloud service providers, allowing users internal observations of cloud service operations. However, like STAR, an essential weakness of CTP is that its information is provided by cloud service provider itself. Dishonest cloud service providers can filter out or change the data. From the point of view of a trust judgement, it raises questions of the data's reliability.

3.4. Trust as a Service

We have already noted the need for employing third-party professionals for QoS monitoring and SLA verification. Independent assessment has utility in other aspects of cloud computing, as well.

RSA announced the *Cloud Trust Authority* (CTA)[27] as a cloud service, called Trust as a Service (TaaS) to provide a single point for configuring and managing security of cloud services from multiple providers. The initial release of the CTA includes: *identity service*, enabling single sign-on among multiple cloud providers, and *compliance profiling service*, enabling a user to view the security profiles of multiple cloud providers against a common benchmark. The CTA is a tool specialized on cloud trust management, and is developed from RSA's philosophy of "trust = visibility + control"[28]. As a cloud-based tool, the CTA could largely simplify cloud users' trust management. However, a cloud user must still make trust judgment about the cloud service assertions streamed in the CTA, because those assertions were made by cloud service providers themselves. Most importantly, a cloud user needs to judge the trustworthiness of the CTA in role as an intermediary.

The essential issue of any TaaS mechanism is about what is the basis of the trust relation between cloud users and those commercial trust brokers. We will discuss the answers later in subsections 'Trust judgement on a cloud broker' and 'Trust judgment on a cloud service provider'.

3.5. Formal Accreditation, Audit, and Standards

Because self-assessment exercises may be compromised by dishonesty, some argue that formal accreditation from a trusted independent authority is necessary for a healthy cloud market; some others argue that formal accreditation "would stifle industry innovation"[2].

External audits, attestations, or certifications for more general purpose (not specific to clouds) have been used in practice. Examples include: the ISO/IEC 27000 series, which are international information security management standards[29]; "Statement on Standards for Attestation Engagements No. 16" (SSAE 16)[30], which is an attestation standard for service organizations, put forth by the Auditing Standards Board (ASB) of the American Institute of Certified Public Accountants (AICPA). SSAE 16 is replacing the older standard "Statement on Auditing Standards No. 70" (SAS 70); "The International Standard on Assurance Engagements 3402" (ISAE 3402)[31], which is a globally recognized standard for assurance reporting on service organizations.

Specific to cloud computing, in addition to CTP and STAR (for self-assessment), CSA also launched the CloudAudit initiative, which provides a common interface and namespace for cloud providers to produce audit assertions, and allows cloud users to automate use of that data in their own audit processes. CloudAudit could facilitate automated cloud audit, conducted by cloud providers (for self-audit), cloud users (for cloud user-audit), and cloud auditors (for formal audit). CloudAudit, CCM, CAIQ, and CTP form the CSA Governance, Risk Management and Compliance (GRC) stack.

To ensure trustworthiness, the International Grid Trust Federation (IGTF) issued GFD-I.169 as guidelines for auditing the cloud/grid assurance bodies—the certification authorities (CAs) issuing X.509 certificates[32].

A formal process for assessment of cloud services and their providers by independent third parties, acceptable to both cloud users and providers, does not yet exist. Formal accreditation specific to independent third-party cloud assessors also does not exist.

3.6. Further Discussion

A reputation-based trust mechanism reflects the overall view of a community towards a cloud service provider. It can help with cloud service selection; but is insufficient for other important purposes.

After establishing an initial trust on a cloud service, a cloud user needs to verify and re-evaluate that trust. QoS monitoring and SLA verification based trust mechanism can help to manage the existing cloud trust relations. The QoS/SLA mechanism can manage "visible" elements of the black box of a cloud service, such as performance; but it cannot help to manage the "invisible" elements inside a cloud service, such as privacy protection.

Cloud transparency mechanisms provide channels for cloud users to "observe" how cloud service providers operate. The mechanisms help to establish trust by making the cloud services more "visible". The essential issue of the transparency mechanisms is that the information is provided by cloud service providers themselves; thus we need to identify the basis for cloud users to trust them.

The TaaS mechanism provides cloud users a solution where the sophisticated tasks of cloud trust management can be delegated to third-party professionals. However similarly the basis for cloud users to trust them needs to be estabulished.

One possible solution to the problems posed in the above mechanisms is formal accreditation and audit. The mechanisms of formal accreditation and audit in the cloud do not exist yet and are still in discussion.

In the rest of this paper, we continue to explore the cloud trust mechanisms by borrowing policy-based trust mechanism from PKI, combined with evidence-based trust and attribute certification and validation.

4. Policy-Based Trust

We earlier identified the need for "formal" trust mechanisms in cloud computing. In a related sphere, PKI is a widely used mature technology that employs "formal" trust mechanisms to support digital signature, key certification and validation, as well as attribute certification and validation. Can we apply trust ideas used in PKI to establish "formal" trust mechanisms to the cloud?

To simplify the discussion, consider the example illustrated in **Figure 1**. Alice has a digital document supposedly signed by Bob using his private key K. To validate, she needs Bob's public key Kb. Assume that Alice trusts only her trust anchor certification authority $CA1$, and she knows only $K1$, her trust anchor's public key. In order for her to verify the signature on the document as being Bob's, she needs to discover a certification path (a chain of certificates) from $CA1$ to $CA3$ who has issued Bob's public key certificate. As shown in the figure, Alice uses $CA1$'s public key $K1$ to validate $CA2$'s public key $K2$; because Alice trusts $CA1$ on public key certification, and $CA2$'s public key is certified by $CA1$, Alice can believe that $CA2$'s public key is $K2$; then Alice uses $K2$ to validate $CA3$' public key $K3$; and finally uses $K3$ to validate Bob's public key Kb. The main issue is *why Alice should believe $K3$ is $CA3$'s public key and Kb is Bob's public key*?

Essentially, to infer belief in a statement "Bob's key is Kb", Alice needs to trust $CA3$, the creator of that assertion, with respect to the truth of the statement;

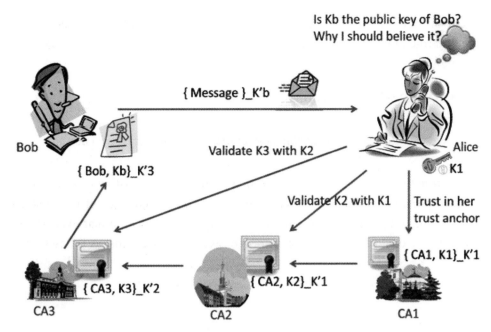

Figure 1. PKI trust example. This example reveals trust relations in public key certification and validation.

however, this raises questions that ask about the foundation of that trust, and how the trust is inferred or calculated. Some research suggests that the trust comes from recommendations along the chain of certificates by those certificate issuers[33]; but the practice of digital certification and validation in real PKI systems suggests that *the trust comes from compliance with certain certificate policies*.

As specified in IETF RFC 5280[34], in addition to the basic statement that binds a public key with a subject, a public key certificate also contains a certificate policy (CP) extension. For a public key certificate issued to a CA, the certificate means that the issuing CA who conforms to the specified CP asserts that the subject CA has the certified public key, and the subject CA also adheres to the specified CP. As a result, to infer Alice's belief in *CA3*'s key and Bob's key, she must trust that CP in the sense that any CA conforming to that CP will generate valid public key certificates. There are more complex and interesting issues in PKI trust[35], but for the purpose of this paper, we will not go further.

In summary, as PKI is currently practiced, trust in a certification authority (CA) with respect to issuing and maintaining valid public key certificates is based on the CA's conformance with certain certificate policies. Certificate policies play a central role in PKI trust. We call this trust mechanism as *policy-based trust*.

5. Evidence-Based Trust

We now discuss using attributes as evidence to make trust decision.

From the definition of trust given in § 'Semantics of Trust', a trustor's belief in the expected behaviour of trustee is based on the evidence about the trustee's attributes of competency, goodwill, and integrity, with respect to that expectation. Formally, we could express a general form of evidence-based trust as follows:

$$believe(u, attr1(s,v1)) \wedge ... \wedge believe(u, attrn(s,vn)) \\ \rightarrow trust\_*(u,s,x,c) \quad (3)$$

which states that if an individual u believes a subject s has attribute *attr1* with value $v1$, ..., attribute *attrn* with value vn, then u trusts (either *trust in belief* or *trust in performance*) s with respect to x, the performance of s or information created or believed by s, in a specific context c.

An entity's belief in an attribute assessment is dependent on whether the entity trusts the entity who makes that attribute assessment. Formally, based on the definition of trust-in-performance, formula (1) in § 'Semantics of Trust', we could have

$$trust\_p(u,a,attr(s,v),c) \wedge madeBy(attr(s,v),a,c) \\ \wedge inContext(c) \rightarrow believe(u, attr(s,v)), \quad (4)$$

which states that if an individual u trusts an attribute authority a to make assertions about a subject s has attribute *attr* with value v in a specific context c, a specific assertion *attr(s, v)* is made by a in context c, and the context c is the case, then u believes that assertion. In the formula, *attr(s, v)* is a reified proposition represented as a term. Since not only the attributes of a cloud service may be assessed and certified, but also the attributes of a cloud entity may be assessed and certified, in the above formula, we may use *attr(e, v)* to state that cloud entity e has attribute *attr* with value *e*. In this way, a logic formula similar to (4) can describe the relation from trust in a cloud auditor to the belief in the certified attribute of a cloud entity such asa service provider.

To use attributes as evidence in trust judgment, we organize the relevant attributes in a two-dimension space: 1) one dimension goes along the domain of the trustor's expectation on the trustee, in the context of cloud computing, including aspects of performance, security, and privacy; 2) another dimension goes along the source of trust, that is, what makes the trustor trust the trustee, including the trustee's competency (capability), integrity (consistency in performance and principles), and goodwill (motivation or intension).

Figure 2 illustrates a spectrum of attributes in cloud computing. Most commonly considered ones fall in the category of competency; attributes that reflect integrity and goodwill are frequently neglected, and should be included in trust judgment. To neglect these is to implicity assume that trust does not depend on them, or if it does, that dependence is satisfied. Characterization and quantification of integrity and goodwill is an interesting research challenge. A trustee's historical behavior might reflect integrity; goodwill might be quantified as performance improvements are measured, and cloud users' feedback.

Domain-specific expectation	Sources of trust		
	Competency	Good will / intention	Integrity / consistency
Performance	Availability; Reliability; ...	<Constant efforts to improve performance>	<Consistency shown by historic data>
Security	Security cert; Security breach rate; ...	<Constant efforts to improve security>	<Consistency shown by historic data>
Privacy	Privacy regulation compliance; Privacy violation rate; ...	<Constant efforts to improve privacy>	<Consistency shown by historic data>

Figure 2. Attributes for evidence-based trust. The attributes used for evidence-based trust judgment can be organized in two dimensions: (1) sources of trust, including competency, goodwill, and integrity; (2) domain-specific expectation.

Different cloud users may have different trust policies, involving different trust attributes. A common trust *framework* supports evidence-based trust judgment for different users and different policies. The connection between evidence-based trust and policy-based trust is that the belief that an entity conforms to a trusted policy implies the belief that the entity has a set of attributes associated with that policy.

6. Attribute Assessment and Certification

When the attributes of a cloud service (or cloud entity) are used as evidence to make trust judgment on the service (or entity), the sources of attribute assessment must be trustworthy, and those attributes need to be distributed in a trustworthy way. In the following, we first discuss the source of attribute assertions and then we discuss attribute certification as a formal approach to deliver cloud attributes.

6.1. Sources of Cloud Attribute Assessment

Assessment of attributes may come from several sources: the cloud user, other peer users, the service provider, cloud auditor/accrediator, and cloud broker. We discuss each of them in turn.

6.1.1. Cloud User Observation

If a cloud user has already interacted with a cloud service or a cloud service provider, then the experience will be the user's direct basis for cloud attribute assessment. Experience is a fundamental factor of trust, and this kind of trust, called "interpersonal trust", has long been studied in both social sciences and computing science.

The advantage of using direct interaction experience is that the data used are first-hand and may be most relevant; the disadvantage is that the data accumulated are limited with respect to the sample size and the range of the usage of the cloud service. A specific user's experience is just one piece of the information revealing the trustworthiness of a cloud service.

6.1.2. Opinions of Other Peer users

When a cloud user has only limited direct experience with a cloud service (or none at all), other peer users' opinions could be an important source of cloud attribute assessment. The major issues are: can those peer reviewers be trusted with respect to their opinions on the cloud service? And how can those different opinions be aggregated?

There are at least two basic approaches to solving the problem: social network based and reputation based.

Social network based approach A cloud user takes one or more trusted friends' opinions, and combines them with that user's personal trust in each of those friends. That user may not have a direct trust relation with a "popular" reviewer, but the user may derive an indirect trust relation with that reviewer through a trust network[17][36], which is a specific form of social networks, comprising of only trust relations. The social network based approach is an analogue of how a person initially trusts an entity, unknown before in the real world. Models in this category are heuristic. Typically, one asks only a small number of trusted friends for their opinions. When a large number of peer users' opinions are involved, the approach becomes reputation based.

Reputation based approach A typical methodology is to aggregate a large number of peer user's ratings, often seen in e-commerce product/service ratings. The advantage is that the data used for assessment may cover many more situations and have a wider time-window of observations; this approach can have a much wider view on the cloud service (or its provider) than a single user does. On the other hand, some weaknesses exist: a large number of raters are required for meaningful and objective ratings; the raters and users should have a common understanding of the attribute semantics and the corresponding measurement; this approach is suitable for the purpose of overall rating, or is limited to rating a small number of attributes; the trustworthiness of individual voter are rarely taken into account; usually, as in e-commerce, the reputation of product/service is calculated by an organization in a centralized manner, so the organization may manipulate the calculation, and the calculating service may become a single point of attack.

6.1.3. Statements from Cloud Service Provider

Some cloud service attributes may be specified, promised, or revealed by its provider. In "service specification" and advertisements, a service provider will specify the featured attributes of a cloud service; the attributes of the service stated in a SLA are the promises of that service provider to that user. Through the CloudTrust Protocol (CTP)[26], cloud users can request and get a response from the provider about "the elements of transparency", the information concerning the compliance, security, privacy, integrity, and operational security history.

However, information about the attributes of a service given by the service provider are usually not directly believed by the first-time users. Sometimes a user may believe a service provider's statements or promises, based on the brand name or reputation of that service provider, or based on the user's past experience of interaction. In any case, the stated attributes are an important part of the watch-list in cloud service monitoring, and they are used to verify whether the service provider behaves as trusted. The conclusion of the verification will be used by the users to build or revise their trust in that service provider.

In general, the statements or promises about the attributes of a cloud service given by a cloud service provider itself need to be verified before used for decision making, and cloud attribute assertions from third party independent professional organizations are expected, which we discuss in the following subsections 'Assessment of cloud auditor/accreditor' and 'Observation of cloud brokers'.

6.1.4. Assessment of Cloud Auditor/Accreditor

NIST identifies a cloud auditor as "a party that can conduct independent assessment of cloud services, information system operations, performance, and security of a cloud implementation. A cloud auditor can evaluate the services provided by a cloud provider in terms of security controls, privacy impact, performance, etc."[14]. Obviously, cloud audit is an important channel of cloud attribute assessment. A limitation of cloud auditing is that the trust assessment reflects only the state at the time of the audit. Trust changes dynamically, as a function of dynamic monitoring of behavior.

A cloud auditor's assessment is usually regarded as a reliable information source for trust judgment. To some cloud users, a cloud auditor as a third-party professional organization may be a satisfactory trust root. However, to some others, the trustworthiness of a cloud auditor also needs to be evaluated by looking into the auditor's attributes and/or policies. Since cloud audit is an important mechanism to ensure trustworthiness of clouds, each cloud auditor should be periodically audited and/or accredited by a professional association such as Auditing Standards Board of AICPA.

In formal accreditation, an entity who provides a professional service is assessed against official standards, and is issued with certification of its competency, authority, or credibility. The certification is provided by an accreditor, who is a third party independent authorized accreditation organization, and who is also accredited by a national standard body or professional association. If formal accreditation is applied to clouds, the cloud attribute assessment from a formal accreditation will be another important information source for cloud trust judgment.

Accreditation is somewhat similar to audit. In both cases an entity is assessed by an independent third party; however, there are subtle differences. First, they may have different focusing aspects of assessment. Accreditation focuses on the qualification of the accredited entity with respect to conducting a specific type of professional services; audit focuses on assessing the performance of the audited entity with respect to the common requirements of a society and/or the professional standards of a professional community. Secondly, audit typically takes place annually or once per half year; accreditation takes place in a longer period (e.g. every 5 years).

In summary, in context of cloud computing, the assessments by audit and accreditation are objective and "formal", but they are not real-time information as from real-time monitoring.

6.1.5. Observation of Cloud Brokers

Cloud brokers play an important role. By the NIST definition[14], a cloud broker is "an entity that manages the use, performance, and delivery of cloud services, and negotiates relationships between Cloud Providers and Cloud Consumers." A cloud broker may provide services in three categories[14]: 1) service intermediation: for a given cloud service, to provide value-added additional services such as performance monitoring and security management; 2) service aggregation: to provide an integrated service by aggregating several cloud services from different providers; 3) service arbitrage: to select proper cloud services in an integrated service, based on the quantified evaluation of the alternative cloud services. The observation of a cloud broker can be an important source of cloud attribute assessment.

The advantages of broker observation include: real-time cloud service performance monitoring; feedback from many peer users; an ability to monitor and evaluate a collection of the same category of cloud services from different providers. A cloud broker potentially has a relatively complete picture of a cloud service.

However, again the question arises whether a cloud broker can be trusted with respect to assessing cloud attributes. This depends on the relationship between broker and providers, and between broker and users. A tight business relation with some cloud providers may make the brokers' opinion be not as objective as the one made in formal audit or accreditation.

From the perspective of cloud market mechanism we imagine that if a cloud broker represents a cloud provider, then the cloud broker may provide information which favors that cloud provider; however, if a broker is independent, and its business depends on the trust relations with users, the broker is more motivated to find and provide information being truly helpful for cloud users. This situation may occur when a cloud broker serves as a gateway for a large number of cloud users in the cloud market. Consistent with the above view, we further imagine that if a cloud broker is highly trusted by some cloud users (especially, end cloud users), the broker may become those cloud users' trust anchor, taking care of trust management for those cloud users.

In order to ensure thata cloud broker behaves as a trustworthy cloud entity, cloud users will expect to learn how a cloud broker works, whether the broker is neutral, what policies the broker follows, and whether the broker has certain attributes that can be used as evidence to judge its trustworthiness. Therefore, essentially a cloud broker is also expected to be formally audited and/or accredited either.

6.2. Attribute Certification

In addition to X.509 identity (public key) certification, there also exists X.509 *attribute certification*[37]. Public key certification is used in authentication; attribute certification is used for both authentication and authorization. An attribute certificate (AC) is a statement digitally signed by the AC issuer to certify that the AC holder has a set of specified attributes. The certified attributes can be access identity, authentication information (e.g. username/password pairs), group membership, role, and security clearance[37]. An AC mainly contains the following fields: unique AC identifier, AC holder, AC issuer, attribute-value pairs, valid period, the Id of the algorithm used to verify the signature of the AC, and extensions, which mainly include AC targeting – a list of specified servers or services where the AC can be used, and CRL (Certificate Revocation List) distribution points.

The current IETF X.509 AC standard[37] might be considered for use in cloud attribute certification, but it has several limitations.

First, the standard does not include important attributes needed in the cloud context. Extensions are possible to deal with this, but still no standards regarding service performance, security, and privacy. Second, with respect to attribute certification, the real authority behind attribute assertion is the entity who really knows the certified entity. For example, with respect to the role or membership of an entity in a specific organization, that organization is naturally the authority to state that attribute. From this point of view, we should discern the difference between "*attribute assertion authority*" (AAA) and *attribute certification authority* (ACA, i.e. AC issuer). We use AA (Attribute Authority) to refer to an entity who is both AAA and ACA. In the context of clouds, who plays the role of AA? From our earlier discussion, it is obvious that the most reliable sources for attribute assertion/ assessment are independent third-party professional organizations such as cloud auditors and accreditors, and even cloud brokers.

Finally, current IETF X.509 AC standard[37] adopts a simple trust structure where "one authority issues all of the ACs for a particular set of attributes". In cloud applications (except for small scale private clouds) an AC issuer may be frequently outside the trust boundary of an AC user. Therefore, mechanisms for cross-domain attribute certification and validation are necessary for both hybrid cloud and public cloud.

7. An Integrated View

Earlier, we envisioned that the attributes of a cloud service (or cloud entity) can be used as evidence for a cloud user to make trust judgment on the service (or entity); we discussed the sources of cloud attribute assessment and *attribute certification*; we also revealed that PKI in practice uses policy-based trust mechanism, which might be used in cloud computing either. In this section, we put together all those mechanisms, including: reputation based, SLA verification based, transparency based, formal accreditation and audit, as well as the suggested policy-based, evidence-based, and cloud attribute certification, to construct an overall framework for analyzing and modeling trust chains among cloud entities.

Figure 3, **Figure 4**, **Figure 5** and **Figure 6** illustrate the dependence between the trust placed in various cloud entities and the sources of evidence for trust judgment. In these figures, the left part illustrates trust placed on different types of cloud entities; the right part illustrates trust mechanisms to be used, which

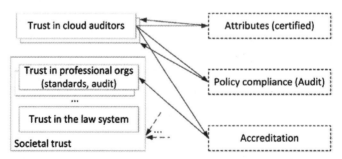

Figure 3. Evidence and chains for trust judgment on a cloud auditor. Trust placed in a cloud auditor is based on one or more of: (1) accreditation, which is further based on the trust placed on professional organizations making standards and audit/accreditation, —a part of societal trust; (2) policy compliance, which is audited by another trusted auditor; (3) attributes, certified by another trusted auditor.

Chapter 18

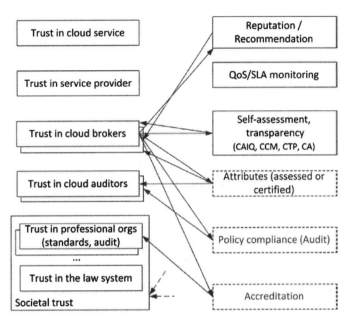

Figure 4. Evidence and chains for trust judgment on a cloud broker. Trust placed in a cloud broker is based on one or more of: (1) accreditation; (2) policy compliance; (3) certified attributes; (4) self-assessment and information revealing, which is based on the trust placed in this broker with respect to telling truth; (5) reputation calculated or recommendation made by another trusted broker.

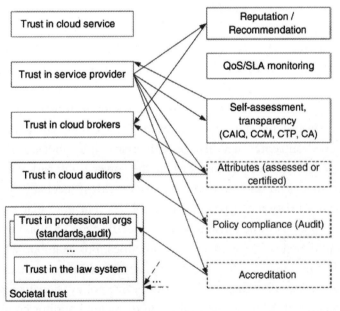

Figure 5. Evidence and chains for trust judgment on a cloud service provider. Similar to the structure of trust judgment on cloud broker showed in **Figure 4**.

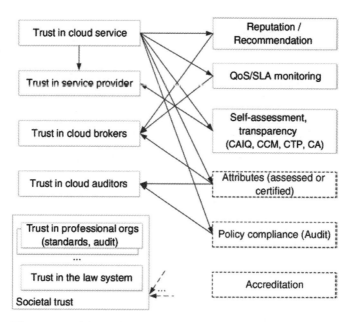

Figure 6. Evidence and chains for trust judgment on a cloud service. Trust placed in a cloud service is based on one or more of: (1) cloud service provider, whom is trusted; (2) policy compliance; (3) certified attributes; (4) QoS monitoring and SLA verification, which are conducted by a trusted party such as a trusted broker; (5) reputation calculated or recommendation made by a trusted broker.

are also the sources of evidence to support trust judgment; the arrows represent dependence relations between them; the dependence relations together form the chains of trust in the cloud. The six mechanisms shown in those pictures are an abstraction of typical mechanisms; a real system support trust judgment in practice may involve several mechanisms. For example, a cloud reputation system may calculate reputation scores, and also provide assessed attributes from brokers and users' reviews. The three mechanisms in the lower-right part with dotted border-lines are suggested ones and do not exist yet. Most mechanisms may support trust judgment on different types of cloud entities, but note that for a same mechanism, the contents to be examined for a specific type of cloud entity could be different from the ones for another types of entities. For example, when applied to a cloud service provider, "policy compliance audit", refers to evaluation of a cloud service provider's conformance to its cloud service policy; however, when applied to a cloud auditor, it refers to the evaluation of a cloud auditor conformance to a cloud audit policy. Now we discuss each trust judgment task in turn.

7.1. Societal Trust

Societal trust is foundational in all trust models that include individuals and organizations; cloud computing is no exception. Each individual in a society has to place trust in some basic parts of the society. Examples include: trust in the law system and government to maintain social order; trust in some professional services; trust in professional organizations with respect to creating and maintaining specific professional services standards. In the cloud context, examples of professional organizations might include AICPA, NIST (National Institute of Standards and Technology), IGTF (International Grid Trust Federation), and CSA (Cloud Security Alliance). We specifically assume that societal trust leads cloud users to put their trust in the accreditation of cloud entities including auditors, brokers, and service providers, with respect to the qualification of a cloud entity on corresponding professional services.

While we recognize societal trust as a root of cloud trust, a deeper treatment of societal trust is beyond the scope of our overview of trust in clouds.

7.2. Trust Judgment on a Cloud Auditor

A cloud auditor is a professional independent assessor of cloud entities. An auditor conforms to professional policies and/or standards in his operations. Cloud auditors should be also externally audited periodically by audit professional organizations, to ensure they comply with established policies and standards.

One cloud user might place a cloud auditor in his trusted societal root, i.e., simply assume the auditor is trustworthy; another user may choose instead to make a trust judgment on a cloud auditor as they do on other cloud entities. By the semantics of trust given in § 'Semantics of Trust', for "trust in a cloud auditor", the **expectancy** of a cloud user on a cloud auditor is the objective and professional assessment on a cloud entity with respect to its cloud services against a specific set of standards; the belief in that expectancy is based on some **evidence** with respect to the auditor's competency, goodwill, and integrity. For this judgment, there may be several sources of information as shown in **Figure 3**, and they are discussed as follows:

- **Accreditation:** A cloud user may check whether a specific cloud auditor is formally accredited by an professional audit organization and/or a cloud computing professional organization. Belief in accreditation is further dependent on whether the cloud user trusts the formal accreditor—an audit professional organization such as ASB of AICPA.

- **Policy compliance audit:** A cloud auditor should conform to professional policies and/or standards in its audit operations, such as SAS 70, SSAE 16, and ISAE 3402; the auditor should assess a cloud entity against widely accepted policies; the quality of the audit operations of an auditor is also assessed through audit, conducted by a different auditor appointed by an professional audit organization. A cloud user may use the audit results as evidence for trust judgment. The cloud user's belief in the audit result is further dependent on the user's trust in the auditor conducting the audit.

- **Certified attributes:** In addition to accreditation and policy compliance, a cloud user may want to check the auditor's other attributes, such as the history of the auditor, experiences of those previously audited by that auditor, the history of the audit applied to the auditor. Some attributes may be contained in audit documents; some others may be certified (or assessed, verified, and digitally signed) by a peer auditor. The cloud user's belief in the certified attributes is dependent on the user's trust in the issuer of the certified attributes.

7.3. Trust Judgment on a Cloud Broker

As discussed in § 'Observation of cloud broker', a cloud broker provides various intermediate services. Any cloud entity offering intermediated services may be regarded as a broker. Examples may include: "market" for cloud services such as SpotCloud[38], and TaaS such as CTA[27]. Note that an online reputation and ranking system for cloud services can also be regarded as a cloud broking service.

For the concept of "trust in a cloud broker", the **expectancy** of a cloud user on a cloud broker includes trustworthy value-added services such as bridging and aggregating services, security and identity management services, objective and precise evaluation of cloud services and their providers. To make evidence-based trust judgments, as illustrated by **Figure 4**, the evidence may include:

- **Accreditation:** Similar to cloud auditors, a cloud broker should be qualified for providing cloud broking services, through formally accreditation by a cloud computing professional organization.

- **Policy compliance audit:** A cloud broker should conform to certain policies and/or standards widely adopted or accepted by the cloud in the broker's operations; the quality of its operations should be audited by a cloud auditor. A cloud user may use the audit result as evidence for trust judgment. The cloud user's belief in the audit result is further dependent on the user's trust in the auditor conducting the audit.

- **Attributes (assessed or certified):** The attributes of a cloud broker on competency, goodwill, and integrity are important evidence for cloud users' trust judgment. In addition to the attributes assessed with respect to policy compliance, other attributes regarding performance, security, and privacy as discussed in § 'Evidence-based trust' may be also audited by a cloud auditor, or assessed and digitally signed by other cloud brokers, or reviewed and digitally signed by some cloud users. The cloud user's belief in the certified/assessed attributes is dependent on the user's trust in the issuer of the certified/assessed attributes.

- Self-assessment and information revealing: Cloud brokers as a special type of intermediated cloud service providers should also adopt the CSA cloud transparency mechanisms to exercise self-assessment such as CAIQ and CCM, and information revealing as does in CTP (discussed in § 'Cloud transparency mechanisms'). The cloud user's belief in the information revealed by the broker is dependent on the user's trust in that broker with respect to telling the truth, which may be verified in a formal audit.

- **Reputation/recommendation:** Reputation and recommendation can be very helpful to new cloud users and/or the users who are planning to recompose their cloud services. The cloud user's belief in the reputation scores and recommendation is dependent on the user's trust in the source of the information, typically, a cloud broker.

7.4. Trust Judgment on a Cloud Service Provider

The trust **expectancy** of a user with respect to a provider is that the provider offers trustworthy cloud services. The evidence for trust judgment on a cloud service provider may include the following sources, as shown in **Figure 5**:

- Accreditation
- Policy compliance audit
- Attributes (assessed or certified)
- Self-assessment and information revealing
- Reputation/recommendation

All of the above mechanisms are similar to the ones applied to cloud brokers, save that the trustee is a cloud service provider rather than a cloud broker.

7.5. Trust Judgment on a Cloud Service

We view a cloud service as an autonomous agent; and that "a cloud user trusts a cloud service" means that the user has the **expectancy** that the cloud service is trustworthy, which means that the cloud service has a set of attributes including reliability, availability, confidentiality, integrity, safety, and privacy; the user believes the expectancy to be true based on some **evidence**, from diverse sources, shown in **Figure 6**:

- **Trust based on the service provider:** by trust in performance, a user trusts a cloud service with respect to performance, security, and privacy, based on the identity of the provider. If the user trusts that the provider gives trustworthy cloud services, then the cloud service is trusted.

- **Policy compliance audit:** A cloud user may examine specific policies and/or standards applied to the service, and investigate the results of formal audits of the provider.

- **Attributes (assessed or certified):** A cloud user may examine the attributes of a cloud service regarding performance, security, and privacy, which may be audited by a cloud auditor, or assessed and digitally signed by cloud brokers, or reviewed and digitally signed by some cloud users. The belief in those attributes is dependent on the trust in the corresponding attribute assessor.

- **Self-assessment and information revealing:** A cloud user may study information about the service which is revealed by the service provider through cloud transparency mechanisms. The user's belief in the information is dependent on the user's trust in the cloud service provider with respect to telling the truth.

- **QoS monitoring and SLA verification:** QoS monitoring and SLA verification (a shorter term "QoS/SLA monitoring" is used in **Figure 6**) is an important source to verify trust and to adjust trust. If the monitoring is conducted by a cloud broker, then the belief in the results of monitoring is dependent on the trust in that broker with respect to objective and professional monitoring.

- **Reputation/recommendation:** a cloud user may trust a cloud service, based on a trusted cloud broker's recommendation. Similar to PKI trust, recommendation may be handled in two ways: one regards the "recommendation" as the broker's trust in that recommended service, and then derives indirect trust on that service through using trust in belief relation with the broker; another is (as in PKI practice) that the broker only certifies that that cloud service has certain attributes or conforms to certain policies, and cloud users to make their own decision whether to trust that service.

7.6. Further Discussion

As seen above, the trust placed on a cloud entity may be dependent on several sources of evidence; however, it is unnecessary to use all of them; a cloud user may use one or more sources of evidence for trust judgment, dependent on the user's trust policy. For example, to decide whether to trust a cloud service provider, a cloud user may simply just check whether the provider passed the formal audit of a widely accepted cloud service policy, conducted by a trusted auditor.

In the discussion above, the trust mechanisms of reputation/recommendation, QoS monitoring and SLA verification, self-assessment and information revealing are already in development; formal accreditation is in discussion, but it does not exist yet; trust mechanisms of attribute assessment/certification, which is used for evidence-based trust judgment, and policy compliance audit, which is used for policy-based trust judgment, are what we suggest, and do not exist in the cloud yet; however policy-based trust has been successfully (more or less) used in PKI practice, and the practice is a proof of feasibility.

The mechanism of using attribute assessment/certification and evidence-based trust judgment could be complex, due to a possibly large set of attributes to consider and a possibly long chains of trust relations. Nevertheless, the policy-based trust judgment can be actually regarded as a simplified version of the attribute/evidence-based mechanism, in the sense that a widely accepted policy captures a set of key attributes.

In the above figures, the trust relations with various cloud entities, shown in the left part of the figures, are dependent on various sources of evidence, shown in the right part of figures; and the derivation of a source of evidence is dependent on some trust relations either. All those dependence relations form the chains of trust. **Figure 7** illustrates some chains of trust focusing on policy-based and attribute/evidence-based mechanisms.

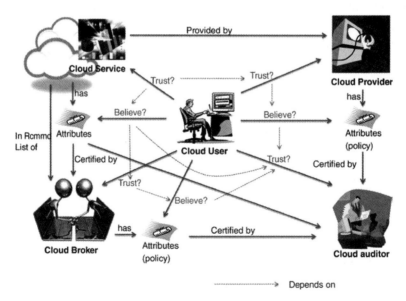

Figure 7. Chains of trust relations in clouds. This figure provides an integrated picture to illustrate the chains of trust relations from a cloud user to a cloud service and related cloud entities, where accreditation is omitted for simplicity.

Chapter 18

Summary and Further Research

Trust is a critical aspect of cloud computing. We examined and categorized existing research and practice of trust mechanisms for cloud computing in five categories—reputation based, SLA verification based, transparency mechanisms (self-assessment and information revealing), trust as a service, and formal accreditation, audit, and standards. Most current work on trust in the cloud focus narrowly on certain aspects of trust; our thesis is that this is insufficient. Trust is a complex social phenomenon, and a systemic view of trust mechanism analysis is necessary. In this paper we take a broad view of trust mechanism analysis in cloud computing and develop a somewhat informal and abstract framework as a route map for analyzing trust in the clouds. In particular, we suggest: 1) a policy-based approach of trust judgment, by which the trust placed on a cloud service or a cloud entity is derived from a "formal" audit proving that the cloud entity conforms to some trusted policies; 2) a "formal" attribute-based approach of trust judgment, by which particular attributes of a cloud service or attributes of a service provider are used as *evidence* for trust judgment, and the belief in those attributes is based on formal certification and chains of trust for validation. To support this mechanism, we propose a general structure of evidence-based trust judgment, which provides a basis to infer the trust in a cloud entity from the belief in the attributes that entity has, and in which, based on the semantics of trust, we define the attributes to be examined are in a space of two-dimensions—domain of expectancy and source of trust including competency, integrity, and goodwill.

Future research will focus on mathematically formal frameworks for reasoning about trust, including modeling, languages, and algorithms for computing trust.

Abbreviations

AA: Attribute Authority; AC: Attribute Certificate; AICPA: American Institute of Certified Public Accountants; ASB: Auditing Standards Board; CA: Certification Authority; CAIQ: Consensus Assessment Initiative Questionnaire; CCM: Cloud Control Matrix; CP: Certificate Policy; CSA: Cloud Security Alliance; CTA: Cloud Trust Authority; CTP: CloudTrust Protocol; GRC: Governance, Risk Management and Compliance; IETF: Internet Engineering Task Force; IGTF: International Grid Trust Federation; ISAE 3402: The International Standard on Assurance Engagements 3402; NIST: National Institute of Standards and Technology; PKI: Public Key Infrastructure; QoS: Quality of Service; SLA: Service Level Agreement, SSAE 16. Statement on Standards for Attestation Engagements No. 16; STAR: Security, Trust & Assurance Registry; TaaS: Trust as a Service.

Competing Interests

The authors declare that they have no competing interests.

Authors' Contributions

JH carried out the study of trust in clouds and drafted the manuscript; DMN helped develop the concepts, reviewed, and revised the manuscript. Both authors read and approved the final manuscript.

Acknowledgements

This material is based upon research sponsored by the U.S. Air Force Research Laboratory (AFRL) and the U.S. Air Force Office of Scientific Research (AFSOR), under agreement number FA8750-11-2-0084. The U.S. Government is authorized to reproduce and distribute reprints for Governmental purposes notwithstanding any copyright notation thereon. We would like to thank Professor Roy Campbell, Professor Ravishankar K. Iyer, Scott Pickard, and many others in ACC-UCoE (Assured Cloud Computing University Center of Excellence, a joint effort of AFSOR, AFRL, and UIUC) for their valuable discussion.

Source: Huang J, Nicol D M. Trust mechanisms for cloud computing [J]. Journal of Cloud Computing Advances Systems & Applications, 2013, 2(1): 1–14.

References

[1] Michael B (2009) In clouds shall we trust? IEEE Security and Privacy 7(5): 3–3.

http://dx.doi.org/10.1109/MSP.2009.124.

[2] Everett C (2009) Cloud computing: A question of trust. Computer Fraud Security 2009(6): 5–7. http://dx.doi.org/10.1016/S1361-3723(09)70071-5.

[3] Garrison G, Kim S, Wakefield RL (2012) Success factors for deploying cloud computing. Commun ACM 55(9): 62–68. http://doi.acm.org/10.1145/2330667.2330685.

[4] Ghosh A, Arce I (2010) Guest editors' introduction: In cloud computing we trust—but should we? Secur Privacy, IEEE 8(6): 14–16. doi:10.1109/MSP.2010.177.

[5] Habib S, Hauke S, Ries S, Muhlhauser M (2012) Trust as a facilitator in cloud computing: a survey. J Cloud Comput Adv Syst Appl 1(1): 19.
doi:10.1186/2192-113X-1-19.
http://www.journalofcloudcomputing.com/content/1/1/19.

[6] Khan K, Malluhi Q (2010) Establishing trust in cloud computing. IT Prof 12(5): 20–27. doi:10.1109/MITP.2010.128.

[7] Michael B, Dinolt G (2010) Establishing trust in cloud computing. IANewsletter 13(2): 4–8. http://iac.dtic.mil/iatac/download/Vol13_No2.pdf.

[8] Park J, Spetka E, Rasheed H, Ratazzi P, Han K (2012) Near-real-time cloud auditing for rapid response. In: 26th International conference on advanced information networking and applications workshops (WAINA), pp 1252–1257. IEEE Computer Society, Washington, DC, USA. doi:10.1109/WAINA.2012.78.

[9] Pearson S (2011) Toward accountability in the cloud. Internet Comput IEEE 15(4): 64–69. doi:10.1109/MIC.2011.98.

[10] Takabi H, Joshi J, Ahn G (2010) Security and privacy challenges in cloud computing environments. Secur Privacy IEEE 8(6): 24–31. doi:10.1109/MSP.2010.186.

[11] Abawajy J (2011) Establishing trust in hybrid cloud computing environments. In: Proceedings of the 2011 IEEE 10th International Conference on Trust, Security and Privacy in Computing and Communications. IEEE Computer Society, Washington, DC, USA. TRUSTCOM '11, pp 118–125. doi:10.1109/TrustCom.2011.18.
http://dx.doi.org/10.1109/TrustCom.2011.18.

[12] Haq IU, Alnemr R, Paschke A, Schikuta E, Boley H, Meinel C (2010) Distributed trust management for validating sla choreographies. In: Wieder P, Yahyapour R, Ziegler W (eds). Grids and service-oriented architectures for service level agreements. Springer, US. pp 45–55. http://dx.doi.org/10.1007/978-1-4419-7320-7_5.

[13] Pawar P, Rajarajan M, Nair S, Zisman A (2012) Trust model for optimized cloud services (Dimitrakos T, Moona R, Patel D, McKnight D, eds.). Springer, Berlin Heidelberg. pp 97–112. http://dx.doi.org/10.1007/978-3-642-29852-3_7.

[14] NIST (2011) NIST cloud computing standards roadmap, NIST CCSRWG-092. first edition. NIST, Gaithersburg, MD, USA.
http://www.nist.gov/itl/cloud/upload/NIST_SP-500-291_Jul5A.pdf.

[15] Blomqvist K (1997) The many faces of trust. Scand J Manage 13(3): 271–286.

[16] Mayer R, Davis J, Schoorman F (1995) An integrative model of organizational trust: Past, present, and future. Acad Manage Rev 20(3): 709–734.

[17] Huang J, Nicol D (2010) A formal-semantics-based calculus of trust. Internet Comput IEEE 14(5): 38–46. doi:10.1109/MIC.2010.83.

[18] Shaoham Y (1987) Temporal logics in ai: Semantical and ontological considerations. Artif Intell 33: 89–104.

[19] Huang J, Fox MS (2006) An ontology of trust: formal semantics and transitivity. In: Proceedings of the ICEC'06, 259–270. ACM, New York, NY, USA. doi:10.1145/1151454.1151499.

[20] Hwang K, Kulkareni S, Hu Y (2009) Cloud security with virtualized defense and reputation-based trust mangement. In: Dependable, Autonomic and Secure Computing, 2009. DASC'09. IEEE Computer Society, Washington DC, USA. Eighth IEEE International Conference on, pp 717–722. doi:10.1109/DASC.2009.149.

[21] CSA (2011) STAR (security, trust and assurance registry) program. Cloud Security Alliance. https://cloudsecurityalliance.org/star/. Accessed on 16 Oct. 2012.

[22] CSA (2011) Security guidance for critical areas of focus in cloud computing v3.0. Cloud Security Alliance. www.cloudsecurityalliance.org/ guidance/csaguide.v3.0.pdf.

[23] CSA (2011) CSA: Security, trust and assurance registry. Cloud Security Alliance. https://cloudsecurityalliance.org/research/initiatives/starregistry/. Accessed on 16 Oct. 2012.

[24] Knode R (2009) Digital trust in the cloud. CSC.COM. http://assets1.csc.com/au/downloads/0610_20_Digital_trust_in_the_cloud.pdf.

[25] CSC (2011) Cloudtrust protocol (CTP). Cloud Security Alliance. https://cloudsecurityalliance.org/research/ctp/. Accessed on 16 Oct. 2012.

[26] Knode R, Egan D (2010) Digital trust in the cloud—A precis for the CloudTrust protocol (V2.0). CSC. https://cloudsecurityalliance.org/wp-content/uploads/2011/05/cloudtrustprotocolprecis_073010.pdf.

[27] RSA (2011) RSA establishes cloud trust authority to accelerate cloud adoption. RSA. http://www.rsa.com/press_release.aspx?id=11320.

[28] EMC (2011) Proof, not promises: Creating the trusted cloud. EMC. http://www.emc.com/collateral/emc-perspective/11319-tvision-wp-0211-ep.pdf.

[29] ISO (2005) ISO/IEC 27001:2005 information technology—security techniques—information security management systems—requirements. ISO. http://www.iso.org/iso/catalogue_detail?csnumber=42103. Accessed on 16 Oct. 2012.

[30] AICPA (2010) SSAE 16. AICPA. http://www.aicpa.org/Research/Standards/AuditAttest/Pages/SSAE.aspx. Accessed on 16 Oct. 2012.

[31] IAASB (2009) ISAE 3402. International Auditing and Assurance Standards Board. http://www.ifac.org/sites/default/files/downloads/b014-2010-iaasb-handbook-isae-3402.pdf. Accessed on 16 Oct. 2012.

[32] IGTF (2010) Guidelines for auditing grid cas version 1.0. IGTF. http://www.ogf.org/documents/GFD.169.pdf.

[33] Maurer UM (1996) Modelling a public-key infrastructure. In: In: ESORICS'96: Proceedings of the 4th European symposium on research in computer security. Springer-Verlag, London, pp 325–350.

[34] Cooper D, Santesson S, Farrell S, Boeyen S, Housley R, Polk W (2008) Internet x.509 public key infrastructure certificate and certificate revocation list (CRL) profile. IETF. http://www.ietf.org/rfc/rfc5280.txt.

[35] Huang J, Nicol D (2009) Implicit trust, certificate policies and formal semantics of PKI. Information Trust Institute, University of Illinois at Urbana-Champaign.

[36] Ziegler CN, Lausen G (2005) Propagation models for trust and distrust in social networks. Inf Syst Front 7(4–5): 337–358. http://dx.doi.org/10.1007/ s10796-005-4807-3.

[37] Farrell S, Housley R, Turner S (2010) An internet attribute certificate profile for authorization. IETF. http://www.ietf.org/rfc/rfc5755.txt.

[38] Enomaly Inc. (2010) SportCloud: Global market for cloud capacity. Enomaly Inc. http://spotcloud.com. Accessed on 18 Jan. 2013.

Chapter 19
A Multi-Level Security Model for Partitioning Workflows Over Federated Clouds

Paul Watson

School of Computing Science, Newcastle University, Newcastle-upon-Tyne, NE1 7RU, UK

Abstract: Cloud computing has the potential to provide low-cost, scalable computing, but cloud security is a major area of concern. Many organizations are therefore considering using a combination of a secure internal cloud, along with (what they perceive to be) less secure public clouds. However, this raises the issue of how to partition applications across a set of clouds, while meeting security requirements. Currently, this is usually done on an adhoc basis, which is potentially error-prone, or for simplicity the whole application is deployed on a single cloud, so removing the possible performance and availability benefits of exploiting multiple clouds within a single application. This paper describes an alternative to adhoc approaches—a method that determines all ways in which applications structured as workflows can be partitioned over the set of available clouds such that security requirements are met. The approach is based on a Multi-Level Security model that extends Bell-LaPadula to encompass cloud computing. This includes introducing workflow transformations that are needed where data is communicated between clouds. In specific cases these transformations can result in security breaches, but the paper describes how these can be detected. Once a set of valid options has been generated, a cost model is used to rank them. The method has been implemented in a tool, which is described in the paper.

1. Introduction

Cloud computing is of growing interest due to its potential for delivering cheap, scalable storage and processing. However, cloud security is a major area of concern that is restricting its use for certain applications: "Data Confidentiality and Auditability" is cited as one of the top ten obstacles to the adoption of cloud computing in the influential Berkeley report[1]. While security concerns are preventing some organizations from adopting cloud computing at all, others are considering using a combination of a secure internal "private" cloud, along with (what they perceive to be) less secure "public" clouds. Sensitive applications can then be deployed on a private cloud, while those without security concerns can be deployed externally on a public cloud. However, there are problems with this approach. Currently, the allocation of applications to clouds is usually done on an adhoc, perapplication basis, which is not ideal as it lacks rigour and auditability. Further, decisions are often made at the level of granularity of the whole application, which is allocated entirely to either a public or private cloud based on a judgment of its overall sensitivity. This eliminates the potential benefits for partitioning an application across a set of clouds, while still meeting its overall security requirements. For example, consider a medical research application in which data from a set of patients' heart rate monitors is analyzed. A workflow used to analyze the data from each patient is shown in **Figure 1**. The input data is a file with a header identifying the patient, followed by a set of heart rate measurements recorded over a period of time. A service (*Anonymize*) strips off the header, leaving only the measurements (this application is concerned with the overall results from a cohort of patients, not with individuals). A second service (*Analyze*) then analyzes the measurements, producing a summary.

Analyzing the heart rate data is computationally expensive, and would benefit from the cheap, scalable resources that are available on public clouds. However, most organizations would be unlikely to consider storing medical records on a public cloud for confidentiality and, in some cases, legal reasons. Therefore, one solution is to deploy the whole workflow on a secure private cloud. However, this may overload the finite resources of the private cloud, resulting in poor performance, and potentially a negative impact on other applications.

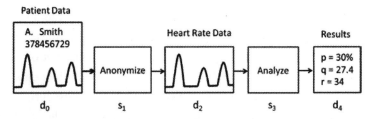

Figure 1. An example medical data analysis workflow.

An alternative solution is to partition the application between the private cloud and an external public cloud in order to exploit the strengths of both. This could be attempted in an adhoc fashion by a security expert but, as this paper describes, there are challenges in working out the set of partitioning options that still preserve the required security of data and services. This paper therefore describes an alternative to adhoc solutions—a method that takes an application consisting of a set of services and data connected in a workflow, and determines the valid set of deployments over a set of clouds, ensuring that security requirements are met. Although the paper is focused on workflows in which services communicate through passing data, the method can be applied to other types of distributed system that are composed of a set of communicating components. The method is based on Multi-Level Security models[2], specifically Bell-LaPadula[3]. The result of the method is the complete set of options that meet the organization's security requirements for the application. The method introduces transformations that need to be performed on the work-flows where data is communicated between clouds; the paper identifies the security issues that can be raised as a result, and the extra security checks that need to be performed to address this. When the method results in more than one valid partitioning option, there is the issue of how to choose the best. The paper shows how a cost model can be introduced to rank the valid options; a model based on price is defined, and applied to the running medical workflow example. The full method, including the cost model, has been implemented in a tool that has been built to automate and explore its application.

The paper is structured as follows. The **Method** section gives a brief introduction to Multi-Level Security models and Bell-LaPadula. It then describes how the Bell-LaPadula rules can be applied to ensure that a workflow meets the security requirements of its constituent services and data. The method is then extended to cloud computing by assigning security levels to clouds, and building on Bell-LaPadula to define a method for determining if security requirements are met in a particular deployment of the constituent parts of a workflow onto a set of clouds. The **Calculating valid deployment options** section then defines a method for enumerating all valid options for deploying a workflow over a set of clouds so as to meet security requirements. It highlights the issues raised when data must flow between clouds, and shows the work-flow transformations and security checks that must be included in the method if security is to be guaranteed. The result is a set of valid options; the **Selecting a deployment option with a cost model** section then introduces a model that can be used to select the best option. The method is then applied to a second, more complex example (in the **A more complex example** section). A tool has been designed and built to implement the method. As described in the **Tooling** Section, it is structured as a set of rules, transforms and a cost model, allowing it to be enhanced to meet other non-functional requirements, including dependability. Following a review of related work, the paper draws conclusions and outlines further work.

2. Method

This section describes how the Bell-LaPadula security model can be applied to workflows, and can then be extended to the deployment of workflows on clouds. Through this section, a workflow is modeled as a directed graph in which services and data are represented as nodes. Services consume zero or more data items and generate one or more data items; the edges in the graph represent the data dependencies.

2.1. Representing Security Requirements

The Bell-LaPadula multi-level access control model[3] is adopted, with services modeled as the subjects (S), and data as the objects (O)[4]. The security model therefore consists of the following:

- a set of actions (A) that subjects (S) can carry out on objects (O). In the case of services operating on data in a workflow, the actions are limited to read and write. Therefore, the set of actions (A) is: A = $\{r, w\}$

- a poset of security levels: L

- a permissions matrix: $M: S \times O \to A$ (the contents of the matrix are determined by the workflow design; *i.e.* if service s_i reads datum d_0 then there will be an entry in the matrix $s_1 \times d_0 \to r$; similarly, if service s_1 writes datum d_2 then there will be an entry in the matrix: $s_1 \times d_2 \to w$

- an access matrix: $B: S \times O \to A$ (this is determined by the execution of the workflow: if there are no choice points then it will equal the permissions matrix, however, if there are choice points then it will equal a subset of the permissions matrix corresponding to the path taken through the workflow when it is executed.

- a clearance map: $C: S \to L$ (this represents the maximum security level at which each service can operate)

- a location map: $l: S + O \to L$ (this represents the security level of each service and datum in the workflow)

In a typical Multi-Level Security scenario, the system moves through a set of states, and the model can have different values for permissions, access, clearance and location in each state. However, here the execution of a workflow is modeled as taking place within a single state. Normally a service would be expected to have a clearance that is constant across all uses of that service in workflows, however the location can be chosen specifically for each workflow, or even (though less likely) for each invocation of a workflow. However, the model itself is general, and makes no assumptions on this.

The Bell-LaPadula model states that a system is secure with respect to the above model if the following conditions are satisfied $\forall subjests\, u \in S$ and $\forall objests\, i \in O$

$$\text{Authorization: } B_{ui} \subseteq M_{ui} \qquad (1)$$

$$\text{clearance: } l(u) \leq c(u) \qquad (2)$$

$$\text{no-read-up: } r \in B_{ui} \Rightarrow c(u) \geq l(i) \qquad (3)$$

$$\text{no-write-down: } w \in B_{ui} \Rightarrow l(u) \leq l(i) \qquad (4)$$

For workflows, the implications of these conditions are:

(1) all actions carried out by services must conform to the permissions granted to those services

(2) a service can only operate at a security level (location) that is less than or equal to its clearance

(3) a service cannot read data that is at a higher security level than its own clearance

(4) a service cannot write data to a lower security level than its own location.

For example, consider a service s_i which consumes datum d_0 and produces datum d_2:

$$\boxed{d_0} \rightarrow \boxed{s_1} \rightarrow \boxed{d_2}$$

(in these diagrams, the $\rightarrow$ is used to show data dependency, and each block—service or datum—is uniquely identified by the subscript). The following rules must be met:

by (3)

$$c(s_1) \geq l(d_0) \qquad (5)$$

and by (4)

$$l(d_2) \geq l(s_1) \qquad (6)$$

The relationship between security levels is captured in **Figure 2**. Arrows represent $\geq$ relationships.

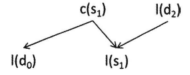

Figure 2. The relationship between security levels for a service that consumes and produces data.

Whilst assigning a security level to a datum in a work-flow is directly analogous to assigning a level to an object (e.g. a document) in the standard Bell-LaPadula model, assigning a security level to a service may be less intuitive. The justification is that an organization may have differing levels of confidence in the set of services they wish to use. For example, they may be very confident that a service written in-house, or provided by a trusted supplier, will not reveal the data it consumes and produces to a third party either deliberately or through poor design; in contrast, there is a risk that a service downloaded from the Internet, of unknown provenance, may do just that. Therefore, the organization can assign a high security level to the former service, and a low level to the latter.

For a specific workflow, when an organization's security experts are assigning locations to services, they may in some cases chose to set the location below that of the clearance level in order to allow a service to create data that is at a lower level than its clearance level; i.e. so that the no write down rule (4) is not violated. This may, for example, take place when the expert knows that the output data will not be sensitive, given the specific data that the service will consume as input in this specific workflow. A concrete example would be a service that summarizes textual data. This has been written to a high standard, and the security expert is confident that it will not leak data to a third party. Therefore, its clearance is high. However, in one particular workflow it is known that this service will only be used to summarise public data downloaded from the World Wide Web, which is also where its output will be published. Therefore, the security expert would set the service's location to an appropriately low level so that the write down rule was not violated.

2.2. Cloud Security

This section describes how the Bell-LaPadula model, as applied to workflows, can be extended to encompass cloud computing.

Let us say that an organization wishes to run a particular workflow. As more than one cloud is available, a decision must be made as to where the data and services should be placed. In current practice, it is typical that a security expert or system administrator would just take a considered view on the overall security level of the work-flow, and that of the clouds on which it could be deployed. For example, let us say that there are two clouds, one a highly trusted private cloud contained within the intranet of the organization, and the other a less trusted public cloud. It may seem obvious in this case that a workflow that operates on sensitive medical data should run only on the internal cloud. Similarly, a workflow that summarises public data could be deployed on the public cloud. However, there are two problems with this approach. Firstly, it is informal, being based on an expert's judgment; a systematic approach is preferable as it will give more consistent, defendable results. Secondly, the approach deploys the whole of a workflow on a single cloud. This rules out other options that may:

- reduce cost: for example by running less sensitive, but computationally intensive, sub-parts of the workflow on a public cloud if that avoids the need to purchase expensive new servers so that the internal cloud can handle the extra load

- increase reliability: for example by having the option to run on a public cloud if the private cloud has an outage

- increase performance: for example by taking advantage of the greater processing capacity of the public cloud for the computationally intensive services in a workflow

Therefore, the rest of this section extends the security model introduced earlier in order to allow systematic decisions to be taken on where the services and data within a workflow may be deployed to ensure security requirements are met.

To do this, the location map is extended to include clouds which we denote by P (to avoid confusion with the C conventionally used to denote the clearance map):

- location map: $l: S + O + P \rightarrow L$

Also, H is added to represent the mapping from each service and datum to a cloud:

$$H: S + O \rightarrow P]$$

We then add a rule that any block (service or datum) must be deployed on a cloud that is at a location that is greater than or equal to that of the block, e.g. for a block x on cloud y:

$$l(p_y) \geq l(b_x) \qquad (7)$$

Returning to the example service introduced in the previous section:

$$\boxed{d_0} \rightarrow \boxed{s_1} \rightarrow \boxed{d_2}$$

if, in H, d_0 is on cloud p_a, s_1 on p_b and d_2 on cloud p_c then the following must be true:

$$l(p_a) \geq l(d_0) \qquad (8)$$

$$l(p_b) \geq l(s_1) \qquad (9)$$

$$l(p_c) \geq l(d_2) \qquad (10)$$

This allows us to extend (6) to:

$$l(p_c) \geq l(d_2) \geq l(s_1) \qquad (11)$$

The complete relationship between security levels blocks and clouds is captured in **Figure 3**.

3. Calculating Valid Deployment Options

Using the above model and rules, it is now possible to automatically enumerate all the valid deployment options for a workflow. These are generated in two stages. Firstly, given the following:

- the set of clouds P
- the set of services S

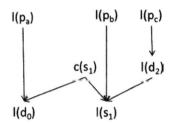

Figure 3. The relationship between cloud and workflow block security levels.

- the set of data O
- the map of security locations l

we can define the valid mappings of services and data onto clouds, using rule (7):

$$V : S + O \rightarrow P$$
$$V = \{b \rightarrow p / b \in S + O, p \in P, l(b) \leq l(p)\}$$

To illustrate this, we use the medical workflow of **Figure 1**, with two clouds. This has two services connected in a pipeline, each with one datum as input and one as output:

$$\boxed{d_0} \rightarrow \boxed{s_1} \rightarrow \boxed{d_2} \rightarrow \boxed{s_3} \rightarrow \boxed{d_4}$$

Table 1 shows an example location and clearance table (while the scheme is general, this example uses only two security levels: 0 and 1). Here, c_1 is a private cloud, which is at a higher security level than the public cloud c_0. The patient data (d_0) is at the highest security level, while the other data is at the lower level as it is not confidential. Service *si* is cleared to access confidential data at level 1, but its location has been set to 0 in this workflow so that it can produce non-confidential output at level 0 without violating the Bell-LaPadula "no-write-down" rule (4).

Based on this mapping of blocks and clouds to locations, **Table 2** then shows the possible valid placement of each block onto the two clouds.

Having determined all valid mappings of services and data to clouds, the set of all valid workflow deployments is given by:

Table 1. Locations and clearances for the medical analysis example.

	Location (l)	Clearance (c)
d_0	1	
s_1	0	1
d_2	0	
s_3	0	0
d_4	0	
c_0	0	
c_1	1	

Table 2. Valid mappings of blocks to clouds.

Block	Cloud c_0	Cloud c_1
d_0		•
s_1	•	•
d_2	•	•
s_3	•	•
d_4	•	•

$$W:(S+O \to P) \to \{(S+O \to P)\} = \{w \in \|V\|, \forall b \in S+O. \exists p \in P.b \to p \in w, |w| = |S+O|\}$$

where $\|V\|$ is the power set of V and $|w|$ is the cardinality of w. Algorithmically, in the implementation of the method, W is computed by forming the cross-product of the block-to-cloud mappings contained in V.

All possible valid workflow deployments—as defined by W—for the running medical workflow example are shown in **Figure 4**. The cloud on which a datum or service is deployed is indicated as a superscript; e.g. d_j^a is datum j deployed on cloud a.

Transferring Data between Clouds

There is still an important issue to be addressed: the approach makes assumptions that are unrealistic for a practical distributed workflow system. It assumes that:

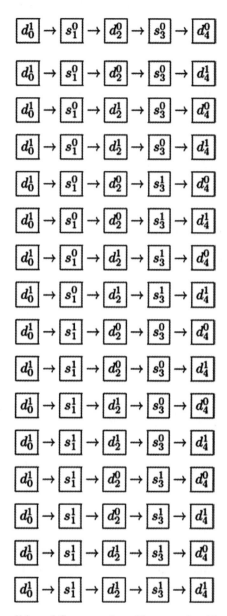

Figure 4. All valid workflows produced by mapping blocks to clouds.

- a service can generate as its output a datum directly on another cloud, without that item being first stored on the same cloud as the service

- a service can consume as its input a datum directly from another cloud, without that item ever being stored on the same cloud as the service

This problem is solved in two stages. Firstly, a new type of service is introduced—sxfer—which will transfer data from one cloud to another (this is analogous to the exchange operator used in distributed query processing[5]). It would be implemented with sub-components running on the source and destination clouds. The sxfer service takes a datum on one cloud and creates a copy on another. All the workflows generated by W are then transformed to insert the transfer nodes whenever there is an intercloud edge in the workflow graph. There are four graph transformations:

$$\boxed{d_j^a} \to \boxed{s_i^a} \Rightarrow \boxed{d_j^a} \to \boxed{s_i^a} \tag{12}$$

$$\boxed{d_j^a} \to \boxed{s_i^b} \Rightarrow \boxed{d_j^a} \to \boxed{sxfer} \to \boxed{d_j^b} \to \boxed{s_i^b} \tag{13}$$

$$\boxed{s_i^a} \to \boxed{d_j^a} \Rightarrow \boxed{s_i^a} \to \boxed{d_j^a} \tag{14}$$

$$\boxed{s_i^a} \to \boxed{d_j^b} \Rightarrow \boxed{s_i^a} \to \boxed{d_j^a} \to \boxed{sxfer} \to \boxed{d_j^b} \tag{15}$$

Transforms (12) and (14) reflect the fact that if both nodes are deployed on the same cloud then no change is needed. In contrast, (13) and (15) introduce new sxfer nodes to transfer data between clouds.

Unfortunately, the creation of new copies of data through transforms (13) and (15) introduces potential security problems. When transform (13) is applied, there is the need to check that cloud b has a sufficiently high security level to store the copy of d_j that would be created on it (the copy inherits the security level of the original). The following rule must therefore be checked to ensure this is true:

$$l(p_b) \geq l(d_j) \tag{16}$$

Similarly, for transform 15:

$$l(p_a) \geq l(d_j) \tag{17}$$

If either is violated then the workflow does not meet the security requirements, and so should be removed from the set W of valid mappings of services and data to clouds. Proof that this violation can only occur in two specific cases now follows.

Firstly, consider (16). By rule (2) we have:

$$c\left(s_i^b\right) \geq l\left(s_i^b\right) \tag{18}$$

First consider the case where:

$$c\left(s_i^b\right) = l\left(s_i^b\right) \tag{19}$$

i.e. the clearance of the object is equal to its location.

Rules (3) and (4) give

$$c\left(s_i^b\right) \geq l\left(d_j\right) \tag{20}$$

and

$$l\left(p_b\right) \geq l\left(s_i^b\right) \tag{21}$$

then, by (19)

$$l\left(p_b\right) \geq l\left(s_i^b\right) \geq l\left(d_j\right) \Rightarrow l\left(p_b\right) \geq l\left(d_j\right) \tag{22}$$

and rule (16) is satisfied. Therefore, in this case there are no violations.

However, if:

$$c\left(s_i^b\right) > l\left(s_i^b\right) \tag{23}$$

i.e. the clearance of the service is strictly greater than its location then combining

(23) with (3) and (4) in a similar way to the above, we get:

$$l(s_i^b) < c(s_i^b) \geq l(d_j) \qquad (24)$$

and

$$l(p^b) \leq l(s_i^b) < c(s_i^b) \qquad (25)$$

so it is possible that

$$l(p^b) < l(d_j) \qquad (26)$$

in which case rule (16) is violated and so that particular workflow deployment does not meet the security requirements.

Turning now to the data produced by services, rule (17) can be violated by transform (15) in the case where the service $s0$ writes up data (4) to a level such that:

$$l(p_a) < l(d_j)$$

The effect of the transformations is to modify the security lattice of **Figure 2** to that of **Figure 5**. The arc from $l(P_b)$ to $l(d_0)$ is introduced by transform (13) which adds a copy of do into the workflow, while the arc from $l(p_b)$ to $l(d_2)$ is introduced by transform (15) which adds a copy of d_2.

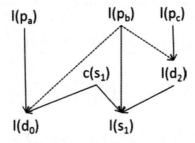

Figure 5. The relationship between security levels after transformation for inter-cloud data transfer.

Applying the transformations to each workflow in **Figure 4**, followed by rules (16) and (17) removes half of the possible deployment options. Removing two duplicates created by the transformations leaves the six valid options shown in **Figure 6**. Another view of the remaining options is shown in **Figure 7** As services can have multiple inputs and outputs, the arcs in the diagrams are labelled with the input/output number. These diagrams were generated automatically by the tool we have built to implement the methods described in this paper. The aim is to provide a security expert with an easy to understand view of the possible options.

Whilst a simple, linear workflow has been used here to illustrate the method, it is applicable to all workflows that can be represented by a directed graph, whatever their structure. The discussion so far does however still leave open the issue of how to choose between these valid options? The next section therefore describes how a cost model (also implemented in the tool) can be used to select the best option based on the charges made by the cloud providers.

4. Selecting a Deployment Option with a Cost Model

Once all valid options for allocating services and data to clouds have been determined, one must be selected, and used to enact the workflow. This decision

$$d_0^1 \to s_1^1 \to d_2^1 \to sxfer \to d_2^0 \to s_3^0 \to d_4^0$$

$$d_0^1 \to s_1^1 \to d_2^1 \to sxfer \to d_2^0 \to s_3^0 \to d_4^0 \to sxfer \to d_4^1$$

$$d_0^1 \to s_1^1 \to d_2^1 \to sxfer \to d_2^0 \to sxfer \to d_2^1 \to s_3^1 \to d_4^1 \to sxfer \to d_4^0$$

$$d_0^1 \to s_1^1 \to d_2^1 \to sxfer \to d_2^0 \to sxfer \to d_2^1 \to s_3^1 \to d_4^1$$

$$d_0^1 \to s_1^1 \to d_2^1 \to s_3^1 \to d_4^1 \to sxfer \to d_4^0$$

$$d_0^1 \to s_1^1 \to d_2^1 \to s_3^1 \to d_4^1$$

Figure 6. The Workflows that remain valid after Transfer Blocks are Added.

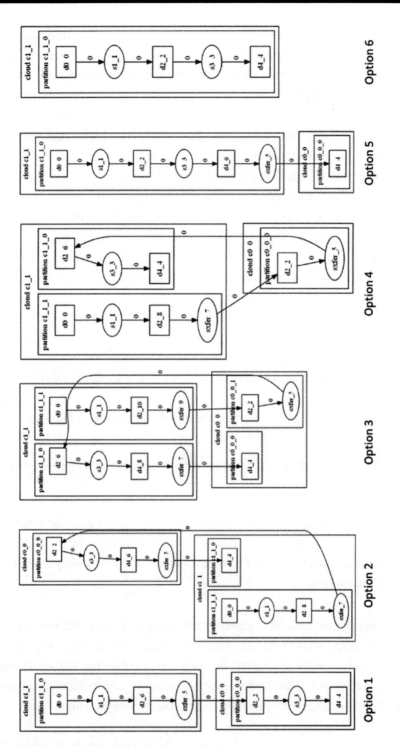

Figure 7. The six valid cloud mappings.

could be made by a deployment expert, but this section describes how it can be achieved automatically through the use of a cost model. Different criteria may be important for different applications (e.g. dependability, performance), but this section illustrates the approach by describing a model that minimizes price.

Cloud pricing is measured using the metrics by which cloud providers allocate charges. For a cloud (*p*) this is represented as:
- volume of data transferred into a cloud: e_{dxi}^{p}
- volume of data transferred out of a cloud: e_{dxo}^{p}
- volume of data stored, per unit of time for which it is stored: e_{ds}^{p}
- time units of cpu consumed in the execution of a service: e_{cpu}^{p}

Cost metrics are characterised for a datum (d) as:
- data size: size (d)
- data longevity—the length of time the datum is stored: longevity(*d*)

Finally, the cost metric for a service (s) is characterised as:
- time units of cpu consumed in the execution of a service: cpu(*s*)

The cost model for a workflow execution can then be defined as:

$$\cos t = \sum_{d=0}^{d=k-1} e_{ds}^{p} \cdot size(d) \cdot longevity(d) + \sum_{s=0}^{s=m-1} e_{cpu}^{p} \cdot cup(s) + \sum_{x=0}^{x=q-1} \left(e_{dxo}^{ps} + e_{dxi}^{pd} \right) \cdot size(d)$$

where *k* is the number of data items in the workflow, and *m* is the number of services, while *q* is the number of inter-cloud data transfers. In the third term that calculates data transfer costs, *ps* represents the source cloud and *pd* the destination cloud for the transfer.

Using the cost model requires estimates of data sizes and cpu costs. This is realistic for many workflows, and producing these estimates is made easier if performance and capacity are logged for each run, so allowing statistical analysis to generate predictions. This is, for example, done by the e-Science Central cloud platform[6] which logs data on all data sizes, and service execution times.

Two examples now highlight the use of the model. Consider the valid mapping options shown in **Figure 7** for the running medical workflow example. In the simplest case, if the performance and cost of both clouds are equal (as in **Table 3**), then the cost difference between options is dependent only on the number of inter-cloud communications. **Table 4** gives example values for the blocks in the workflow. The size of d_0 will be known as it is the input to the workflow, while that for d_2 and d_4 are estimates, perhaps based on the results of previous runs. To set the longevity values, it is assumed that the input (d_0) and output data (d_4) is stored for a year, while intermediate data (d_2) (along with any intermediate copies of data created by transforms) is immediately discarded once it has been consumed by a service: in this case s_2.

Table 5 shows the results when the cost model is applied. Each row represents the cost of an option in **Figure 6**. The final column of the table gives the order of the options (from lowest to highest cost). This confirms that the cheapest is option 6, in which all the blocks are deployed on the same cloud, and so there are no inter-cloud transfer costs.

While it may seem that an option in which all services and data are deployed on a single cloud will always be the cheapest, if CPU costs vary between clouds, then inter-cloud transfers may be worthwhile. **Table 6** shows clouds with a different set of cost parameters. Here, a private cloud (c_1) has higher security, but higher CPU and data costs, compared to a public cloud (c_0). The effect of plugging these values into the cost model is shown in **Table 7**. The result is that the best option is

Table 3. Cloud costs: Example 1.

Cloud	Storage (GB / Month)	Transfer in (/GB)	Transfer out (/GB)	CPU (/s)
c_0	10	10	10	10
c_1	10	10	10	10

Table 4. Block info.

Block	Size (GB)	Longevity (months)	CPU (s)
d_0	10	12	
s_1			100
d_2	5	0	
s_3			50
d_4	1	12	

Table 5. Workflow deployment options costs: Example 1.

Option	Storage	Transfer	CPU	Total	Order
1	1320	100	1500	2920	3
2	1320	120	1500	2940	4
3	1320	220	1500	3040	6
4	1320	200	1500	3020	5
5	1320	20	1500	2840	2
6	1320	0	1500	2820	1

Table 6. Cloud costs: Example 2.

Cloud	Storage (GB / Month)	Transfer in (/GB)	Transfer out (/GB)	CPU (/s)
c_0	5	5	5	5
c_1	10	5	5	10

Table 7. Workflow deployment options costs: Example 2.

Option	Storage	Transfer	CPU	Total	Order
1	1260	75	1250	2585	1
2	1320	90	1250	2660	2
3	1260	165	1500	2925	5
4	1320	150	1500	2970	6
5	1260	15	1500	2775	3
6	1320	0	1500	2820	4

now the one that allocates as much work as possible to the public cloud, which has lower CPU costs.

5. A More Complex Example

The medical example used to date consists to a purely linear workflow: each service reads and writes only one data item. However, the method supports arbitrary work-flows, and this is now demonstrated through the example of a more complex workflow, which is based on that introduced in[4]. It is shown in **Figure 8**.

With security settings shown in **Table 8**, the workflow meets the Bell-LaPadula criteria, and produces the three workflow partitionings shown in **Figure 9**.

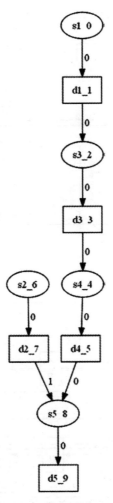

Figure 8. A more complex workflow.

Table 8. Block locations and clearances.

	Location (*l*)	Clearance (*c*)
$s1_0$	0	1
$s2_6$	0	0
$s3_2$	1	1
$s4_4$	0	1
$s5_8$	1	1
$d1_1$	1	
$d2_7$	0	
$d3_3$	1	
$d4_5$	1	
$d5_9$	1	

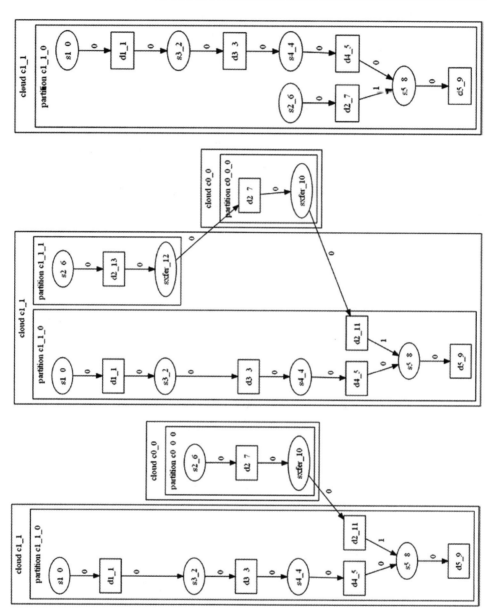

Figure 9. Valid workflow partitionings for the second example.

With block costs from **Table 9** and cloud costs from **Table 6** (as in the previous example), the costs of the three partitioning options are shown in **Table 10**. This example illustrates the importance of the use of the cost model in allowing an expert or automatic deployment system to filter out valid but inefficient workflow partitionings such as option 2.

Table 9. Block Costs.

	Size	Longevity	CPU
d1	10	12	
d2	5	12	
d3	1	0	
d4	20	0	
d5	20	12	
s1			100
s2			50
s3			10
s4			40
s5			60

Table 10. Workflow partitioning option costs.

Option	Storage	Transfer	CPU	Total	Rank
1	4500	50	2350	6900	2
2	5100	100	2600	7800	3
3	4200	0	2600	6800	1

6. Tooling

The method described in this paper has been implemented in a tool which takes the workflow and security requirements as input, and generates as output the set of valid partitions with costs. The tool is implemented in the functional language Haskell[7], with workflows represented as directed graphs. This section explains how the tool implements the multilevel security method that is the focus of this paper, and then goes on to explain how it can be used to process other classes of nonfunctional requirements.

The tool is structured so that three types of functions are used to process the initial workflow:

1) **Transformation** functions take a workflow as input and generate a set of workflows derived from it. In the multilevel security method they are used for two purposes: to generate the candidate workflow partitionings from the initial workflow (**Figure 4**); and to insert the inter-cloud transfers according to (13) and (15).

2) **Rules** are implemented as filter functions that take a workflow as input and return a boolean indicating whether the workflow meets the rule. Workflows that do not meet a rule are removed from the set of valid workflows. In the methods described in this paper, they are used to check that the initial workflow conforms to Bell-LaPadula (**Figure 2**), and that the candidate workflow partitionings conform to **Figure 3**.

3) **Cost Model** functions take in a candidate workflow and assign a cost. They are used to implement the model described in Section "Selecting a deployment option with a cost model".

Overall, the tool takes the initial workflow (e.g. that of **Figure 1**) and security requirements, applies the set of transformations and rules, and then uses the cost model function to rank the valid workflows.

The tool also automatically generates the diagrams that are shown in **Figure 7** using the GraphViz software library[8]; these visualisations have proved to be a useful way to review the available options. It can also generate an html report for a security review, containing all the information generated by the method, including the diagrams, security tables and ranked cost tables. Finally, it can automatically generate LATEX tables. Therefore all the tables and diagrams used in this paper have been automatically generated by the tool. Automatically generating diagrams and tables eliminates the risk of transcription errors[9] when conveying results through reports and papers. These reports can be used by system administrators to configure the partitions manually onto the clouds, but we are also currently developing a tool to do this automatically, as described in the next section.

Generalising the Approach

While this paper focuses on meeting security requirements, the structure of the tool, with its three types of functions described above, allows it to be utilised as a general system for generating partitions of workflows over clouds that meet nonfunctional requirements. As well as security, those requirements can include dependability and performance. For example, the tool is currently being extended to encompass dependability requirements. One example requirement came out of discussions with the designer of an application in which the key workflow was very similar to that of **Figure 1**, but with the additional constraint that the input and output data for the workflow as a whole (e.g. d_0 and d_4 in **Figure 1**) must not be stored on the same cloud. We were able to meet this requirement simply by adding one extra rule into the tool. This is a function *apart* that takes three arguments:

- the set of blocks that need to be kept apart on different clouds

- the mapping of blocks onto clouds

- the workflow to be checked against the rule

The result of the function is true if every block contained in the first argument is deployed on a different cloud. One subtlety is that the function must take into account the fact that transformations may create one or more copies of the blocks that have been specified by the user as needing to be kept apart. For example, transforms (13) and (15) can create such copies. It is important that those copies are also included in the set of blocks that need to be kept apart. This is is achieved by exploiting the fact that the tool assigns blocks a name and identifier pair, e.g. '(xfer, 14)': While the identifier of each block is unique, the name can be shared by multiple blocks. When a block is copied by a transformation, it is given the same name but a new, unique identifier (similarly, separate deployments of the same service share the same name, but have a different unique identifier), e.g. a transform copying data item (d, 14) could generate (d, 27) where 27 is a unique identifier, not shared with another block. The *apart* function therefore works by comparing whether two blocks with the same name are stored on the same cloud, irrespective of their identifiers. For the running medical workflow example, adding the rule that d_0 and d_4 in **Figure 1** must be deployed on different clouds results in only allowing the single partitioning option shown in **Figure 10**.

Figure 11 shows the architecture of the generalised tool. The Deployment Manager takes in the workflow and the user's nonfunctional requirements. It uses the rules and transformations to generate a set of valid partitionings over the available clouds, and then applies a cost model to rank them. The 'best' workflow is then executed across the set of clouds.

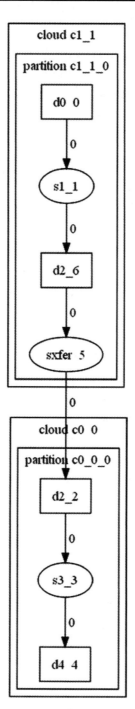

Figure 10. The only valid workflow partitioning if d_0 and d_4 must not be deployed on the same cloud.

To achieve this, each cloud must run software that can store data and execute workflows. In this work the e-Science Central cloud platform[6] is used. This has the advantage of providing a portable platform that can be deployed on a range of clouds including public clouds (Amazon and Windows Azure) but also private clouds. **Figure 12** shows the e-Science Central Architecture.

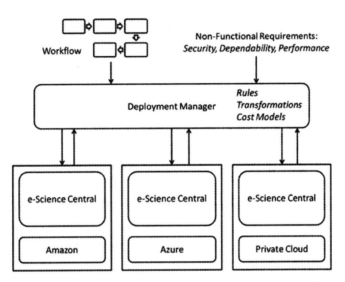

Figure 11. The Architecture of the General Cloud Workflow Partitioning Tool.

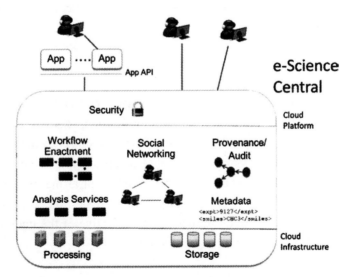

Figure 12. The Architecture of the portable e-Science Central Cloud Platform.

This is a cloud Platform-as-a-Service that provides users with the ability to store, share and analyse data in the cloud. Data can be uploaded and tagged with structured or unstructured metadata. Tools are provided to allow services, written in a variety of languages, to be packaged and loaded into the platform. Users can then create workflows from those services and execute them in the cloud; a graphical, in-browser editor is provided to allow users to create new workflows. e-Science Central provides scalable cloud computing as it can distribute the execution of a set of workflows across a set of nodes. This is mainly used to scale throughput (to increase the number of workflow executions per second), but it can also be used to reduce response time where there are opportunities to execute sub-workflows in parallel. The system has achieved scalability at over 90% efficiency, when running on up to 190 cloud nodes, each with two cores.

All accesses to e-Science Central system are policed by a Security service. This ensures that access to all data, services and workflows can be controlled by users. The basis for this is a social networking system that allows users to connect to each other, as well as to create and join groups. Users can chose to keep data, services and workflows private, or can share them with other users, or with groups to which they are connected.

All actions carried out in e-Science Central are recorded in an audit/provenance log. Users can view and query the subset of this log that the security system permits them to see. This allows users to determine exactly how each data item was created (e.g. the graph of service/workflow executions) and how it has been used.

e-Science Central has a Software-as-a-Service interface that allows users to perform all actions through a web browser. In addition, there is an API so that programs can drive the system. This includes all the actions needed to deploy and execute partitioned workflows over federated clouds: storing data, creating and executing workflows. The portability of e-Science Central means that the 'Deployment Manager' in **Figure 11** can create and trigger the execution of the workflow partitions in exactly the same way, irrespective of the underlying cloud on which the partition is deployed and executed.

Figure 11 shows two way communication between the 'Deployment Manager' and 'e-Science Central'. The above description of the tool describes the flow of information from the 'Deployment Manager' to the 'e-Science Central' instances in the form of workflow partitions to be executed. Information flowing in the reverse direction can include cost updates. If cloud providers start to vary their pricing models frequently, and provide APIs to access it, this information could also be fed back into the Deployment Manager so that the cost models can be kept up-to-date. e-Science Central also has a sophisticated provenance capture and analysis system[10]; this includes collecting information on data sizes and service execution times that can be used to improve the accuracy of the estimated block costs that feed into the cost models.

The Deployment Manager could also monitor the availability of the clouds, for example by sending regular "ping" requests to the e-Science Central deployments, or by including timeouts in the calls to execute a workflow partition on a cloud. This would then allow it to select dynamically between possible workflow partitionings based on which clouds are currently available. Of course, there is a danger that all valid options depend on the availability of one particular cloud (as in the running medical workflow example). Rather than discover this at runtime, it is possible to determine this statically using the tool. This can be done by running the tool multiple times, each time omitting one cloud. This will determine whether the execution of a workflow is critically dependent on a single cloud (in this case the tool will generate no valid options). Having done this, an organization for whom the workflow is business-critical could ensure that the cloud in question has sufficiently high levels of availability, or identify (or create) a second cloud with a sufficiently high security clearance that could also be used by the workflow if the other failed.

7. Related Work

The motivation for this paper came from the author's experience of cloud applications with security constraints (e.g. healthcare applications in the 'Social Inclusion through the Digital Economy' (SiDE) project[11]). However, the general concern that security was a barrier to use of the cloud for many organizations and applications has been widely discussed[1]. The general issues associated with security and clouds are discussed in[12]. A high-level approach to deciding where an application could be deployed is discussed in[13]. Another approach to eliciting and exploiting information on the security and other properties of clouds is described in[14]. These methodologies could be valuable in assigning security levels to clouds, services and data: something which is orthogonal to the scheme described in this paper.

In[4], Bell-LaPadula is also applied to workflow security. Petri Nets are used to model the workflow, rather than the approach taken in this paper. However, the key difference is that its scope does not extend to considering the deployment ofblocks within a workflow across a set of computational resources, as this paper does. It also differs in including *clearance* but not *loation* in its embodiment of Bell-LaPadula.

There has been a large body of work on using cost models to predict execution times in order to select between options for deploying workflows over grids and clouds[15][16]. However, perhaps due to the relatively recent introduction of pay-as-you-go cloud computing, there is much less work on using price-based cost models. In[17], both execution time and price-based models are used to compare a set of options for allocating a workflow over local resources and a public cloud. The work in[18] uses non-linear programming to generate options for using clouds to execute a workflow. Security is not a consideration in any of these papers.

Once the partitioning of a workflow over a set of clouds has been decided, a distributed workflow enactment engine is needed to actually run the workflow. The issues around this are discussed in[19] and a solution is proposed.

8. Conclusions

This paper has described a new method for automatically determining valid options for partitioning a workflow of services and data across a set of clouds based on security requirements. A cost model is then used to choose between the available options. The main contribution is to show how multilevel security, which has previously been applied to workflows, can be extended to encompass the allocation of workflow services and data to clouds. This has demonstrated that the need for inter-cloud data transfers raises interesting potential security violations that need to be addressed; in the running medical workflow example, this ruled out over half of the possible partitioning options. Although the paper focuses on workflows, the method can be applied to other distributed systems whose components are to be partitioned over a set of clouds.

The tool developed to implement the method has proved invaluable in two ways. Firstly, it removes the chance of human error in applying the various stages of the method to the input workflow. To reinforce this advantage, it can automatically generate both html and LTEXtables, diagrams and reports (the LTEXtables were used in this paper). Secondly, developing the tool forced us to think about how best to structure the implementation of the method, which resulted in a very general system that operates on rules, transforms and cost models. Whilst the focus of this paper is on the multilevel security rules and transforms, as described in subsection "Generalising the approach", we have been exploring the extension of the approach to other non-functional requirements such as dependability. Whilst this work is ongoing, even in its current form the method can illuminate dependability issues; for example, analysing the set of valid options can highlight the dependency of the workflow on a specific cloud (as in the running medical workflow example). This will allow an organization which is dependent on the workflow to ensure that this cloud has sufficiently high levels of availability, or encourage it to identify a second cloud with a sufficiently high security clearance that could also be used by the workflow.

Overall, the hope is that the approach described in this paper can move the process of partitioning workflows over federated clouds from one in which a human administrator makes an informed but adhoc choice, to one in which a tool, such as the one built to implement this method, can determine the valid options based on a rigorous underlying set of rules, and then suggest which is the best, based on a cost model. The approach therefore has the advantage that it can reduce both security violations and execution costs.

Competing Interests

The author declare that he have no competing interests.

Acknowledgements

The author would like to thank Leo Freitas, John Mace, Paolo Missier,

Chunyan Mu, Sophie Watson, Feng Hao, Zhenyu Wen, Simon Woodman and Hugo Hiden for their comments and suggestions. This work was funded by the Research Councils UK "Social Inclusion through the Digital Economy" project EP/G066019/1.

Source: Watson P. A multi-level security model for partitioning workflows over federated clouds [J]. Journal of Cloud Computing, 2011, 1(1): 180–188.

References

[1] Armbrust M, Fox A, Griffith R, Joseph AD, Katz RH, Konwinski A, Lee G, Patterson DA, Rabkin A, Stoica I, Zaharia M (2009) Above the clouds: a berkeley View of Cloud Computing, Technical Report UCB/EECS-2009-28, EECS Department, University of California, Berkeley.
http://www.eecs.berkeley.edu/Pubs/TechRpts/2009/EECS-2009-28.html.

[2] Landwehr CE (1981) Formal models for computer security. ACM Comput Surveys 13: 247–278.

[3] Bell DE, LaPadula LJ (1973) Secure Computer Systems: Mathematical Foundations. Tech. rep., MITRE Corporation.

[4] Knorr K (2001) Multilevel Security and Information Flow in Petri Net Workflows. In: Proceedings of the 11th Conference on Advanced Information Systems Engineering.

[5] Graefe G (1990) Encapsulation of parallelism in the Volcano query processing system. In: Proceedings of the 1990 ACM SIGMOD international conference on, Management of data, SIGMOD '90. ACM New York, pp 102–111.
http://doi.acm.org/10.1145/93597.98720.

[6] Watson P, Hiden H, Woodman S (2010) e-Science Central for CARMEN: science as a service. Concurr Comput: Pract Exper 22: 2369–2380.
http://dx.doi.org/10.1002/cpe.v22:17.

[7] Jones SP (2003) Haskell 98 language and libraries: the Revised Report. Cambridge University Press. ISBN 0521826144.

[8] Ellson J, Gansner E, Koutsofios L, North SC, Woodhull G (2002) Graphviz—open source graph drawing tools. Graph Drawing 2265: 483–484.
http://www.springerlink.com/index/bvdkvy7uj1plml8n.pdf.

[9] Kelly MC (1989) The Works of Charles Babbage. William Pickering, London.

[10] Woodman S, Hiden H, Watson P, MissierP(2011)Achieving reproducibility by combining provenance with service and workflow versioning. In: 6th Workshop on Workflows in Support of Large-Scale Science.

[11] Social Inclusion through the Digital Economy. www.side.ac.uk.

[12] Mace J, van Moorsel A, Watson P (2011) The case for dynamic security solutions in public cloud workflow deployments, dsnw. IEEE/IFIP41st International Conference on Dependable Systems and Networks Workshops. pp 111–116. http://csdl2.computer.org/csdl/proceedings/dsnw/2011/0374/00/05958795-abs.html.

[13] Capgemeni (2010) Putting Cloud security in perspective. Tech. rep., Capgemeni.

[14] Pearson S, Sander T (2010) A mechanism for policy-driven selection of service providers in SOA and cloud environments. In: New Technologies of Distributed Systems (NOTERE), 2010 10th Annual International Conference on. pp 333–338.

[15] Yu J, Buyya R, Tham CK (2005) Cost-based scheduling of scientific workflow applications on utility grids. In: First International Conference on e-Science and Grid Computing (e-Science'05). pp 140–147.

[16] Singh G, Kesselman C, Deelman E (2007) A provisioning model and its comparison with best-effort for performance-cost optimization in grids. In: Proceedings of the 16th international symposium on High performance distributed computing—HPDC'07. ACM Press, New York, pp 117–126.

[17] Deelman E, Singh G, Livny M, Berriman B, Good J (2008) The cost of doing science on the cloud: the Montage example. In: Proceedings of the 2008 ACM/IEEE conference on Supercomputing, SC'08. IEEE Press, Piscataway, pp 50:1–50:12. http://dl.acm.org/citation.cfm?id=1413370.1413421.

[18] Pandey S, Gupta K, Barker A, Buyya R (2009) Minimizing Cost when Using Globally Distributed Cloud Services: A Case Study in Analysis of Intrusion Detection Workflow Application. Tech. rep., Cloud Computing and Distributed Systems Laborato, The University of Melbourne, Australia, Melbourne, Australia.

[19] Woodman S (2008) A programming system for process coordination in virtual organisations. PhD thesis, School of Computing Science, Newcastle University.

Chapter 20

Enhancing the Security of LTE Networks against Jamming Attacks

Roger Piqueras Jover[1], Joshua Lackey[1], Arvind Raghavan[2]

[1]AT&T Security Research Center, New York, NY 10007, USA
[2]Blue Clover Devices, San Bruno, CA 94066, USA

Abstract: The long-term evolution (LTE) is the newly adopted technology to offer enhanced capacity and coverage for current mobility networks, which experience a constant traffic increase and skyrocketing bandwidth demands. This new cellular communication system, built upon a redesigned physical layer and based on an orthogonal frequency division multiple access (OFDMA) modulation, features robust performance in challenging multipath environments and substantially improves the performance of the wireless channel in terms of bits per second per Hertz (bps/Hz). Nevertheless, as all wireless systems, LTE is vulnerable to radio jamming attacks. Such threats have security implications especially in the case of next-generation emergency response communication systems based on LTE technologies. This proof of concept paper overviews a series of new effective attacks (smart jamming) that extend the range and effectiveness of basic radio jamming. Based on these new threats, a series of new potential security research directions are introduced, aiming to enhance the resiliency of LTE networks against such attacks. A spread-spectrum modulation of the main downlink broadcast channels is combined with a scrambling of the radio resource allocation of the uplink control channels and an advanced system information message encryption scheme. Despite the challenging implementation on commercial networks, which would require inclusion of these solutions in future releases of the LTE standard, the security solutions could strongly enhance the security of LTE-based national emergency response communication systems.

Keywords: LTE, Jamming, Security, OFDMA

1. Introduction

As mobile phones steadily become more powerful and bandwidth demands skyrocket, cellular operators are rapidly deploying broadband data services and infrastructure to enhance capacity. The long-term evolution (LTE) is the recently deployed standard technology for communication networks, offering higher data speeds and improved bandwidth. This new cellular communication system is the natural evolution of 3rd Generation Partnership Project (3GPP)-based access networks, enhancing the Universal Mobile Telecommunications System (UMTS).

LTE provides capacity to user equipments (UEs) by means of a centralized assignment of radio resources. A newly enhanced physical (PHY) layer is implemented based on orthogonal frequency division multiple access (OFDMA) and substantially improves the performance of the former wideband code division multiple access (W-CDMA)[1]. The new modulation scheme provides a large capacity and throughput, potentially reaching a raw bit rate of 300 Mbps in the downlink with advanced multiple input multiple output (MIMO) configurations[1].

Due to its spectrum efficiency and great capacity, LTE is planned to be adopted as the basis for the next-generation emergency response communication system, the Nationwide Interoperable Public Safety Broadband Network[2]. In this context, the characteristics of such LTE-based public safety networks are already under consideration in the industry[3]. Note that, specially in the case of this application, the security requirements of LTE communication networks are of paramount importance.

Despite the tremendous capacity and system enhancements implemented by LTE, cellular networks are known to be, as any kind of wireless network, vulnerable to radio jamming. Although it is a simple and well-known attack, radio jamming is the most common way to launch a localized denial of service (DoS) attack against a cellular network[4]. The impact of such attacks is very local and mainly constrained by the transmitted power of the jamming device. The attacker is only able to deny the service locally to UEs located in its vicinity. However, more sophisticated attacks have been discovered as a potentially more effective way to jam LTE networks[5][6]. These smart jamming attacks aim to saturate specifically the main downlink broadcast channel of LTE networks in order to launch a local DoS attack that requires less power, making it stealthier. Further complex attacks, such as low-power smart jamming, identify the actual physical resource blocks (PRBs) assigned to essential uplink control channels by capturing the unprotected broadcast messages sent from the base station (eNodeB). The interception of such unencrypted network configuration data allows the attacker to selectively saturate uplink control channels in order to extend the range of the attack to an entire cell or sector. Note that network configuration contained in the broadcast channel can also be leveraged to deploy an effective rogue base station and other kinds of attacks.

Although radio jamming attacks have a rather local range, they become highly relevant in the current cybersecurity scenario. Reports of very targeted and extremely sophisticated attacks have emerged over the last 2 years[7]. These attacks, popularly known as advanced persistent threats (APTs), span over months or even years and target large corporations and government institutions with the goal of stealing intellectual property or other valuable digital assets[8]. The advent of APTs has substantially changed the set of assumptions in the current threat scenario. When it comes to very well-planned and funded cyber attacks, the scale of the threat is not important anymore. Instead, achieving a very specific and localized goal for economic benefit or military advantage is the key element. In this context, scenarios such as a local DoS attack against the cell service around, for example, a large corporation's headquarters or the New York Stock Exchange becomes very relevant. DoS is also often a tool used to knock a phone off a secure network and force it down to an insecure radio access network (RAN) to pursue further attacks and data exfiltrations[9].

The goal of this proof of concept paper is to raise awareness on the traditionally overlooked threat of radio jamming and to propose a combination of potential research directions and LTE RAN enhancements against sophisticated jamming attacks. This theoretical enhanced security architecture relies on a boost of the jamming resiliency of the main downlink broadcast channels and the encryption of the data broadcasted in it. The potential results would be twofold. On one hand, an attacker would not be able to easily jam the downlink broadcast channels and, therefore, deny the service to UEs in its vicinity. On the other hand, no network configuration information could be intercepted and decoded, preventing an attacker to gain knowledge on each cell's specific configuration, which could be leveraged in security attacks. Finally, a proactive smart jamming multi-antenna cancelation technique is presented.

Some of the proposed security solutions involve substantial changes at the PHY layer of LTE networks which could be very challenging to implement on a commercial network and would require collaboration within the industry. Nevertheless, such security architecture could substantially increase the reliability and resiliency against security attacks of the Nationwide Interoperable Public Safety Broadband Network[2]. Anti-jamming enhancements could be included to the list of requirements for LTE-based public safety networks that are not in the scope of current releases of the LTE standard, such as direct communication and group communication[3].

The remainder of the paper is organized as follows. Section 2 briefly overviews the cell selection procedure in LTE networks, the main downlink broadcast channels, and the feasibility of eavesdropping unprotected broadcasted network configuration messages. Three attacks against LTE networks are described in Section 3. The proposed research directions and theoretical architecture to mitigate radio jamming attacks is introduced in Section 4. Finally, related work is reviewed in Section 5, and the concluding remarks are presented in Section 6.

2. Initial Access to LTE Networks

This section overviews the basic procedures necessary for a phone to synchronize with and connect to an LTE network. Any UE willing to access the network must first perform a cell selection procedure. After this procedure, the UE decodes the physical broadcast channel (PBCH) to extract the basic system information that allows the other channels in the cell to be configured and operated. The messages carried on this channels are unencrypted and can be eavesdropped by a passive radio sniffer. Once at this point, the UE can initiate an actual connection with the network by means of a random access procedure and establish a radio access bearer (RAB) in order to send and receive user traffic. The whole process is portrayed in **Figure 1**.

2.1. Cell Search Procedure

The cell search procedure consists of a series of synchronization steps that allow the UE to determine time and frequency parameters required to detect and demodulate downlink signals as well as to transmit uplink signals with the right timing. The three major steps in this procedure are symbol timing acquisition, carrier frequency synchronization, and sampling clock synchronization. To achieve full synchronization, the UE detects and decodes the primary synchronization signal (PSS) and the secondary synchronization signal (SSS), which are fully described in[10]. The mapping of the PSS and SSS in the central subcarriers of the LTE frame as well as the main functions of these synchronization signals is shown in **Figure 2**.

The PSS enables the UE to acquire the time slot boundary independently from the cyclic prefix configuration of the cell, which at this point is unknown to the UE. Based on the downlink frame structure, the PSS is transmitted twice per radio frame. This enables the UE to get time synchronized on a 5-ms basis, which simplifies the required inter-frequency and inter-RAT measurements. The PSS is transmitted occupying the six central PRBs of the LTE frequency configuration[11].

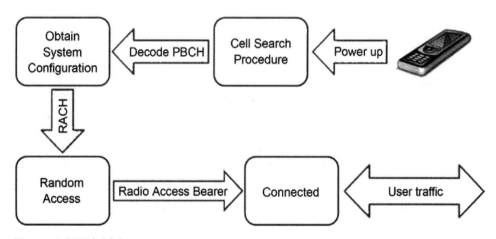

Figure 1. LTE initial access.

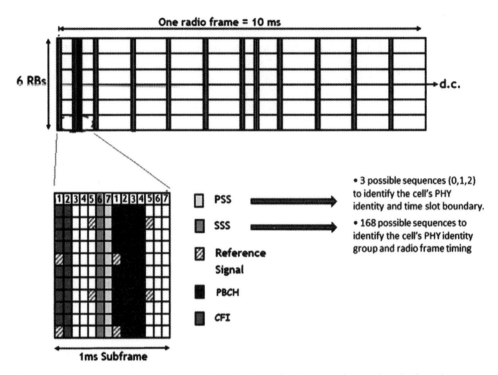

Figure 2. Primary and secondary synchronization signals mapping and main functions.

With 12 subcarriers per PRB, this results in 1.08 MHz of bandwidth (BW). This way, independently of the BW configuration of the cell, the UE is able to decode it.

The next step is to obtain the radio frame timing and the group identity of the cell, which is found in the SSS. In the time domain, the SSS is transmitted in the preceding symbol to the PSS. The SSS also has a 5-ms periodicity and occupies 62 of the 72 central subcarriers so it can be decoded without knowledge of the system BW configuration.

Decoding this signal, the device determines the unique identity of the cell. At this point, the terminal can get fully synchronized with the eNodeB because the reference signals are transmitted in well-defined resource elements and the current synchronization allows locating them. A reference symbol from the generated reference signal pattern is transmitted on every sixth subcarrier. In the time domain, every fourth OFDM symbol holds a reference symbol. This results on four reference symbols per PRB.

2.2. LTE Physical Broadcast Channel

The LTE PBCH is crucial for the successful operation of the LTE radio interface. Therefore, its transmission has to be optimized so it can be reliably decoded by cell edge users with low latency and low impact on battery life. This is achieved by means of low system overhead (the effective data rate is of just 350 bps) and transmission with the lowest modulation and coding scheme (MCS) in order to minimize the bit error rate (BER) for a given signal-to-noise ratio (SNR)[1].

The main LTE system information is transmitted over the PBCH within the master information block (MIB). This message contains the most frequently transmitted parameters, essential for an initial access to the cell, such as the system BW, the physical hybrid ARQ indicator channel (PHICH) structure and the most significant eight bits of the system frame number (SFN).

The remainder of the system configuration is encoded in the system information blocks (SIBs), which are modulated on the physical downlink shared channel (PDSCH). These messages can be mapped on the PDSCH based on their broadcast id, the system information RAN temporary identifier (SI-RNTI), which is fixed in the specifications and therefore known *a priori* to all UEs and potential attackers. The SIB-1 message contains transport parameters necessary to connect to the cell as well as scheduling information, and the SIB-2 message contains information on all common and shared channels. Subsequent SIB messages define multiple parameters, such as the power thresholds for cell re-selection and the list of neighboring cells.

2.3. MIB and SIB Message Eavesdropping

The MIB and SIB messages are broadcasted on PRBs known *a priori* and transmitted with no encryption. Therefore, a passive sniffer is able to decode them. This simplifies the initial access procedure for the UEs but could be potentially leveraged by an attacker to craft sophisticated jamming attacks, optimize the configuration of a rogue base station or tune other types of sophisticated attacks. **Figure 3(a)**, **Figure 3(b)** presents our lab system configuration eavesdropped with a

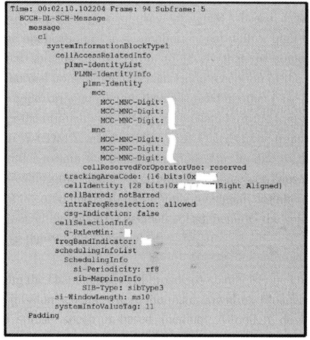

Figure 3. MIB (a) and SIB-1 (b) LTE messages eavesdropped with a commercial traffic sniffer.

commercial of-the-shelf LTE wireless traffic sniffer. Note that details such as the system BW, the cell identity, and the MCC and the mobile network code (MNC) of the eNodeB are broadcasted in the clear. These values have been faded out on purpose in the figures.

Similarly, using the same commercial traffic sniffer, the subsequent SIB messages can be intercepted. For example, the SIB-2 messages contains the PRB mapping of other control channels, such as the uplink (UL) resources reserved for the UE random access procedure on the random access channel (RACH).

Note that a commercial traffic sniffer is not necessary to obtain this information. A skilled programmer could design a PBCH traffic sniffer implemented on a cheap software-defined radio (SDR) platform such as the universal software radio peripheral (USRP)[12], which is commonly used as radio transceiver in GSM (Global System of Mobile Communications) open source projects[13].

3. Attacks against Cellular Networks

Radio jamming is the deliberate transmission of radio signals to disrupt communications by decreasing the SNR of the received signal. This attack essentially consists of blasting a high-power constant signal over the entire target band of the system under attack[4][14].

This attack is broadly known as a simple and common way to attack a wireless network and has been widely studied in the literature in the context of wireless local area networks (WLAN)[4], sensor networks[15], and cellular networks[14]. Despite the attack's simplicity, often, the only solution is to locate and neutralize the attacker, specially in the situation where the entire band of the system is being jammed. The very large amount of transmitted power, though, results in a reduced stealthiness so more elaborated schemes to jam cellular networks are being proposed in the literature.

It has been shown that a standard barrage jamming attack is the optimal jamming strategy when the attacker has no knowledge of the target signal[16]. This section overviews specific derivations of radio jamming attacks against cellular networks based on the knowledge of the target LTE signal that an attacker can obtain from publicly available documents and standards. A popular new threat vector that can be exploited as a result of such attacks is also described.

3.1. Downlink Smart Jamming

Downlink smart jamming consists of generating malicious radio signals in order to interfere with the reception of essential downlink control channels. A recent report introduces the potential theoretical results of jamming the PBCH of LTE networks[5]. The authors of the original study expanded the details of this study in a recent paper[6]. This attack, which could be applied to both 2G and 3G networks as well, targets this channel because, as described in Section 2.1, its assigned PRBs are known *a priori* and always mapped to the central 72 subcarriers of the OFDMA signal. Given that this channel is required to configure and operate the other channels in the cell, this jamming attack is characterized by a low duty cycle and a fairly low bandwidth.

The range of the jammer in this case is still rather small, with a very localized impact. The transmission and modulation characteristics of the PBCH still require a fairly high-power interfering signal to deny the service to noncell edge users. Note that, in order to outpower the legitimate signal, the attacker is bounded by the large transmitted power at the eNodeB and the potentially low transmitted power of the jamming device.

More sophisticated versions of this attack have been proposed, targeting the downlink pilot signals used by the UE to estimate the channel for signal equalization[17]. However, Release 10 of the LTE standard covers the concepts of heterogeneous networks (HetNets), with strong enhancements in the pilot signals to avoid strong interference between the pilots sent by different overlaying cells (macrocells and pico/femto/metrocells)[18]. As a consequence of the inter-cell interference coordination (ICIC) efforts of Release 10, the downlink (DL) pilot signals might experience an enhancement in their resiliency against jamming.

3.2. Uplink (Low-Power) Smart Jamming

Low-power smart jamming takes a step further by targeting essential uplink control channels. Note that, as depicted in **Figure 4**, the range of an uplink smart jamming attack is less local and covers the entire cell or sector. This is because the attacker jams UL control channels, preventing the eNodeB from receiving essential UL signaling messages required for the correct operation of the cell. By overwhelming reception at the eNodeB by means of a jamming signal, the attacker is effectively preventing the base station to communicate with every UE in the cell, thus extending the range of the attack to the entire cell.

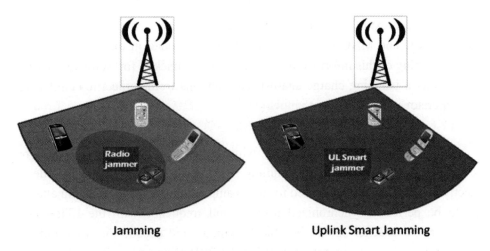

Figure 4. Impact range of radio jamming vs UL smart jamming.

Moreover, the attacker is not bounded by the high power of downlink signals transmitted by the eNodeB (often in the range of 48 dBm), but by the maximum power, a legitimate UE can transmit, which is fixed at 23 dBm in the case of LTE[19]. In this case, an attacker sitting in the vicinity of the eNodeB transmitting at the same power level as any legitimate smartphone could potentially jam the uplink control messages of all the UEs within a given cell or sector. Furthermore, the attacker could use a very directive antenna pointed towards the eNodeB and substantially enhance the effectiveness of the attack.

This type of attack has been previously demonstrated in the context of GSM networks targeting the uplink RACH[20].

The first message exchange on this channel allows the UE to synchronize in the uplink and, after the initial access procedure, radio resources can be allocated to the UE.

In order to target a specific LTE uplink control channel, the attacker would need to know the actual PRBs assigned to it at the PHY layer. This PRB assignment can be obtained from publicly available documentation. Nevertheless, as it will be presented in Subsection 4.4, if the actual location of this signals in the time-frequency LTE frame was randomized or scrambled, such radio resource assignment information could still be obtained from the SIB unprotected messages carried by the PBCH and PDSCH.

In the context of a sophisticated and highly targeted attack, one should note that the MCC and MNC of an eNodeB are also encoded in the SIB-1 message. Eaves-dropping of this information would allow an attacker, for example, to selectively target a jamming charge against base stations from a specific cellular network operator.

Note that uplink smart jamming, while being much more effective than basic jamming or downlink smart jamming, is a more complex attack. In order to selectively jam the PRBs assigned to, for example, the RACH channel, an attacker should be perfectly synchronized in time and frequency with the LTE signal. Moreover, the attacker should be able to capture and decode the MIB and SIB messages in order to extract the actual RACH PRB allocation information. Therefore, a skilled attacker and moderate development work on, for example, software-defined radio would be required.

3.2.1. LTE Link Budget

To illustrate the gain in transmitted power and, therefore, range of uplink smart jamming, we compute the link budget for a typical 10 MHz LTE system in both the uplink and downlink. The main parameters of this LTE configuration are described in **Table 1**. Such calculations can be done with multiple open access tools available online such as[21].

Given an eNodeB transmitting at the standard power (48 dBm), the received power at a UE located at the edge of a cell is of −100.80 dBm for the largest possible cell with radius 0.4 Km (i.e., −100.80 dBm is the receiver sensitivity in the DL). Although the maximum size of the cell is limited by the UL link, for the sake of comparison, we compute the received power at the eNodeB from a UE located at the cell edge, 0.4 km away from the UL receiver. This link budget results in −132.26 dBm received at the eNodeB. This difference of 31.46 dB between the received power in the UL and DL indicates that the power the eNodeB receives from a UE at the cell edge is 1000 times lower than the power that the same UE at the cell edge receives from the eNodeB in the dowlink. This gives a clear indication on the much lower jamming signal power requirements for an UL smart jamming strategy.

Table 1. LTE link budget parameters from a standard 10 MHz deployment.

System parameters	Values
System BW	10 MHz
Subchannel reuse	One-three
Carrier frequency	2.5 GHz
#TX antennas	2
#RX antennas	2
Path loss model	Cost 231
BS antenna height	30 m
UE antenna height	1.5 m
MCS	QPSK 1/2
SNR_{min} for MCS	8.5 dB [22]
Thermal noise density	-174 dBm/Hz
Log-normal fading margin	6 dB
Downlink	
eNodeB max power	43 dBm
Multi-antenna gain	3 dB
TX antenna gain	17 dBi
Noise figure	4 dB
Uplink	
UE max power	23 dBm
Multi-antenna gain	0 dBm
TX antenna gain	-1 dBm
Noise figure	9 dB

3.2.2. Attack Complexity

Based on the characteristics of UL smart jamming, the attacker would require full synchronization in time and frequency with the LTE signal to be able to, for example, selectively jam the RACH. This raises the complexity of the attack as compared to DL smart jamming. Nevertheless, there are numerous of-the-shelf and open access tools that could be leveraged in this context.

The USRP is commonly used for GSM-related projects, but there are certain ongoing open source projects that could be used to write software radio applications that synchronize with an LTE signal. One example is the openLTE project[23]. Leveraging these tools, a skilled attacker could potentially implement an advanced jammer at a very low cost. Moreover, there are other off-the-shelf applications and tools that allow a user to synchronize with an LTE signal such as specialized LTE sniffing hardware and commercial software-based LTE base stations.

3.3. Rogue Base Station Attacks

Rogue base station attacks have been proposed in the literature as a means to, for example, steal credentials or invade the privacy of mobile users[9][24]. These attacks are based on the deployment of a GSM rogue base station combined with jamming the UMTS and/or LTE network in order to force as many UEs as possible to camp on the fake GSM cell. Many security features of GSM have been defeated over the last few years[25]. Given that the authentication algorithm is not symmetric, the network is not required to authenticate, so the UE believes it is connected to a real base station.

An efficient technique to maximize the potential number of devices camping on the fake cell is by advertising the id of the rogue base station based on the list of neighboring cells broadcasted by legitimate base stations. Note that such information can be extracted from the unprotected downlink broadcast MIB and SIB messages. From the data sniffed from such broadcast messages, one can efficiently tune the transmitted power of the rogue cell as well such that the UEs will handoff to the rogue base station.

Note that both techniques leveraged to optimize a rogue base station attack (jamming of the LTE network and obtaining information from the unencrypted MIB and SIB messages) leverage the vulnerabilities introduced in Section 2.

4. LTE Security Solutions against Jamming Attacks

One of the goals of this proof of concept paper is to propose research directions to enhance the resilience of LTE against smart jamming threats. We introduce a set of security research directions at the PHY layer of LTE networks, aiming to enhance the resiliency of data communications against jamming. The envisioned security system would protect communication systems by mitigating the radio jamming attacks discussed in Section 3. These solutions would also minimize the system configuration information that an attacker can easily eavesdrop in order to leverage a jamming charge or the deployment of a rogue base station.

The proposed theoretical security system is based on an enhancement of the resiliency against radio jamming of the PBCH by means of a spread spectrum transmission. This can be combined with scrambling of the PRB allocation of UL control channels and a distributed encryption scheme for downlink control broadcast messages. On one hand, the system protects its most vulnerable resources, downlink control channels, which are the target of DoS attacks[6]. On the other hand, MIB and SIB messages are protected so an attacker cannot learn any information on the PRB allocation for the other control channels, which are now randomly allocated in time and frequency. Only with the information encoded and encrypted in the MIB and SIB messages an attacker would be able to aim to the UL control channels with a jamming charge. Full application of such security solutions render a jamming attack to be only as effective as basic barrage jamming. Note that in jamming mitigation studies, the goal is precisely to force any sophisticated jamming attack to be just as efficient as standard jamming[17].

Note that the full implementation of the proposed techniques would not be trivial, as it will be discussed throughout this section. For example, the scrambling of the PRB allocation of UL control channels will challenge the SC-FDMA scheduling in the uplink because it could potentially break up the continuity of user allocations. The successful implementation of some of these solutions would be very challenging in commercial networks. Nevertheless, such modifications at the PHY layer could be aimed for the development on the Nationwide Interoperable Public Safety Broadband Network, with a PHY layer based on LTE[2][3]. Such next-gen- eration communication systems for emergency response present strict security requirements and should be protected against potential jamming attacks.

4.1. Spread-Spectrum Jamming Resiliency

By means of jamming the central 1.08 MHz of any LTE signal, an attacker would deny the service to all UEs in its vicinity. Therefore, it is important to enhance the protection of the main broadcast channels at the PHY layer. The goal is to counteract the advantage in bandwidth and transmitted power the jammer has due to this LTE vulnerability[6].

Newly deployed LTE networks implement a completely redesigned modulation scheme that substantially maximizes the performance of the wireless channel in terms of bits per second per Hertz (bps/Hz). However, the implementation of an OFDMA-based PHY layer lacks of the inherent interference resilience features of code division multiple access (CDMA)-based networks. While OFDMA is often the choice because of its robust performance in challenging multipath environments, it is not optimal for scenarios where adversarial entities intentionally attempt to jam communications, such as in tactical scenarios[17].

The strong interference resiliency of CDMA-based net-works is well known[26][27]. The application of a scrambling signal with a high chip rate to the transmitted signal spreads the spectrum to levels that, in some cases, can be masked by the thermal noise at the receiver. Upon reception of the signal, application of the same code, orthogonal with the code used in other base stations or UEs, allows to recover the original signal. Due to the nature of the transmitted signal in UMTS, based on W-CDMA, an interfering signal needs to be transmitted at a very high power in order to jam the communication. This is due to the fact that the process of despreading the signal spectrum at the receiver causes, assuming an interfering signal uncorrelated with the scrambling signal, an inherent reduction of the interference power by $\log 10(G)$ decibels (dBs), being G the spreading factor or processing gain of the W-CDMA signal[26].

Considering the characteristics of broadcast channels, one could envision an alternative transmission scheme where the main downlink broadcast channels are protected by a spread spectrum-based method. Although downgrading from OFDMA could potentially decrease the available throughput for broadcast messages, such control channels are known for having very low overhead and a low throughput of, in the case of the PBCH, just 350 bps[1].

4.1.1. System Description

The proposed security solution applies a spread spectrum-based modulation to the downlink control channels in order to extend their spectrum over the available BW. This could be done by just expanding the BW of the downlink broadcast signals or by applying an actual CDMA-based modulation on this portion of the LTE signal.

This solution by itself would prevent a downlink jamming attack to be launched with a simple radio transmitter or jammer, which substantially increases the attack complexity and cost. To perform such attack, full synchronization in time and frequency would be required in order to apply the same CDMA spreading code to the jamming signal. In the case that an attacker does incur this cost, a further enhancement to this solution is described in Subsection 4.3.

Assuming a scrambling or spreading sequence with a rate of $Rb \cdot G$, with Rb being the rate of the PBCH messages, a jammer would theoretically require an extra $\log 10(G)$ dBs of transmitted power in order to achieve the same result. With the transmitted power kept constant, the BW of the jamming signal would be reduced by a factor of up to G times. With both power and BW kept constant, the range of the attack would be reduced.

4.1.2. Limitations and Potential Implementation

The main limitation of the solution is that the UE requires a finer synchronization with the DL signal. In addition to that, the effectiveness of the defense is directly proportional to the spreading factor of the broadcast signal. Therefore, either extra BW should be allocated for the PBCH or its PRB allocation should be modified and spread over the available 1.08 MHz. Nevertheless, with an effective throughput of just 350 bps, there is potentially room for improvement.

In order to be implemented in commercial cellular, this technique would require changes in the LTE standards. Moreover, it would not be backwards compatible with current LTE terminals unless the PBCH and broadcasting messages were transmitted both within the central subcarriers and with the spread spectrum enhancement. Nevertheless, this solution is feasible and could be implemented in the context of an anti-jamming security-enhanced LTE-based military or tactical network, which would use custom wireless devices and eNodeBs.

4.2. LTE, MIB, and SIB Message Encryption

As introduced in Subsection 2.2, the MIB and SIB messages broadcasted by an eNodeB contain essential network configuration parameters that aid the UE to synchronize and establish a connection with the network. Nevertheless, all these messages are transmitted in the clear with obvious security implications.

Assuming a hypothetical scenario with control broadcast messages encrypted but no specific protection for the PSS and SSS, an attacker could not obtain any configuration information by means of a commercial traffic sniffer. A skilled attacker could still synchronize with the network by means of a SDR platform. Extraction of network configuration information, though, would be impossible, assuming a strong encryption scheme. Therefore, the only way an attacker could obtain the network configuration details would be by using a legitimate wireless device and hijacking it to extract data from the baseband chip, which is unreachable from the user space.

4.2.1. Initial Limitations

A simple encryption scheme cannot be applied to the system information messages. If the information in the PBCH was encrypted with a secret key, the mobile terminal must know that key *a priori*. Assuming the case of a mobile terminal being turned on or roaming to a new network, the device must still be able to decode the PBCH to establish a connection. In parallel, key exchange algorithms cannot be executed with the network at this stage because the device is not connected and authenticated yet. Therefore, the key that encodes the MIB and SIBs must be hard-coded in the UE.

Relying on one common key for all users and cells is not possible either. If this key was compromised by any means, the whole system would be useless. Therefore, the system must be able to operate with a large number of different keys or able to generate a large number of keys.

4.2.2. Assumptions

The main assumption for this security architecture is a global collaboration of all mobile operators. Subscriber identity module (SIM)-based authentication schemes allow to provide cellular services to a mobile terminal independently of the network being roamed (assuming, of course, that the user has roaming activated and the phone and network are compatible). Encryption of the MIB and SIB messages would require a similar collaboration among carriers.

It is important to note that, in the case of such encryption scheme being applied exclusively to a national LTE-based emergency response broadband network, such as the Nationwide Interoperable Public Safety Broadband Network, this limitation does not apply.

The proposed solution assumes that the UE is equipped with a trusted hardware (HW) platform which is secure and able to both securely store data and perform cryptographic operations. Note that all UEs are already provisioned with such element, the SIM card. Basing the encryption scheme on the SIM, though, could potentially allow an attacker to capture LTE wireless traffic and decode it afterwards with a legitimate SIM on a standard card reader connected to a personal computer. Therefore, such encryption scheme would only be effective if it was implemented on a protected trusted platform module (TPM) that can be only operated by, for example, the baseband of a cell phone. Note that there are initial plans in the industry to equip UEs with a TPM[28]. In the context of a national emergency response network, equipping mobile devices with a secure TPM is feasible. Finally, a strong private key encryption scheme is assumed.

4.2.3. System Description

The proposed solution is depicted in **Figure 5**. On the network side, either each base station or a node in the EPC stores a set of N secret keys in a secure location. In the case of storing the keys in a centralized way, this secure storage could be the home subscriber server (HSS) or any other newly implemented network node. On the other hand, the TPM (or SIM card) in each mobile phone stores securely the same set of N keys. The value of N can be arbitrarily large.

Note that, in practice, only one secret key K would be required. Based on this initial secret key, each sub-key K_j $j = 1, ..., N$ would be generated as $K_j = H(K|j)$, being H a hash function and '|' the concatenation operation. Assuming a robust hash function, eventual leakage of a sub-key K_j would not provide an attacker any information on the actual secret key K. However, leakage of the main secret key K would relent this method useless.

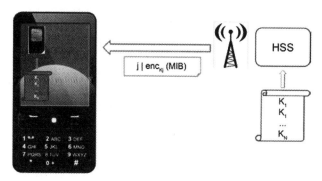

Figure 5. LTE MIB message encryption.

The operation of the system is described as follows. The eNodeB selects a key K_j with id j. The broadcast MIB and SIB messages are encrypted with the key K_j and transmitted over the air. Along with the encryption of the broadcast message, for example $encK_j(MIB)$, the system transmits the id j in plain text. This way, any UE knows what key to use to decrypt the messages. Note that an attacker would learn the id j but would not be able to know the key K_j that is being used to encrypt the broadcast messages.

If at any point a key was compromised, the network would be able to switch to a different key K_i and continue operating normally. A broadcast message would be sent to the mobile devices to alert them of this change. Incoming connections, either via handovers from other cells or new devices being turned on, would just receive the updated broadcast messages $[i|encK_i(MIB)]$ and continue operating normally.

Note that the network could choose to use a different key at each cell/sector. This way, if an attacker managed to compromise a key, a potential attack during the time it would take the network to change to a new key would be localized and only impact one cell or sector.

In this section, the MIB message is used as example, but the same scheme could be applied to SIB messages as well.

4.2.4. Limitations and Potential Implementation

The main limitations for this security enhancement of LTE networks is the requirement for a global partnership of both cellular operators, device manufacturers (in the case of the deployment of a TPM), and SIM card providers. Moreover, a substantial editing of the standards would be necessary. Therefore, such a security solution, when framed in the context of commercial wireless networks, would be more appropriate for the upcoming fifth generation of mobile networks, the standards of which are just being started.

In the context of high security demanding LTE-based communications, such as military networks or first responders, the encryption of broadcast messages is more feasible. Based on custom hardware equipped with state-of-the-art TPM and encryption schemes, such a system could be deployed.

It is important to note that, by itself, the encryption of the broadcast messages would not prevent an attack from jamming the LTE central subcarriers (downlink smart jamming) and block users from detecting and decoding the PSS and SSS and, therefore, connect to the cell.

4.3. Spread-Spectrum and Encryption Combination

As introduced in Subsection 4.1, if an attacker obtained the sequence used to protect the central subcarriers of the LTE signal, the spread spectrum anti-jamming solution would be useless against a skilled attacker using a fully synchronized jamming device. In this section, we introduce an enhancement of the spread spectrum protection based on the encryption scheme described in Subsection 4.2.

4.3.1. Assumptions

In order to prevent an attacker from encoding the jamming signal with the right sequence and, therefore, bypass the protection, this spreading sequence must be only known by the UEs and the eNodeB. In parallel, the entire system cannot depend on a single spreading sequence because, if it was compromised, the protection would be bypassed. Therefore, the spreading sequence selection can be implemented as follows.

4.3.2. System Description

The proposed enhanced security architecture is depicted on **Figure 6**. On the network side, either each eNodeB or the HSS stores a set of M secret spreading sequences. The value of M can be arbitrarily large. The eNodeB selects a sequence S_i with id i and scrambles the PBCH signal with it prior to broadcasting it. Along with the scrambled signal $PBCH(t) \cdot S_i(t)$, the eNodeB broadcasts the id i, either on the same channel or on a separate resource. This allows the UE to despread the PBCH with the right sequence. An attacker would just learn the id i but would not be able to extract the sequence C_i used to protect the DL signal.

Note that, when enhancing the resiliency against jamming of the MIB messages following this scheme, the sequence $S_i(t)$ used to scramble the signal can be seen as the equivalent of the key k_i used to encrypt the MIB.

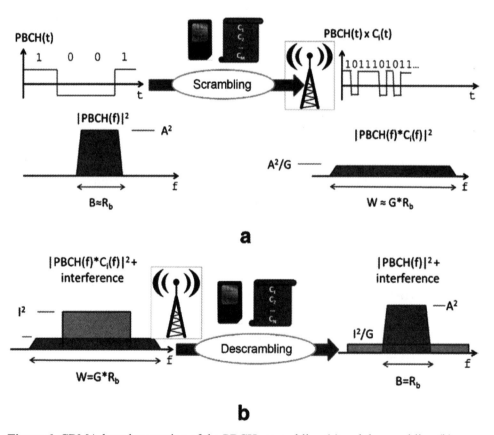

Figure 6. CDMA-based protection of the PBCH: scrambling (a) and descrambling (b).

4.3.3. Limitations and Potential Implementation

With the combination of the MIB and SIB message encryption and the spread spectrum protection against jamming, an attacker cannot leverage the knowledge of the id i, broadcasted in the clear, to optimize the attack. However, if the attacker was able to jam the actual broadcast transmission of the id i, the wireless system would be inoperative as no legitimate user would be able to determine what spreading code is being used in a given cell.

This could be prevented in different ways. In the case of LTE-based systems with a specific application, such as military ad-hoc networks or first responder wireless systems, the id i could be distributed through an out of band secondary channel or known a priori before deploying the ad-hoc network. Alternatively, the id i could be broadcasted in a more complex yet secure manner, such as the primary scrambling code detection in UMTS networks[26]. In this case, a receiver detects which one of the 512 possible primary scramble codes is being used within a cell by applying a correlation receiver. The necessary signals for the primary scrambling code detection are often transmitted in a way that they can be received and decoded at a low SNR regime. A similar procedure could be implemented to, for example, broadcast which one of 512 possible sequences C_i with $i = \{1, 2, ..., 511\}$ is being used.

Finally, the sequence id could be broadcasted repeatedly in a frequency hopping scheme over the entire system bandwidth. Although this would require substantial changes to the LTE standards, it would force an attacker to jam the entire band to jam the id broadcasting operation. Therefore, the goal of counteracting the advantage a jammer has in LTE networks would be achieved as the only feasible option would be barrage jamming.

4.4. LTE UL Control Channel PRB Scrambling

The PRB allocation of the Physical Uplink Control Channel (PUCCH) is known a priori as defined by the standards. The UL control signaling on this channel is transmitted in a frequency region on the edges of the system BW. In parallel, the PRB allocation of other essential UL control channels, such as the RACH, can be extracted from the SIB messages.

4.4.1. System Description

The proposed security architecture scrambles the PRB allocation of UL control channels so they cannot be the target of an uplink smart jamming attack. Based on the encryption scheme described in Section 4.2, a legitimate UE would be able to decode the system configuration and normally operate on the UL control channels. An attacker, though, would not be able to locate any UL control channel and its best UL jamming strategy would be equivalent to a basic barrage jamming.

4.4.2. Limitations and Potential Implementation

Periodically modifying the PRB allocation of certain UL control channels, such as the RACH, would not be challenging given the multiple possible configurations of the RACH in current LTE networks. However, the allocation of the PUCCH away from the edges of the spectrum would generate new limitations. The frequency diversity achieved through frequency hoping would not be maximized anymore. In parallel, the maximum achievable PUSCH data rate would decrease due to the fact that uplink allocations must be contiguous in frequency to maintain the single-carrier nature of the uplink LTE signal[1]. Random allocation of the UL control channels could potentially pose a challenge to SC-FDMA scheduling because it could break up the continuity of user allocations.

In this case, the implementation of this security solution would require changes in the LTE standards. However, a potential application for security-demanding military and first responder LTE-based networks could be implemented using nonstandard hardware on both transmitting and receiving sides.

4.5. Selective Uplink Smart Jamming Interference Cancelation

In order to enhance their coverage range and apply diversity techniques, cell towers are equipped with multiple antennas. The number of antennas at the cell tower is commonly three but some network operators are pushing this number to five in LTE networks to expand further the system's capacity[29]. The cell range in the DL is often bounded by the base station's transmitted power, which is significantly higher than that of a UE. Therefore, the multiple antennas implement spatial diversity in the UL to extend the cell's range limited by the UE.

The proposed security architecture includes a further application that exploits the availability of multiple antennas to suppress the interfering signal of the smart UL jammer, defined in Subsection 3.2. Similar methods have been proposed in the literature aiming to mitigate the effects of a jamming signal in a wireless system by means of jointly mechanically adjusting an array with two antennas and applying an interference cancelation algorithm[30].

4.5.1. System Description

The proposed security scheme implements beam-forming techniques at the eNodeB that leverage the availability of up to five antennas in reception. By means of a configurable signal feed, with variable delays and gains, the radiation pattern of an antenna array can be molded to achieve either enhanced directivity or strongly attenuate the signal coming from a specific direction[31]. Assuming the location of the jammer was known, a null in the antenna radiation pattern of the eNodeB could be generated to selectively block the interference. **Figure 7** depicts an example of such application.

The attenuation of the interfering signal will depend on the null of the radiation pattern. In order to locate the source of the interference, a narrow directive

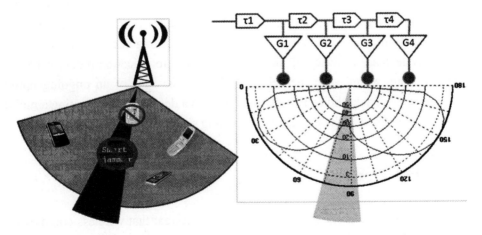

Figure 7. Beam-forming scheme for selective cancelation of UL smart jamming.

radiation pattern can be shifted while monitoring SNR and traffic congestion metrics on the UL control channels, scanning this way the entire cell or sector. This would allow to determine the angle of arrival of an incoming UL jamming signal.

4.5.2. Limitations and Potential Implementation

Note that the proposed architecture requires the multiple antennas of the eNodeB to perform both spatial diversity and beam-forming. The spatial separation between antennas required to optimize the diversity receiver substantially increase the phase or delay between each antenna element. In terms of the beam-forming, this could result in a suboptimal radiation pattern with considerable side lobes and a wider null in the radiation pattern. A trade-off should be found between the performance of the array in terms of diversity/MIMO and the ability to generate a narrow beam.

In parallel, all the UEs located in line with and in the close vicinity of the jammer would not be able to access the networks. Nevertheless, the range of the jammer would be significantly reduced, efficiently mitigating any uplink smart jamming attack.

It is important to note that this particular security enhancement is completely independent and both the hardware (multiple antennas at each eNodeB) and the technology (beamforming) already exist. A potential implementation could be framed within the concept of self-organizing networks (SON), as an automatic smart jamming attack detection plus self-healing security function. Upon detection of an anomaly in a given eNodeB, in the shape of a strong decrease in the load or anomalous decrease in the SNR, the cell would go into detection mode. A narrow reception beam would scan the cell or sector. In the case of an ongoing uplink smart jamming attack, the cell would then go into a defense state, creating a null in reception and blocking the malicious interfering signal.

5. Related Work

Jamming attacks are the main basic type of threat that wireless communication networks face given the fact that the threat vector exploited is inherent to the actual technology. There is no way to prevent an attacker from broadcasting high-power signals on the frequency band allocated to a commercial mobility network. The goal of this attack is often to prevent users to access communication networks, which catalogues this threat as a DoS attack. Several attacks proposed in the literature use radio jamming as a first step in order to force UEs to an insecure access network[24].

Jamming attacks have been in the scope of network and security research for several years already[16]. As new network standards arise, jamming attacks spread their threat over new technologies such as wireless sensor networks (WSNs)[15] and WLANs[4]. Mobility networks, the main commercial wireless networks, have also been considered in radio jamming studies[14].

In parallel, the potential of this kind of attack has lead to improvements and refinements, resulting in more sophisticated jamming techniques. Over the years, authors have proposed ways to launch DoS attacks against mobility networks by overloading the system at the paging channel[32] or with a spike in core network signaling messages[33]. Some other sophisticated jamming techniques have been proposed for UMTS networks[34].

The author of[20] was the first to implement an actual smart jamming attack against an UL control channel in a GSM network, opening a new simple but very effective attack vector to be leveraged in a radio jamming attack. The same idea has recently been proposed as a potential way to jam LTE networks[6].

Despite the prevalence and effectiveness of jamming in the context of wireless networks, there is a clear lack of security strategies to mitigate the impact of such attacks, specially in current mobility networks and upcoming LTE-based emergency response broadband systems. Current standardization bodies do not consider any jamming resiliency requirements for the next planned release of the LTE advanced standard. Nevertheless, some work has been done in addressing jamming attacks in WLANs[35] and WSNs[15].

6. Conclusions

Jamming attacks are one of the main types of security attack that mobility networks face. This threat is inherent to the actual wireless technology employed in this type of network, and in its most basic implementation (barrage jamming), there is no means to prevent an attacker from broadcasting a high power interfering signal on a commercial frequency band.

Despite that jamming attacks are well known and have been widely studied in the literature, no actual security and mitigation strategies have been proposed to enhance the resiliency against jamming attacks in mobility networks. This has resulted on a constantly growing list of new proposals for sophisticated DoS attacks against cellular networks based on jamming principles. However, standardization bodies do not include any anti-jamming guidelines or requirements for the upcoming new releases of LTE advanced. Nevertheless, the forecasted application of LTE-based technologies to implement national emergency response networks make the reliability and security requirements of LTE of paramount importance.

In this proof of concept paper, we overview a series of simple but effective jamming attacks that extend the range of basic jamming while requiring less power. Based on these new threats, classified as smart jamming, we propose a series of potential security research directions that could protect LTE cellular networks, forcing a potential attacker to rely on just basic jamming to attempt a DoS charge. The goal is to raise awareness on this traditionally overlooked threat and spark security research work in this area. We are, in parallel, implementing smart jamming in the lab as well as some of the proposed security solutions.

A potential enhancement of the anti-jamming properties of the main DL broadcast channels, importing concepts from spread spectrum modulations, protects the wireless interface from a smart jamming attack aimed to such control channels. In parallel, a randomization of the PRB allocation of UL control channels plus a sophisticated encryption method for DL system configuration messages, backed up by the deployment of a TPM in the UE, prevent an attacker from launching a smart jamming attack against these essential UL channels. Finally, a method that leverages the current availability of antennas at the eNodeB is proposed to filter out an UL smart jamming signal in order to block an UL smart jamming attack. The limitations for all these solutions have been discussed as well.

Such enhancements, or similar proposals, should be considered in the scope and requirements of the upcoming releases for wireless cellular networks, specially for the Nationwide Interoperable Public Safety Broadband Network. Mobility networks, providing mobility services to billions of customers over the world, were never designed with a security perspective. The evolution from GSM to UMTS and finally LTE has addressed encryption and authentication issues, aiming to enhance the overall system security. The same kind of proactive approach should be taken in order to mitigate potential DoS jamming attacks against mobility networks.

Competing Interests

The authors declare that they have no competing interests.

Authors' Information

Dr. Raghavan participated in this work while being a member of the AT&T Radio Access and Devices team.

Source: Jover R P, Lackey J, Raghavan A. Enhancing the security of LTE networks against jamming attacks [J]. Eurasip Journal on Information Security, 2014, 2014(1): 1–14.

References

[1] S Sesia, M Baker, I Toufik, *LTE, The UMTS Long Term Evolution: From Theory to Practice*. (Wiley, New York, 2009).

[2] Nationwide Public Safety Broadband Network. US Department of Homeland Security: Office of Emergency Communications (2012). http://goo.gl/AoF41.
Accessed Feb 2014.

[3] T Doumi, M Dolan, S Tatesh, A Casati, G Tsirtsis, K Anchan, D Flore, LTE for public safety networks. IEEE Comm. Mag. **51**(2), 106–112 (2013).

[4] W Xu, Y Zhang, T Wood, The feasibility of launching and detecting jamming attacks in wireless networks, in *ACM MOBIHOC; Urbana-Champaign* (ACM New York, 2005), pp. 46–57.

[5] D Talbot, *One simple trick could disable a city 4G phone network*. (MIT Technology Review, 2012). http://goo.gl/jROMe2.

[6] M Lichtman, JH Reed, TC Clancy, M Norton, Vulnerability of LTE to hostile interference, in *Proceedings of the IEEE Global Conference on Signal and Information Processing*, GlobalSIP'13, Austin, TX (IEEE New York, 2013), pp. 285–288.

[7] When advanced persistent threats go mainstream. Emc corporation: security for business innovation council (2011). http://www.emc.com/collateral/industry-overview/sbic-rpt.pdf.

[8] D Alperovitch, Revealed: operation shady RAT. Threat research, mcafee (2011). http://www.mcafee.com/us/resources/white-papers/wp-operation-shady-rat.pdf.

[9] D Perez, J Pico, A practical attack against GPRS/EDGE/UMTS/HSPA mobile data communications, in *In BlackHat DC*, (2011). http://goo.gl/KGN3j.

[10] A Rossler, Cell search and cell selection in UMTS LTE. White paper, Rhode & Schwarz (2009). http://goo.gl/ntWJPA.

[11] 3rd Generation Partnership Project; Technical Specification Group Radio Access Network, LTE; Evolved Universal Terrestrial Radio Access (E-UTRA); Physical channels and modulation. 3GPP TS 36.211 vol. v9.1.0 (2010).

[12] Ettus Research. USRP. http://www.ettus.com/ Accessed Mar 2014.

[13] Kestrel Signal Processing, Inc. The OpenBTS Project. http://openbts.sourceforge.net/.

[14] M Stahlberg, Radio jamming attacks against two popular mobile networks, in *Helsinki University of Technology. Seminar on Network Security. Mobile Security*, (2000). Accessed Apr 2014.

[15] W Xu, K Ma, W Trappe, Y Zhang, Jamming sensor networks: attack and defense strategies. IEEE Netw. **20**(3), 41–47 (2006).

[16] T Basar, The Gaussian test channel with an intelligent jammer. IEEE Trans. Inform. Theor. **29**, 152–157 (1983).

[17] T Clancy, Efficient OFDM denial: pilot jamming and pilot nulling, in *Communications (ICC), 2011 IEEE International Conference on* (IEEE New York, 2011), pp. 1–5.

[18] P Bhat, S Nagata, L Campoy, I Berberana, T Derham, G Liu, X Shen, P Zong, J Yang, LTE-advanced: an operator perspective. IEEE Comm. Mag. **50**(2), 104–114 (2012).

[19] 3rd Generation Partnership Project; Technical Specification Group Radio Access Network, LTE; Evolved Universal Terrestrial Radio Access (E-UTRA); User Equipment (UE) radio transmission and reception. 3GPP TS 36.101 vol. fv10.3.0 (2011).

[20] D Spaar, A practical DoS attack to the GSM network, in *In DeepSec*, (2009). http://tinyurl.com/7vtdoj5.

[21] LTE/WiMAX link budget calculator (2010). http://goo.gl/phn2we. Accessed Mar 2014.

22. K Ramadas, R Jain, WiMAX system evaluation methodology, in *Wimax Forum, Jan*, (2007). http://goo.gl/sNIj70.

23. B Wojtowicz, OpenLTE. An open source 3GPP LTE implementation. http://sourceforge.net/projects/openlte/. Accessed Apr 2014.

24. K Nohl, S Munaut, Wideband GSM sniffing. In 27th Chaos Communication Congress (2010). http://goo.gl/wT5tz.

25. E Gadaix, GSM and 3G security, in *In BlackHat Asia*, (2001). http://tinyurl.com/85plhlv.

26. J Pérez-Romero, O Sallent, Agustí R, MA Diaz-Guerra, *Radio Resource Management Strategies in UMTS*. (John Wiley & Sons, New York, 2005). http://books.google.com/books?id=581gFV8abl4C.

27. AJ Viterbi, *CDMA: Principles of Spread Spectrum Communication, Volume 129*. (Addison-Wesley Boston, MA, 1995).

28. P Vig, Trusted platform module. Microsoft secret weapon in the mobile arena. Zunited (2012). http://goo.gl/Iqldu.

29. S Marek, AT&T's Rinne: using SON helps improve throughput and reduce dropped calls. FierceBroadband Wireless (2012). http://goo.gl/xV70k.

30. TD Vo-Huu, EO Blass, G Noubir, Counter-jamming Using mixed mechanical and software interference cancellation, in *Proceedings of the Sixth ACM Conference on Security and Privacy in Wireless and Mobile Networks,* WiSec '13 (ACM New York, 2013), pp. 31–42.

31. C Balanis, *Antenna Theory: Analysis and Design*. (Wiley, New York, 1982).

32. J Serror J, Impact of paging channel overloads or attacks on a cellular network, in *Proceedings of the ACM Workshop on Wireless Security (WiSe)* (IEEE New York, 2006), pp. 1289–1297.

33. P Lee, T Bu, T Woo, On the detection of signaling DoS attacks on 3G wireless networks, in *INFOCOM 2007. 26th IEEE International Conference on Computer Communications. IEEE*, (2007).

34. G Kambourakis, C Kolias, S Gritzalis, J Park, DoS attacks exploiting signaling in UMTS and IMS. Comput. Commun. **34**(3), 226–235 (2011).

35. S Khattab, D Mosse, R Melhem, Jamming mitigation in multi-radio wireless networks: reactive or proactive? In *Proceedings of the 4th International Conference on Security and Privacy In Communication Netowrks, SecureComm'08* (ACM New York, 2008), pp. 27:1–27:10.

Chapter 21
Combating Online Fraud Attacks in Mobile-Based Advertising

Geumhwan Cho[1], Junsung Cho[1], Youngbae Song[1], Donghyun Choi[2], Hyoungshick Kim[1]

[1]Department of Computer Science and Engineering, Sungkyunkwan University, Seobu-ro 2066, 16419 Suwon, Republic of Korea
[2]Samsung Electronics, Samsung-ro 129, 16677 Suwon, Republic of Korea

Abstract: Smartphone advertisement is increasingly used among many applications and allows developers to obtain revenue through in-app advertising. Our study aims at identifying potential security risks of mobile-based advertising services where advertisers are charged for their advertisements on mobile applications. In the Android platform, we particularly implement bot programs that can massively generate click events on advertisements on mobile applications and test their feasibility with eight popular advertising networks. Our experimental results show that six advertising networks (75%) out of eight are vulnerable to our attacks. To mitigate click fraud attacks, we suggest three possible defense mechanisms: 1) filtering out program-generated touch events; 2) identifying click fraud attacks with faked advertisement banners; and 3) detecting anomalous behaviors generated by click fraud attacks. We also discuss why few companies were only willing to deploy such defense mechanisms by examining economic misincentives on the mobile advertising industry.

Keywords: Advertising Network, Click Fraud, Android

1. Introduction

As smartphones become more popular, the mobile advertisement market is also growing rapidly[1]. Mobile advertisement is a primary business model that offers the financial incentives for developers to distribute free applications. In the mobile advertisement market, advertising networks serve as a single vendor for advertisers and pay a developer according to the numbers of *impressions* (the number of times an advertisement has been served) and/or *clicks* generated by users[2]; application developers expect users to "pay" for their applications by viewing (*i.e.*, impressions) or clicking advertisements (i.e., generating clicks) as many as possible.

In those business models, the most important security problem is to detect and prevent (artificially created) fraudulent events, which have no intention of generating value for advertising[3]. Although this security issue has been extensively studied, most studies have focused on preventing click fraud attempts housed on web pages rather than mobile platforms[4][5].

Recently, there have been a few attempts to analyze the security risks of smartphone advertisement. Crussell *et al.*[6] particularly analyzed the prevalence of fraudulent advertisement behaviors generated by real Android apps.

In this paper, we extend their work by implementing independent bot programs to generate fraudulent click events in an automatic manner. Unlike the previous study[6], which is based on emulation results, we applied our bot programs to the eight real advertising networks (AdMob, Millennial Media, AppLovin, AdFit, MdotM, LeadBolt, RevMob, and Cauly Ads) and found that artificially generated click events were successfully approved in the six advertising networks (out of eight networks). We highlight our key contributions as follows:

- We design and develop bot programs capable of automatically generating fraud events to mimic users' activities on advertisements.

- We particularly show the feasibility of automatic click generation attacks on eight popular mobile advertising networks to evaluate their security risks. Seventy-five percent of the systems that we experimented (Millennial Media,

AppLovin, AdFit, MdotM, RevMob, and Cauly Ads) did not detect our anomalous click attempts.

- We suggest three possible defense mechanisms: 1) filtering out program-generated touch events (at client side); 2) identifying click fraud attacks with faked advertisement banners (at client side); and 3) detecting anomalous behaviors generated by click fraud attacks (at server side).

- We discuss the issue of economic misincentives on the mobile advertising industry to discover an inherent problem in using countermeasures against click fraud attacks.

The rest of this paper is organized as follows. The related work is reviewed in Section 2. In Section 3, we provide some background on the mobile advertisement (particularly for the Android platform). In Section 4, we present how bot programs can be implemented for online fraud in mobile-based advertising. Then, we evaluate the feasibility of bot programs against click-based advertisements through intensive experiments with real advertisement services in Section 5. In Section 6, we discuss three practical defense mechanisms to detect and prevent automated fraudulent clicks. In Section 7, we explore the misincentive problem that can inherently corrupt ad networks through false clicking. Our conclusions are in Section 8.

2. Related Work

Over the last few years, online advertisement has been widely studied because it has become a significant source of revenue for web-based businesses. However, it also introduces a new type of cyber criminal activities called "click fraud". Click fraud is the practice of deceptively clicking on advertisements with the intention of either increasing third-party website revenues or exhausting an advertiser's budget[7]. Kshetri[8] examined the mechanisms and processes associated with the click fraud industry from an economics viewpoint. Miller *et al.*[9] analyzed the characteristics of real-world click fraud by examining the operations and underlying economic models of two modern malware families, Fiesta and 7cy, which are typically used for click fraud.

To prevent those click frauds, several defense techniques have been introduced. The simplest solution is to use threshold-based detection. If a website is receiving a high number of click events with the same device identifier (e.g., IP address) in a short time interval, those events can be considered as fraud. However, click fraud detection is not trivial—clicks can sophisticatedly be generated to bypass such naive defense schemes. For example, attackers are behind proxies or globally distributed[10]. Also, device identifiers such as IP address can easily be modified.

To mitigate such sophisticated attacks, Kitts et al.[11] discussed how to design a data mining system to detect large-scale click fraud attacks. Metwally et al.[12] developed a technique based on the traffic similarity analysis to discover a type of fraud called *coalitions* performed by multiple fraudsters. Another interesting approach is to use bait advertisements[10][13]. Xu et al.[14] proposed a systematic approach by introducing additional tests to check whether visiting clients are clickbots.

Only a few studies have analyzed the security of mobile advertising networks although many applications use one or more advertising services as a source of revenue for the Android developers. Those studies were mainly focused on discussing the security concerns about unnecessary permissions required by advertisement libraries. Pearce et al.[15] showed that 49% of Android applications contain at least one advertisement library, and these libraries overprivilege 46% of advertising-supported applications. Shekhar et al.[16] proposed an approach called *AdSplit* to separate applications from its advertisement libraries that might request permissions for sensitive privileges.

When it comes to click fraud in mobile platforms, Crussell et al.[6] raised the issue about click fraud in the context of mobile advertising. However, their study results may not be sufficient to show the real impacts of click fraud attacks in mobile platforms because their study mainly focused on analyzing existing mobile applications' fraudulent behaviors that could be used for advertisement fraud. In contrast, we studied how vulnerable real mobile advertising networks are to click fraud attacks by implementing bot programs and testing their feasibility with real advertisement networks. We particularly extend our preliminary work[17] to generalize click fraud attacks with various revenue models and develop practical defense mechanisms for mitigating click fraud attacks on mobile devices. We also discuss the economic aspects of security failure that might be an inherent problem of click fraud in mobile advertising.

3. Background

In this section, we explain definitions of the terminologies used in the remaining of the paper. To provide a better understanding of click fraud attacks, we present how ad networks typically work between entities and explore business models popularly used for mobile advertising networks.

3.1. Terminology

We define the following definitions.

- Publisher is an entity which deploys a mobile application with advertisements.

- Advertiser is an entity which pays the advertising networks for their advertisements being displayed on applications.

- Advertising network (Ad network) is an entity which manages publishers and advertisers. They can buy and sell advertisement traffic through trusted partner networks.

- Impressions are a metric on counting the number of times an advertisement has been deployed.

- Clicks are another metric on counting the number of events that are generated when users click on an advertisement.

- Advertisement request (Ad request) is the form of HTTP traffic that is generated from impressions or clicks. Whenever a valid request message is generated (i.e., advertisement is shown to a user or clicked by a user), advertisers have to pay the ad network and a percentage of amount is also paid to the publisher.

3.2. How ad Networks Work

As we have already explained in Section 3.1, an ad network acts as a moderator between publishers and advertisers. In Android, a *jar* file acts as a moderator, which should be included in an application to embed advertisements into the publisher's applications. When a publisher wants to display advertisements as a part of its application, she must sign up with the ad network and download the advertisement library. That library typically provides an API for embedding advertisements into the UI of the publisher's application and fetching, rendering, and tracking advertisements. The device identifier is generally used to uniquely identify the publisher, who wanted to embed those advertisements.

As shown in **Figure 1**, we illustrate how ad networks manage publishers, applications, and advertisers with advertisement library. Since an advertiser A wishes to send advertisements to many Android users, she requests to distribute her advertisement via an ad network N. After receiving the request, the ad network adds the advertiser's advertisement to the ad network's software development kit (SDK) library. Imagine that there is a publisher P who wants to make money with

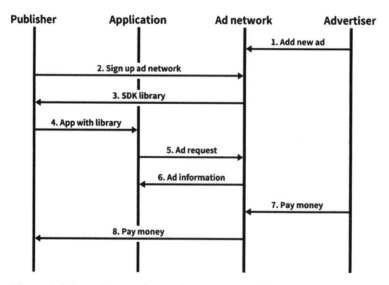

Figure 1. How ad networks work to manage publishers, applications, and advertisers with advertisement library.

mobile advertising services. When the publisher signs up with the ad network, she downloads the latest ad network's SDK library and release her application *App* with this library.

When a user clicks on the advertisement related to the advertiser A, the application App delivers the *ad request* message containing the user device's identifier to the ad network N. The ad network N then sends the response to the application App running on the user's device. For that click-on-the-advertisement banner, the advertiser A pays the ad network N and the publisher P.

3.3. Revenue Models

Mobile advertisement services are dramatically growing up as the number of smartphone users is also increasing. In particular, the Android market offers the opportunity for advertisers and publishers who are interested in the mobile advertisement business. A publisher (i.e., application developer) releases a (free) mobile application with advertisements so that she can make money through those advertisements where revenue is typically determined by the amounts of *impressions* and/or *clicks*. The following revenue models are generally used:

- Cost per click (CPC) is that advertisers charge per click which is generated by users. It is used for the number of times a website visitor or a user for an application clicks on a banner. This measurement is also popularly used since it can be implemented in a simple manner.

- Cost per mile (CPM) is that advertisers charge per a thousand impressions to publishers with ad networks. It is also referred as the cost per thousand (CPT) since it has estimated the cost per thousand views of the advertisements. This measurement is widely available for advertisements for Android developers.

- Cost per action (CPA) is that advertisers charge per specific action such as filling a form, signing up for an offer, completing a survey, or downloading software. This model seems advantageous to advertisers since they only pay for specific actions which are directly related to their advertisements. However, it is not easy to implement when it comes to complex actions.

4. Implementation of Online Fraud Attacks

In this paper, we particularly focus on the implementation of bot programs against the CPC revenue model in order to test their feasibility with real ad networks (see Section 4.1). The test results were presented in Section 5. However, similar automatic attacks can also be implemented for the other revenue models. We will briefly introduce such attack designs in Section 4.2 and 4.3.

4.1. Attack against CPC

As we can see in Section 3, in the CPC revenue model, a publisher makes money whenever a user clicks on the advertisement. Therefore, a malicious publisher may involve in the act of generating such click events on the advertisement with the publisher's profit. In theory, those events can be simply generated. However, it is still questionable whether real advertisement networks are vulnerable to those attacks since they may provide some countermeasures to defeat such attacks.

To analyze the risk of real mobile ad networks, we implemented an independent Android app that can automatically generate click events for advertisements displayed on Android apps. We note that a malicious publisher can implement such an app (*i.e.*, bot program), and the publisher can also install the app on his (or her) own device to generate fake click events within a short time interval in order to make its own profit. This is a very effective economic activity for attackers. For example, in the case of Cauly, a publisher receives only US$0.02 per click event—if an attack sequence is generated within 7 s on average, the attacker can earn US$1480.72 in a week by running a bot program on the attacker's device.

For simplicity, we used Android Debug Bridge (ADB) which is a debug support tool to communicate between a host and an Android device. In other words, with ADB, we can control an Android device without any restrictions. To generate a click event on an advertisement embedded in our prototype with a victim ad network's SDK library, we send a sendevent command to the Android app via ADB. Finally, a virtual click event is generated on an advertisement in the Android app. Surely, such a bot program can also be implemented as a stand-alone Android app without ADB support.

Chapter 21

However, the generation of click events alone is not enough. We empirically found that some ad networks (e.g., Cauly Ads) have checked the requesting device's identifiers such as international mobile equipment identity (IMEI) and/or "Device ID" (also known as android_id) to limit the number of possible ad requests from a device during a specific period (e.g., a day). In our empirical experiments, Cauly Ads limited the number of click events to 19 per day for a device (based on the "Device ID" information). If a device generates click events more than the threshold number (e.g., 19 per day in Cauly Ads), the device might be black-listed by an ad network. Therefore, we tried to generate click events which resembles events generated by multiple devices associated with multiple users. **Figure 2** illustrates how a bot program pretends to be a normal one with multiple devices. Bot programs can generate not only click events but also (faked) *device identifiers* to disable device detection for limiting the number of ad requests from a single device.

4.1.1. Generation of Device IDs

In Android, Device ID (or android_id) is a 64-bit number (as a hex string) that is randomly generated when the user first sets up the device and should remain constant for the lifetime of the Android device.[1] This information can be used for identifying or tracking a device (particularly for tablet devices that do not have IMEI). However, we can observe that Device ID is independently (and randomly) created from an Android device itself. Therefore, we can generate new Device IDs for an Android device without any restrictions.

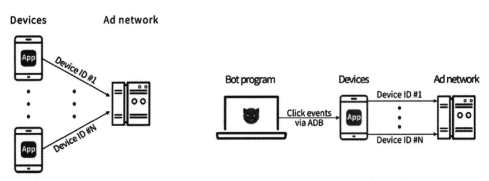

(a) Normal situation with multiple devices (b) Pretending to be normal by a bot program

Figure 2. Bot programs can generate click events with different device identifiers, respectively, such that those look like click events generated by multiple devices.

Our main concern here is how to replace the existing Device ID with newly generated ones. The simplest solution is to perform the "factory" reset which is a procedure that securely erases all user data on the Android device and returns the device to its initial state. However, the whole process is time consuming; an Android app with the victim's SDK library should be installed again even when the Android device is successfully initialized.

Alternatively, we can update the Device ID value without performing the factory reset. Android saves all device settings including Device ID in a SQLite database file.[2] This database file can be managed by a program named sqlite3 at runtime. We assume that sqlite3 is already installed on the Android device. A bot program can open the settings.db file and modify the Device ID value via ADB (see **Figure 3**) before generating ad requests.

4.1.2. Generation of ad Requests

In order to successfully send ad requests to an ad network, a bot program sequentially generates a series of pre-defined click events via ADB: (1) the bot program first starts to run an Android app with the victim's SDK library, (2) generates a series of click events on the advertisement in the app, and then (3) updates Device ID with a new random 64-bit number. Whenever Device ID is successfully changed, the bot program repeats this procedure from (2). The reason we use a new Device ID for every request is to avoid detection by the ad network. That is, when the ad network monitors ad request traffic to detect click fraud by counting the number of ad requests from the same device, the bot program can trick the ad network with new Device IDs into believing as if those requests came from different users (or Android devices) individually.

4.2. Attack against CPM

To automatically increase the number of impressions on a mobile application,

```
$ sqlite3 /data/data/com.android.providers.settings/databases/settings.db
sqlite> update secure set value='ce8946b96e22354e' where name='android_id';
```

Figure 3. Example of SQLite commands to update Device ID (also known as android_id).

the most straightforward approach is to view new advertisements as many as possible.

A typical ad network's SDK library updates its advertisements periodically (e.g., every 30 s). To reduce this delay time for updating advertisements, an attacker can make a targeted app with ad library terminate and restart in an automatic manner. In fact, a similar technique can also be needed for attacking CPC when multiple clicks are not allowed on the same advertisement.

However, since advertisers generally charge per thousand impressions, the effectiveness of such attacks is not comparable with the attacks against CPC in terms of efficiency. For example, in the case of Cauly, a publisher receives only $0.09 per thousand impressions.

4.3. Attack against CPA

The CPA model is a generalization of CPC; advertisers can consider various activities (e.g., watching a video clip, installing another app, signing up for a website) including click events.

In general, it is necessary to learn prior knowledge about specific event sequences for CPA activities in order to automatically generate those sequences against CPA. Surely, it seems to be trickier than implementing the attacks for CPC consisting of touch events with fixed screen positions.

5. Experiments

We implemented several independent Android apps with real ad networks' SDK library for evaluating their potential security risk against click fraud attacks. We performed click generation attacks described in Section 4 on eight popular ad networks (AdMob, Millennial Media, AppLovin, AdFit, MdotM, LeadBolt, RevMob, and Cauly Ads), respectively. Those ad networks were selected by "AppBrain" website that offers the lists of top 500 most installed ad networks.[3] We selected those ad networks which are suitable for testing the CPC revenue model with a banner. For each ad network, we created a user account to receive payments and implemented an Android app with the SDK library for advertisements of the ad network. We note that a malicious publisher (as also the app developer) uses his (or her) own device in order to use those apps on the device without any restriction.

In our threat model, the attacker's (i.e., publisher's) goal is to successfully deliver ad request messages to a victim ad network to receive payments for those messages from the ad network. Therefore, we tested whether ad request messages can be successfully delivered without any trouble whenever we had attempted to generate a click event (with a different Device ID) on the victim's advertisements. Specifically, we assume that the attack is successful if our bot program generates over 100 ad request messages without any disruption to receive payments for the generated clicks.

In our experiments, 75% (Millennial Media, AppLovin, AdFit, MdotM, RevMob, and Cauly Ads) of the ad networks that we analyzed failed to prevent the automated click generation attacks conducted by bot programs. Probably, this implies that many real-world ad networks seem to be significantly dangerous due to click fraud. In this paper, our attack attempts are not sophisticated but straightforward. However, 75 % of the ad networks are vulnerable to those attempts.

Fortunately, the other two mobile ad networks (AdMob and LeadBolt) are secure against automated click generation attacks. When we generate click events even with uniquely different Device IDs, those networks detect our attacks only with a small number of attack attempts and then finally blocked our accounts. For example, our account used in the experiments for LeadBolt was temporarily paused due to traffic abnormality (see **Figure 4**). We surmise that those networks might have their own defense mechanisms to detect such abnormal request patterns. We will discuss such defense mechanisms in detail in the next section.

The main motivation of our experiments is to analyze potential risks of mobile advertisement services and suggest reasonable countermeasures to mitigate such risks. Therefore, we only checked ad networks' responses for our click fraud attempts; however, actual money is not withdrawn from our bank accounts. We also reported the discovered design flaws to the ad networks, which acknowledged them. Another ethical concern is related to Device ID. When we replace Device ID,

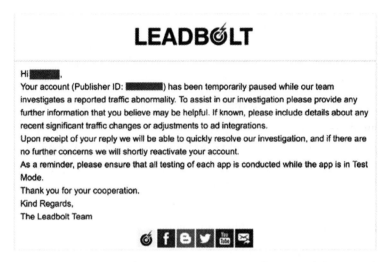

Figure 4. Our LeadBolt account was temporarily paused due to traffic abnormality.

the used Device ID may belong to some legitimate user that might be potentially blamed for our attack experiments. However, we note that this possibility is very unlikely because the used Device IDs are randomly generated ones rather than real Android devices' Device ID.

6. Countermeasures

In this section, we describe several defense mechanisms for preventing click fraud (i.e., automatic click generation on Android's banner advertisements) attacks.

We extend our previous work[17] with more detailed implementation designs by fixing incorrect representation and clarifying some ambiguous parts in the previously suggested models.

6.1. Distinguishing Human-Generated Touch Events from Program-Generated Touch Events

To prevent touch events generated by bot programs for click fraud attacks, the most straightforward defense mechanism is to effectively distinguish such events from human-generated touch events and filter out them. To achieve this goal, we first analyze how Android handles with touch events and then suggest a possible reference implementation to filter out program-generated touch events according to security policies.

We propose a modification of the existing Android architecture to trace physically generated touch events on the device screen (see **Figure 5**). Android is built on the top of a Linux kernel and includes a middleware framework and an application layer. Touch events are gathered at the /dev/input/event# node called Device Input Event files regardless of either human-generated or program-generated events. Next, the InputReader framework reads those events from Device Input Event files and the collected events are dispatched by Input-Dispatcher to a proper app.

Before touch events are recorded at the Device Input Event files, human-generated touch events are processed through Input Driver while program-generated touch events are handled with the evdev_write function in the evdev.c file. We can develop a filter for program-generated events with this difference—the evdev_write function should be modified. We can accept or reject program-generated touch events before writing them on Device Input Event files depending on the target program's security policy defined in its Security Policy files. For

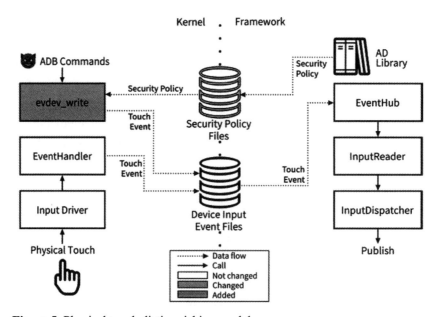

Figure 5. Physical touch distinguishing model.

example, program-generated touch events on a specific rectangle region (for displaying advertisements) can be ignored during the target program is actively running. We suggest that those files can be included in the ad network's SDK library and installed with the program itself together. We expect that the existing architecture can be slightly modified to filter out automatically generated touch events by programs without significant loss in efficiency.

For using this filtering mechanism at the kernel level, however, the integrity of the evdev_write function and security policy files should be protected. In the legacy Android architecture, if a bot program with root privileges can be installed, it is not possible to satisfy this requirement because those files might be easily tampered by the bot program. Therefore, secure environments for the suggested technique should also be provided to protect those files from a malicious root. Hardware-assisted solutions (e.g., TrustZone-based Real-time Kernel Protection[18]) might be used to achieve this security goal.

We implemented a prototype of our defense framework to filter out program-generated touch events and tested its effectiveness against click fraud attacks on real mobile devices. We modified the Android operating system 5.1.1 (i.e., the evdev_write function). To simplify implementation, we used a naive security policy to ignore all touch events generated by programs. We tested its feasibility with a Nexus 5 smartphone running the modified operating system against our own CPC bot implementation and a popular automatic click event generation tool called DummySprite (http://www.dummysprite.com) on Android. Our prototype implementation successfully prevented their attack attempts without incurring significant performance degradation.

6.2. Honey Advertisement

"Honey" is the traditional term used to indicate a "decoy" or "bait" for attackers in the field of security. For example, a honeypot is a security resource, which is intended to be attacked and compromised to gain more information about an attacker and his attack techniques[19].

To mitigate click fraud, we suggest an advertising system for automatic click generation software. We call this approach "honey advertisement". Unlike the existing advertisement systems, ad networks' SDKs often display *transparent* advertisement banners (in a random manner) to deceive malicious bot programs that automatically generate click events on those banners. A human cannot see those banners while a machine cannot distinguish transparent banners from normal banners. Therefore, if an ad request for a transparent banner was generated, this request might be triggered by automatic click generation software rather than a human user and can finally be used to set off an alarm of an automatic click fraud attack on the advertising infrastructure. A similar idea was introduced in the previous work[10].

We implemented a proof-of-concept app to show the feasibility of honey advertisement (see on **Figure 6**). As shown in this figure, some advertisement banners can be transparently displayed in a random manner. In a normal situation, a visible banner image file with a link to the advertiser's server is displayed while a transparent image file is used with a link to the ad network's server in honey advertisement. When a transparent advertisement is clicked, the ad request is delivered to the ad network's server. Basically, such events are likely to be triggered by a bot program rather than a human user because human users cannot see transparent advertisements. Therefore, we may detect the bot program's existence with a low chance of false alarms.

6.3. Detecting Anomalous Behaviors

We can see that ad requests generated by bot programs have significantly

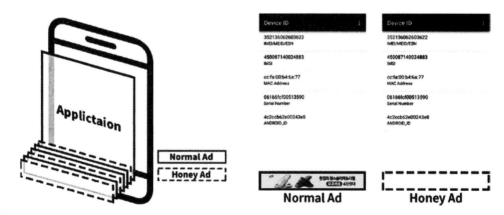

Figure 6. Our proof-of-concept implementation for honey advertisement.

different patterns compared with those generated by human users—for example, the automatically generated requests would be periodically repeated during a relatively short time. Oentaryo et al.[20] and Kitts et al.[11] introduced systems, respectively, to detect click fraud patterns in online advertisement using server-side event models. Since the existing Android platform can be used without modifications to support this approach, we highly recommend deploying similar systems at the server side.

As presented in Section 5, we believe that some of ad networks (e.g., AdMob and LeadBolt) might already use such systems to detect abnormal request patterns for click fraud in mobile platforms.

7. Economic Aspects of Security Failure

In this section, we discuss why a small number of ad networks have only detected our straightforward click fraud attacks.

This might be explained from the economic aspects of security failure. When we examine the incentives of market players for mobile advertising, we might discover an inherent problem of click fraud.

As discussed in Section 6, several defense mechanisms fighting against online click fraud can be deployed. But, who should pay for those mechanisms? We note that click fraud may directly incur significant losses in advertisers instead of ad networks.

In theory, those solutions could be deployed by ad networks who are running mobile advertising platforms. However, we claim that many ad networks might not actually be interested in mitigating click fraud.

In the mobile advertising industry, ad networks manage publishers and advertisers as a moderator (see Section 3). In general, advertisers pay money to both ad networks and publishers for their advertisements. For example, in the CPC revenue model, whenever advertisements on mobile applications are clicked, publishers receives money; ad networks also earn money. That is, ad networks would rather profit from click fraud attacks than from defenses to mitigate such attacks.

With those misincentives between players, ad networks are not motivated enough to detect online fraud attacks. Perhaps this is another example of "negative externality"[21] in the field of information security, which is an economic activity that imposes a negative effect on an unrelated third party.

8. Conclusions

This paper evaluates the potential risk of automated online fraud attacks in mobile advertising. Our experimental results show that 75% of those networks (Millennial Media, AppLovin, AdFit, MdotM, RevMob, and Cauly Ads) are vulnerable to click fraud attacks.

To mitigate such automated attacks, we suggest three possible defense mechanisms: (1) filtering out program-generated touch events; (2) identifying click fraud attacks with faked advertisement banners; and (3) detecting anomalous behaviors generated by click fraud attacks. We also discuss why few companies were only willing to deploy such defense mechanisms with the economic aspects of security failure in the mobile advertising industry.

However, our current results are not enough to generalize our observations because of the limited number of tested ad networks. As an extension to this paper, we need to consider performing our tests on a large sample of ad networks.

In this paper, we only considered the simplest attack pattern where click events were successively generated within a fixed time interval and finally failed to successfully attack two ad networks. Therefore, as another extension to this work, we also plan to test other attack sequences patterns against secure ad networks.

Endnotes

1) http://developer.android.com/reference/android/provider/Settings.Secure.html.
2) /data/data/com.android.providers.settings/databases/settings.db.
3) http://www.appbrain.com/stats/libraries/ad?list= top500.

Competing Interests

The authors declare that they have no competing interests.

Acknowledgements

This work was supported by the National Research Foundation of Korea (NRF) grant funded by the Korea government (No. 2014R1A1A1003707), ITRC (IITP-2015-H8501-15-1008) studentships, and ICT R&D program (2014-044-072-003, "Development of Cyber Quarantine System using SDN Techniques") of MSIP/IITP.

Source: Cho G, Cho J, Song Y, *et al.* Combating online fraud attacks in mobile-based advertising [J]. Eurasip Journal on Information Security, 2016, 2016(1): 1–9.

References

[1] S Dhar, U Varshney, Challenges and business models for mobile location-based services and advertising. Commun. ACM. **54**(5), 121–128 (2011).

[2] I Leontiadis, C Efstratiou, M Picone, C Mascolo, in *Proceedings of the Twelfth Workshop on Mobile Computing Systems & Applications*. Don't kill my ads!: balancing privacy in an ad-supported mobile application market (ACM, 2012).

[3] N Immorlica, K Jain, M Mahdian, K Talwar, in *Proceedings of the Workshop on Internet and Network Economics*. Click fraud resistant methods for learning click-through rates (Springer Berlin Heidelberg, 2005).

[4] N Daswani, C Mysen, V Rao, S Weis, K Gharachorloo, S Ghosemajumder, Online advertising fraud. Crimeware Underst. New Attacks Defenses. **40**(2), 1–28 (2008).

[5] B Stone-Gross, R Stevens, A Zarras, R Kemmerer, C Kruegel, G Vigna, in *Proceedings of the 2011 Conference on Internet Measurement Conference*. Understanding fraudulent activities in online ad exchanges (ACM, 2011).

[6] J Crussell, R Stevens, H Chen, in *Proceedings of the 12th Annual International Conference on Mobile Systems, Applications, and Services*. MAdFraud: investigating ad fraud in android applications (ACM, 2014).

[7] KC Wilbur, Y Zhu, Click fraud. Mark. Sci. **28**(2), 293–308 (2009).

[8] N Kshetri, The economics of click fraud. IEEE Secur. Priv. **8**(3), 45–53 (2010).

[9] B Miller, P Pearce, C Grier, C Kreibich, V Paxson, in *Proceedings of Detection of*

Intrusions and Malware, and Vulnerability Assessment. What's clicking what? Techniques and innovations of today's clickbots (Springer Berlin Heidelberg, 2011).

[10] H Haddadi, Fighting online click-fraud using bluff ads. ACM SIGCOMM Comput. Commun. Rev. **40**(2), 21–25 (2010)

[11] B Kitts, JY Zhang, G Wu, W Brandi, J Beasley, K Morrill, J Ettedgui, S Siddhartha, H Yuan, F Gao, *et al.*, Click fraud detection: adversarial pattern recognition over 5 years at Microsoft. Real World Data Min. Appl. **17**(1), 181–201 (2015).

[12] A Metwally, D Agrawal, A El Abbadi, in *Proceedings of the 16th International Conference on World Wide Web*. Detectives: detecting coalition hit inflation attacks in advertising networks streams (ACM, 2007).

[13] V Dave, S Guha, Y Zhang, Measuring and fingerprinting click-spam in ad networks. ACM SIGCOMM Comput. Commun. Rev. **42**(4), 175–186 (2012).

[14] H Xu, D Liu, A Koehl, H Wang, A Stavrou, in *Proceedings of 19th European Sysposium on Reseach in Computer Security*. Click fraud detection on the advertiser side (Springer International Publishing, 2014).

[15] P Pearce, AP Felt, G Nunez, D Wagner, in *Proceedings of the 7th Symposium on Information, Computer and Communications Security*. AdDroid: privilege separation for applications and advertisers in android (ACM, 2012).

[16] S Shekhar, M Dietz, DS Wallach, in *Proceedings of USENIX Security Symposium*. AdSplit: separating smartphone advertising from applications (USENIX, 2012).

[17] G Cho, J Cho, Y Song, H Kim, in *Proceedings of the International Workshop on Cyber Crime*. An empirical study of click fraud in mobile advertising networks (IEEE, 2015).

[18] AM Azab, P Ning, J Shah, Q Chen, R Bhutkar, G Ganesh, J Ma, W Shen, in *Proceedings of the ACM SIGSAC Conference on Computer and Communications Security*. Hypervision across worlds: real-time kernel protection from the ARM trustzone secure world (ACM, 2014).

[19] L Spitzner, *Honeypots: Tracking Hackers*. (Addison-Wesley Longman Publishing Co., Inc., Boston, MA, USA, 2002).

[20] R Oentaryo, E-P Lim, M Finegold, D Lo, F Zhu, C Phua, E-Y Cheu, G-E Yap, K Sim, MN Nguyen, *et al.*, Detecting click fraud in online advertising: a data mining approach. J. Mach. Learn. Res. **15**(1), 99–140 (2014).

[21] T Moore, R Clayton, R Anderson, The economics of online crime. J. Econ. Perspect. **23**(3), 3–20 (2009).

Chapter 22

Fingerprint-Based Crypto-Biometric System for Network Security

Subhas Barman[1], Debasis Samanta[2], Samiran Chattopadhyay[3]

[1]Department of Computer Science and Engineering, Govt. College of Engineering and Textile Technology, 4, Cantonment Road, Berhampore-742101, West Bengal, India
[2]School of Information Technology, Indian Institute of Technology Kharagpur, Kharagpur-721302, West Bengal, India
[3]Department of Information Technology, Jadavpur University, Kolkata-700098, India

Abstract: To ensure the secure transmission of data, cryptography is treated as the most effective solution. Cryptographic key is an important entity in this process. In general, randomly generated cryptographic key (of 256 bits) is difficult to remember. However, such a key needs to be stored in a protected place or transported through a shared communication line which, in fact, poses another threat to security. As an alternative to this, researchers advocate the generation of cryptographic key using the biometric traits of both sender and receiver during the sessions of communication, thus avoiding key storing and at the same time without compromising the strength in security. Nevertheless, the biometric-based cryptographic key generation has some difficulties: privacy of biometrics, sharing of biometric data between both communicating parties (*i.e.*, sender and receiver), and generating revocable key from irrevocable biometric. This work addresses the above-mentioned concerns. We propose an approach to generate cryptographic key from cancelable fingerprint template of both communicating parties. Cancelable fingerprint templates of both sender and receiver are securely transmitted to each other using a key-based steganography. Both templates are combined with concatenation based feature level fusion technique and generate a combined template. Elements of combined template are shuffled using shuffle key and hash of the shuffled template generates a unique session key. In this approach, revocable key for symmetric cryptography is generated from irrevocable fingerprint and privacy of the fingerprints is protected by the cancelable transformation of fingerprint template. Our experimental results show that minimum, average, and maximum Hamming distances between genuine key and impostor's key are 80, 128, and 168 bits, respectively, with 256-bit cryptographic key. This fingerprint-based cryptographic key can be applied in symmetric cryptography where session based unique key is required.

Keywords: Symmetric Cryptography, Cryptographic Key Generation, Biometric Security, Crypto-Biometric System, Network Security

1. Introduction

Information security and a secure transmission of data become very important in information and communication technology. A third party can trap data or steal important data stored in a computer. To prevent this, it is advocated to encrypt the messages to provide information security. This type of protection is usually provided using cryptography. In cryptography, a key ($K1$) is used to encrypt a message (called plaintext P) with encryption algorithm (E) into ciphertext (C). The ciphertext is converted into plaintext using a key (K_2) and decryption algorithm (D). There are two types of cryptography: symmetric cryptography and asymmetric cryptography. In symmetric cryptography (e.g., Data Encryption Standard (DES)[1], Advanced Encryption Standard (AES)[2]), same key (*i.e.*, $K_1 = K_2 = K$) is used in encryption ($C = E_K(P)$) and decryption algorithm ($P = D_K(C)$). In asymmetric cryptography (e.g., Rivest-Shamir-Adleman (RSA) algorithm[1]), two different keys are used (*i.e.*, $K_1 \neq K_2$), the public key is used to encrypt a message ($C = E_{K1}(P)$), and private key is used to decrypt the ciphertext into plaintext ($P = D_{K2}(C)$)[1]. Strength of cryptography in respect to security depends on the strength of keys used in encryption and decryption algorithms. A key is said to be strong if it is not easily guessed and not feasible to break within a real time. So, the issue that arises is the selection of cryptographic key. If the key is simple or very short in length, then for the attacker it is easy to guess the key. If the key is very long (128, 192, or 256 bits, for example, in AES algorithm[2]), then it is very difficult to memorize the key by a user. As a consequence, user should store it in a smart card or hardware token which can be misplaced or stolen out by an attacker. Moreover, the token or smart card is protected by password-based authentication mechanism to control the access of cryptographic key. Nevertheless, password can be forgotten or guessed by social engineering[3] and dictionary attack[4]. Both knowledge-based (e.g., password) and possession based (e.g., token) authentication systems are unable to assure non-repudiation property in traditional cryptography.

Of late, biometric is being integrated with cryptography (called crypto-biometric system) to alleviate the limitations of the above-mentioned systems[5][6]. Biometric is the unique measure of the identity of individuals with their behavioral and physiological traits like face, fingerprint, iris, retina, palm-print, speech, etc.[7]. Many researchers are trying to use biometric traits in the authentication component of cryptography to remove the requirement of password-based authentication. The integration of biometric with cryptography, deals with either cryptographic key release[8]–[10] or cryptographic key generation[6][11]–[17], is promising in many aspects. As biometric is directly linked with the owner, it removes the problem to memorize the cryptographic key and confirms the non-repudiation of users.

Crypto-biometric system, however, has some issues. Any biometric system needs to provide biometric template protection which confirms the privacy and security of biometric data[18]. The biometric data used in a biometric system should not leak any information about the biometric features. It is also required to provide revocability to the irrevocable biometric data. In password-based authentication systems or token-based authentication systems, passwords or tokens are easy to change while it is compromised. But, biometric traits are inherent and fixed forever, that is, the biometric data is irrevocable[7]. The owner of biometric traits is not able to revoke her biometric when it is compromised. As a result, the biometric data become useless forever[6]. To overcome this problem, it demands a cancelable transformation[18][19] of biometric template to provide revocability to the irrevocable biometric. Simultaneously, it would ensure the privacy of biometric data [5], so that the transformed template does not leak any information about the original template. Moreover, biometric data is required to be transmitted over non-secure communication channels for remote use. Therefore, there is a need to generate cryptographic key, which is revocable and non-invertible from the biometrics of two different users without compromising the privacy and security of the biometrics involved in key generation process.

This work aims to address the above-mentioned concerns and proposes a solution to develop a cryptobiometric system. Our proposed solution includes the following: 1) how to generate cancelable fingerprint template so that biometric features of neither communicators are never disclosed to anyone, 2) how to generate a unique cryptographic key for encryption (decryption) of messages using the cancelable fingerprint templates of both sender and receiver, and 3) how to generate revocable session key from irrevocable biometric traits prior to each session. In this paper, we propose an approach to generate, share, and update cryptographic key for symmetric cryptography from the fingerprints of sender and receiver at their sites for encryption and decryption, respectively. Initially, sender shares two secret keys namely stego key (K_g) and shuffle key (K_{shuf}) with receiver. Stego key is generated from a password (*pwd*) by sender and receiver using pseudo random number generator (PRNG)[1]. Shuffle key (Kshuf) is generated randomly, which is a binary stream of bits and stored in token. In this work, sender shares K_{shuf} and *pwd* with receiver using public key cryptography. With our proposed approach, asymmetric cryptography is proposed to exchange an initial shuffle key K_{shuf} and a password *pwd* between sender and receiver. For session keys, we propose biometric-based cryptographic key generation to establish a link of users biometric with cryptographic key. In our approach, biometrics of both communicating parties are integrated to generate cryptographic keys so that we can avoid the complex random number generation and alleviate the issue of storing the random cryptographic keys in the custody of sender and receiver. Moreover, revocable key generation in every session and protecting the privacy of biometric templates are the challenge which has been addressed in this work. Both sender and receiver exchange their cancelable fingerprint template with each other using key-based steganography. Both cancelable templates are then merged together using concatenation-based feature level fusion technique[20] to generate a combined template. Shuffle key is used to randomize the elements of the combined template. Finally, cryptographic key is generated from this shuffled template using a hash function. In our approach, fingerprint identity of sender is not disclosed to receiver and vice versa as cancelable template is exchanged between them to derive cryptographic key. Moreover, in our approach, cryptographic key or fingerprint template or both can be revoked easily if required. The revocability is provided to the cryptographic key with cancelable template and or with updated shuffle key.

The rest of the paper is organized as follows. A brief review of related research is given in Section 2. The proposed approach for cryptographic key generation from the fingerprints of sender and receiver is given in Section 3. The experimental results and security analysis are discussed in Sections 4 and 5, respectively. Finally, the paper is concluded in Section 6.

2. Literature Survey

Our work consists of mainly three sub-tasks: i) transformation of biometric template, ii) secure transmission of biometric data, and iii) crypto-biometric system. There exist few work in the literature related to each sub-task, which are discussed in this section.

2.1. Biometric Template Transformation

Biometric systems require a transformation of biometric template to ensure privacy, security, and revocability of biometric data. The technique which can meet this requirement is called cancelable or revocable biometric. This privacy enhancement problem is identified, and conceptual frameworks of biometric templates are presented in[21][22]. Ratha et al.[23] formally defined the problem of cancelable biometric. Uludag et al.[6] provides a comprehensive review on privacy and revocability of biometrics with some corrective measures. Recently, Ratha et al.[19] proposed three practical solutions to cancelable biometrics and generate cancelable fingerprint templates. These three template transformation approaches are Cartesian, polar, and functional transformations on feature domain. In Cartesian transformation approach, the minutiae space is divided into rectangular cells which are numbered with sequence. A user-specific transformation key (*i.e.*, matrix) is used to shift the cells to a new location, and the minutiae points are relocated to the new cells. In polar transformation, coordinate space is divided into polar sectors that are numbered in sequence. The sector position is changed with the help of translation key, and it changes the minutiae location also. In functional transformation, Ratha et al.[19] model the translation using a vector-valued function $\vec{F}(x,y)$ which is an electric potential field parameterized by a random distribution of charges. The phase angle of the resulting vector decides the direction of translation and the magnitude $|\vec{F}|$ of this vector function parameterizes the extent of movement. In an alternate formulation, Ratha et al.[19] use the gradient of a mixture of Gaussian kernels to determine the direction of movement and the extent of movement is determined by the scaled value of the mixture. Some researchers proposed shuffling-based transformation to generate cancelable templates using a user-specified random key[24][25]. In these works, the iris code is divided into blocks and then the blocks are shuffled with a user-specified random shuffling key to generate cancelable iris template.

Jain et al.[18] reviewed the existing work of fingerprint template protection such as encryption, template transformation, and crypto-biometric systems. They analyzed the practical issues involved in applying these techniques for biometric template protection. They compared the existing solutions of template protection on the basis of template security and matching accuracy of biometrics in transformed domain.

2.2. Biometric Data Transmission

There are many work reported in the current literature where the biometric data is transmitted over communication channels for the purpose of remote authentication. Existing work[26]–[28] consider hiding of biometric data within another media called cover media using data hiding technique. Different types of data hiding techniques are used for secure transmission of biometric data using steganography. Minutiae points of fingerprint are hidden within face or synthetic fingerprint using watermarking technique and sent to other user via insecure communication channel[26]. Similarly, fingerprint is also used as cover media to hide other biometric data (i.e., face) in watermarking technology and used as carrier image for secure transmission of biometric data[27]. Note that in the data hiding concept, use of real biometric as cover media is risky as it reveals sender's biometric identity to receiver. Agrawal and Savvides[29] propose a biometric data hiding approach where a biometric (iris and fingerprint) data is encrypted with a key and the encrypted biometric data is encoded with error correcting code. The encoded biometric data is embedded bit by bit using the sign of discrete cosine transform (DCT) coefficients of a random cover image.

2.3. Crypto-Biometric Systems

Biometric-based cryptosystems are classified into two types, namely key release and key generation. In the first approach, a randomly created cryptographic key is protected from unauthorized access with users' biometric data. Fuzzy vault[8]-[10] and fuzzy commitment scheme[24][25][30] fall under this category. In fuzzy vault scheme, biometric data (e.g., minutiae points) is considered as an unordered set $s^E = x_1, x_2, ..., x_r$ of r elements. The secret key (k_v) of k bits is transformed into a polynomial of degree k. All elements of s^E are evaluated on the polynomial and the polynomial evaluation value $P(x_i)$ and x_i, that is, $(x_i, P(x_i))$ points are secured with some randomly generated chaff points $(a_j, b_j)_{j=1}^{q}$ which do not lie on the polynomial P (i.e., $b_j \neq P(a_j)$ and $a_j \notin s^E$, $\forall j = 1, 2, ..., q$). The genuine $((x_i, P(x_i)))$ and chaff (a_j, b_j) points constitute the fuzzy vault. The security of this vault depends on the computational difficulties of solving polynomial reconstruction problem. Now, the secret key is released only when the query biometric is close to the set s^E. Most of the existing fingerprint-based fuzzy vault use the (x, y) coordinate values of minutiae points[9][10] whereas, Nandakumar et al.[8] propose a fingerprint-based fuzzy vault where both (x, y) coordinates and orientation (θ) of minutiae points are used. On the other side, in fuzzy commitment scheme, biometric data is represented in a binary vector b^E and the vector is locked by a random secret key of less or equal bits of b^E with XOR operation. In[24][25], iris code is combined with a random key using XOR operation and using the query iris code, the secret key is extracted from the combined iris code.

Few approaches have been proposed to generate cryptographic key from the biometric traits[14]-[16][31][32]. Monrose et al.[14] propose an approach to generate cryptographic key from user's voice while speaking a passphrase. Feng et al.[15] propose a cryptosystem, that is, BioPKI, where user's online signature is used to generate a private key. In[16][32], face biometric is used to extract a suitable length cryptographic key. Iris is also used for cryptographic key generation from iris texture[12][13][31]. Rathgeb et al.[31] analyze the iris feature vector and detected the most stable or reliable bits in the binary iris code to construct cryptographic key. Fingerprint, the most universal and acceptable biometric, is also used to derive a cryptographic key from cancelable fingerprint template[17][33][34]. Main problem of the approaches[17][33][34] is that it is not able to generate revocable key for session based communication.

In recent research, multimodal or multiple biometrics are used in crypto-biometric systems[11]–[13]. A Jagadeesan et al. proposed a method[12] where multimodal biometrics (fingerprint and iris) are used. They applied the feature level fusion of minutiae points and texture properties of iris to generate the multimodal biometric templates and the key is generated from this template. In another work[13], Jagadeesan et al. use the same biometrics (fingerprint and iris) but different method to generate the transformed template. The exponentiation operation is performed where iris texture values are used as base numbers and minutiae coordinates are used as exponent. Then, the next prime number is calculated for each exponentiation result and multimodal template is generated using multiplication of two resultants prime numbers to generate a key of 256 bits. This approach is, however, not free from key sharing problem of traditional symmetric cryptography.

Dutta et al. reported a method of fingerprint-based cryptography and network security[11]. In this method, they work with the fingerprints of sender and receiver. The fingerprint of receiver is transmitted to sender, and it is merged with sender's fingerprint to generate cryptographic key (of 128 bits) using standard hash function (MD5). In their approach, cryptographic key along with a random vector are watermarked into the genuine fingerprint and watermarked image is sent to the recipient. This method is not secure as genuine biometric is used as the cover image for data hiding. As the key and random sequence vector are transmitted over insecure channel with data hiding technique, it causes security threats to the message transmission if the fingerprint of the user is compromised to a third party, anyway. In this approach, fingerprint of receiver is sent to sender and fingerprint of sender is used as cover image and the watermarked image containing the master key and random vector is sent to receiver. Thus, fingerprint of sender is known to receiver and vice versa. A third party (man-in-middle) can generate the key using cryptographic hash function and with the knowledge of fingerprints of sender and receiver. Further, this approach is silent about the revocability issue.

3. Proposed Methodology

In this section, we discuss our proposed approach in details. An overview of our approach is shown in **Figure 1**. In our approach, both sender and receiver extract minutiae points from their own fingerprints. The minutiae points are transformed into a cancelable form called cancelable template. The cancelable templates are exchanged between them using steganography. The stego key (K_g) is used by both parties for secure steganographic use. The stego key is generated from a password (*pwd*) using PRNG. Both cancelable templates are combined together and shuffled using shuffle key (K_{shuf}) and finally the cryptographic key is generated from it following a hash function. In this scheme, initial shuffle key K_{shuf} and password *pwd* both are selected by sender. Sender uses asymmetric cryptography to share the concatenated shuffle key and password, that is, K_{shuf}||pwd to receiver. Sender uses the public key K_{pub} of receiver to encrypt the (K_{shuf}||pwd) and sends EK_{pub} (K_{shuf}||pwd) to receiver. Receiver can decrypt the shuffle key and password using his own private key K_{prv}, and they are used for key generation and template sharing, respectively. The above-mentioned steps in our approach are stated in details in the following.

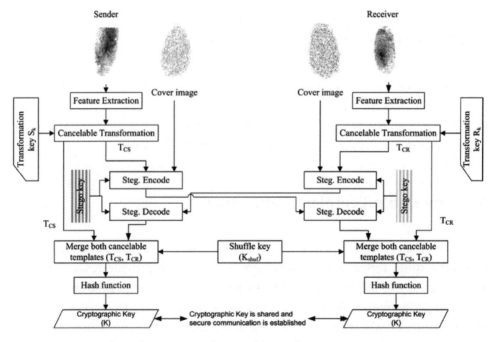

Figure 1. An overview of our proposed crypto-biometric system.

3.1. Feature Extraction from Fingerprint Image

We consider minutiae points (ridge ending and bifurcations) as the biometric features. The features are stored as (x, y, θ) form, where (x, y) denotes coordinate value and θ is the orientation of a minutiae point. In fingerprint authentication systems, inclusion of more features increase matching scores and hence the angle information is preferred. The objective of our work is to use fingerprint data as the source of randomness rather than the authentication of a user. It is observed that x and y coordinate values are enough to provide randomness in data. Therefore, we have considered only (x, y) coordinate values as the minutiae points in our work. We extract minutiae-based features from the fingerprint images of both sender and receiver. For the reference in our subsequent discussion, we denote them as follows.

- FS = Set of minutiae points extracted from sender's fingerprint.
  $= \left[m_1^s, m_2^s, ..., m_{N_s}^s \right]$; where $m_i^s = (x_i, y_i)$, m_i^s is the ith minutiae points of sender's fingerprint, $i = 1$ to N_s and N_s is the size of F_S

- FR = Set of minutiae points extracted from receiver's fingerprint.
  $= \left[m_1^r, m_2^r, ..., m_{N_r}^r \right]$; where $m_i^r = (x_i, y_i)$, m_i^r is the ith minutiae points of receiver's fingerprint, $i = 1$ to N_r and N_r is the size of F_R.

Note that, for all fingerprint images, it is a general observation that number of minutiae points for a person lies within 50. However, in case, if there are more than 50 minutiae points (i.e., $N_s, N_r \geq 50$), then first 50 minutiae points according to their quality value would be selected and the rest be discarded.

3.2. Cancelable Template Generation

The fingerprint templates of both sender and receiver are transformed into a non-invertible forms, called cancelable templates, to provide revocability as well as privacy to the fingerprint data of both users before transmitting them to their counter partners. The position of minutiae features are changed using Cartesian transformation with the help of a user-specified transformation key (i.e., S_k for sender and R_k for receiver). The overall process of transformation is discussed below.

1) The coordinate system is divided into N cells of same size $h \times w$, where h is the height and w is the width of the cells. The total number of cells N can be calculated with equation given below

$$N = \frac{(H \times W)}{(h \times w)} \qquad (1)$$

where H and W are the height and width of the fingerprint image.

2) Each cell is denoted by $C_{i,j}$ where $i = 1$ to n (n cells in each row) and $j = 1$ to m (m cells in each column). The cells can be represented in a one-dimensional vector, and cell $C_{i,j}$ can be represented by c_t and the value of t can be computed in the following way.

$$t = \{(j-1) \times n\} + i \qquad (2)$$

where for any value of i ($1 \leq i \leq n$) and j ($1 \leq j \leq m$), there will be a unique t such as $1 \leq t \leq N_{cell}$. Each cell contains either no minutiae or a set of minutiae points. It depends on the cell size, distribution of minutiae points in fingerprint image. For sender's fingerprint, if the tth cell c_t contains n_{mt} number of minutiae then the ith minutiae point of tth cell of senders fingerprint can be represented as m_i^{st} and the value of i is defined as $1 \leq i \leq n_{mt}$. For an illustration, we divide the image space into n vertical cells and m horizontal cells, and the cells are shown in **Figure 2** where the two-dimensional cells are shown in left and the sequence number is shown in right side table.

3) Generate a user-specific (0, 1) matrix (M) of size $N \times N$ (where $N = n \times m$) to map the cells with their new positions as per the following equation.

$$C'_{i'j'} = C_{i,j} M \qquad (3)$$

where $C_{i,j}$ is the original cell which is replaced by cell $C'_{i'j'}$. Here, one cell can replace zero or multiple cells. The number of minutiae points (say n_{mt}) in the new cell can be the same or different.

								31	32	33	34	35	36
m	C_{1m}	C_{2m}	...	C_{im}	...	C_{nm}		25	26	27	28	29	30
:	:	:	...	:	...	:		19	20	21	22	23	24
j	C_{1j}	C_{2j}	...	C_{ij}	...	C_{nj}		13	14	15	16	17	18
:	:	:	...	:	...	:		7	8	9	10	11	12
2	C_{12}	C_{22}	...	C_{i2}	...	C_{n2}		1	2	3	4	5	6
1	C_{11}	C_{21}	...	C_{i1}	...	C_{n1}							
	1	2	...	i	...	n			$i = j = 4;\ n = m = 6$				

Figure 2. (x, y) coordinate space is divided into cells, and the equivalent cell numbers are shown in right.

4) Modify the (x, y) coordinates according to their new cell locations, and cancelable template T_C is generated as follows.

$$T_C = F\left(C_{i,j}, C'_{i',j'}\right) \tag{4}$$

The (x, y) coordinate values of all minutiae points belong to $C'_{i',j'}$ will be placed to the cell $C_{i,j}$ according to the mapping function (F).

For the replacement of cell $C_{i,j}$ by $C'_{i',j'}$ ($i.e., C_{i,j} \leftarrow C'_{i',j'}$), the (x, y) coordinate value of minutiae point of cell $C'_{i',j'}$ can be computed for new location as follows.

(a) If $i' = i$ then $x_i^d = x_i$
(b) If $i' > i$ then $x_i^d = x_i + (i' - i) * w$
(c) If $i' < i$ then $x_i^d = x_i - (i - i') * w$

where x_i is the x coordinate value of a minutiae points and x_i^d is the displaced x coordinate value of the same minutiae points and w is the width of a cell of size $h \times w$. Similarly, depending on the value of j', the y_j will be changed to y_j^d.

For example, a simplified coordinate value of minutiae points are divided by four cells and numbered as $C_{1,1} = 1$, $C_{1,2} = 2$, $C_{2,1} = 3$, $C_{2,2} = 4$. The Cartesian transformation is given in the following equation.

$$\begin{bmatrix} 1 & 2 & 3 & 4 \end{bmatrix} \begin{bmatrix} 1 & 0 & 0 & 0 \\ 1 & 1 & 0 & 0 \\ 0 & 0 & 0 & 1 \\ 0 & 0 & 1 & 0 \end{bmatrix} = \begin{bmatrix} 3 & 2 & 4 & 3 \end{bmatrix} \tag{5}$$

where, cells 1 and 4 are replaced by cell 3 while cell 3 is replaced by cell 4. The cell-wise coordinate values of minutiae points before and after transformation are shown in **Figure 3**. A pictorial representation of the cancelability transform, using real minutiae sets is shown in **Figure 4**. Here, we have divided the height H and width W of the fingerprint image by 8 to make $8 \times 8 = 64$ cells of size $\frac{H}{8} \times \frac{W}{8}$. Also, it may be noted that the height and width of the cells in last row and last column may vary.

After transformation of fingerprint template using this transformation, the transformed template (T_C), also known as cancelable template, contains modified minutiae points denoted by $m_i^{s'}$, which represents ith modified minutiae point or ith elements of the cancelable template. In the cancelable template, we consider that it contains 50 modified minutiae points. If it exceeds 50, then we consider only first 50 elements and if it contains less than 50 elements, then we augment sufficient numbers of zero elements at the end to make it of intended size. In the subsequent discussion, the cancelable templates of sender and receiver are denoted by T_{CS} and T_{CR}, respectively, where $T_{CS} = \{m_1^{s'}, m_2^{s'}, ..., m_{N_{TCS}}^{s'} | N_{TCS} = |T_{CS}|\}$ and $T_{CR} = \{m_1^{r'}, m_2^{r'}, ..., m_{N_{TCR}}^{r'} | N_{TCR} = |T_{CR}|\}$.

3.3. Steganographic Encoding

In this work, cancelable template of one party (say sender) needs to be sent through a shared communication channel to other party (say receiver) and vice versa. Sender (and receiver) uses steganography-based data hiding technique to

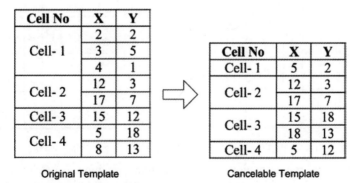

Figure 3. Original fingerprint template and cancelable template.

Fingerprint-Based Crypto-Biometric System for Network Security

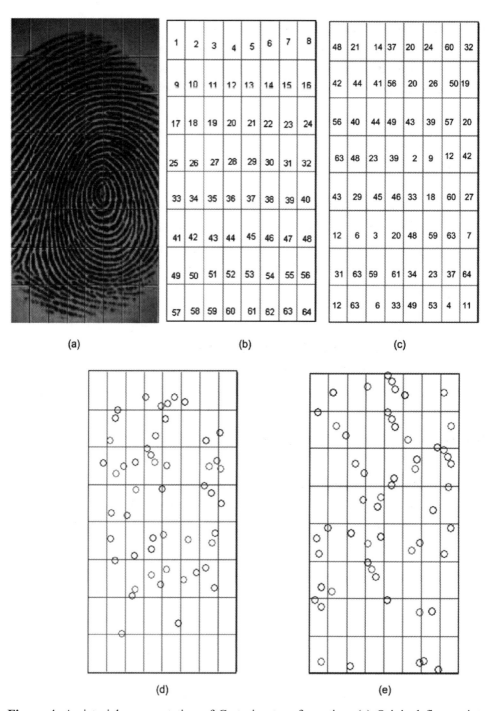

Figure 4. A pictorial representation of Cartesian transformation. (a) Original fingerprint image, (b) Cells, (c) Cells which replaces the original cells, (d) Original minutiae points, (e) Minutiae points in cancelable template.

hide the cancelable template data into a cover image (I) (of size MI pixels, say). The cancelable template is converted into binary stream $(s_1, s_2, s_3, ..., s_L)$, where L is the number of bits in cancelable template after the conversion, which is to be hidden into cover image using LSB steganography[35]. A secret key, called stego key K_g, is generated from a password (*pwd*) using pseudo random number generator where the password is used as the seed value. The stego key ($K_g = [k_1, k_2, ... k_{Nkg}]$; where $L \leq N_{kg} \leq M_I$) is used to select the pseudo random path of pixel locations in cover image (I) to hide the cancelable template bit by bit in cover image. The stego image (I_{stego}) is sent to the recipient from sender and *vice versa*.

3.4. Steganographic Decoding

In this phase, hidden data are extracted from the stego image (I_{stego}) using the decoding function. The stego key (K_g) is used to locate the pixels where data embedding take place. The extracted binary stream is then used to reconstruct the cancelable template.

3.5. Merging Cancelable Templates T_{CS} and T_{CR}

After receiving the cancelable template of the counter partner, receiving party merges his own cancelable template with the received cancelable template. Say, sender has its own cancelable template T_{CS} and received the cancelable template T_{CR} from receiver. Both T_{CS} and T_{CR} consist of modified minutiae points ($m_i^{s'}$, $m_i^{r'}$) of sender and receiver, respectively. These two cancelable templates are fused using the feature level fusion[20][36] of modified minutiae features.

Feature level fusion is achieved with concatenation of two feature sets[20] $T_{CS} = \left[m_1^{s'}, m_2^{s'}, ..., m_{N_{TCS}}^{s'} \right]$ and $T_{CR} = \left[m_1^{r'}, m_2^{r'}, ..., m_{N_{TCR}}^{r'} \right]$. ($|T_{CS}| = N_{TCS}$ and $|T_{CR}| = N_{TCR}$) to generate a combined template T_f ($|T_f| = N_{TCS} + N_{TCR}$). For this purpose, all x coordinate values of $m_i^{s'}$ from TCS are stored in vector X and the x coordinate values of $m_i^{r'}$ from T_{CR} are *augmented to the same vector*. Similarly, the y coordinate values of TCS and TCR are combined and stored in another vector Y. These two vectors (X, Y) generate a new (x, y) coordinates of T_f. The size of T_f is the total size of T_{CS} and T_{CR}. That is

$$T_f = T_{CS} \| T_{CR} ; |T_f| = |T_{CS}| + |T_{CR}|. \tag{6}$$

where $\|$ denotes the augmentation operation, $|T_f|$ is the size of T_f. In the combined template, redundancy may exist. The redundancy, if it exists, is removed, and only unique modified minutiae points are selected from T_f. As an example, sample feature level fusion of two feature sets is shown in **Figure 5(c)**.

The elements of vectors X and Y are shuffled separately using the shuffle key (K_{shuf}). The initial shuffle key (of 200 bits and 100 bits are required for each template) can be generated randomly. Our proposed shuffling method is illustrated in **Figure 6**. In this shuffling method, the vector elements where corresponding key bits are 1 are sorted starting at the beginning, and the remaining elements where the key bits are 0 are placed starting from the end. In this way, all elements of vector X and Y are shuffled and the shuffled X, Y vectors (denoted as XS and YS) result a modified F. For example, a sample shuffled F is shown in **Figure 5(d)**.

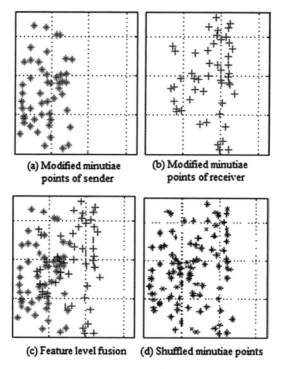

(a) Modified minutiae points of sender
(b) Modified minutiae points of receiver
(c) Feature level fusion
(d) Shuffled minutiae points

Figure 5. A snapshot of fused features and shuffled features. (a) Modified minutiae points of sender. (b) Modified minutiae points of receiver. (c) Feature level fusion. (d) Shuffled minutiae points.

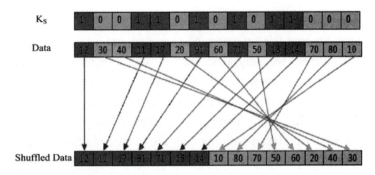

Figure 6. Shuffling method.

Each corresponding element of shuffled vectors (X^S and Y^S) is merged using XOR operation. For this purpose, x_i and y_i ($x_i \in X^S$ and $y_i \in Y^S$) are converted into binary numbers and bitwise XOR operation is followed for all elements of X^S and Y^S. The results of bitwise XOR operation are stored in a vector F_{code}.

$$F_{code} = \int F_{code_i} = \int bitwiseXOR(x_i, y_i) \tag{7}$$

Finally, the cryptographic key is generated from this F_{code} using a hash function which is as follows. The F_{code} is divided into blocks (i.e., $F_{code} = B_1 \| B_2 \| \ldots \| B_{nb}$; say total n_b blocks) of size 256 bits each. A vector (K) of size 256 bits of all zeros is generated as the initial hash value. Now, block $B1$ is XORed with initial K and the output is stored in K_2. The K_2 is XORed with the next block B_2 and the result is stored in K_3 and so on. Finally, the hash value K_{nb+1} is the cryptographic key (K). That is

$$K_{i+1} = (K_i \oplus B_i); \quad \text{where } i = 1 \text{ to } n_b \text{ and } |K_i| = |B_i| = 256 \text{ bits} \tag{8}$$

This way, sender and receiver both derive the same secret key which establishes a secure communication between the sender and receiver for a session. For a new session, a new session key can be generated from the same fingerprints using an updated shuffle key, which is discussed in the next sub-section.

3.6. Shuffle Key Update

For better security measure, we propose to change the cryptographic key in each session. In other words, if there is a chance to compromise cryptographic key, then it is desirable that the key must be canceled and a new key be used for the next session. Further, we may note that if both cancelable templates become known to a third party then with the knowledge of key generation algorithm, key can be derived by the third party. But, in our approach, cryptographic key generation depends on another factor namely shuffle key (Kshuf). The revocability of the cryptographic key is achieved with not only the cancelable fingerprint template but also with shuffled key. Our protocol is also able to update the shuffle key time to time using the fingerprint data of both users. To realize this, we propose to update the shuffle key from one session to another. Session-wise shuffle key update procedure is shown in **Figure 7**. Initiation to update shuffle key can be taken by sender or receiver.

The steps followed in our shuffle key update process when it is initiated by the sender are given below.

1) Both sender and receiver share their cancelable fingerprint data (Sections 3.3 and 3.4) and generate F_{code} following the method discussed in Section 3.5.

2) Shuffle key (K_{shuf}) is XORed with the first $|K_{shuf}|$ bits of F_{code} to obtain a new shuffle key (K_{SS}), that is, $K_{SS} = K_{shuf} \oplus F_{code}$.

3) Sender computes the hash value of the new shuffle key ($h(K_{SS})$) and sends it along with update request to receiver. We have used XOR-based hash function (h), as discussed in Section 3.5.

4) Similarly, receiver also generates a new shuffle key ($K_{SR} = K_{shuf} \oplus F_{code}$) and computes the hash of the new shuffle key ($h(K_{SR})$) using the same one way hash function (h) which is used in sender side.

5) Receiver compares the computed hash ($h(K_{SR})$) with received hash ($h(K_{SS})$), if both are same, then receiver replaces old shuffle key (K_{shuf}) with new one (K_{SR}) and sends success message to the sender.

6) On the basis of receiver's report, sender also replaces the old shuffle key (K_{shuf}) with new shuffle key (K_{SS}).

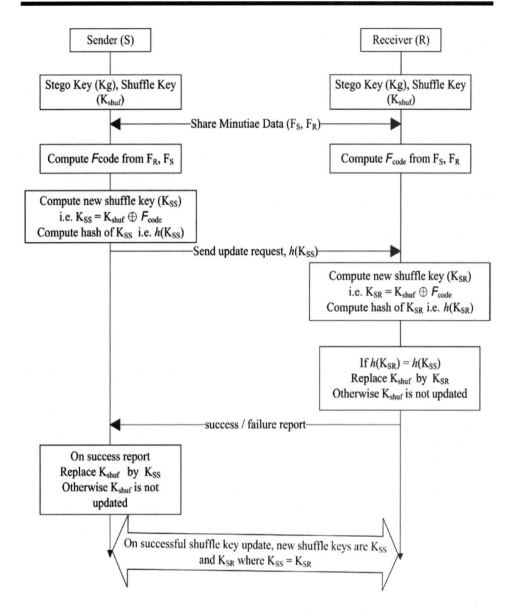

Figure 7. Shuffle key update.

In this way, both sender and receiver are able to update their shuffle key. In every session, a unique shuffle key is generated from the fingerprint data of sender and receiver. After the session is over, old shuffle key is destroyed and modified one is used for next session of communication. In this way, our approach can provide the diversity of cryptographic key of fingerprint-based symmetric cryptography.

4. Experiment and Experimental Results

There are mainly two objectives in our experiment. First, we investigate the impact of data encoding into cover images (*i.e.*, synthetic fingerprints from DB4 of each database) and accuracy of data decoding from stego images. In the next part of our experiment, we measure the randomness of cryptographic key generated from fingerprints of genuine users with respect to the key generated from impostor's fingerprints. In this regard, the Hamming distances between the genuine and impostor's keys are measured and corresponding histograms are plotted. Here, we consider all possible cases of attacks when different entities are compromised and the Hamming distances are computed for each case.

4.1. Database

We have tested our work using the fingerprint images from publicly available fingerprint databases, FVC2000[37], FVC2002[38], and FVC2004[39]. Each FVC database has four subsets labeled as DB1 to DB4. Each subset has a development set (B) (from fingers 101 to 110) with ten people and a test set (A) with 100 peoples (fingers from 1 to 100). There are eight samples for each person. The details of fingerprints used in our experiment are shown in **Table 1**. In these databases, DB1, DB2, and DB3 are real fingerprint databases and DB4 is synthetic fingerprint[40] database. In set B of each databases, total number of fingerprint images are (10 × 8 × 4) = 320; and in set A of each database, there are 100 × 8 × 4 = 3200 fingerprint instances.

Table 1. Fingerprints used in our experiments.

	FVC2000 (sensors, image size)	FVC2002 (sensors, image size)	FVC2004 (sensors, image size)
DB1	Optical (KeyTronic), 300 × 300	Optical (Identix) (CrossMatch V300), 388 × 374	Optical sensor, 640 × 480
DB2	Capacitive (ST Microelectronics), 256 × 364	Optical (Biometrika), 296 × 560	Optical Sensor (Digital Persona U.are.U 4000), 328 × 364
DB3	Optical (Identicator Technology), 448 × 478	Capacitive (Precise Biometrics), 300 × 300	Thermal sweeping sensor (Atmel FingerChip), 300 × 480
DB4	SFinGe v2.0, 240 × 320	SFinGe v2.51, 288 × 384	SFinGe v3.0, 288 × 384

4.2. Experimental Setup

A unique pair of fingerprints is taken (as fingerprint of sender and receiver) from a subset (*i.e.*, DB1 or DB2 or DB3 or DB4) of a specific FVC to generate a genuine key. All remaining pairs of fingerprints in the same subset of the same FVC database are taken to generate impostor keys with respect to that genuine key. This way, all pairs of genuine fingerprints and corresponding pairs of impostor fingerprints are chosen. Initially, all unique combination of two person's fingerprints from 110 person's fingerprints of each subset of the fingerprint database is computed. Thus, the total number of pairs for genuine users is $\frac{110}{2} = 55$ for each subset (*i.e.*, DB1, DB2, DB3, and DB4 for both sets A and B). For each FVC database (considering all four subsets of set (A+B), the total number of genuine keys is $4 \times 55 = 220$. Similarly, for every genuine key, remaining 54 keys are impostor keys. In our experiment, the duplicate pairs of genuine and impostor fingerprints are avoided carefully. As a result, we get $220 \times 3 = 660$ impostor's keys per FVC database and a total number of $3 \times 4 \times \frac{55 \times 54}{2} = 17{,}820$ impostor's keys.

Now, the minutiae points from fingerprint images are extracted using NBIS software (MINDTCT)[41]. The MINDTCT tool takes fingerprint image as input and returns minutiae points set in the format of (x, y, θ, q) where q is the quality of that minuitae point. Average number of minutiae points is found as 50. We consider only (x, y) coordinates of first 50 minutiae points according to the quality of minutiae reported by MINDTCT in our experiment. The minutiae points are transformed into cancelable template with a user-specified transformation key (Section 3.2). We have divided the fingerprint images into 64 cells. The cell size is computed by dividing the height (H) and width (W) with 8 (*i.e.*, cell size = $h \times w$ where $h = \frac{H}{8}$ and $w = \frac{W}{8}$). The cancelable template converted into binary stream (Section 3.5). Most of the sizes of fingerprint images are within the range of 256 to 511. The maximum values of x and y coordinate points thus can be represented with 9 bits binary numbers. However, as exceptions, the y coordinate value of the fingerprint images in some FVC2002 database (it is A and B sets in DB2), and the x coordinate values of fingerprint images in FVC2004 database (e.g. A and B sets in DB1) exceed the range of 511. As the x and y coordinate values of minutiae points in almost all fingerprint images are less than 512, we represent them by 9 bits (maximum decimal value with 9 bits is $2^9 - 1 = 511$). Of course, the coordinates exceeding 511 are with a less approximation, which do not affect the results adversely. The values of x and y coordinate of modified minutiae points (*i.e.*, the elements of cancelable templates) also lie within this range. Therefore, each element of the cancelable template is converted into 18 bits binary number (9 bits for x coordinate and 9 bits for y coordinate) and binary conversion of all elements of cancelable template produces a binary bit stream of $18 \times 50 = 900$ bits.

We propose to generate the transformation keys (S_k, R_k) randomly. In our experiment, these keys are generated using pseudo random number generator (PRNG) available in MATLAB. We have used a unique transformation key to generate a cancelable template from fingerprint of a specific person. In our approach, stego key (K_g) is also generated randomly (using PRNG in MATLAB) from a password (*pwd*) and a unique stego key is assigned to each cover image of the synthetic fingerprints (*i.e.*, the fingerprints in DB4). A unique shuffle key (pseudo random number) is used for a unique pair of fingerprint images.

We use synthetic fingerprint[40] from DB4 database (of FVC2000, FVC2002, FVC2004) as the cover image to hide cancelable fingerprint template of genuine users using LSB steganography. The cover image is picked up at random by both sender and receiver. A stego key (K_g) is used to locate the pixels of cover image where the cancelable template (C_{TS}, C_{TR}) bits are hidden during steganographic encoding (Section 3.3). Similarly, the same stego key is used to decode the cancelable template bits from stego image during steganographic decoding (Section 3.4).

Note that in our experiment, both sender and receiver exchange their own cancelable template between them using the same stego key (K_g) but different cover images (say I_S, I_R).

4.3. Experimental Results

The results of our experiment are stated below with respect to different scenarios.

4.4. Case 1: Impact of Steganography

In our experiment, effect of data encoding over cover image is investigated and we follow the evaluation method as given in[26][27]. The stego key K_g is fixed for each unique pair of users and it differs when the pair of sender and receiver is changed.

Few observations are summarized in **Table 2**. The first column is the average pixel value for cover images. The second column is the average pixel value for the stego images. The third column is the pixel change with respect to cover image. The last column represents the absolute pixel change of the total encoded pixels. It is also observed that 100% of the encoded message is extracted using the same stego key from stego image and nearly 0.47% pixels of the cover image is changed due to data hiding.

4.5. Case 2: Both Fingerprints and shuffle Keys are Unknown

For this purpose, we consider that the attacker has no knowledge about either the genuine fingerprints or shuffle key (K_{shuf}) to compromise the cryptographic key. In this case, unique S_k and R_k are used to transform each fingerprint and unique K_{shuf} is used for generation of each key. In this condition, we compute the Hamming distance between genuine and impostors' keys and the Hamming distances are plotted using histogram. The histogram in **Figure 8(a)** shows the distribution of Hamming distances of 17,820 comparisons between genuine and impostors' keys. It is observed from the histogram that mean Hamming distance is 49.95% which means that the average hamming distance between the genuine and impostors' key is 128 bits. In this case, the Hamming distances are spreaded between the range of 34.38% to 62.11% with a standard deviation of 0.032. According to the quantity of impostor's key, it is observed that 40% to 60% bits of the genuine keys are different from 99.89% impostor's keys. There is a small number (0.04%) of impostor's key whose unmatched bits are below 40%.

Table 2. Effect of steganography.

Cover pixel average	Stego pixel average	Overall pixel change	Absolute pixel change
173.26	173.48	0.47%	50.45%

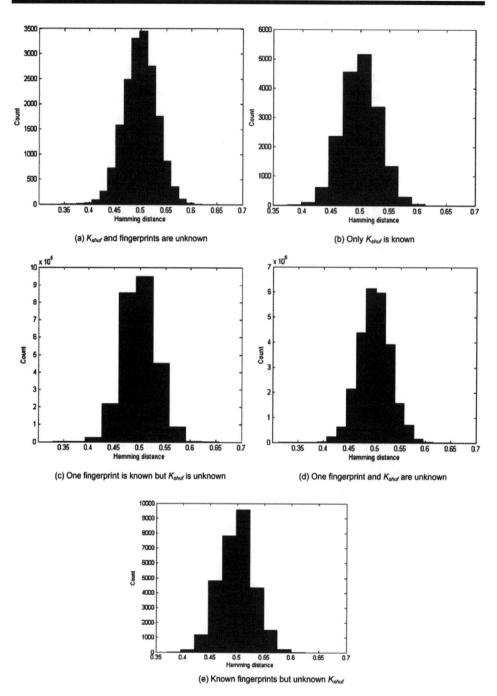

Figure 8. Hamming distances between genuine and impostor's keys. (a) K_{shuf} and fingerprints are unknown. (b) Only K_{shuf} is known. (c) One fingerprint is known but K_{shuf} is unknown. (d) One fingerprint and K_{shuf} are unknown. (e) Known fingerprints but unknown K_{shuf}.

4.6. Case 3: Shuffle Key Is Known

Now, we consider that the shuffle key (K_{shuf}) is compromised by the attacker and an attacker tries to generate the same cryptographic key using this shuffle key from the fingerprints other than genuine fingerprints. In this case, same shuffle key (Kshuf) is used to generate genuine and impostor keys. The transformation keys (Sk and Rk) are distinct for each fingerprint. In our experiment, Hamming distances are computed to measure the similarity or dissimilarity of the genuine keys and impostor's keys. The observation is shown using histogram in **Figure 8(b)**. In this case, the Hamming distances are distributed from 37.50% (minimum) to 61.33% (maximum) with a mean of 50% and standard division of 0.0309. Maximum impostors' keys (*i.e.*, 99.85% impostors' keys) differ from genuine key with the range of 40% to 60% Hamming distances. Even when the similarity of the impostor's key is maximum (*i.e.*, 62.5% bits are similar), the attacker needs to guess 96 bits (*i.e.*, 296 trials in brute force attacks) to crack the genuine key.

4.7. Case 4: Only One Genuine Fingerprint/Cancelable Template Is Known

In the proposed approach, we also consider that one of the genuine fingerprints (either the fingerprint of sender or receiver) along with transformation key is compromised by the impostor and cryptographic key is generated using one genuine cancelable template and one impostor's cancelable fingerprint template. In this case, shuffle key K_{shuf} is unknown to the attacker and a different K_{shuf} is used to generate impostor keys. Whereas, we have used one fixed transformation key (S_k) to generate genuine and impostor templates. The dissimilarity of the trial key with respect to genuine key is shown in **Figure 8(c)**. In this case, maximum 67.19% bits of the genuine key is unchanged for impostor's key and up to 65.63% bits of the genuine key is changed in impostor's key. Whereas, mean value of the Hamming distances and standard deviation of the distributions are almost similar as in other cases.

4.8. Case 5: One Genuine Fingerprint and Shuffle Key are Known

In this case, we consider that the impostor knows the cancelable template of either sender or receiver and the shuffle key (K_{shuf}) to be used in the key generation. In this case, one fingerprint along with transformation key (either S_k or R_k) and K_{shuf} are common in genuine and impostor keys. Hamming distances are computed and plotted using histogram. The histogram in **Figure 8(d)** shows that the knowledge of fingerprint and shuffle key helps the attacker slightly. In this case, minimum 31.25% bits of the genuine key is changed with respect to impostor's key and maximum 65.23% bits of the impostor's key are different from genuine key.

4.9. Case 6: Both Genuine Fingerprints Are Known But Shuffle Key is Unknown

Let us consider that attacker has complete knowledge about both genuine fingerprints or cancelable templates but no knowledge about shuffle key K_{shuf}. In this case, we have used same cancelable templates (*i.e.*, same fingerprints with same S_k and R_k) with ten different shuffle keys to generate ten impostors keys and compare the similar ity between genuine key and these impostors keys with respect to Hamming distance. The histogram in **Figure 8(e)** shows that an attacker is not able to compromise the cryptographic key even when both fingerprints/templates are known but the shuffle key is unknown. The average Hamming distance between a geneuine and impostor key is 49.99% of the length of cryptographic key. The most of the impostor's keys (99.86% of impostor's keys) are reported a dissimilarity between the range of 40% to 60% with respect to genuine key.

From our experimental results, we may conclude that an attacker is not able to generate the cryptographic key without the complete knowledge about both shuffle key and fingerprints of sender and receiver. According to our experiment, for all cases, minimum Hamming distance is 31.25% (case 5), that is, 80 bits of the impostor's key are mismatched and maximum Hamming distance is about 65.63% (case 4) that is, 168 bits of the impostor's key are needed to correct to break the genuine key. It is also observed that average Hamming distance is about 50% which means 128 bits among 256 bits of the genuine key are dissimilar on an average case. If we consider the average case, then at least 2^{128} trials are required in brute force attack to break the cryptographic key.

In order to evaluate the proposed method on the basis of execution time, we have consider the time to extract features from fingerprint, steganographic encoding, and decoding time and time for cryptographic key generation from two minutiae data sets. The computation time of our approach is given in **Table 3**, and it is observed that maximum time is required for steganographic encoding and decoding and minimum time is required for key generation from cancelable template.

5. Security Analysis

In this section, security efficacy of the proposed approach is analyzed under different conditions. In our work, both communicating parties exchange their fingerprint data after transformation and mutually agree on two secrets like stego key (K_g) and shuffle key (K_{shuf}).

5.1. Privacy of Fingerprints

We use least significant bit (LSB)-based steganography to exchange fingerprint data between sender and receiver. In our proposed approach, stego key (K_g) is used to hide fingerprint data in the pixels which are randomly chosen from cover image. For each communication, a unique synthetic fingerprint image (I) is used as cover image. An eavesdropper does not suspect the existence of the genuine fingerprint data (*i.e.*, cancelable template) due to high imperceptible stego image (Istego). The cover image is not required to decode the hidden data from stego image. An adversary, with sufficient knowledge of decoding methods, is not able to extract the correct fingerprint data from the stego images I_{stego}^S and I_{stego}^R of sender and receiver, without possessing the K_g. Our experimental result shows that only 0.47% pixels of cover image is modified, which assures that the conventional *Targeted Steganalysis* is not able to detect the existence and meaning of the hidden

Table 3. Computation time of our approach.

Operations	Time (in sec)[a]
Feature extraction	0.05
Cancelable template generation	0.002
Steganographic encoding and decoding	0.15
Cryptographic key generation	0.002
Total time	0.204

[a]The experiment is conducted with Intel® Core™ 2 Duo processor with 2.4 GHz clock speed running with Windows 7 OS.

data[42]. The statistical steganalysis-like *histogram attack*[43] detects only sequential embeddings but it does not detect the random embedding of small size message (*i.e.*, size <50% of available LSB of *I*). Another steganalysis-like *sample pair analysis*[44] is able to detect message embedding of size up to 5% of the available embedding space (LSB) of cover image while *RS analysis*[45] can detect embedding of message of size 2% to 5%.

Both sender and receiver share their fingerprint data in a transformed format, that is, cancelable template. Receiver is not able to derive the minutiae points of sender's fingerprint from cancelable template of sender and vice versa. It assures that fingerprint identity of one user is not disclosed to other user. Even if the transformation process and transformation key are disclosed, the attacker will not be able to compute the entire set of original minutiae points from the cancelable template. This can be argued as follows.

- According to the nature of transformation key, some cells may replace multiple cells and some cells may not replace any cell but they may be replaced by other cells. For example, say, there are four cells (1, 2, 3, 4) and they are replaced by (3, 2, 4, 3). Now it is found that cell 3 replaces cells 1 and 4, and cell 1 does not replace any cell. If transformation key and cancelable template both are compromised, then attacker may know the minutiae points belongs to cell 2, 3, and 4 but not of the cell 1.

- In fact, an attacker needs to know the complete information about the fingerprint image size, that is, height and width of the image to compute the minutiae points accurately.

The cancelable templates (T_{CS}, T_{CR}) are fused, and the resultant template (T_f) of fusion is shuffled and hashed to generate cryptographic key (K). Therefore, the cryptographic key is non-invertible, which confirms that the cryptographic key does not leak any information about the fingerprints of users.

5.2. Security of Cryptographic Key

In our approach, the cryptographic key is generated from the combination of two fingerprints of sender and receiver. The key is not shared by them but generated at their end separately. There is no need to store the key for the use in decryption. So the key is secured from any attack. An attacker needs to compromise either (stego key K_g, sender's stego image I_{stego}^S, receiver's stego image I_{stego}^R shuffle key K_{shuf}) or (sender's fingerprint F_S, receiver's fingerprint F_R, transformation key of sender S_K, transformation key of receiver R_K, K_{shuf}) to generate the genuine cryptographic key (K). Otherwise, the following conditions arise for different attacks.

5.2.1. Known Stego Key Attack

It means that the stego key Kg is compromised by eavesdropper who eavesdrops both stego images I_{stego}^S and I_{stego}^R. Next, as the shuffle key K_{shuf} is unknown, he has to guess the K_{shuf} which is of 200 bits. It requires a trial of 2200 to break the K_{shuf} using brute force attacks. Indeed, without K_{shuf}, it is almost impossible to compute the original cryptographic key even if both fingerprints data are known.

5.2.2. Known Shuffle Key Attack

It happens when the token is stolen or lost and the attacker gets access to the shuffle key (K_{shuf}) but he has no knowledge about the fingerprint data of genuine sender and receiver. In this case, the attacker has to guess 50 + 50 = 100 minutiae points for two fingerprints. Otherwise, he has to compute the key using impostor's fingerprints which may be available to him. As the fingerprint is unique for a person, there is a rare chance to get the same fingerprint from impostor's fingerprint database. Even in our experiment, there is no collision between genuine cryptographic key and impostor's cryptographic key when the K_{shuf} is known. The shuffle key is updated in every session, which guarantees that it is impractical to compute the K_{shuf} of previous session or the same of future sessions.

5.2.3. Known Fingerprint Attack

In this attack, there are two scenarios that may occur. In the first scenario, only the fingerprints of both parties are compromised but transformation key is not known by attacker. In the second scenario, the attacker has complete knowledge about the fingerprints along with transformation key, that is, the attacker is able to generate cancelable templates from fingerprints (of both sender and receiver) using the transformation keys and algorithm of transformation.

In the first case, attacker is not able to generate even cancelable template from the knowledge of genuine fingerprint. An adversary should know two parameters (*i.e.*, fingerprint and transformation key) along with transformation function (f_c), to generate cancelable template from fingerprint template. Now according to the condition, adversary knows F_S, F_R, and f_c but does not have any idea about transformation keys (S_K, R_K). Similarly, to generate the cryptographic key, an adversary should know another parameter which is shuffle key (K_{shuf}) also. The key generation function f_k is known as it is public, but all other three parameters are to be compromised by the attacker.

$$K = f_k\{f_c(F_S, S_K), f_c(F_R, R_K), K_{shuf}\}$$
$$= f_k\{f_c(F_S, ?), f_c(F_R, ?), ?\}$$

where K is the cryptographic key and '?' marks represent those parameters which are unknown to the adversary.

In the second case,

$$K = f_k\{f_c(F_S, S_K), f_c(F_R, R_K), K_{shuf}\}$$
$$= f_k\{T_{CS}, T_{CR}, ?\}$$

only one parameter (K_{shuf}) which is unknown to the adversary. Then, the attacker has to break the K_{shuf} which needs 2200 trials in *brute force* attack.

5.2.4. Known Key Attack

In this case, we consider that the shuffle key K_{shuf} and or cryptographic key (K) are compromised by the attacker. Our proposed approach assures that compromise of (K_{shuf_i} or K_i) of *i*th session does not affect the previous or future session as $K_{shuf_i} \neq K_{shuf_{i+1}} \neq K_{shuf_{i-1}}$ or $K_i \neq K_{i+1} \neq K_{i-1}$. In our approach, both K_{shuf} and

K are updated session wise. Initially, K_{shuf} is randomly generated and time to time it is updated with the help of F_{code}. Cryptographic key (K) is revocable and is used as a session key for symmetric cryptography. When the session is over, the current session key is destroyed.

5.2.5. Resists Replay Attack

Our approach can prevent replay attack using session key. In every session of communication, a unique session key is used to establish a secure communication between sender and receiver and the session key is destroyed after the session. Our proposed approach is able to generate 100! × 100! different cryptographic keys from two finger-print biometric traits. If an eavesdropper wants to make replay attack using a message previously transmitted by legal users, then it will make no sense to the legal user as the cryptographic key is changed. Even when stego image is used for replay attack, eavesdropper is not able to decode the stego image without the stego key.

5.2.6. Resists Man-in-Middle Attack

In our approach, fingerprints of communicating parties are transmitted over communication channel using data hiding scheme. If the man-in-middle (MiM) eavesdrops the stego image and is able to decode the hidden data by any means, then the MiM requires the perfect knowledge of secret shuffle key (K_{shuf}). Otherwise, he is not able to generate the genuine key (K) and, as a result, is not able to decrypt the ciphertext sent by genuine sender.

Case 1: The *MiM* is able to receive messages exchanged between sender and receiver. In the worst case, we also consider that the *MiM* also knows the stego key K_g.

$$T'_{CS} = D_s\left(I^S_{stego}, K_g\right)$$
$$T'_{CR} = D_s\left(I^R_{stego}, K_g\right)$$
$$K_{est} = f_k\left(T'_{CS}, T'_{CR}, K'_{shuf}\right)$$

where Ds is the data unhiding function, T'_{CS} and T'_{CR} are the cancelable templates of sender and receiver, respectively, extracted by *MiM*, whereas K'_{shuf} is the estimated shuffle key by the *MiM*. If $K_{shuf} \neq K'_{shuf'}$, then the estimated cryptographic key Kest does not match with the genuine key K.

Case 2: The *MiM* receives the messages I^S_{stego} and I^S_{stego} sent by sender and receiver, respectively, but sends a stego image, where some random data is encoded by the *MiM* to both parties. The *MiM* acts like receiver to genuine sender and behaves like sender to genuine receiver. The *MiM* is able to receive the encrypted message sent by any genuine party.

$$K^S = f_k\left(T_{CS}, T_{\min}, K_{shuf}\right)$$
$$K^R = f_k\left(T_{CR}, T_{\min}, K_{shuf}\right)$$
$$K^S_{est} = f_k\left(T'_{CS}, T_{\min}, K'_{shuf}\right)$$
$$K^R_{est} = f_k\left(T'_{CR}, T_{\min}, K'_{shuf}\right)$$

where the cancelable template of *MiM* is T_{mim} and K^S, K^R are the keys generated by sender and receiver, respectively, and used in encryption. K^S_{est} and K^R_{est} are the estimated key by the *MiM* and used for decryption.

Now, the *MiM* tries to decrypt the encrypted messages (*i.e.*, Cipher$S$ of sender, Cipher$R$ of receiver) received from sender and or receiver in the following ways,

$$Cipher_S = E_{K^S}\left(message_S\right)$$
$$Cipher_R = E_{K^R}\left(message_R\right)$$
$$message^m_S = D_{K^S_{est}}\left(Cipher_S\right)$$
$$message^m_R = D_{K^R_{est}}\left(Cipher_R\right)$$

In the estimation of cryptographic keys $\left(K^S_{est}, K^R_{est}\right)$, MiM trials the k_{shuf} which produces different keys, (*i.e.*, $\left(K^S_{est} \neq K^S\right)$, and $\left(K^S_{est} \neq K^R\right)$), which do not decrypt the ciphertexts correctly (*i.e.*, $message_S \neq message^m_S$ and $message_R \neq message^m_R$). In this way, our approach resists man-in-middle attack.

6. Conclusions

Cryptographic key generation and subsequently its maintenance are the two important issues in traditional cryptography. A cryptographic key should be generated in such a way that it is hard enough to guess and then it should be managed without any overhead of users. This work addresses these issues and propose a novel approach to generate random cryptographic key using fingerprint biometric of sender and receiver.

In our works, the privacy and security of fingerprint data are provided with cancelable template. Also, we propose a protocol with which key can be revoked thus addressing the limitation of irrevocability property of biometric trait. More significantly, there is no need to store the key, prior to communication. In fact, our protocol adds more security allowing to generate different keys in different sessions. The proposed crypto-biometric system is resilient to many attacks such as known key attacks, replay attack, man-in-middle attacks, etc. Our proposed approach thus provides an effective solutions where we need a session-based cryptographic key during message transmission over an insecure network channel.

Competing Interests

The authors declare that they have no competing interests.

Source: Barman S, Samanta D, Chattopadhyay S. Fingerprint-based crypto-biometric system for network security [J]. Eurasip Journal on Information Security, 2015, 2015(1): 1–17.

References

[1] W Stallings, *Cryptography and Network Security: Principles and Practice, 5e.* (Prentice Hall, 2010).

[2] Advance Encryption Standard (AES), *Federal Information Processing Standards Publication 197.* (United States National Institute of Standards and Technology (NIST), November 26, 2001).

[3] K Mitnick, W Simon, S Wozniak, *The Art of Deception: Controlling the Human Element of Security*. (Wiley, New York, 2002).

[4] DV Klein, in *Proceedings of the 2nd USENIX Security Workshop (Portland)*. Foiling the cracker: A survey of, and improvements to, password security, (1990), pp. 5–14.

[5] F Hao, R Anderson, J Daugman, Combining Crypto with Biometrics Effectively. IEEE Trans. Comput. **55**(9), 1081–1088 (2006).

[6] U Uludag, S Pankanti, S Prabhakar, AK Jain, Biometric Cryptosystems: Issues and Challenges. Proc. IEEE. **92**(6), 948–960 (2004).

[7] D Maltoni, D Maio, AK Jain, S Prabhakar, *Handbook of Fingerprint Recognition*. (Springer-Verlag, New York, 2003).

[8] K Nandakumar, A Jain, S Pankanti, Fingerprint-based fuzzy vault: Implementation and performance. IEEE Trans. Inf. Forensics Secur. **2**(4), 744–757 (2007).

[9] N CT Charles, DJ Kiyavash, in *Proceedings of the 2003 ACM SIGMM workshop on Biometrics methods and applications*. Lin, Secure smartcardbased fingerprint authentication (ACM New York, NY, USA, 2003), pp. 45–52.

[10] S Yang, I Verbauwhede, in *Proceedings (ICASSP'05). IEEE International Conference on Acoustics, Speech, and Signal Processing*, Vol. 5. Automatic secure fingerprint verification system based on fuzzy vault scheme (IEEE Philadelphia, Pennsylvania, USA, 2005), pp. v/609–v/612.

[11] S Dutta, A Kar, BN Chatterji, NC Mahanti, in *Proc. Adv. Concepts Intell. Vis. Syst.*, LNCS 5259. Network Security Using Biometric And Cryptography (Springer Berlin Heidelberg, 2008), pp. 38–44.

[12] A Jagadeesan, K Duraiswamy, Secured Cryptographic Key Generation from Multimodal Biometrics: Feature Level Fusion of Fingerprint and Iris. Int. J. Comput. Sci. Inform. Secur. **7**(2), 28–37 (2010).

[13] A Jagadeesan, T Thillaikkarasi, K Duraiswamy, Cryptographic Key Generation from Multiple Biometrics Modalities: Fusing Minutiae with Iris Feature. Int. J. Comput. Appl. **2**(6), 16–26 (2010).

[14] F Monrose, MK Reiter, Q Li, S Wetzel, in *Proceedings of IEEE Symposium on Security and Privacy*. Cryptographic key generation from voice (IEEE Computer Society Washington, DC USA, 2001), pp. 202–213.

[15] H Feng, CC Wah, Private key generation from on-line handwritten signatures. Inform Manag. Comput. Secur. **10**(4), 159–164 (2002).

[16] B Chen, V Chandran, in *Proceedings of 9th Biennial Conference of the Australian Pattern Recognition Society on Digital Image Computing Techniques and Applications*. Biometric Based Cryptographic Key Generation from Faces (Glenelg Australia, 2007), pp. 394–401.

[17] SVK Gaddam, M Lal, Efficient Cancellable Biometric Key Generation Scheme for

Cryptography. Int. J. Netw. Secur. **11**(2), 57–65 (2010).

[18] AK Jain, K Nandakumar, A Nagar, in *Security and privacy in biometrics*. Fingerprint Template Protection: From Theory to Practice (Springer London, 2013), pp. 187–214.

[19] NK Ratha, S Chikkerur, JH Connell, RM Bolle, Generating Cancellable Fingerprint Templates. IEEE Trans. Pattern Anal. Mach. Intell. **29**(4), 561–572 (2007).

[20] A Ross, AK Jain, Information fusion in biometrics. Pattern Recognit. Lett. **24**, 2115–2125 (2003).

[21] A Bodo, Method for Producing a Digital Signature with Aid of Biometric Feature. German Patent DE 4243908A1 (1994).

[22] JH RM Bolle, S Connell, NK Pankanti, AW Ratha, *Senior, Guide to Biometrics.* (Springer-Verlag, New York, 2003).

[23] NK Ratha, JH Connell, R Bolle, Enhancing Security and Privacy in Biometric-Based Authentication System. IBM Syst. J. **40**(3), 614–634 (2001).

[24] S Kanade, D Camara, E Krichen, D Petrovska-Delacre˜ taz, B Dorizzi, Evry F, in *Proceedings of 6th Biometrics Symposium (BSYM 2008).* Three Factor Scheme for Biometric-Based Cryptographic Key Regeneration Using Iris (Tampa, Florida, USA, 2008), pp. 59–64.

[25] S Kanade, D Petrovska-Delacre˜ taz, B Dorizzi, in *Proceedings of Fourth IEEE International Conference on Biometrics: Theory Applications and Systems (BTAS), Washington, DC, USA, 2010.* Generating and sharing biometrics based session keys for secure cryptographic applications, (2010), pp. 1–7.

[26] A Jain, U Uludag, Hiding Fingerprint Minutiae in Images. in Proceedings of Third Workshop on Automatic Identification Advanced Technologies (AutoID), (Tarrytown, New York. USA, 97–102 (2002).

[27] A Jain, U Uludag, in *Proceedings of 16th International Conference on Pattern Recognition,* vol. 3. Hiding a Face in a Fingerprint Image (Canada, 2002), pp. 756–759.

[28] A Jain, U Uludag, Hiding Biometric Data. IEEE Trans. Pattern Anal. Mach. Intell. **25**, 1494–1498 (2003).

[29] N Agrawal, M Savvides, in *Proceedings of IEEE Computer Society Conference on Computer Vision and Pattern Recognition*. Biometric data hiding: A 3 factor authentication approach to verify identity with a single image using steganography, encryption and matching (Miami Beach Florida, 2009), pp. 85–92.

[30] A Juels, M Wattenberg, in *Proc. 6th ACM Conf. Computer and Communications Security*, ed. by G Tsudik. A fuzzy commitment scheme (ACM New York, NY, USA, 1999), pp. 28–36.

[31] C Rathgeb, A Uhl.Context-based biometric key generation for Iris. IET Comput. Vis. **5**(6), 389–397 (2011).

[32] C Yao-Jen, W Zhang, T Chen, in *Proceedings of IEEE International Conference on*

Multimedia and Expo (ICME'04), Taipei, 2004, Vol. 3. Biometrics-based cryptographic key generation (IEEE, 2004), pp. 2203–2206.

[33] N Lalithamani, KP Soman, in *the 2nd International Conference on Computer Science and Information Technology*. Towards Generating Irrevocable Key for Cryptography from Cancelable Fingerprints (Beijing China, 2009), pp. 563–568.

[34] N Lalithamani, KP Soman, Irrevocable Cryptographic Key Generation from Cancelable Fingerprint Templates: An Enhanced and Effective Scheme. Eur. J. Sci. Res. **31**(3), 372–387 (2009).

[35] V Lokeswara Reddy, A Subramanyam, P Chenna Reddy, Implementation of LSB Steganography and its Evaluation for Various File Formats. Int. J. Adv. Netw. Appl. **02**(05), 868–872 (2011).

[36] A Ross, K Nandakumar, AK Jain, *Handbook of Multibiometrics*. (Springer-Verlag, Berlin, Germany, 2006).

[37] Fingerprint Verification Competition FVC2000, [Online]. Available: http://bias.csr.unibo.it/fvc2000.

[38] Fingerprint Verification Competition FVC2002, [Online]. Available: http://bias.csr.unibo.it/fvc2002.

[39] Fingerprint Verification Competition FVC2004, [Online]. Available: http://biometrics.cse.msu.edu/fvc04db/index.html.

[40] R Cappelli, A Erol, D Maio, D Maltoni, in *Proceedings of 15th International Conference on Pattern Recognition, 2000,* vol. 3. Synthetic Fingerprint Image Generation, (2000), pp. 475–478.

[41] C Watson, M Garris, E Tabassi, C Wilson, M McCabe, S Janet, Ko K, *User's Guide to NIST Biometric Image Software (NBIS)*. (National Institute of Standards and Technology, Gaithersburg, MD, 2007).

[42] A Rocha, W Scheirer, T Boult, S Goldenstein, Vision of the unseen: Current trends and challenges in digital image and video forensics. ACM Comput. Surv. **43**(4), 26:1–26:42 (2011).

[43] A Westfeld, A Pfitzmann, in *Proc. Information Hiding, 3rd Int'l Workshop*. Attacks on Steganographic Systems (Springer Verlag, 1999), pp. 61–76.

[44] S Dumitrescu, Wu Xiaolin, N Memon, in *International Conference on Image Processing, (ICIP 2002)*, vol.3, no. 24-28. On steganalysis of random LSB embedding in continuous-tone images (Rochester, New York, USA, 2002), pp. 641–644.

[45] J Fridrich, M Goljan, Du Rui, Detecting LSB steganography in color, and gray-scale images. MultiMedia IEEE. **8**(4), 22–28 (2001).

Chapter 23
A Secure User Authentication Protocol for Sensor Network in Data Capturing

Quan Zhou[1,2], **Chunming Tang**[1,2], **Xianghan Zhen**[3], **Chunming Rong**[4]

[1]Key Laboratory of Mathematics and Interdisciplinary Sciences of Guangdong Higher Education Institutes, Guangzhou University, Guangzhou, China
[2]School of Mathematics and Information Science, Guangzhou University, Guangzhou, China
[3]College of Mathematics and Computer Science, Fuzhou University, Fuzhou, China
[4]Faculty of Science and Technology, University of Stavanger, Stavanger, Norway

Abstract: Sensor network is an important approach of data capturing. User authentication is a critical security issue for sensor networks because sensor nodes are deployed in an open and unattended environment, leaving them possible hostile attack. Some researchers proposed some user authentication protocols using one-way hash function or using biometric technology. Recently, Yel *et al.* and Wenbo *et al.* proposed a user authentication protocols using elliptic curves cryptography. However, there are some security weaknesses for these protocols. In the paper, we review several proposed user authentication protocols, with a detail review of the Wenbo *et al.*'s user authentication protocol and a cryptanalysis of this protocol that shows several security weaknesses. Furthermore, we propose a secure user authentication protocol using identity-based cryptography to overcome those weaknesses. Finally, we present the security analysis, a comparison of security, computation, and performance for the proposed protocols, which shows that this user authentication protocol is more secure and suitable for higher security WSNs.

Keywords: Data Capturing, Wireless Sensor Networks, User Authentication, Identity-Based Cryptography

1. Introduction

With the application of big data, there are some base manipulation processes: data capturing, data transport, data storage, data extraction & integration, data analysis & interpretation and data application. In the data capturing, using all kinds of devices and methods to collect data, such as smart devices, sensors, Web. So there are three important approaches of data capturing: Internet, Internet of Things (IoT) and sensor network[1]. Wireless Sensor networks (WSNs) is an open environment distributed network, which is an important approach of data capturing for big data. Nevertheless, with the application of dig data, the requirement of real-time data from WSNs is increasing highly. In some situations the gateway impossibly does force a user to access the sensor node directly. In such case the security and reliability to inquire and data disseminate are very important. Only when every client (remote sensor node, remote user) in the WSNs proves his/her identity can he/she be allowed to join the WSNs and access to resource, such as real time data. Thus, a key security requirement for WSNs is user authentication[2]-[5].

In 2004, Sastry *et al.*[2] proposed a security scheme using access control lists (ACL) for IEEE 802.15.4 networks in the gateway node. An ACL would be maintained in gateway node and sensor nodes. Watro *et al.*[6] proposed a user authentication protocol using RSA and Differ-Hellman algorithm, but which was open to hostile attack by a user masquerading.

In 2005, Benenson *et al.*[7] proposed a user authentication protocol based on elliptic curve discrete logarithm problem (ECDLP) to handle the sensor node capture attack, which relied on a trusted third party.

In 2006, Wong et al.[8] proposed a dynamic user authentication scheme for WSNs based on a light-weight strong password using hash function, which included three phases: registration phase, login phase and authentication phase. Nonetheless, Tseng et al.[9] and Das[10] pointed out that this protocol had some weaknesses in protecting against replay attack, forgery attack, stolen-verifier attack, sensor node revealing and exposing the password to the other node and no updating user's password. In 2007, Tseng et al.[9] proposed an enhanced user authentication protocol by adding an extra phase (password changing phase) on Wong et al.'s phases. However, in 2008 Ko[11] showed the Tseng et al.'s protocol was still insecure and did not provide mutual authentication.

In 2009, Das[10] proposed a two-factor user authentication protocol based on password and smart card against stolen-verifier attack. Nevertheless, Nyang et al.[12] showed there were some security weaknesses in offline-password guessing attacks.

In 2010, Vaidya et al.[13] demonstrated the Tseng et al.'s protocol, Wong et al.'s protocol and Ko's protocol were still not strong enough to protect again replay attack, stolen-verifier attack and man-in-the-middle attack. Khan et al.[14][15] pointed out the Das's protocol did not provide mutual authentication, and against by passing attack and privileged insider attack. Moreover, Chen et al.[16] also demonstrated the Das's protocol did not provide mutual authentication between the gateway node and the sensor node. And Chen et al. proposed a more secure and robust two-factor user authentication scheme for WSNs.

In 2011, Yeh et al.[17] found that the Chen et al.'s protocol failed to provide a secure method for updating password and insider attack. And Yeh et al. proposed a new user authentication scheme for WSNs using elliptic curve cryptography (ECC). Unfortunately, Han[18] found this protocol still had some weaknesses: no mutual authentication, no key agreement between the user and the sensor node, and no prefer forward security. Meanwhile, Yuan et al.[19] proposed a biometric-based user authentication for WSNs using password and smart card in 2010. Unfortunately, in 2011 Yoon et al.[20] showed the integrity problem of the Yuan et al.'s protocol and proposed a new biometric-based user authentication scheme without using password for WSNs.

In 2012, Ohood et al.[21] pointed out Yoon et al.'s scheme still had some drawbacks, such as no key agreement, no message confidentiality service, no providing against DoS and node compromise attack. Moreover, Ohood et al.[22] proposed an efficient biometric authentication protocol for WSNs.

Recently, Wenbo et al.[23] in 2013 proposed a new user authentication protocol for WSNs using elliptic curve cryptography to overcome the security weaknesses of Yeh et al.'s protocol. Although they suggested security improvements of Yeh et al.'s protocol, there were some security weaknesses in their protocol, e.g. no mutual authentication between the user and sensor node, no protecting against insider attack, forgery attack and DoS (denial of service) attack.

To address all of the issues raised in the above studies, we propose a secure user authentication protocol using identity-based cryptography on the basis of our previous studies to trusted management and trusted architecture of WSNs[24]–[26]. Our proposal addresses the key security issues.

The remainder of this paper is organized as follows: in Section Related works, we review the Wenbo et al.'s protocol and a detail cryptanalysis; next we present our user authentication protocol based on identity-based encryption in Section Proposed protocol; in Section Security and performance analysis, a security and performance analysis of the related protocol is presented; in Section Conclusion, we provide some conclusion remarks.

2. Related Works

2.1. Notation

In **Table 1**, some notations used throughout this paper and their corresponding definitions are shown.

2.2. Review of Wenbo's Scheme

In the Wenbo's protocol, the gateway GW held two master keys (x and y). And it was assumed that the gateway and the sensor nodes shared a long-term common secret key, $SK_{GS} = h(S_n\|y)$. The Wenbo's protocol involves the registration

Table 1. Notations.

Symbol	Define
p	A big prime number
F_p	A finite field
E	An elliptic curve in F_p with a large order
P	A point on elliptic curve E with order q that is a big prime number
U	A remote user
ID	An identity
PS	A user password
GW	Gateway of WSNs
S_n	Sensor node of WSNs
Q_{id}	Public key of id
d_{id}	Private key of id
P_{set}	A system parameter set of PKG
$h(.)$	A public secure one-way hash function
$H_1(.)$	A public function: $\{0,1\}^* \to G_1$, the G_1 is a group $G_1 = \{NP | n \in \{0,1...q-1\}\}$
$H_2(.)$	A public function $G_2 \to \{0,1\}^*$, G_2 is subgroup with an order q of $GF(p^2)^*$
$f(.)$	A public function: $G_1 \to \{0,1\}^*$
$ê(.)$	An admissible pairing: $G_1 \times G_1 \to G_2$
$E_k(m)$	Encrypt message m with key k
$D_k(c)$	Decrypt message c with key k
$\|$	A string concatenation operation
$\oplus$	A XOR operation

phase, login phase, authentication phase and password update phase, which can be briefly described as follows.

2.2.1. Registration Phase

In this phase, a user U submits his/her ID_u and a hash of his/her password to GW via a secured channel. Then, GW issues a license to U. The steps are described as follows.

Step 1: U → GW: $\{ID_u, PS'\}$.

U enters an identity, selects a random number br and a password PS. And U computes $PS' = h(PS \oplus br)$. Then U sends message $\{ID_u, PS'\}$ to GW via a secured channel.

Step 2: GW → a smart card of U: $\{Bu, Wu, h(.)\}$.

GW computes $Ku = h(ID_u \| x) \times P$, $Bu = h(ID_u \oplus PS')$, and $Wu = Bu \oplus Ku$, where x is a master key of GW. Then the GW stores (Bu, Wu) into a smart card and sends it to U.

2.2.2. Login Phase

When U access Sn, U needs enter his ID_u and PS. And the smart card must confirm the validity of U via the following steps.

Step 1: Validate U.

The smart card check whether $Bu = h(ID_u \oplus h(PS \oplus br))$ hold. If the answer is no, the U's identification validation fails and the smart card will terminate this request.

Otherwise, the smart card continues to execute the next step.

Step 2: U's smart card generates a random number r_u, calculates X and a. $X = r_u \times P$, $X' = r_u \times (Bu \oplus Wu)$, and $a = h(ID_u \| X \| X' \| T_u)$, where T_u is the curren time of U's system.

Step 3: U → Sn: $\{ID_u, X, T_u, a\}$.

The $\{ID_u, X, T_u, a\}$ is submitted to Sn via public channel.

2.2.3. Authentication Phase

The authentication phase includes: Sn checking the validity of the request message of U, GW authenticating Sn and U, Sn authenticating GW and U, U authenticating Sn and GW.

Sn checks the validity of the request message of U When receiving the login message $\{ID_u, X, T_u, a\}$ at time T', Sn checks and generates request message which is sent to GW for authentication. Sn executes the following steps.

Step 1: Checks T_u.

Sn checks if $(T'-T_u) \leqq \Delta T$ holds, where ΔT denotes the expected time interval for transmission delay. If the an-swer is yes, the validity of T_u can be assured, and Sn executes the next step. Otherwise Sn rejects the login request.

Step 2: Picks a random number r_s and calculates Y and b.

$Y = r_s \times P$, $b = h(SK_{GS}||ID_u||X||T_u||a||ID_{Sn}||Y||T_s)$, where Ts denotes the current request time of the Sn system.

Step 3: Sn → GW: $\{ID_u, X, T_u, a, ID_{Sn}, Y, T_s, b\}$.

The $\{ID_u, X, T_u, a, ID_{Sn}, Y, T_s, b\}$ is submitted to GW via public channel.

GW authenticates Sn and U When receiving the request message that sent by Sn at time T'', GW checks and validates Sn and U, and generates the response message that will be sent to Sn. GW executes the following steps.

Step 1: Validates if T_s and T_u.

GW checks whether $(T''-T_s) \leqq \Delta T$ and $(T''-T_u) \leqq \Delta T$ hold. If the answer is yes, the validity of T_s and T_u can be assured and GW executes the next step. Otherwise GW rejects this request message.

Step 2: Calculates b^*.

$$b^* = h\left(SK_{GS}||ID_u||X||T_u||a||ID_{Sn}||Y||T_s\right)$$

A Secure User Authentication Protocol for Sensor Network in Data Capturing

Step 3: Confirms whether $b = b^*$ and validates Sn.

GW checks if $b = b^*$ holds. If the answer is yes, GW accepts this request message and executes the next step. Otherwise, GW rejects this request message.

Step 4: Calculates X' and a^*.

$X' = h(ID_u||x) \times X$, $a^* = h(ID_u||X||X'||T_u)$, where x denotes a master key of GW.

Step 5: Confirms whether $a = a^*$.

GW checks if $a = a^*$ holds. If the answer is yes, GW accepts this request message and executes the next step. Otherwise, GW rejects the request message.

Step 6: Calculates y and l.

$$y = h\left(SK_{GS}\|ID_u\|X\|T_u\|a\|ID_{Sn}\|Y\|T_G\right)$$

$$l = h\left(ID_u\|X\|X'\|T_u\|Y\|T_s\right)$$

where TG denotes the current response time of GW.

Step 7: $GW \rightarrow S_n: \{T_G, y, l\}$

The $\{T_G, y, l\}$ is submitted to Sn via public channel.

Sn authenticates GW When receiving the response message that sent by GW at time T''', Sn checks and validates GW, and generates the message that will be sent to U. Sn executes the following steps.

Step 1: Validates T_G.

Sn checks if $(T'''-T_G) \leqq \Delta T$ holds. If the answer is yes, the validity of T_G can be assured and Sn executes the next step. Otherwise Sn rejects the response message.

Step 2: Calculates y*.

$$y^* = h\left(SK_{GS} \| ID_u \| X \| T_u \| a \| ID_{Sn} \| Y \| T_G\right)$$

Step 3: Validates y.

Sn checks if y = y* holds. If the answer is yes, Sn accepts this response and executes the next step. Otherwise, Sn rejects this response message.

Step 4: Calculates KSU, g and session key sk

$$k_{SU} = r_s \times X, g = h\left(Y \| T_s \| 1 \| K_{SU}\right), sk = h\left(X \| Y \| K_{US}\right)$$

Step 5: $S_n \to U$: {Y, T_s, 1, g}

The {Y, T_s, 1, g} is submitted to U via public channel.

U authenticates GW and Sn When receiving the response message that sent by Sn at time T'''', U checks and validates GW and Sn. U executes the following steps.

Step 1: Validates Ts.

U checks if (T''''-Ts) ≦ ΔT holds. If the answer is yes, the validity of TS can be assured and U executes the next step. Otherwise, U rejects the response message.

Step 2: Calculates KUS, l* and g*.

$$\begin{aligned}K_{SU} &= r_u \times Y, l^* \\ &= h\left(ID_u \| X \| X' \| T_u \| Y \| T_s\right), \text{and } g^* \\ &= h\left(Y \| T_s \| 1 \| K_{SU}\right).\end{aligned}$$

Step 3: Confirms l and g.

U checks if $l = l^*$ and $g = g^*$ hold. If the answer is yes, U accepts the response message and executes the next step. Otherwise, U rejects the response message.

Step 4: Calculates session key sk.

$$sk = h\left(X\|Y\|K_{US}\right).$$

2.2.4. Password Update Phase

When U wants to update his/her old password, U and the smart card execute the following steps.

Step 1: U inserts his/her smart card into the smart terminal and enters his/her identify ID_u, the old password PS and the new password PSn.

Step 2: The smart card calculates PS′ = h(PS⊕br), and checks whether Bu = h(ID_u⊕PS′) holds. If it does not hold, the smart card stops U's request. Otherwise, the smart card continues to compute Ku = h(ID_u∥PS′)⊕Wu, PSn′ = h(PS_n⊕br), Bu′ = h(ID_u⊕PSn′) and Wu′ = Bu′ ⊕Ku. Finally, the smart card replaces (Bu, Wu) with (Bu′, Wu′).

2.5. Cryptanalysis of Wenbo's Protocol

2.5.1. Security Requirements in WSNs

(1) Secure user authentication in WSNs should be based on full mutual authentication.

(2) Secure user authentication in WSNs should resist masquerade, replay, forgery and DoS attacks.

(3) Secure user authentication in WSNs should resist internal attack (com-

promise attack).

(4) Secure user authentication in WSNs with smart card should reject Virus Injection attack.

2.5.2. No Full Mutual Authentication

Because Wenbo's protocol does not authenticate U during the authentication phase (Sn checks the validity of the request message of U), a malicious user can attack Sn and GW by means of forging. The attack could be accomplished as follows:

(1) The attacker sends a forging message $\{ID_a, X_a, Tu_a, a_a\}$ to Sn.

(2) Sn sends a message $\{ID_a, X_a, Tu_a, a_a, ID_{Sn}, Y, T_s, b\}$ to GW for authenticating the user when receiving the forging message.

During the above process, since Sn does not authenticate the user, Sn directly generates authenticating request message for GW to authenticate the user. When GW receives this request message, GW can finish the process from Step 1 to Step 4 of authentication phase (GW authenticates Sn and U). This is because there is no mechanism for Sn to be assured that U is real user of WSNs. Thus, the Wenbo's protocol does not provide mutual authentication between U and Sn. There is no full mutual authentication between Sn and U. This protocol cannot reject DoS attack to Sn and GW.

2.5.3. No Protection against Forgery Attack

Because the confidential information (Bu, Wu) is not encrypted to be stored, the attacker can masquerade as a legal user U. In the case that an attacker steals the (Bu, Wu) from the smart card via some a Virus or a Trojan in the user terminal, he/she maybe try to impersonate user U to access resource in WSNs. The attack can be accomplished via the following means.

(1) The attacker steals the (Bu, Wu)} via some methods, such as Virus soft-

ware, Trojan.

(2) The attacker could compute Ku = Bu⊕Wu and gain the secret Ku.

(3) The attacker picks a random number Ru.

(4) The attacker could computes $X_a = R_u \times P$, $X_a' = R_u \times K_u$, and $a_a = h(ID\|X_a\|X_a'\|T_a)$ because the point P on elliptic curve E is public.

(5) The attacker sends the request message $\{ID_u, X_a, T_a, a_a\}$ to the Sn via public channel.

(6) Sn can finish the authentication phase processes. And GW also can accomplish the authentication phase processes.

After GW and Sn finish to authenticate, the attacker can gains the session key sk. The attacker continues to access Sn. Thus, the Wenbo's protocol does not provide sufficient protection against forgery attack.

2.5.4. No Protection against Insider Attack

In the Wenbo's protocol, U uses a single password for accessing Sn. It is convenient for a user. Nevertheless, if the system manager or a privileged user of GW obtains (Bu, Wu) of U during U registration phase, he/she maybe try to impersonate U to access the resource in WSNs. The attacking processes are the same as the forgery attack. Thus, the Wenbo's protocol does not provide sufficient protection against an insider attack on GW by a privileged user.

2.5.5. No Protection against Compromise Attack

In the Wenbo's protocol, the gateway and the sensor nodes shared a long-term common secret key SKGS. If an attacker captures some a sensor node, he/she can attain the shared secret key SKGS via some methods since the SKGS is not encrypted. So it is very easy to impersonate a sensor node in WSNs. Even the attacker may make many sensor nodes to impersonate the sensor nodes of in WSNs.

3. Proposed Protocol

To solve the security weaknesses of the Wenbo's protocol, we propose a new user authentication protocol for WSNs using identity-based cryptography. First, we review the fundamentals of identity-based cryptography, and then survey the identity-based cryptography which is suitable for our design of a secure authentication protocol for WSNs. In the proposed protocol, GW integrates the trusted and reputation scheme[24][26]. The proposed five phases are described in detail later.

3.1. Identity-Based Cryptography

Identity-based cryptography is a kind of public-key based scheme. The public key is the unique identity of the user. The private key is generated by a third party called a Private Key Generator (PKG) with its master secret and user's identity. In the identity-based cryptography system, firstly, the PKG must create a master public key and a master private key. Then any user may use this master public key and also use the user's identity to generate the user's public. The user's private key is created by the PKG with the user's identity.

For every two parties using in identity-based cryptography, it is easy to calculate a shared secret session key between them using its own private key and public key of another party. For example, a sensor node Sn with public key Q_{Sn} and private key d_{Sn}, and a user U with public key Q_u and private key du can calculate their shared secret session key by computing key = $ê(Q_u, d_{Sn})$ = $ê(d_u, Q_{Sn})$.

In the proposed protocol, GW is the PKG. GW selects a random number s ∈ Z* that is kept secret. GW computes K_{pub} = s × P. This public-private key pair < K_{pub}, s > is the master key pair of GW. And GW computes Q_{GW} = H1 (ID_{GW}), d_{GW} = s × Q_{GW}. QGW is the authentication public key of GW. d_{GW} is the authentication private key of GW.

3.2. Registration Phase

In the registration phase, Sn and U register to GW. The processes are the follow as.

3.2.1. Sensor Node Registration

In the WSNs, all sensor nodes must register to GW before being deployed. GW creates a private key for every sensor node. And the system parameters P_{set}, the public functions and the private key are stored in the sensor node. GW completes the following steps.

Step 1: Creates the public key Q_{Sn}

GW uses the identity ID_{Sn} of Sn to generate the public key Q_{Sn}, $Q_{Sn} = H_1(ID_{Sn})$.

Step 2: Generates the private key d_{Sn}.

GW uses the master key s and the public key Q_{Sn} to create the private key d_{Sn}, $d_{Sn} = s \times Q_{Sn}$

Step 3: Installs system parameters, public functions and private key of Sn.

GW installs the system parameters P_{set}, d_{Sn} and other public functions into Sn. That is to say, $\{P_{set}, d_{Sn}, h(.), f(.), H_1(.), e(.)\}$ is stored into the Sn.

3.2.2. User Registration Phase

Before accessing a sensor node in WSNs, any user must register to GW and gains a set P_{set} and other parameters. The registration phase is shown in the **Figure 1**.

Step 1: U → GW: $\{ID_u, Reg\text{-}inf, T_1\}$.

U sends the register request message $\{ID_u, Reg\text{-}inf, T_1\}$ to GW at the time T1.

Step 2: GW → U: $\{ID_{GW}, P, xP, h(.), a_1, T_2\}$.

When receiving the register request message of U at the time T', firstly GW checks whether $(T'-T_1) \leq \Delta T$ holds. If the answer is no, GW rejects the register

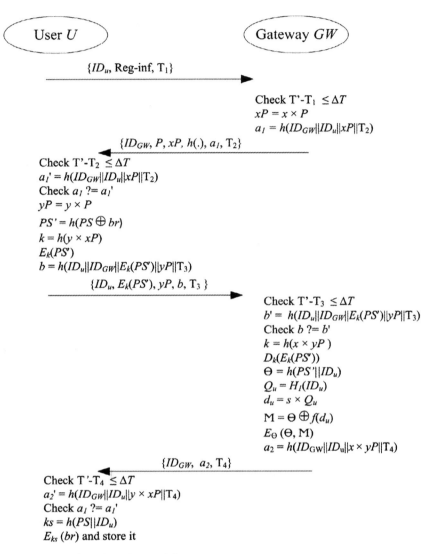

Figure 1. Registration phase of the user.

request message of U. Otherwise, GW selects a random number $x \in Z_q^*$ and computes $xP = x \times P$. Then GW calculates $a_1 = h(ID_{GW}\|ID_u\|xP\|T_2)$, where T_2 is the current time of GW. Finally, GW sends the register response message {ID_{GW}, P, xP, h(.), a1, T_2} to U.

Step 3: U → GW: {ID_u, Ek (PS′), yP, b, T_3}.

When receiving the register response message {ID_{GW}, P, xP, h(.), a1, T_2} at

A Secure User Authentication Protocol for Sensor Network in Data Capturing

the time T', U checks whether (T'-T2) ≤ ΔT holds. If the answer is no, U rejects the register response message. Otherwise, U computes a_1' = $h(ID_{GW}\|ID_u\|xP\|T_2)$ and checks whether a_1' = a_1 holds. If the answer is no, U rejects the register response message. Otherwise, U picks a random number $y \in Z_q^*$ and computes $yP = y \times P$. And U selects a password $PS \in Z_q^*$ and a random number $br \in Z_q^*$. U calculates $PS' = h(PS \oplus br)$ and $k = h(y \times xP)$. Then U encrypts PS' with the session key k, $E_k(PS')$. Finally, U computes $b = h\left(ID_u \| ID_{GW} \| E_k(PS') \| yP \| T_3\right)$ where T_3 is the current times of U. And U sends a message $\{ID_u, E_k(PS'), yP, b, T_3\}$ to GW.

Step 4: $GW \rightarrow U : \{ID_{GW}, P_{set}, E_\Theta(\Theta, M), a_2, T_4\}$.

Receiving the message $\{ID_u, E_k(PS'), yP, b, T_3\}$ at the time T', GW firstly checks whether (T'-T_3) ≤ ΔT holds. If the answer is no, GW rejects this message. Otherwise, GW computes b' = $h(ID_u\|ID_{GW}\|E_k(PS')\|yP\|T_3)$ and checks whether b' = b holds. If the answer is no, GW rejects this message. Otherwise, GW generates the session key k and decrypts $E_k(PS')$, $k = h(x \times yP)$, $D_k(E_k(PS'))$ to gain PS'. Then GW computes $\Theta = h(PS'\|ID_u)$, $Q_u = H_1(ID_u)$ and $d_u = s \times Q_u$. And GW also calculates $M = \Theta \oplus f(du)$. GW encrypts the (Q_u, M), $E_\Theta(\Theta, M)$ and computes a_2 = $h(ID_{GW}\|ID_u\|xyP\|T_4)$. At last GW stores (Pset, $E\Theta(\Theta, M)$, h(.), f(.), H1(.), ê(.)) into a smart card that is sent to U. Moreover GW sends the register acknowledge message $\{ID_{GW}, a_2, T_4\}$ to U.

Step 5: U encrypts and stores br.

When receiving the register acknowledge message $\{ID_{GW}, a_2, T_4\}$ at the time T', U firstly checks whether (T'-T_4) ≤ ΔT holds. If the answer is no, U rejects this mes-sage. Otherwise U computes a_2' = $h(ID_{GW}\|ID_u\|yxP\|T_4)$ and checks whether a_2' = a_2 holds. If the answer is no, U rejects this message. Otherwise, U computes $ks = h(PS\|ID_u)$ and encrypts br, $E_{ks}(br)$. Finially U stores $E_{ks}(br)$.

3.3. Login Phase and Authentication Phase

Accessing the data in Sn, U must login Sn and be authenticated by GW and Sn. And U must complete the login phase and authentication phase. Login phase and authentication phase are shown in **Figure 2**.

Figure 2. Login phase and authentication phase.

3.3.1. Login Phase

U must enter his ID_u and password PS firstly. Then, after the smart card validates U via the following steps, the smart card sends the access request message to Sn.

Step 1: Gains br.

U enters his identity ID_u and password PS to the smart terminal. And the smart terminal computes $ks = h(PS\|ID_u)$, and $D_{ks}(E_{ks}(br))$ to gain br.

Step 2: Validate U.

The smart card computes $PS' = h(PS \oplus br)$, $\Theta' = h(PS'\|ID_u)$ and $D\Theta'(E\Theta(\Theta, M))$ to gain the (Θ, M). The smart card checks whether $\Theta = \Theta'$ holds. If the answer is no, the smart card stops and alarms. Otherwise, the smart card continues to execute the next step.

Step 3: Computes Q_{Sn}, Q_{GW}, d_u, X and Y.

$$Q_{Sn} = H_1(ID_{Sn}), Q_{GW} = H_1(ID_{GW}), d_u = H_1(M \oplus \theta),$$
$$X = \hat{e}(d_u, Q_{Sn}) \text{ and } Y = \hat{e}(d_u, Q_{GW}).$$

Step 4: Generates a, b and encrypts (a, b).

The smart card calculates $a = h(ID_u\|ID_{GW}\|Y\|T_u)$, $b = h(ID_u\|ID_{Sn}\|X\|a\|T_u)$ and $E_X(a, b)$, where T_u is the current time of the smart terminal system.

Step 5: U → Sn: $\{ID_u, ID_{Sn}, E_X(a, b), T_u\}$.

The smart card sends the login request message $\{ID_u, ID_{Sn}, E_X(a, b), T_u\}$ to the Sn.

3.3.2. Authentication Phase

The authentication phase includes Sn authenticating U and GW, GW authenticating Sn and U, U authenticating Sn and GW. The authentication phase can complete the mutual authentication.

Sensor node Sn authenticates user U When receiving the login request message $\{ID_u, ID_{Sn}, E_X(a, b), T_u\}$ sent by U at time T', Sn firstly checks the validity of the request message. Then Sn authenticates U.

Step 1: Validates login request message.

Sn checks whether $(T'-T_u) \leq \Delta T$ holds. If the answer is no, Sn rejects the login request of U. Otherwise, it continues to perform the next step.

Step 2: Decrypts $E_X(a, b)$.

Sn computes $Q_u = H_1(ID_u)$, $X' = \hat{e}(Q_u, d_{Sn})$ and $DX'(E_X(a, b))$ to gain (a, b).

Step 3: Computes $b' = h(ID_u\|ID_{Sn}\|X'\|a\|T_u)$.

Step 4: Validates U.

Sn checks if $b = b'$ holds. If the answer is yes, the validity of U can be assured and Sn continues to perform the next step. Otherwise, it rejects the login request message of U.

Step 5: Computes QGW, Z, β and encrypts.

$Q_{GW} = H_1(ID_{GW})$, $Z = \hat{e}(d_{Sn}, Q_{GW})$ and $\beta = h(ID_u\|ID_{Sn}\|Z\|a\|T_u\|T_{Sn})$, where T_{Sn} is the current time of Sn system. And Sn encrypts (a, β), $E_Z(a, \beta)$.

Step 6: Sn → GW: $\{ID_{Sn}, ID_{GW}, ID_u, E_Z(a, \beta), T_u, T_{Sn}\}$

Sn sends a request message $\{ID_{Sn}, ID_{GW}, ID_u, E_Z(a, \beta), T_u, T_{Sn}\}$ to GW.

Gateway GW authenticates sensor node Sn When receiving the request message $\{ID_{Sn}, ID_{GW}, ID_u, E_Z(a, \beta), T_u, T_{Sn}\}$ at time T'', GW checks the validity of this message firstly. And GW authenticates Sn and U. Finally, GW creates a response message for Sn and U.

Step 1: Validates request message of Sn.

GW checks whether $(T''-T_u) \leq \Delta T$ and $(T''-T_{Sn}) \leq \Delta T$ hold. If the answer is no, GW rejects the request message. Otherwise, GW continues to perform the next step.

Step 2: Computes Q_{Sn}, Z' and gains (a, β).

GW computes $Q_{Sn} = H_1(ID_{Sn})$, $Z' = \hat{e}(Q_{Sn}, d_{GW})$ and $D_{Z'}(E_Z(a, \beta))$ to gain (a, β).

Step 3: Computes $\beta' = h(ID_u \| ID_{Sn} \| Z' \| a \| T_u \| T_{Sn})$.

Step 4: Validates Sn.

GW checks if $\beta' = \beta$ holds. If the answer is yes, the validity of Sn can be assured and GW continues to perform the next step. Otherwise, it rejects the request message.

Step 5: Computes Q_u, Y' and a'.

GW computes $Q_u = H_1(ID_u)$, $Y' = \hat{e}(Q_u, d_{GW})$ and $a' = h(ID_u \| ID_{GW} \| Y' \| T_u)$.

Step 6: Validates U.

GW checks if $a' = a$ holds. if the answer is yes, the validity of U can be assured and GW continues to perform the next step. Otherwise, GW rejects the request message.

Step 7: GW $\to$ Sn: $\{ID_{GW}, ID_{Sn}, ID_u, E_{Z'}(\gamma, \delta), T_{GW}\}$.

GW generates the response message for Sn and U. GW calculates: $\gamma = h(ID_u \| ID_{Sn} \| ID_{GW} \| Y' \| T_u \| T_{Sn})$ and $\delta = h(ID_u \| ID_{Sn} \| ID_{GW} \| Z' \| \gamma \| T_{Sn} \| T_{GW})$, where TGW is the current time of GW's system. And GW encrypts (γ, δ) with the key Z', $E_{Z'}(\gamma, \delta)$, and sends the response message $\{ID_{GW}, ID_{Sn}, ID_u, E_{Z'}(\gamma, \delta), T_{GW}\}$ to Sn.

Sensor node Sn authenticates gateway GW When receiving the response message $\{ID_{GW}, ID_{Sn}, ID_u, E_{Z'}(\gamma, \delta), T_{GW}\}$ sent by GW at time T''', Sn checks and authenticates GW via the following steps.

Step 1: Validates the response message.

Sn checks if $(T'''-T_{GW}) \leq \Delta T$ holds. If the answer is no, Sn rejects this response message. Otherwise, Sn continues to perform the next step.

Step 2: Gains (γ, δ).

Sn decrypts the $E_{Z'}(\gamma, \delta)$ with the key Z, $D_Z(E_{Z'}(\gamma, \delta))$, to gain (γ, δ).

Step 3: Computes δ'.

$$\delta' = h\left(ID_u \| ID_{Sn} \| ID_{GW} \| Z \| \gamma \| T_{Sn} \| T_{GW}\right)$$

Step 4: Validates GW.

Sn checks if $\delta' = \delta$ holds. If th answer is yes, the validity of GW can be assured and Sn continues to execute the next step. Otherwise, it rejects the response message.

Step 5: Generates $K_{session}$, ζ and encrypts.

Sn computes $K_{session} = h(X \| T_u \| T_{Sn})$,

$$\zeta = h\left(ID_u \| ID_{Sn} \| K_{session} \| \gamma \| T_{Sn}\right) \text{ and } E_{X'}(\zeta, \gamma)$$

Step 6: Sn → U: $\{ID_{Sn}, ID_u, ID_{GW}, E_{X'}(\zeta, \gamma), T_{Sn}\}$.

Sn sends the response message $\{ID_{Sn}, ID_u, ID_{GW}, E_{X'}(\zeta, \gamma), T_{Sn}\}$ to U.

User U authenticates sensor node Sn When U receives Sn's response message $\{ID_{Sn}, ID_u, ID_{GW}, E_{X'}(\zeta, \gamma), T_{Sn}\}$ at time T'''', U checks this message and authenticates Sn and GW. U performs the following steps.

Step 1: Validates the response message.

U checks whether (T''''-T_{Sn}) ≤ ΔT holds. If the answer is no, U rejects this response message. Otherwise, it continues to perform the next step.

Step 2: Gains (ζ, γ).

U computes $D_X(E_{X'}(ζ, γ))$ to decrypt $E_{X'}(ζ, γ)$ with the key X to gain (ζ, γ).

Step 3: Generates Ksession and ζ'.

U computes Ksession = $h(X||T_u||T_{Sn})$, and ζ' = $h(ID_u||ID_{Sn}||K_{session}||γ||T_{Sn})$

Step 4: Validates Sn.

U checks whether ζ = ζ' holds. If the answer is yes, the validity of Sn can be assured and U continues to execute the next step. Otherwise, U rejects the response message.

Step 5: Computes γ' = $h(ID_u||ID_{Sn}||ID_{GW}||Y||T_u||T_{Sn})$.

Step 6: Validates GW.

U checks whether γ' = γ holds. If the answer is yes, the validity of GW can be assured and U accepts this response message. Otherwise, U rejects this response message.

After U authenticates Sn and GW, U will access the data of the Sn with the session key Ksession.

3.4. Password Update Phase

When U updates his password, U enters his ID_u, old password PS and news password PSn to the smart terminal or a update password program. The smart card must compute a new password value, which is encrypted and stored in the smart card. The user password update phase includes the following steps.

Step 1: U enters his ID_u, old password PS and news password PSn to the smart terminal or a update password program.

Step 2: The smart terminal computes ks = $h(PS\|ID_u)$ and $D_{ks}(E_{ks}(br))$ to gain br firstly. Then it computes PS' = $h(PS \oplus br)$, PSn' = $h(PSn \oplus br)$. The smart terminal sends $\{ID_u, PS', PSn'\}$ to the smart card.

Step 3: The smart card computes Θ' = $h(PS'\|ID_u)$ and $D\Theta'(E\Theta(\Theta, M))$ to gain (Θ, M).

Step 4: The smart card checks whether $\Theta' = \Theta$ holds. If the answer is no, the smart card rejects the password update and alarms. Otherwise, the smart card continues to perform the next step.

Step 5: The smart card calculates Θ_n' = $h(PSn'\|ID_u)$ and $M' = \Theta_n' \oplus (\Theta \oplus M)$.

Step 6: The smart card encrypts the new sensitive password value $(\Theta n', M')$ with the key Θ_n', $E_{\Theta n}'(\Theta n', M')$, and replaces the $E_\Theta(\Theta, M)$ with $E_{\Theta n}'(\Theta n', M')$.

4. Security and Performance Analysis

4.1. The Proposed Protocol Provides Message Confidentiality Service

Proof

Message confidentiality service against eavesdropping attack is performed by data encryption service. Our proposed protocol can provide sufficient confidentiality for sensitive data stored and transmitted with encrypting data (e.g. $E_k(PS')$, $E_\Theta(\Theta, M)$, $E_X(a, b)$, $E_Z(a, \beta)$, $E_Z'(\gamma, \delta)$ and $E_X'(\zeta, \gamma)$. More specifically, these sensitive information are confidential against the attacker. If the sensitive data is stored or transmitted without encryption in the public channel, the attacker maybe view the plaintext data. This attack maybe occur in Wenbo's protocol and Yoon and Yoo's protocol[15]. Moreover, in Wenbo's protocol the sensitive (Bu, Wu) that was not encrypted was stored in the smart card and the long-term shared secret key SKGW was not also encrypted in the Sn. In the[15] Sn's response message that was not encrypted was sent to U by a public channel directly.

4.2. The Proposed Protocol Resists an Integrity Attack

Proof

The data integrity attack includes data modification attack, data corruption attack and data insertion attack. The integrity service assures the transmitted data is not modified by an unauthorized entity.

In our proposed protocol, Sn can guarantee the login request message $\{ID_u, ID_{Sn}, E_X(a, b), T_u\}$ from U has not been modified by an unauthorized entity via decrypting $E_X(a, b)$, recomputing and checking b. GW can also guarantee the authentication request message $\{ID_{Sn}, ID_{GW}, ID_u, E_Z(a, \beta), T_u, T_{Sn}\}$ from Sn has not been modified by an unauthorized entity via decrypting $E_Z(a, \beta)$, recomputing and checking a, β. Similarly, Sn can guarantee the authentication response message $\{ID_{GW}, ID_{Sn}, ID_u, E_{Z'}(\gamma, \delta), T_{GW}\}$ from GW has not been modified by an unauthorized entity via decrypting $E_{Z'}(\gamma, \delta)$, recomputing and checking δ. Moreover, U uses the same way to guarantee the authentication response message $\{ID_{GW}, ID_{Sn}, ID_u, E_{X'}(\zeta, \gamma), T_{Sn}\}$ from Sn has not been modified.

4.3. The Proposed Protocol Resists a Denial Attack

Proof

This type of attack is that the participating entity denies in all of the operations or part of its. However, in our proposed protocol, we assume that GW is a trusted party. And GW creates the unique private key for every entity (sensor node, user). Although GW does not store the private key of an entity, it can trace the entity operations with the entity's public key and HMAC. Therefore, the entity cannot deny that he/she performed all participation.

4.4. The Proposed Protocol Resists a DoS Attack

Proof

The DoS attack can be occurred by the attacker who transmitting the large number of request messages to Sn or GW in the login phase or in the authentication phase. In our proposed protocol, since every message associates with a timestamp T and is authenticated, the unauthenticated message or the timeout message is rejected. So the proposed protocol can reject DoS attack.

4.5. The Proposed Protocol Resists a Sensor Node Compromise Attack

Proof

Since WSNs is normally deployed in an open environment, the attacker is easy to capture a sensor node and may attempt to get some information stored in the sensor node. When the attacker gets the secret from the capturing sensor node, he/she can attack the WSNs. If the authenticating user and data access from the sensor node are allowed directly to the user without the license of gateway, this attack is very high, which occurs in Watro *et al.*'s scheme[19].

In our proposed protocol, And U does not access data from Sn until it is authorized by GW and Sn. And U's request message must be authenticated by Sn firstly, and the request message must be authenticated by GW. After that GW sends the license of U's to Sn and U. Only U can access the data of sensor node when his/her license from GW is the same as Sn's from GW. Moreover, in our proposed protocol GW can monitor whether a sensor node is captured with the trusted and reputation management scheme[24][26]. If some a sensor node is captured by an attacker, GW can detect and isolate it.

4.6. The Proposed Protocol Resists a Replay Attack

Proof

The replay attacks are impossible if the previous information is not reused again. In our proposed protocol, the login message and the authentication message are validated by checking timestamps. When an attacker eavesdrops the communication between U and Sn or between Sn and GW, he/she does not reusable again. We assume if an adversary intercepts a login request message $\{ID_u, ID_{Sn}, E_X(a, b),$

T_u} and attempts replaying the same message for login to Sn. The verification of the login request fails because of $(T_a-T_u) > \Delta T$, where Ta denotes the time when Sn receives the replaying message. Similarly, if an adversary intercepts {ID_{Sn}, ID_{GW}, ID_u, $E_Z(\alpha, \beta)$, T_u, T_{Sn}} and attempts to replay it to GW, he/she cannot pass the verification of GW because the time expires [i.e. $(T_b-T_{Sn}) > \Delta T$ and $(T_b-T_u) > \Delta T$], where Tb denotes the time when the replaying message is received by GW. Also if an adversary intercepts {ID_{GW}, ID_{Sn}, ID_u, $E_{Z'}(\gamma, \delta)$, T_{GW}} and attempts replaying the same message to Sn, he/she cannot pass the verification of Sn because of $(T_c - T_{GW}) > \Delta T$, where T_c denotes the time when Sn receives the replaying response message. Moreover, if an adversary intercepts {ID_{GW}, ID_{Sn}, ID_u, $E_{X'}(\zeta, \gamma)$, T_{Sn}} and attempts replaying the same message to U, he/she also cannot pass the verification of U because of $(T_d-T_{Sn} > \Delta T)$, where Td denotes the time when U receives the replaying response message.

4.7. The Proposed Protocol Resists an Impersonation Attack

Proof

In our proposed protocol, all sensitive information that is transmitted is encrypted with some a key. Additionally, the messages are validated and authenticated. Only when an attacker knows the master key s or solves Bilinear Differ-Hellman Problem can he/she attain the private key. It is impossible for an attacker.

In the login phase, only when an attacker knows U's private key du can he/she generate a legal login request message {D_u, ID_{Sn}, $E_X(a, b)$, T_u} to impersonate the U. Moreover it is impossible that an attacker gains the sensitive key material (Θ, M) that is encrypted to only be stored in the smart card without the user U's password. Thus it is not possible to compute X without du for an attacker. And as long as an attacker does not possess Sn's private key dSn, he/she cannot generate a legal authentication request message {ID_{Sn}, ID_{GW}, ID_u, $E_Z(\alpha, \beta)$, T_u, T_{Sn}} and {ID_{GW}, ID_{Sn}, ID_u, $E_{X'}(\zeta, \gamma)$, T_{Sn}} to impersonate Sn. This is because that the attacker cannot compute the key Z and the key X' without dSn. Similarly, an attacker also cannot generate a legal response message {ID_{GW}, ID_{Sn}, ID_u, $E_{Z'}(\gamma, \delta)$, T_{GW}} to impersonate GW. This is due to that an attacker does not know the private key dGW of GW.

4.8. The Proposed Protocol Resists a Stolen Verifier Attack

Proof

An attacker who has stolen U's private key materials $E_\Theta(\Theta, M)$ from the smart terminal or the smart card via the Trojan or other intruding methods cannot obtain any useful information. This is due to that the private key materials are encrypted. The attacker cannot decrypt $E_\Theta(\Theta, M)$ to gain (Θ, M) without U's password PS. And the attacker also cannot attain any useful private key information of U from GW because U's private key materials are not stored in the GW database.

4.9. The Proposed Protocol Resists a Stolen Smart Card Attacks

Proof

The attacker who has stolen U's smart card cannot impersonate this user to access Sn. Because the attacker does not know U's password, the smart card does not validate the login request and rejects the access request of the attacker.

4.10. The Proposed Protocol Resists an Insider Attack

Proof

The insider attack is intentionally misused by authorized entities. In our proposed protocol, the gateway manager or system administrator cannot attain U's password PS because in the registration phase U transmits $E_k(PS')$ to GW instead of the plain password PS, and any sensitive key material information of U and any verifier table are not stored in GW. Additionally, the smart terminal manager or administrator also cannot attain the useful information of U's key from the smart card and the smart terminal because of the sensitive key material encrypted. Therefore, the proposed protocol can resist the privileged insider attacks.

4.11. The Proposed Protocol Resists a Man-in-the-Middle Attack

Proof

The man-in-the-middle attack is that an attacker intercepts the communication between the legal user and other entity (e.g. sensor node, gateway) and successfully masquerades as the user or other entity by some methods. In our proposed protocol, U is authenticated by Sn in the login phase, Sn and U are authenticated by GW in the authentication request phase, and Sn also authenticates GW in the authentication response phase, U validates Sn and GW in the authentication response phase. That is to say, our proposed protocol can provide complete mutual authenticate among entities and resists the man-in-the-middle attack.

Table 2 shows the security functionality comparisons between our proposed protocol and the related protocols. According to the **Table 2**, although the Ohood *et al.*'s protocol presents the same security as ours, the Ohood *et al.*'s protocol needs some complicated biometric equipments. Compared against each other, our protocol provides is more security services than the other protocols.

Performance analysis The section summarizes the performance results of the proposed protocol. We define the notation Th as the hash function computation

Table 2. Security comparison.

	Benenson et al. [7]	Das [10]	Chen and Shih [16]	Yuan et al. [19]	Yeh et al. [17]	Yoon and Yoo [20]	Ohood et al. [21]	Wenbo and Peng [22]	Ours
Data Confidentiality	NP	NP	NP	NP	NP	NP	P	NP	P
Data Integrity	NP	P	P	NP	NP	P	P	P	P
Password Update	NR	NP	R	NP	P	NR	NR	P	P
Key Agreement	NP	NP	NP	NP	NP	NP	P	P	P
Mutual Authentication	NP	NP	P	NP	NP	P	P	NP	P
Denial Attack	No	No	No	Yes	No	Yes	Yes	No	Yes
DoS Attack	No	No	No	No	No	No	Yes	No	Yes
Compromise Attack	Yes	No	No	No	No	No	Yes	Yes	Yes
Replay Attack	Yes	Yes	Yes	Yes	No	Yes	Yes	No	Yes
Impersonation Attack	No	Yes	Yes	No	No	Yes	Yes	No	Yes
Insider Attack	Yes	No	No	No	No	Yes	Yes	Yes	Yes
Forgery Attack	Yes	No	Yes	Yes	Yes	Yes	Yes	No	Yes
Stolen-Verifier Attack	Yes	Yes	Yes	Yes	No	Yes	Yes	Yes	Yes
Guessing Attack	Yes	Yes	Yes	Yes	Yes	Yes	Yes	Yes	Yes
Man-in-the-Middle Attack	No	No	Yes	No	No	Yes	Yes	No	Yes

Yes: Resist Attack, No: Not Resist Attack, P: Provided, NP: Not Provided, R: Required, NR: Not Required.

cost, Texp as the modular exponential computation cost, Tpm as the elliptic curve point multiply cost, Tpa as the elliptic curve point addition cost, Tpair as pairing computation cost, Trc as RC5 computation cost, Taes as AES computation cost, Te as the elliptic curve polynomial computation cost. The comparison of related protocols is illustrated in the **Table 3**.

According to **Table 3**, Chen *et al.*'s protocol needs eight hash function computations, Yoon el at.'s needs thirteen hash function computations, Yuan *et al.*'s also need thirteen hash function computations, Das's protocol needs six hash function computations. And Benenson *et al.*'s protocol needs 2n hash function computations and 3n + 1 modular exponential computations[22]. Ohood *et al.*'s biometric authentication protocol needs four RC5 computations and ten hash function computations. Yeh *et al.*'s protocol needs fifteen hash function computations, four elliptic curve point addition computations , ten elliptic curve point multiply computations and two elliptic curve polynomial computations. Wenbo *et al.*'s protocol needs eighteen hash function computations and seven elliptic curve point multiply computations. Our proposed protocol needs eighteen hash function computations, four elliptic curve point multiply computations, eleven AES computations and six pairing computations. Although our protocol needs more computations than their protocols, their protocols suffer from security issues or need complicated biometric equipments. Our protocol addressed these issues and provides better security and more security services than the other related protocols.

5. Conclusion

In the paper, we discussed an approach of data capturing for big data that is data collecting via sensor networks and its user authentication protocol. We have analyzed Wenbo *et al.*'s user authentication protocol for WSNs. The Wenbo's protocol, which does not provide mutual authentication between user and sensor node

Table 3. Computation performance comparison.

	Benenson et al. [7]	Das [10]	Chen and Shih [16]	Yeh et al. [17]	Yoon and Yoo [20]	Ohood et al. [21]	Yuan et al.[19]	Wenbo et al. [22]	Ours
Registration Phase	1Texp	1Th	1Th	4Th +2Tmp	3Th	2Th	4Th	3Th +1Tpm	4Th + 4Tpm + 3Taes
Login and Authentication Phase	2nTh +3nTexp	5Th	7Th	11Th + 4Tpa + 8Tpm + 2Te	10Th	4Trc +8Th	9Th	15Th +6Tpm	14Th + 6Tpair + 8Taes
Total	2nTh +3nTexp +1Texp	6Th	8Th	15Th + 4Tpa + 10Tpm + 2Te	13Th	4Trc + 10Th	13Th	18Th +7Tpm	18Th + 4Tpm + 11Taes + 6Tpair

and confidentiality service, is susceptible to insider, replay, denial, compromise, forgery, man-in-the-middle and DoS attacks. We have also reviewed the protocols of Yeh *et al.*, which does not provide mutual authentication and protect against insider, denial, compromise, man-in-the-middle and DoS attacks, of Das, which is vulnerable to forgery, denial, compromise, DoS, man-in-the-middle attacks, of Benenson *et al.*, which susceptible to denial, compromise, DoS, man-in-the-middle attacks, of Chen *et al.* which is vulnerable to denial, insider, compromise and DoS attacks, of other biometric authentication protocols. Since WSNs need more secure mutual authentication method in an insecure network environment, we use the IBE mechanism to design a new user authentication protocol. Our protocol can prevent all the problems of the former schemes. Furthermore, it enhances the WSNs authentication with higher security than the other protocol. Therefore, the protocol is more suited to open and higher security WSNs environment in despite of more computation cost.

Competing Interests

The authors declare that they have no competing interests.

Authors' Contributions

ZQ designed the user authentication protocol and analyzed performance. TC implemented the security analysis for protocol. ZX analyzed the security of Wenbo's protocol. RC coordinated the whole study. All authors read and approved the final manuscript.

Acknowledgements

This work was supported in part by the National Natural Science Foundation of China under Grant No. 11271003, the National Research Foundation for the Doctoral Program of Higher Education of China under Grant No.20134410110003, High Level Talents Project of Guangdong, Guangdong Provincial Natural Science Foundation under Grant No.S2012010009950 and No. S2012040007370, the Project of Department of Education of Guangdong Province under Grant No

2013KJCX0146, and the Natural Science Foundation of Bureau of Education of Guangzhou under Grant No. 2012A004. We sincerely thank all the researchers in our references section for the inspiration they provide.

Source: Zhou Q, Tang C, Zhen X, et al. A secure user authentication protocol for sensor network in data capturing [J]. Journal of Cloud Computing, 2015, 4(1): 1–12.

References

[1] Kenneth NC, Viktor MS (2013) The rise of big data: how It's changing the way we think about the world. Fortuna's Corner in cloud computing, Cybersecurity, Dow, Intelligence Community, Internet, Markets, national security, S & P, Uncategorized, US Military, April 24, 2013.

[2] Sastry N, Wagner D (2004) Security considerations for IEEE 802.15.4 networks. In Proceedings of the ACM Workshop on Wireless Security (WiSe'04). 32–42.

[3] WG802.15 (2003) IEEE Standards for 802.15.4, Part 15, Amendment 4. Wireless medium access control and physical layer specifications for low-rate wireless personal area networks. IEEE, Washington DC, USA.

[4] Das ML, Saxena A, Gulati VP (2004) A dynamic ID-based remote user authentication scheme. IEEE Trans Consum Electron 50(2):629–631.

[5] Leung KC, Cheng LM, Fong AS, Chan CK (2003) Cryptanalysis of a modified remote user authentication scheme using smart cards. IEEE Trans Consum Electron 49(4):1243–1245.

[6] Watro R, Kong D, Cuti S F, Gardiner C, Lynn C, Kruus P (2004) TinyPK: securing sensor networks with public key technology. In Proceedings of the ACM Workshop on Security of Ad Hoc and Sensor Networks (SASN'04), 59–64.

[7] Benenson Z, Gedicke N, Raivio O (2005) Realizing robust user authentication in sensor networks. In Real-World Wireless Sensor Networks (REALWSN), 14.

[8] Wong KHM, Yuan Z, Jiannong C, Shengwei W (2006) A dynamic user authentication scheme for wireless sensor networks. In Proceedings of the IEEE International Conference on Sensor Networks, Ubiquitous, and Trustworthy Computing, 244-251.

[9] Tseng HR, Jan RH, Yang W (2007) An improved dynamic user authentication scheme for wireless sensor networks. In Proceedings of the 50th Annual IEEE Global Telecommunications Conference (GLOBECOM'07), 986–990.

[10] Das ML (2009) Two-factor user authentication in wireless sensor networks. IEEE Trans Wirel Commun 8(3):1086–1090.

[11] Ko KC (2008) A novel dynamic user authentication scheme for wireless sensor net-

works. In Proceedings of the IEEE International Symposium on Wireless Communication Systems, ISWCS'08, 608–612.

[12] Nyang DH, Lee MK (2009) Improvement of Das's two-factor authentication protocol in wireless sensor networks. Available via DIALOG. http://eprint.iacr.org/2009/631.pdf Accessed 15 Jan 2014.

[13] Vaidya B, Rodrigues JJ, Park JH (2010) User authentication schemes with pseudonymity for ubiquitous sensor network in NGN. Int J Communication Syst 23 (9–10): 1201–1222.

[14] Khan MK, Alghathbar K (2010) Cryptanalysis and security improvements of 'two-factor user authentication in wireless sensor networks'. Sensors 10(3):2450–2459.

[15] Khan MK, Alghathbar K (2010), Security Analysis of Two-Factor Authentication In Wireless Sensor Networks. In Proceedings of Advances in Computer Science and Information Technology: AST/UCMA/ISA/ACN 2010 Conferences, 55–60.

[16] Chen TH, Shih WK (2010) A robust mutual authentication protocol for wireless sensor networks. ETRI J 32(5):704–712.

[17] Yeh HL, Chen TH, Liu PC, Kim TH, Wei HW (2011) A secured authentication protocol for wireless sensor networks using elliptic curves cryptography. Sensors 11(5): 4767–4779.

[18] Han W (2013), Weakness of a secured authentication protocol for wireless sensor networks using elliptic curves cryptography. Available via DIALOG. http://eprint.iacr.org/2011/293 Accessed 15 May 2014.

[19] Yuan J, Jiang C, Jiang Z (2010) A biometric-based user authentication for wireless sensor networks. Wuhan Univ J Nat Sci 15(3):272–276.

[20] Yoon EJ, Yoo K Y (2011) A new biometric-based user authentication scheme without using password for wireless sensor networks. In Proceedings of the 20th IEEE International Workshops on Enabling Technologies: Infrastructure for Collaborative Enterprises, 279–284.

[21] Ohood A, Mznah AR, Abdullah AD (2013) An efficient biometric authentication protocol for wireless sensor networks. Int J Distributed Sensor Networks 4:1–13.

[22] Wenbo S, Peng G (2013) A new user authentication protocol for wireless sensor networks using elliptic curves cryptography. Int J Distrib Sens Netw 3:1–7.

[23] Quan Z (2011) Trusted transmission model of wireless sensor networks, Ph.d. Theis. South China Agricultural University, China.

[24] Quan Z, Gui F, Deqin X, Jiuhao L (2010) trusted transport model based cluster-merkle-tree for WSNs. in Processing of 2010 IEEE International Conference on Computer Application and System Modeling V1, 564–568.

[25] Quan Z, GUI F, Deqin X, Yi T (2012) Trusted architecture for farmland WSNs. in

Processing of 2012 Forth IEEE International Conference on Cloud Computing Technology and Science, 782–787.

[26] Boneh D, Franklin M (2001) Identity based encryption from the Weil pairing. in processing of Advances in Cryptology. Lect Notes Computer Sci 2139:213–229.

Chapter 24

An Economic Perspective of Message-Dropping Attacks in Peer-to-Peer Overlays

Kevin W. Hamlen[1], William Hamlen[2]

[1]Computer Science Department, The University of Texas at Dallas, USA
[2]Department of Finance and Managerial Economics, School of Management, State University of New York at Buffalo, USA

Abstract: Peer-to-peer networks have many advantageous security properties, including decentralization, natural load-balancing, and data replication. However, one disadvantage of decentralization is its exclusion of any central authority who can detect and evict malicious peers from the network. It is therefore relatively easy to sustain distributed denial-of-service attacks against these networks; malicious peers simply join the network and fail to forward messages. This article shows that peer-to-peer message-dropping attacks can be understood in terms of a well-established category of economic theory: the theory of the second best. In particular, peers who wish to continue service during an attack seek a second best solution to a utility optimization problem. This insight reveals useful connections between economic literature on the second best and computer science literature on peer-to-peer security. To illustrate, we derive and test an economics-inspired modification to the Chord peer-to-peer routing protocol that improves network reliability during message-dropping attacks. Under simulation, networks using the modified protocol achieve a 50% increase in message deliveries for certain realistic attack scenarios.

1. Introduction

Peer-to-peer networks are an increasingly popular vehicle for highly fault-tolerant, light-weight, and low-cost distributed computing across heterogeneous hardware. Cloud computing[1], digital music exchange[2], digital libraries[3], secure data management systems[4], and bioinformatics databases[5] are just a few of the venues where this technology is being used today. Unlike traditional networks, peer-to-peer networks lack any centralized server; every agent acts as both server and client. This provides a natural resistance to many attacks, since adversaries must compromise a large number of peers instead of just a few central servers to corrupt data integrity or disrupt its availability.

Unfortunately, decentralization can leave peer-to-peer networks vulnerable to a different sort of denial-of-service attack wherein malicious agents join the network and misroute or drop messages, thereby disrupting communication. Constraining this message-dropping behavior is difficult since victims typically learn only that their messages weren't delivered, not who was at fault. Even when malicious behavior is localized to a particular peer, there is no central authority who can evict the peer from the network, leaving it free to perpetuate the attack. A small number of malicious agents can amplify their message-dropping with a Sybil attack[6], in which each joins the network many times under various false identities to occupy a greater percentage of the overlay space. Since dropping messages is computationally inexpensive for the attacker, a relatively small number of attackers can significantly disrupt overlay traffic.

We gain insight into this problem by connecting it to economic theories of the second best[7] (cf.,[8]). Second best economic solutions are required when one or more units contravene the first best solution, and there is no way to remove the misbehaving units from the system. The second best solution incorporates (takes as given) the misbehavior of the deviate units to obtain a second best optimum. In the case of peer-to-peer networks, we find that application of second best theory to the secure routing problem yields recommendations that advise non-malicious peers on how to route messages optimally given observed misbehavior of their neighbors. By taking the network topology as given, such recommendations can be implemented atop any given topology to achieve better performance during a message-dropping attack.

Adopting the piecemeal approach of Davis and Whinston[9] further allows each peer to optimize its own behavior with only local knowledge, and without requiring cooperation from other peers. Thus, the recommendations can be implemented fully automatically by individual peer clients without a central authority that has global knowledge, and without requiring other agents to adopt the new protocol.

We show how these theoretical insights can be used in practice by deriving a simple modification to the Chord peer-to-peer networking protocol[10] that improves overlay performance during message-dropping attacks. When faced with message-dropping agents, peers approximate a second best alternative to optimal overlay routes. Simulation of our modified protocol shows that message delivery rates improve by over 50% with a corresponding increase in overall system utility under realistic attack scenarios. The modified protocol is simple to implement since it does not require any change to the network topology and uses only information that is already available in a standard Chord network.

We begin by summarizing related work in §2. Section 3 reviews the economic theory of the second best and argues its relevance to peer-to-peer systems. The individual and social welfare perspectives are each elaborated in §4, yielding a general framework for expressing peer-to-peer message-dropping attacks as second best utility optimization problems. The framework is general enough to include many different definitions of utility, including those that incorporate both reliability and risk. Section 5 applies this framework to Chord, showing its effectiveness in resisting both non-coordinated and coordinated, distributed, message- dropping attacks. Finally, §6 concludes with a summary and suggestions for future work.

2. Related Work

Over the past decade there has been an explosion of research devoted to peer-to-peer security (cf.,[11]). Relevant issues include robust search[12], pollution prevention (i.e., inhibiting the spread of unwanted objects), secure data storage, and data confidentiality. Most work on these assumes a secure routing framework that facilitates reliable, robust communication between peers. For example, reputation-based trust managers such as EigenTrust[13], Credence[14], and Penny[15], aggregate local reputation information whose exchange requires a secure routing framework.

Secure routing is divisible into three sub-problems[16]: secure identifier assignment, routing table maintenance, and message forwarding. Secure identifiers prevent attackers from misrepresenting or abusing their overlay positions to intercept more than their share of traffic. Secure routing tables maintain peers' connections to appropriate sets of neighbor peers. Finally, secure message forwarding delivers messages using the secure identifiers and routing tables. This last sub-problem is the focus of our work.

Secure message forwarding has been studied from an economic perspective in the context of selfish routing[17]. Selfish routers enjoy the message-forwarding services provided by fellow peers, but fail to forward the messages of others. Unlike message-droppers, selfish routers desire services from the network. This has led to incentive-based solutions such as pricing networks[18], negotiating contracts for mutual message delivery[19], rewarding message delivery with increased reputation[20], and rewarding message delivery with increased quality of service[21]. In contrast, message-droppers only desire to disrupt service, making incentivization inapplicable.

Economists have also studied peer-to-peer networks in the context of free-riding. Free-riding peers obtain shared resources from the network but fail to share their own resources. Krishnan *et al.*[22] observe that the shared resources resemble public goods[23]—an insight that has generated a growing body of work devoted to incentivizing free-riding peers to share[24]. However, unlike public goods, the connectivity that is threatened during a message-dropping attack is not equally available to all peers; it is influenced by the position of the attackers relative to each peer. This leads us to a different economic model for message-dropping.

Message forwarding in peer-to-peer networks can be divided into protocols for structured and unstructured networks. Structured networks route queries using a multi-hop protocol that passes the message from peer to peer. Examples include CAN[25], Pastry[26], Chord[10], and Tapestry[27]. Unstructured peer-to-peer networks like Gnutella broadcast queries using a multicast protocol that floods the query to all peers within a given radius. The unstructured approach tends to be more robust against message-dropping because of its high redundancy, but it does not always scale well[28]. Although our work is potentially applicable to unstructured topologies, we focus on structured ones since message-dropping tends to be a more significant threat in those contexts.

A large body of prior work improves the robustness of structured networks by augmenting their topologies with redundant routing paths (e.g.,[29]–[31]). Route redundancy increases the probability that at least one message replica reaches its destination. Such topologies can be improved further via adaptive techniques that adjust the topology dynamically in response to observed failures (e.g.,[32][33]) or based on the interaction history of peers (e.g.,[34][35]). In contrast to these approaches, our work adapts routing flow rates without modifying the topology. Our work therefore complements the above by optimizing routing behavior once a (possibly dynamic) topology is chosen.

Adaptive techniques, including ours, require a means of detecting message-dropping behavior. Section 5 adopts a sampling approach in which non-malicious peers test for malicious behavior by periodically sending probe messages. Such sampling has been effectively used in unstructured networks to detect and identify message-droppers[36]. Introduction graphs[37] and flow rate histories[38] can further help to identify and isolate malicious peers, but sampling has the advantage of being easy to implement atop networks that do not collect this extra information.

3. Peer-to-Peer and the Second Best

3.1. Overview of the Second Best

The theory of the second best has had a significant impact on economics throughout the past half-century, being applied to such subjects as health care, antitrust, and international trade[8]. In the case of peer-to-peer overlays, the original design of the overlay topology can be thought of as a first best solution that optimizes message delivery when all peers are non-malicious. Adapting individual peer routing behavior to the presence of malicious peers can therefore be viewed as a second best solution to this optimization problem.

Economic theory generally begins with economic subunits such as consumers, households, or firms that attempt to optimize their own objective functions subject to various constraints. These economic subunits work within a larger administrative perspective that is defined by the set of rules governing the system and the degree of decentralization. This leads to a larger fundamental problem first formulated by Nobel laureate Paul Samuelson[39], who observed that every society acts as if it is attempting to maximize a (known or unknown) social welfare function subject to constraints.[a] The welfare function is so named because it incorporates the goals of all economic subunits. For any viable system, the optimal conditions derived from the fundamental problem must also be consistent with the optimal behavior of the economic subunits.

Following Samuelson and the later literature[7], we write this problem in its most general form as:

$$\text{maximize } F(x_1,...,x_n) \text{ subject to } G(x_1,...,x_n) = 0 \qquad (1)$$

The objective function is F, the binding constraint is G, and the choice variables are xi with $i \in 1...n$. In a decentralized, consumer-oriented economy these choice variables are selected by independent economic subunits (e.g., consumers). The objective function is related to the maximization of the consumers' objective functions (utility functions) subject to constraints reflecting consumer income limitations. In a peer-to-peer network the choice variables are selected by individual peers and the constraints are those imposed by the routing protocol and the overlay topology.

The first best solution to the problem formalized by Equation (1) has the following necessary conditions for an optimal solution[7]:

$$F_i + \lambda G_i = 0 \quad \forall i \in 1...n \qquad (2)$$

where F_i and G_i are partial first derivatives and λ is a Lagrange multiplier[b] associated with the constraint in Equation (1). To obtain the first best solution, individual subunits solve their respective constrained optimization problems, and these must be consistent with the necessary conditions given by Equation (2).

Samuelson was the first to recognize that when some economic subunits fail to adhere to Equation (2), the ramifications extend beyond that deviate sector:

"First, what is the best procedure if for some reason a number of optimum conditions are not realized? What shall we do about the remaining ones which are in our power? Shall we argue that 'two wrongs do not make a right' and attempt to satisfy those we can? Or is it possible that failure of a number of conditions necessitates modifying the rest? Clearly the latter alternative is the correct one." ([39], p. 252).

In other words, if the first best solution is unobtainable because one sector behaves suboptimally, then the second best solution does not necessarily imply that the remaining sectors should continue to satisfy their first best optimization conditions; many subunits may need to change their optimal behaviors.

Lipsey and Lancaster[7] formalized the second best solution with an additional constraint containing the first-order behavior of the deviating sector. We use the concise presentation of Henderson and Quandt [[40], p. 316], who append the following constraints to the problem given by Equation (1):

$$F_j = h_j G_j, \quad h_j \neq \lambda \quad h_j \neq 0 \qquad (3)$$

The second best optimal condition that replaces Equation (2) isc

$$F_i + \lambda G_i + \mu\left(F_{ji} - h_j G_{ji}\right) = 0 \quad \forall i \in 1...n \qquad (4)$$

where F_{ji} and G_{ji} are the partial cross-derivatives between subunits i and j and μ is the new Lagrange variable associated with constraint 3. Thus, unless the cross-derivatives are zero, each sector $i \in 1...n$ has optimal behavior that deviates from Equation (2).

Negishi[41] (cf.,[42]) further found that under standard economic assumptions of a competitive, decentralized economy (viz., concave functions and at least one interior solution) it is as if society seeks to maximize a specific social welfare function expressible as a weighted sum of the consumer utility functions $U_i(x_i)$. The final form of the society's objective function is therefore

$$F(x_1,...,x_n) = \sum_{i=1}^{n} \alpha_i U_i(x_i) \qquad (5)$$

The coefficients α_i are the final weights given by the implicit social welfare to the utility of the economic subunits.[d]

These weights have a specific economic interpretation: Each consumer or household's weight is the reciprocal of its marginal utility of income. thus, if a wealthy family places relatively low value on the last dollar earned, it receives a relatively high weight in the objective function of the society. In essence, the system tends to weight the more successful households (in terms of income).

We find a similar result in the case of peer-to-peer systems. During a message-dropping attack, non-malicious peers can be defined as those that derive utility from message deliveries. This leads to a social welfare (administrative objective) function that weights peers by the reciprocal of their relative reliability–i.e., their message delivery success rates. When all weights are equal (a standard assumption when applying utility theory to information systems problems[43]), this corresponds to the special case where all peers in the network are equally reliable.

Davis and Whinston[9] used Negishi's result to provide a piecemeal approach to the second best problem. They concluded that the difficulties of implementing a complex set of second best conditions is often overestimated since, based on Equation (5), many of the second derivatives are zero in a decentralized system and can therefore be dropped. In the context of a peer-to-peer network, this implies that it is possible to approach the second best solution without the need to share global information amongst peers. Instead, each peer adapts its own behavior individually, based on its own knowledge.

Based on the economic development above, there are two separate aspects of the problem: the mathematical method proposed by Lipsey and Lancaster, and the welfare and utility considerations discussed by Samuelson and Negishi. Section 3.2 examines the former aspect and §4.2 the latter.

3.2. Example Application of the Second Best to a Peer-to-Peer System

Application of the second best approach to resisting message-dropping attacks in peer-to-peer overlays can be illustrated by a simple example. We begin by considering a single peer p of degree k in a network of size n (with k ≪ n). Peer p periodically receives messages for other peers, each of which it must forward to one of its k neighbors. We assume a flat identifier space for this example, and that peer p may forward each message to any one of its k neighbors (though for any given message, certain neighbors are better positioned than others to deliver it).

The first best solution to this problem is he one addressed when designing the overlay topology, which assigns p an optimal set of k neighbors given an assumed distribution of message destinations seen by p. Specifically, we seek the k neighbors that minimize the absolute distance between each message's final destination and the nearest of the k neighbors to that destination:

$$\underset{N_1,\ldots,N_k}{\text{minimize}} \sum_{i=1}^{k} \int_{(N_i+N_{i-1})/2}^{(N_i+N_{i+1})/2} |x - N_i| D_p(x) \, dx \qquad (6)$$

where $N_i \forall i \in 1\ldots k$ are the desired neighbor identifiers in ascending order, $N_0 = 0$ and $N_k + 1 = n$ are the limits of the identifier space, and D_p is the probability density of message destinations seen by p. For example, if D_p is a uniform distribution, the optimal first order conditions (derived by setting the derivative of Equation (6) with respect to Ni to zero) are:

$$N_i = \frac{1}{2}(N_{i-1} + N_{i+1}) \quad \forall i \in 1\ldots k \qquad (7)$$

That is, a uniformly distributed set of random messages is delivered most effectively when the k neighbors of p have identifiers that are evenly spaced along the interval [0, n).

The above implicitly assumes that peers behave optimally, forwarding each message to the neighbor closest to its destination. A second best solution is needed when some peers behave suboptimally and there is no way (short of centralizing the system) to force optimal behavior. One option in this case is to implement a new first best solution, but this requires global information about the overlay topology, which isn't typically available to peers once the network has been deployed and malicious behavior becomes evident. The second best solution takes the topology as given and re-solves the optimization problem to obtain a new recommended optimal behavior for peer p given the suboptimal behavior of its neighbor(s).

For example, suppose that peer p discovers that of the am messages it forwarded to peer $m \in 1...k$ during some sampling period, only a_m^* of them were ultimately delivered to their final destinations. In the first best solution given by Equation (7), peer p forwards an average of w/k of its messages to each of its neighbors, where w is the total number of messages; thus, when $a^* < w/k$, neighbor m is behaving suboptimally (possibly due to the suboptimal behavior of its neighbors). The second best solution with respect to peer m's actions requires appending the following constraint to Equation (6), with new Lagrange multiplier β:

$$\beta\left(a_m^* - w/k\right) \geq 0 \qquad (8)$$

When not all messages that peer p forwards to peer m are ultimately delivered (*i.e.*, am <w/k), we obtain a different optimal identifier value for the neighbor on each side of neighbor m (obtained by summing Equations (6) and (8), setting the derivaive to zero, and solving for the unknowns):

$$N_{m-1}^* = \frac{1}{2}\left(N_{m-2} + N_m\right) + \frac{1}{2}\beta \qquad (9)$$

$$N_{m+1}^* = \frac{1}{2}\left(N_{m+2} + N_m\right) - \frac{1}{2}\beta \qquad (10)$$

Even though peer p cannot change the identifiers of its neighbors, it should forward its messages as if neighbor $m-1$ had identifier N_{m-1}^* and neighbor $m + 1$ had identifier N_{m+1}^*.

As peer m's reliability decreases, β increases and N_{m-1}^* and N_{m+1}^* approach N_m. Peer p therefore forwards fewer messages to neighbor m since fewer destinations are closer to Nm than to N^*_{m-1} or N^*_{m+1}. In the limiting case where $a^*_m = 0$ (*i.e.*, neighbor m is completely unreliable), peer p only forwards to m those messages whose final destinations are m itself.

We next generalize this approach to a larger class of topologies and message-dropping attacks, and we examine the second best approach from both the perspective of individual peers and that of the system as a whole.

4. Attack Resistance as Utility Optimization

Economic formulations of the second best typically have two aspects: an individual perspective in which individuals in the society seek to maximize their own utility subject to individual constraints, and an administrative perspective in which the society as a whole acts as if it is seeking to maximize an objective function subject to administrative constraints. In this section we develop the each of these aspects as they relate to message-dropping attacks in peer-to-peer overlays. The analysis of the individual perspective yields a piecemeal approach[9] to resisting message-dropping attacks, wherein each peer individually adjusts its optimal behavior to account for the suboptimal behavior of its immediate neighbors. The analysis of the administrative perspective yields a measure of the network's success in resisting message-dropping attacks, providing a means to evaluate defense effectiveness. We initially consider only peer reliability; risk is added in §4.3.

4.1. The Individual Perspective

The optimization problem presented in §3.2 can be cast in the more general framework of a utility maximization problem. Following the economic literature, we assume that all decision-makers seek to maximize their own utility when making choices. The utility $U_i(P_i)$ of a non-malicious peer $i \in 1...n$ is a function of its message-delivering reliability $P_i \in [0,1]$. Following the standard assumptions on utility functions, we assume that utility increases with reliability but at a decreasing rate:

$$\frac{dU_i}{dP_i} > 0 \quad \text{and} \quad \frac{d^2 U_i}{dP_i^2} < 0 \quad \forall i \in 1...n \tag{11}$$

The first derivative of utility is the marginal utility and the second assumption above is the law of diminishing marginal utility.

Since each peer in the overlay forwards messages through its neighbors, its reliability P_i can be expressed in terms of the reliabilities of its $k \leq n$ neighbors:

$$P_i = \sum_{j=1}^{k} w_{ij} R_{ij}(w_{ij}) \quad \text{and} \quad \sum_{j=1}^{k} w_{ij} = 1 \quad \forall i \in 1...n \tag{12}$$

where $w_{ij} \in [0,1]$ is the relative share of its messages that peer i forwards to neighbor j, and $R_{ij}(w_{ij})$ is the reliability that peer i estimates for neighbor j. R_{ij} is a function of w_{ij} because the observed reliability of neighbor j typically varies with the share of messages it receives from i. As j receives a larger share, a greater portion of their destinations are farther from j, making those messages harder for j to deliver. Thus, $\partial R_{ij}/\partial w_{ij} \leq 0$.

The optimization problem confronting each peer i is that of maximizing utility subject to the constraint above. Written in the Lagrange (Kuhn-Tucker[44]) format it is:

$$\underset{w_{i1},...,w_{ik}}{\text{maxmize}} U_i(P_i) + \Phi_i \left(1 - \sum_{j=1}^{k} w_{ij} \right) + \lambda_i \left(P_i - \sum_{j=1}^{k} w_{ij} R_{ij}(w_{ij}) \right) \tag{13}$$

where φ_i and λ_i are the Lagrange multipliers.

For example, adopting reliability function $R_{ij}(w_{ij}) = d_{ij} w_j^{-c}$ (generalizing the example in §3.2 in which $d_{ij} = 1$ and $c = 1$), using the natural logarithmic utility function $U_i = \log$, and solving the resulting Kuhn-Tucker conditions (see the appendix), we find that if peer i deems neighbor m to be a fraction $\varepsilon \in [0,1]$ less reliable than neighbor j (i.e., $R_{im}(w) = \varepsilon R_{ij}(w) \quad \forall w \in [0,1]$, then peer i's optimal relative use of m compared to j is

$$w_{im} = \sqrt[c]{\varepsilon} w_{ij} \tag{14}$$

Hence, we advise peer i to use neighbor m less than it uses neighbor j by a factor of $\sqrt[\varepsilon]{\varepsilon}$.

4.2. The Administrative Perspective

In this section we derive a measure of the social welfare (i.e., verall system utility) of a peer-to-peer system using the results from the previous sections. This provides a general measure of the performance of the system during a message-dropping attack in terms of peer utilities. In particular, networks that attain higher social welfare can be characterized as more robust against message-dropping attacks. Section 4.3 illustrates the generality of this metric by showing how it can incorporate risk as well as reliability, and §5 uses this measure to evaluate the performance of our method when implemented in an actual peer-to-peer network.

The peer-to-peer system acts as though it has an administrator who seeks to maximize some vector $\left(U_1\left(\hat{P}_1\right),\cdots,U_n\left(P_n\right)\right)$ of the utility functions of all peers in the system[41][42], where $\hat{P}_i = P_i / \sum_{j=1}^{n} P_j$ denotes the relative reliability of peer i. Such an administrator might not actually exist, but the system behaves as if it does. Recall that this equates to maximizing the weighted sum in Equation (5). The administrator's optimization problem can therefore be written as:

$$\underset{\hat{P}_1,...\hat{P}_n}{\text{maximize}} \sum_{i=1}^{n} \alpha_i U_i\left(\hat{P}_i\right) \quad \text{subject to} \quad \sum_{i=1}^{n} P_i = 1 \quad (15)$$

Thus, the system behaves as if an administrator determines the relative reliabilities of all members, knowing that the selection must ultimately account for the various utility functions of the members as well as their respective optimization behaviors.

A peer-to-peer network's success in resisting a message-dropping attack can therefore be measured by computing the objective function in Equation (15). Combining the optimality result of the individual behavior described by Equation (13) with the administrative optimal behavior described by Equation (15), we find (see the appendix) that consistency between the individual behavior and the administrator's optimal solution requires that $\alpha_i = (\delta/\Phi_i)\hat{P}_i$ where δ is the Lagrange multiplier associated with the constraint in Equation (15) and $\Phi_i = \left(\partial U_i / \partial \hat{P}_i\right) P_i$

This implies that weight α_i is directly related (up to common factor δ) to peer i's relative reliability and inversely related to its marginal utility. Using $U_i = \log$ yields $\varphi_i = 1$ for all $i \in 1...n$. Since constant factor δ has no effect on optimization problem 15, this simplifies to $\alpha_i = \hat{P}_i$

Thus, a peer-to-peer network's success in resisting a message-dropping attack is measurable via the following social welfare function:

$$\sum_{i=1}^{n} \hat{P}_i \log(P_i) \tag{16}$$

where $\hat{P}_i$ is the fraction of the messages forwarded by peer i that were ultimately delivered to their destinations. This can be interpreted as a familiar result from information theory. It is the negation of the Shannon entropy of the peer-to-peer system, and the administrative problem therefore reduces to the problem of minimizing the entropy subject to the constraints. When there are no additional constraints, the optimal solution is obviously one in which all peers are equally reliable–i.e., $\hat{P}_i = P_j$ $\forall i, j \in 1...n$. In the case where there are additional constraints (e.g., some peers are malicious and therefore have constrained reliabilities), the optimal solution is nontrivial, as we see in §5.

4.3. Risk

In coordinated, distributed, message-dropping attacks, malicious peers vary their behavior over time, dropping some but not all messages they receive in an effort to evade detection. Malicious peers may even coordinate their behavior changes so as to keep each individual peer's reliability relatively high while keeping overall availability of network services low. Peer reliability alone is not an adequate measure of malicious behavior during such an attack; one must also consider variance or risk.

Following the work of von Neumann and Morgenstern[45], there has been strong agreement that individuals tend to maximize expected utility in the presence of risk. The negative exponential utility function is most commonly used to examine effects of both mean and variance in such contexts:

$$U_i(P_i) = -\exp(-a_i P_i) \quad \forall i \in 1...n \qquad (17)$$

where parameter $a_i > 0$ is a measure of the risk aversion of peer i. (Equation (17) satisfies the assumptions about utility given by Equation 11, and approximates the natural logarithmic utility function used in §4[46].e) When random variable Pi ($i \in 1...n$) has an approximately normal distribution with mean E[Pi] and variance Var[Pi], the expected utility is

$$E[U_i(P_i)] = -\exp\left(-a_i E[P_i] + \frac{1}{2}a_i^2 Var[P_i]\right) \qquad (18)$$

With risk, the administrative perspective differs from §4.2. There, utility U_i is a function of only one argument P_i and increases monotonically with P_i; but with risk, expected utility is a function of two arguments $E[P_i]$ and $Var[P_i]$ and is concave. Assuming there is at east one interior, feasible solution, we can again write the administrative objective function as a weighted sum of the expected utilities of the peers. Substituting these into Equation (5), the administrative optimization problem becomes

$$\text{maximize} \sum_{i=1}^{n} \alpha_i E\left[U_i(\hat{P}_i)\right] \qquad (19)$$

$$\text{subject to} \sum_{i=1}^{n} E\left[\hat{P}_i\right] = 1 \text{ and } \sum_{i,j=1}^{n} Cov\left[P_i, P_j\right] = 0 \qquad (20)$$

where $\hat{P}_i$ is the relative reliability of peer i as defined in §4.2 and Cov[X, Y] denotes the covariance of random variables X and Y.

The first best solution is obviously one in which all peers are invariably equally reliable. Since zero variation implies $E[\hat{P}_i] = P_i$, this reduces to the same optimization problem as derived in §4.2; we therefore conclude that $\alpha_i = \hat{P}_i$ as before. When the variance is non-zero for some peers, we seek a second best solution. In that case the form of the social welfare function stays the same but is evaluated at the second best solution. Thus, $\alpha_i = \hat{P}_i$ in that case as well, and we conclude that social welfare can be measured by weighting each peer's expected utility by its relative reliability. The optimal conditions for Equations (19)–(20) are derived in the appendix.

We next consider how individual peers make optimal decisions in the presence of risk in a way that is consistent with the administrative perspective above. Using Equation (12), the expected reliability and variance of each peer $i \in 1...n$ is

$$E[P_i] = \sum_{j=1}^{k} w_{ij} E[P_j] \qquad (21)$$

$$Var[P_i] = \sum_{j=1}^{k} w_{ij}^2 Var[P_j] + 2\sum_{j=1}^{k} \sum_{h=j+1}^{k} w_{ij} w_{ih} Cov[P_j, P_h] \qquad (22)$$

where $w_{ij} \in [0,1]$ is again the relative share of messages that peer i forwards to neighbor j. Substituting these into the Kuhn-Tucker conditions for Equations (19)–(20) (see the appendix) yields the following system of linear equations that must be solved to find peer i's optimal relative use w_{ij} of each of its neighbors $j \in 1...k$:

$$\frac{1}{a_i}\left(E[P_b] - E[P_j]\right) = \sum_{h=1}^{k}\left(Cov[P_h, P_b] - Cov[P_h, P_j]\right) \qquad (23)$$

Since the shares w_{ij} are all relative, we choose some arbitrary benchmark neighbor $b \in 1...k$ in terms of which peer i computes the other optimal shares.

One approach to implementing the above in actual peer-to-peer client software is with a linear constraint solver. The system of linear equations given by 23 can be expressed as a matrix computatio of the form

$$(\mathbf{C} - \mathbf{D})\mathbf{w} = \mathbf{E} \qquad (24)$$

where $C_{hj} = Cov[P_h, P_j]$ is the covariance matrix; $D_{hj} = Cov[P_h, P_b]$ if $h \neq b$ and $D_{bj} = 0$ otherwise; $E_j = (E[P_b]-E[P_j])/a_i$ if $j \neq b$ and $E_b = 1$ otherwise; and w is the unknown length—k vector of relative shares for which the system must be solved. The problem can be further simplified if we assume that under normal conditions most of the covariance terms for any individual peer are likely to vanish. We can therefore approximate the above solution by setting them to zero, which simplifies to the following formula for computing the optimal relative use of neighbor $j \in 1...k$ by peer $i \in 1...n$:

$$w_{ij} = \frac{E[P_j] - E[P_b]}{a_i Var[P_j]} + \frac{Var[P_b]}{Var[P_j]} w_{ib} \quad (25)$$

Peers that use Equation (25) to guide their relative usage of their neighbors tend to maximize expected reliability and minimize risk as they forward messages. This can have some interesting ramifications for peer behavior. For example, depending on their risk aversion a_i, they may sometimes forward messages through less reliable peers to avoid a more reliable but much riskier one. Risk-averse peers also tend to diversify their message-forwarding behavior similar to an investor's diversification of a portfolio. This can result in a better outcome when resending a dropped message since there is a higher chance that the message will not take the same route to its destination even when the overlay topology remains static.

5. Implementation

To put our approach into practice, we implemented it within a Chord network[10]. We begin with a review of Chord's overlay structure and routing protocol in §5.1. Section 5.2 then formulates the Chord protocol as a utility optimization problem using the second best. Finally, §5.3 describes our experimental methodology and results.

5.1. The Chord Potocol

Chord[10] is a structured peer-to-peer protocol with a ring-shaped overlay. Each peer's ring position is defined by an integer identifier. Identifiers are derived via secure hash functions so that attackers cannot easily choose their positions. Each peer is directly connected to k = $\lfloor \log_2 n \rfloor$ neighbors, where n is the size of the identifier space. For example, in a Chord network that can accommodate 2^{160} peers, each peer has 160 neighbors.

The neighbor set of peer i is densest near i and thins farther away. Specifically, the jth neighbor of peer i is the peer whose identifier is closest to (but no less than) $(id_i + 2^{j-1}) \mod n \, (\forall j \in 1...k)$. Thus, peer i's first neighbor is its successor in the ring, each subsequent neighbor is approximately twice as far from i as its previous neighbor, and peer i's last neighbor is approximately halfway around the ring. To send a message to peer h, peer i forwards it to the neighbor whose identifier is closest to but no greater than h's identifier (modulo n). When all peers adhere to this protocol, messages are delivered to their final destinations in at most $O(\log_2 n)$ hops because each hop at least halves the distance from the message's current position to its destination. Without malicious peers, the topology is naturally load-balancing in that a uniform distribution of message sources and destinations tends to solicit equal relative use of each peer's k neighbors.

During a message-dropping attack, however, malicious peers drop the messages they receive instead of forwarding them. Since Chord is deterministic, a single malicious peer on the route from i to h can thereby prevent i from sending any messages to h until the topology changes (e.g., due to churn). With multiple attackers, the identifier assignment process tends to distribute attackers approximately uniformly across the identifier space. As a result, attackers can intercept a significant portion of the overlay traffic. For example, even when malicious peers comprise only 10% f the network they can intercept about 40% of the messages on average[47].

5.2. Applying the Second Best Approach to Chord

Peers can forward messages via different neighbors than the ones prescribed by the Chord protocol at the expense of longer message delivery paths. This flexibility allows peers to potentially improve message delivery rates in the presence of malicious peers via a second best routing strategy. Specifically, a peer can potentially forward each message to any neighbor between itself and the message's intended destination, not just the closest one to the destination.

However, this flexibility must be exercised in moderation to avoid unacceptably long routing paths, since forwarding messages in very small hops greatly increases the worst-case path length bound given in §5.1. For example, if each peer forwards messages to its nearest neighbor, the worst-case path length is $O(n)$, which is clearly unreasonable when $n \approx 2160$. More generally, when peers forward messages to their rth-closest neighbors, the worst-case path length increases by a factor of $(r - \log_2(2^r - 1))^{-1}$. Hence, forwarding to the 2nd-closest neighbor multiplies the worst-case path length by a factor of about 2.4, and forwarding to the 3rd-closest multiplies it by a factor of over 5.

To keep the worst-case path length reasonable, we therefore modify the Chord protocol to allow (non-malicious) peers to forward each message only to the closest neighbor or 2nd-closest neighbor to the message's intended final destination. In our experiments we found that allowing peers to forward to other neighbors is seldom useful, since during a message-dropping attack greatly increased path lengths almost always include at least one malicious peer.

Given this restriction, the reliability P_i of any peer $i \in 1...n$ can be expressed in terms of the reliabilities of its neighbors as follows:

$$P_i = s_1 P_1 + \sum_{j=2}^{k} s_j \left(w_{ij} P_j + d(1 - w_{ij}) P_{j-1} \right)$$

where $s_j \in [0,1]$ is the fraction of the messages seen by peer i whose destinations are closest to neighbor j, $w_{ij} \in [0,1]$ is the share of those sj messages that peer i chooses to forward to neighbor j (instead of to neighbor j–1), and $d \in [0,1]$ models a istance penalty for forwarding messages to the 2nd-closest neighbor instead of to the closest one. (In our implementation we used d = 0.8, but other values in the interval [0.1, 0.9] performed similarly.) The expected value and variance of P_i are

$$E[P_i] = s_k w_{ik} E[P_k] + \sum_{j=1}^{k-1} \left(s_{j+1} d(1 - w_{ij+1}) + s_j w_{ij} \right) E[P_j] \qquad (26)$$

$$Var[P_i] = s_k^2 w_{ik}^2 Var[P_k] + \sum_{j=1}^{k-1} \left(s_{j+1} d(1 - w_{ij+1}) + s_j w_{ij} \right)^2 Var[P_j] \qquad (27)$$

respectively, for all $i \in 1...n$.

Individual utility is modeled by Equation (17) with the addition of a unit constant to force non-negative utilities[48]; hence, $U_i(P_i) = 1 - \exp(-a_i P_i)$. Solving the resulting optimization problem given by Equation (23) (and zeroing the covariance terms as in §4.3), yields a system of linear equations of the form A = 0, where A is the n × k matrix defined by

$$A_{ij} = -s_{ij}\left(E[P_{ij}] - dE[P_{ij-1}]\right) + as_{ij}\left[\left(s_{ij}w_{ij} + s_{ij+1}d(1-w_{ij+1})\right)Var[P_{ij}]\right.$$
$$\left. - d\left(s_{ij-1}w_{ij-1} + s_{ij}d(1-w_{ij})\right)Var[P_{ij-1}]\right] \quad (28)$$

for all $j \in 2...k-1$, and for $j \in \{1, k\}$ we have $A_{i1} = 0$ and

$$A_{ik} = -s_{ik}\left(E[P_{ik}] - d\,E[P_{ik-1}]\right)$$
$$+ as_{ik}\left[s_{ik}w_{ik}Var(P_{ik}) - d\left(s_{ik-1}w_{ik-1} + s_{ik}d(1-w_{ik})\right)Var[P_{ik-1}]\right] \quad (29)$$

Fully solving the above system subject to constraint $w_{ij} \in [0,1]$ requires a mixed integer programming algorithm. However, the implementation can be significantly simplified by approximating the solution iteratively. We used a quasi-Newtonian approximation obtained by computing the Hessian matrix of Equations (28)–(29), zeroing the cross-derivatives, and solving for w. This yields the following rule for updating share w_{ij}:

$$\Delta w_{ij} = \frac{-(A_{ij})}{(A_{ij})^2 + a_i s_{ij}^2\left(Var[P_{ij}] + d^2 Var[P_{ij-1}]\right)}\psi \quad \forall j \in 2...k \quad (31)$$

$$w_{i1} = 1 \quad (32)$$

where A is defined by Equations (28)–(29) and ψ controls the rate of convergence. (In our implementation we used ψ = 1.) Equation (31) reflects the inflexibility of traffic forwarded to a peer's first neighbor (since no neighbors fall between a peer and its first neighbor).

In summary, non-malicious peers in our modified system continuously adjust their relative usage of neighbors in small increments, based on the most current information available concerning neighbor reliability and riskiness. These adjustments are made so as to maximize reliability and minimize risk. That is, each peer optimizes its own expected utility subject to the constraints imposed by the routing protocol.

5.3. Experimental Results

To test our solution, we simulated a Chord network in which non-malicious peers maximize expected utility by adapting their relative use of their neighbors according to Equation (30). Malicious peers drop some or all messages they are asked to forward. To assess the network's success in resisting the attack, we computed the social welfare [Equation (16)] that it attained over each simulation. We also measured the total percentage of messages that were successfully delivered. Each simulation involved sending a total of one million randomly generated messages through the overlay, and simulation results were averaged over 50 trials each.

We assume that senders learn whether their messages were ultimately delivered, but not who dropped undelivered messages. This is consistent with networks in which delivered messages solicit unforgeable, direct responses from recipients. For example, object lookups in Chord solicit a direct response that does not use the overlay, and that can be authenticated via cryptographic message signing. This information allows each peer i to estimate a running mean $E[P_{ij}]$ and running variance[f] $Var[P_{ij}]$ for each neighbor $j \in 1...k$. Each peer also tracks its own relative usage s_{ij} of each neighbor.

Non-malicious peers begin the simulation with $w_{ij} = 1$ and $s_{ij} = 1$ for all $i \in 1...n$ and $j \in 1...k$. That is, each peer initially behaves as in a traditional Chord network, forwarding each message to the closest neighbor and using all neighbors approximately equally. At regular intervals, non-malicious peers modify w_{ij} according to Equation (30). (If w_{ij} rises above 1 or descends below 0, it is truncated down to 1 r up to 0, respectively.) In our simulation, peers recomputed w_{ij} after every 1000 messages they sent. We used a convergence rate of $\psi = 1$, a distance penalty of $d = 0.8$, and a risk aversion of $a_i = 1$ to strike a roughly even balance between reliability maximization and risk minimization.

Figure 1 shows the overall utilities (i.e., social welfare) attained by a traditional Chord network, an adaptive Chord network that uses our utility optimization

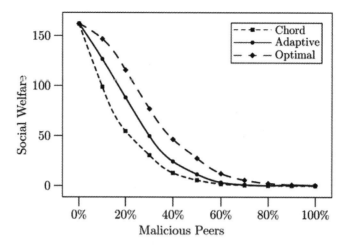

Figure 1. Social welfare attained during message-dropping attacks.

procedure, and an optimal Chord network in which peers have global knowledge of the overlay topology and know the identifiers of all malicious peers. For each attack, malicious peers dropped all messages they received (other than the messages intended for themselves) and were roughly uniformly distributed throughout the overlay.

The curve for the optimal Chord network was computed by exhaustively deciding for each possible source-destination pair whether there exists a route through the overlay that does not include any malicious peers (subject to the constraint that non-malicious peers must not route messages farther away than their 2nd-closest neighbor to the message's intended destination). We simulated networks with up to 10K peers, but the number of peers did not influence any of our results (except that computing the optimal curve for very large networks was not feasible). The curves shown in **Figure 1** are for a network with 256 peers.

As **Figure 1** illustrates, once malicious peers comprise over half the network, our adaptive approach is unable to provide substantial improvement; however, we see significant gains in the more typical scenarios where malicious peers comprise 10%–30% of the network. Under those conditions the adaptive approach resulted in about a 60% increase in social welfare compared to a traditional Chord network—about 55% of the gain that was possible even with global information. **Figure 2** verifies that this increase in social welfare translates to a corresponding increase in message deliveries. Message delivery success rates were increased by

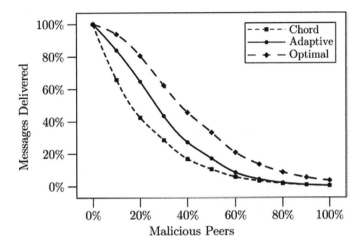

Figure 2. Message delivery success rates during message-dropping attacks.

about 52% when malicious peers comprised 10%–30% of the network–about 58% of what was possible with global information.

We next considered a more sophisticated message-dropping attack in which the attackers vary their behavior over time in an effort to avoid detection. Malicious peers coordinate these behavior changes to keep each malicious peer's observed reliability relatively high while keeping overall network connectivity low. In our simulation, attackers chose their reliabilities from a normal distribution of mean 0.3 and variance 0.12. A coordinated, distributed, denial-of-service attack of this kind can be quite effective against defense mechanisms that rely on average reliability as the sole indicator of maliciousness. Our protocol's inclusion of risk as a secondary indicator was therefore important for resisting this attack.

Figure 3 shows that our adaptive approach continues to provide effective recommendations to non-malicious peers during the attack. The three curves are closer than in the non-coordinated attack since malicious peers in this simulation deliver at least some messages. Nevertheless, the adaptive network is still able to achieve a 30% increase in social welfare (58% of what was possible with global information) when attackers comprise 10%–30% of the network. Likewise, the message delivery rates reported in **Figure 4** show a 28% increase (60% of what was possible with global information). Most importantly, a point-by-point comparison reveals that introducing attacker behavior variations only increased social welfare and message deliveries compared to when they dropped all messages.

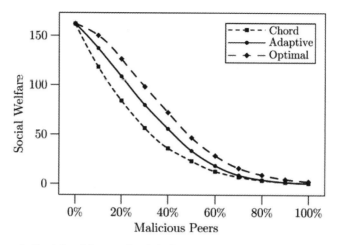

Figure 3. Social welfare attained during coordinated, distributed attacks.

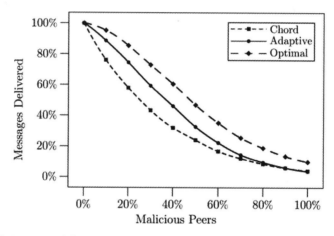

Figure 4. Message delivery success rates during coordinated, distributed attacks.

Thus, behavior variations were ineffective when attacking Chord networks equipped with our adaptive routing protocol.

During testing, our utility optimization strategy demonstrated little sensitivity to parameter changes and implementation details. For example, different convergence rates ψ, different distance penalties d, different approximation methods for Equations (28)–(29), and different refresh rates for recomputing shares w_{ij} resulted in little or no change to the results reported here (except when parameters were set to extreme values). This seems to indicate that our method is does not require much manual tuning to perform well.

6. Conclusion

The theory of the second best has played a significant role over the past several decades in solving numerous important problems in economics. In this article we have shown that it also applies to the problem of resisting message-dropping attacks in peer-to-peer overlay networks. If one views the design of the underlying overlay topolgy as an optimization problem, the second best solution yields recommendations on how to make optimal use of that existing topology in the presence of malicious peers who drop messages.

This has implications both for individual peers and for the peer-to-peer system as a whole. For individual peers, the second best solution provides peer-specific recommendations on how to forward messages so as to maximize each peer's individual reliability and minimize its risk. For the system as a whole, it maximizes the overall objective of the system given the misbehavior of the attackers. We found that in this context the overall objective can be expressed as a weighted sum of the utility functions of the individual peers, where the weights are the relative reliabilities of the peers. When individual utility functions are standard Bernoulli logarithmic functions, this equates to minimizing the Shannon entropy of the peer-to-peer system.

As a practical application of our work, we solved the above optimization problem for the Chord peer-to-peer network protocol[10] and implemented it in a simulator. Non-malicious peers in our modified network forward messages according to the recommendations prescribed by the second best solution. Rather than compute the second best solution directly, each peer approximates it iteratively using an efficient quasi-Newtonian algorithm. We simulated simple message-dropping attacks in which attackers drop all messages, as well as coordinated, distributed message-dropping attacks in which attackers vary their behavior to avoid detection. The modified protocol achieves a 50% increase in message deliveries and a 60% increase in social welfare when malicious peers comprise about 20% of the network. Behavior variations are not effective as a means of disguising the attack; they only result in higher message delivery rates and higher social welfare in networks equipped with our adaptive protocol.

In future work we intend to apply our approach to other distributed computing paradigms, such as clouds. Tapestry[27] networks are more densely connected than Chord networks, incorporating extra routing links for improved fault tolerance. These extra links could provide more opportunities for second best optimization. CAN[25] poses interesting mathematical challenges for our method since it uses a multidimensional identifier space.

In addition, much prior work on adaptive overlay routing has focused on adapting the overlay topology in response to observed peer behavior and performance. Since our second best optimization approach takes the topology as given, it could be implemented atop one of these adaptive topologies. Future work should investigate the interaction between these two approaches.

Finally, we plan to investigate second best optimization approaches to protecting these networks from other forms of attacks, such as message misrouting, message integrity and confidentiality violations, and reputation mismanagement. These are all significant current-day threats to large, distributed data management systems, and would likely benefit from second best optimization.

Appendix

We here sketch derivations of the solutions of the three main optimization problems presented throughout the paper. We begin with the problem of individual utility optimization presented in §4.1. The necessary (Kuhn-Tucker) conditions for the problem given by Equation (13) are

$$\frac{\partial U_i}{\partial P_i} + \lambda_i \geq 0 \qquad P_i\left(\frac{\partial U_i}{\partial P_i} + \lambda_i\right) = 0 \qquad (32)$$

$$\Phi_i - \lambda_i d_{ij} w_{ij}^{-c}(1-c) \geq 0 \qquad w_{ij}\left(\Phi_i - \lambda_i d_{ij} w_{ij}^{-c}(1-c)\right) = 0 \qquad (33)$$

$$1 - \sum_{j=1}^{k} w_{ij} \geq 0 \qquad \Phi_i\left(1 - \sum_{j=1}^{k} w_{ij}\right) = 0 \qquad (34)$$

$$P_i - \sum_{j=1}^{k} d_{ij} w_{ij}^{(1-c)} \geq 0 \qquad \lambda_i\left(P_i - \sum_{j=1}^{k} d_{ij} w_{ij}^{(1-c)}\right) = 0 \qquad (35)$$

for all $i \in 1...n$, where φ_i is the Lagrange multiplier associated with the constraint.

Equation (33) implies that $\varphi_i > 0$ for any positive share $w_{ij} > 0$ with $d_{ij} > 0$, and that for any two positive shares $w_{ij}, w_{im} > 0$ and $d_{ij}, d_{im} > 0$ with $c > 0$ the ratio of shares is

$$\frac{w_{im}}{w_{ij}} = \left(\frac{d_{im}}{d_{ij}}\right)^{1/c}$$

The assumption in §4.1 that peer m is a factor ε less reliable than peer j implies that $d_{im} = \varepsilon d_{ij}$. This yields the result given by Equation (14).

We next consider the administrative optimization problem presented in §4.2. The optimal (Kuhn-Tucker) conditions associated with Equation (15) are[g]

$$\alpha_i \frac{\partial U_i}{\partial \hat{P}_i} - \delta \leq 0 \qquad \hat{P}_i\left(\alpha_i \frac{\partial U_i}{\partial \hat{P}_i} - \delta\right) = 0 \qquad \forall i \in 1...n \qquad (36)$$

$$1 - \sum_{i=1}^{n} \hat{P}_i \geq 0 \qquad \delta\left(1 - \sum_{i=1}^{n} \hat{P}_i\right) = 0 \qquad (37)$$

where δ is the Lagrange multiplier associated with the constraint in Equation (15). By combining the optimality result of the individual behavior given by Equations (32)–(35) with the optimal behavior from the administrative perspective given by Equations (36)–(37), we see that whenever $\hat{P}_i > 0$, consistency between the individual behavior and the administrator's optimal solution requires that

$$\alpha_i = \frac{\delta}{\Phi_i} \hat{P}_i \quad \text{and} \quad \Phi_i = \frac{\partial U_i}{\partial \hat{P}_i} \hat{P}_i$$

Section 4.2 describes how this result leads directly to the conclusion that $\alpha_i = \hat{P}_i$.

Finally, we consider the optimization problem given in §4.3, which introduces risk. The necessary (Kuhn-Tucker) conditions for the administrative optimization problem given by Equations (19)–(20) are

$$\alpha_i a_i E\left[U_i\left(\hat{P}_i\right)\right] + \delta \geq 0$$
$$E\left[\hat{P}_i\right]\left(\alpha_i a_i E\left[U_i\left(\hat{P}_i\right)\right] + \delta\right) = 0 \quad (38)$$

$$\frac{1}{2}\alpha_i a_i^2 E\left[U_i\left(\hat{P}_i\right)\right] Var\left[P_i\right] - \gamma \leq 0$$
$$\left(\frac{1}{2}\alpha_i a_i^2 E\left[U_i\left(\hat{P}_i\right)\right] Var\left[P_i\right] - \gamma\right) Var\left[P_i\right] = 0 \quad (39)$$

for all $i \in 1 \ldots n$, and

$$\delta\left(1 - \sum_{i=1}^{n} E\left[\hat{P}_i\right]\right) = 0 \quad (40)$$

where δ and γ are the Lagrange multipliers. Recall that $\alpha_i = \hat{P}_i$ (see §4.3); therefore the individual optimization problem in Equations (21)–(22) contributes the following additional (Kuhn-Tucker) conditions:

$$w_{ij}\left(\frac{\partial E[U_i]}{\partial w_{ij}} - \Phi_i\right) = 0 \quad (41)$$

$$1 - \sum_{j=1}^{k} w_{ij} \geq 0 \quad \Phi_i\left(1 - \sum_{j=1}^{k} w_{ij}\right) = 0 \quad (42)$$

Equation (41) can be reexpressed as $w_{ij}\left(E[U_i] z_{ij} - \varphi_i\right)$, where z_{ij} is defined by

$$z_{ij} = -a_i E\left[P_j\right] + a_i^2 \sum_{h=1}^{k} w_{ih} Cov\left[P_j, P_j\right] \quad (43)$$

In the above, φ_i is the Lagrange multiplier associated with the constraint requiring that each peer i's relative usage of its neighbors sums to 1. In most situations the constraint would be binding and $\varphi_i > 0$. This complicates the solution for any single w_{ij}. We can, however, examine the ratio $E[U_i]z_{ij} = \varphi_i$ to $E[U_i]z_{ib} = \varphi_i$ for any two shares w_{ij}, $w_{ib} > 0$. The benchmark neighbor b is arbitrarily chosen by peer i as the standard by which its other neighbors are assessed. From the conditions above we see that $z_{ij} = z_{ib}$, leading to the system of linear equations given in Equation (23).

Notes

[a]This "as if" approach is attributed to Nobel laureate Milton Friedman[49], who recognized that participants need not know or understand a model for it to adequately explain an outcome. Kenneth Arrow[50], another Nobel laureate, later proved that there are limits on what any welfare function can achieve. [b]Samuelson[39] and others used the Lagrange method, which requires binding constraints. Subsequent work used the Kuhn-Tucker conditions[44], which allow non-binding constraints. [c]If λ = hj in Equation 4, the Lagrange multiplier μ is zero, re-attaining the first best solution. Only when $\lambda \neq$ hj does the added constraint change the solution. [d]Hamlen and Hamlen[51] show that Negishi's solution, while important in defining the social welfare function, can be evaluated only at the equilibrium solution and therefore does not aid in obtaining the solution. [e]This follows from isoelasticity of natural logarithm[46]. [f]Neither running mean nor running variance require maintaining any message history. We computed running variance via $Va[rP_{ij}] = E[P_{ij}^2] - E[P_{ij}]^2$. [g]Partial derivatives ∂ are used here to remind us that the administrator's objective function contains all $P_i \forall i \in 1...n$.

Acknowledgements

This work includes material supported by the National Science Foundation under award NSF-0959096. Any opinions or findings expressed are those of the authors and not of the NSF.

Authors' Contributions

KWH conceived of the research, implemented the Chord simulator, carried out the experiments, and analyzed the results. WH provided economic interpretations and solved the optimization problems. Both authors drafted and approved the final manuscript.

Competing Interests

The authors declare that they have no competing interests.

Source: Hamlen K W, Hamlen W. An economic perspective of message-dropping attacks in peer-to-peer overlays [J]. Security Informatics, 2012, 1(1): 1–22.

References

[1] Marozzo F, Talia D, Trunfio P: A Peer-to-peer Framework for Supporting MapReduce Applications in Dynamic Cloud Environments. In Cloud Computing: Principles, Systems and Applications. Edited by: Antonopoulos N, Gillam L. Springer; 2010.

[2] Alexander PJ: Peer-to-Peer File Sharing: The Case of the Music Recording Industry. Review of Industrial Organization 2002, 20:151–161.

[3] Amrou A, Maly K, Zubair M: Freelib: Peer-to-peer-based Digital Libraries. Proceedings of the 20th International Conference on Advanced Information Networking and Applications (AINA) Vienna, Austria; 2006, 9–14.

[4] Hamlen KW, Thuraisingham B: Secure Peer-to-peer Networks for Trusted Collaboration. Proceedings of the 3rd International Conference on Collaborative Computing: Networking, Applications and Worksharing White Plains, New York; 2007, 58–63.

[5] Chou PH, Ortega RB, Borriello G: The Chinook Hardware/Software Co-synthesis System. Proceedings of the 8th International Symposium on System Synthesis (ISSS) Cannes, France; 1995, 22–27.

[6] Douceur JR, Donath JS: The Sybil Attack. Proceedings of the 1st International Workshop on Peer-to-Peer Systems (IPTPS) Cambridge, MA; 2002, 251–260.

[7] Lipsey RG, Lancaster K: The General Theory of Second Best. The Review of Economic Studies 1956, 24:11–32.

[8] Hamlen W: Gleanings into the Second Best Debate. Business Modelling: Multidisciplinary Approaches–Economics, Operational, and Information Systems Perspectives (In Honor of Andrew B. Whinston) 2002.

[9] Davis OA, Whinston AB: Piecemeal Policy of the Second Best. The Review of Eco-

nomic Studies 1967, 34(3):323–331.

[10] Stoica I, Morris R, Karger D, Kaashoek MF, Balakrishnan H: Chord: A Scalable Peer-to-peer Lookup Service for Internet Applications. IEEE/ACM Transactions on Networking 2003, 11:17–32.

[11] Wallach DS: A Survey of Peer-to-Peer Security Issues. Proceedings of the International Symposium on Software Security—Theories and Systems, Mext-NSF-JSPS (ISSS) Tokyo, Japan; 2002, 42–57.

[12] Risson J, Moors T: Survey of Research towards Robust Peer-to-peer Networks: Search Methods. Computer Networks 2006, 50(17):3485–3521.

[13] Kamvar SD, Schlosser MT, Garcia-molina H: The EigenTrust Algorithm for Reputation Management in P2P Networks. Proceedings of the 12th International World Wide Web Conference (WWW) Budapest, Hungary; 2003, 640–651.

[14] Walsh K, Sirer EG: Experience with an Object Reputation System for Peer-to-Peer Filesharing. Proceedings of the 3rd Symposium on Networked System Design and Implementation (NSDI) San Jose, California; 2006.

[15] Tsybulnik N, Hamlen KW, Thuraisingham B: Centralized Security Labels in Decentralized P2P Networks. Proceedings of the Annual Computer Security Applications Conference (ACSAC) Miami, Florida; 2006, 315–324.

[16] Castro M, Druschel P, Ganesh A, Rowstron A, Wallach DS: Secure Routing for Structured Peer-to-peer Overlay Networks. Proceedings of the 5th Symposium on Operating Systems Design and Implementation (OSDI) Boston, Massachusetts; 2002, 299–314.

[17] Roughgarden T, Tardos E: How Bad is Selfish Routing? Journal of the ACM 2002, 49(2):236–259.

[18] Cole R, Dodis Y, Roughgarden T: Pricing Network Edges for Heterogeneous Selfish Users. Proceedings of the 35th Annual ACM Symposium on Theory of Computing (STOC) San Diego, California; 2003, 521–530.

[19] Feldman M, Chuang J, Stoica I, Shenker S: Hidden-action in Multi-hop Routing. Proceedings of the 6th ACM Conference on Electronic Commerce (EC) Vancouver, British Columbia; 2005, 117–126.

[20] Blanc A, Liu YK, Vahdat A: Designing Incentives for Peer-to-Peer Routing. Proceedings of the 24th Annual Joint Conference of the IEEE Computer and Communications Societies (INFOCOM) Miami, Florida; 2005, 374–385.

[21] Ngan TW, Wallach DS, Druschel P: Incentives-Compatible Peer-to-Peer Multicast. Proceedings of the 2nd Workshop on the Economics of Peer-to-peer Systems (IPTPS) Berkeley, California; 2003, 149–159.

[22] Krishnan R, Smith M, Telang R: The Economics of Peer-to-peer Networks. Journal of Information Technology Theory and Application (JITTA) 2003, 5(3):31–44.

[23] Hardin G: The Tragedy of the Commons. Science 1968, 162:1243–1248.

[24] Feldman M, Chuang J: Overcoming Free-Riding Behavior in Peer-to-Peer Systems. ACM SIGecom Exchanges 2005, 5(4):41–50.

[25] Ratnasamy S, Francis P, Handley M, Karp R, Schenker S: A Scalable, Content-addressable Network. Proceedings of the ACM Conference on Applications, Technologies, Architectures, and Protocols for Computer Communication (SIGCOMM) San Diego, California; 2001, 161–172.

[26] Rowstron A, Druschel P: Pastry: Scalable, Decentralized Object Location and Routing for Large-scale Peer-to-peer Systems. Proceedings of the IFIP/ACM International Conference on Distributed Systems Platforms (Middleware) Heidelberg, Germany; 2001, 329–350.

[27] Zhao BY, Huang L, Stribling J, Rhea SC, Joseph AD, Kubiatowicz JD: Tapestry: A Resilient Global-scale Overlay for Service Deployment. IEEE Journal on Selected Areas in Communications (JSAC) 2004, 22:41–53.

[28] Lv Q, Cao P, Cohen E, Li K, Shenker S: Search and Replication in Unstructured Peer-to-peer Networks. Proceedings of the 16th International Conference on Supercomputing New York; 2002, 84–95.

[29] Artigas MS, López PG, Skarmeta AG: A Novel Methodology for Constructing Secure Multipath Overlays. IEEE Internet Computing 2005, 9(6):50–57.

[30] Fiat A, Saia J, Young M: Making Chord Robust to Byzantine Attacks. Proceedings of the 13th Annual European Symposium on Algorithms (ESA) Mallorca, Spain; 2005, 803–814.

[31] Naor M, Wieder U: A Simple Fault Tolerant Distributed Hash Table. Proceedings of the 2nd International Workshop on Peer-to-peer Systems (IPTPS) Berkeley, California; 2003, 88–97.

[32] Ren S, Guo L, Jiang S, Zhang X: SAT-Match: A Self-adaptive Topology Matching Method to Achieve Low Lookup Latency in Structured P2P Overlay Networks. Proceedings of the 18th International Symposium on Parallel and Distributed Processing (IPDPS) Santa Fe, New Mexico; 2004, 83–91.

[33] Hong F, Li M, Yu J, Wang Y: PChord: Improvement on Chord to Achieve Better Routing Efficiency by Exploiting Proximity. Proceedings of the 25th IEEE International Conference on Distributed Computing Systems Workshops (ICDCSW) Columbus, Ohio; 2005, 806–811.

[34] Condie T, Kamvar AD, Garcia-molina H: Adaptive Peer-to-Peer Topologies. Proceedings of the 4th International Conference on Peer-to-Peer Computing (P2P) Zurich, Switzerland; 2004, 53–62.

[35] Gatani L, Re GL, Gaglio S: An Adaptive Routing Protocol for Ad Hoc Peer-to-peer Networks. Proceedings of the 6th IEEE International Symposium on a World of Wireless, Mobile and Multimedia Networks (WoWMoM) Taormina, Italy; 2005, 44–50.

[36] Xie L, Zhu S: Message Dropping Attacks in Overlay Networks: Attack Detection and Attacker Identification. ACM Transactions on Information and Systems Security (TISSEC) 2008, 11(3).

[37] Danezis G, Lesniewski-Laas C, Kaashoek MF, Anderson R: Sybil-resistant DHT Routing. Proceedings of the 10th European Symposium on Research in Computer Security (ESORICS) Milan, Italy; 2005, 305–318.

[38] Stavrou A, Locasto ME, Keromytis AD: Pushback for Overlay Networks: Protecting against Malicious Insiders. Proceedings of the 6th International Conference on Applied Cryptography and Network Security (ACNS) New York, New York; 2008, 39–54.

[39] Samuelson PA: Foundations of Economic Analysis. Harvard University Press; 1947.

[40] Henderson JM, Quandt RE: Microeconomic Theory: A Mathematical Approach. Economics Handbook Series, McGraw Hill, 3, 1980.

[41] Negishi T: Welfare Economics and Existence of an Equilibrium for a Competitive Economy. Metroeconomica 1960, 12:92–97.

[42] Takayama A: Mathematical Economics. Cambridge University Press, 2, 1985, 117.

[43] Varian H: Economic Mechanism Design for Computerized Agents. Proceedings of the 1st Usenix Workshop on Electronic Commerce New York, 1995, 13–21.

[44] Kuhn HW, Tucker AW: Nonlinear Programming. Proceedings of the 2nd Berkeley Symposium on Mathematical Statistics and Probability 1951, 481–492.

[45] von Neumann J, Morgenstern O: Theory of Games and Economic Behavior. Princeton University Press, 2, 1944.

[46] Merton RC: Lifetime Portfolio Selection Under Uncertainty: The Continuous-time Case. Review of Economics and Statistics 1969, 51:247–257.

[47] Cooke J: A Comparison of the Effectiveness of Message Dropping Attacks against Structured Peer-to-peer Networks. Senior undergraduate honors thesis, University of Texas at Dallas, Richardson, TX 2008.

[48] Panzar JC, Sibley DS: Public Utility Pricing Under Risk: The Case of Self-Rationing. American Economic Review 1978, 68(5):888–895.

[49] Friedman M: Essays in Positive Economics. University of Chicago Press, 1953.

[50] Arrow KJ: A Difficulty in the Concept of Social Welfare. Journal of Political Economy 1950, 58(4):328–346.

[51] Hamlen W, Hamlen KW: A Closed System of Production Possibility and Social Welfare. Computers in Higher Education Economics Review 2006, 18:15–18.

Chapter 25

Fluency of Visualizations: Linking Spatiotemporal Visualizations to Improve Cybersecurity Visual Analytics

Zhenyu Cheryl Qian[1], Yingjie Victor Chen[2]

[1]Interaction Design, Purdue University, 552 W. Wood Street, 47907 West Lafayette IN, USA
[2]Computer Graphics Technology, Purdue University, 402 S. Grant Street, 47907 West Lafayette IN, USA

Abstract: This paper adopts the metaphor of representational fluency and proposes an auto linking approach to help analysts investigate details of suspicious sections across different cybersecurity visualizations. Analysis of spatiotemporal network security data takes place both conditionally and in sequence. Many visual analytics systems use time series curves to visualize the data from the temporal perspective and maps to show the spatial information. To identify anomalies, the analysts frequently shift across different visualizations and the original data view. We consider them as various representations of the same data and aim to enhance the fluency of navigation across these representations. With the auto linking mechanism, after the analyst selects a segment of a curve, the system can automatically highlight the related area on the map for further investigation, and the selections on the map or the data views can also trigger the related time series curves.

This approach adopts the slicing operation of the Online Analytical Process (OLAP) to find the basic granularities that contribute to the overall value change. We implemented this approach in an award-winning visual analytics system, SemanticPrism, and demonstrate the functions through two use cases.

Keywords: Representational Fluency, Cybersecurity Analysis, Spatiotemporal Visualization, Interaction Design

1. Introduction

One of the biggest challenges the information security society faces is analyzing large-scale spatiotemporal datasets. In most organizations and companies, their computer networks are routinely capturing huge volumes of historical data describing the network events. Most of these events are recorded as spatiotemporal data because every event takes place at a certain time and in a certain location. The location could be either a physical location (e.g., an office) or a virtual space (e.g., an Internet IP address)[1]. Different kinds of events have more detailed information, such as operations, products, targets, and human involvement, which can add more dimensions to the spatiotemporal database. As a result, such a dataset is usually both high-dimensional and very large.

Peuquet[2] identified three components in spatiotemporal data: space (where), time (when), and objects (what). Foresti et al.[3] also labeled when, where, and what (W3) as the three attributes of cybersecurity alerts and events because of their very nature. According to their definitions, when refers to the point in time where the event happened, where to the location of the event that happened, and what to the type of the event. The space of what and where are finite, and the when space is semi-infinite[3]. Finding the relations among these components and answering related questions are essential to analysis[4].

The two most popular methods to visualize and analyze the cybersecurity spatiotemporal data are geospatial visualization and time series curves. (1) The geospatial visualization is usually integrated with a time slider to adjust the time frame. The high-dimensional data are often displayed on the geospatial map with multiple views and layers overlaid with numerous data points, connections, and details. The visualization can easily overwhelm the display space on a single monitor. These types of visualizations challenge human cognition to remember what was seen previously, where it was, and its potential relationship to current information[5]. (2) The time series curves focus on providing the analyst an overview of how the data change over time. Significant value changes can be clearly reflected on the curve as peaks or valleys, which hint for the analyst to pay attention to these significant situations. This type of visualization is clean and easy to read, but it skips the context of spatial information.

In a survey of cybersecurity visualization techniques, Shiravi et al.[6] argued that user experience should be one of the key issues a successful visual analytics system should consider. The user experience is not only about elegant appearance or powerful functions, but also, and more importantly, about a smooth and fluent analysis process. Heer and Shneiderman also stressed that "visual analytics tools must support the fluent and flexible use of visualizations at rates resonant with the pace of human thought"[7]. However, in most cases the complex data and multiple visualizations lead to poor user experience.

This paper aims to promote the fluency of navigating in the spatiotemporal visualizations and to enhance the user experience of cybersecurity analysis. It demonstrates a solution to link the time series curves, geospatial visualizations, and data views together and to help the user achieve situational awareness through comprehension of the what, where, and when attributes of cybersecurity issues. This paper was originated from our previous work[8] that attempted to link the user from the temporal time series curve to geospatial visualizations. At this paper, we were able to extend the approach and its application to link the user in multiple directions among temporal visualization, geospatial visualization, and data view. We borrow the term "representational fluency"[9] from psychology and pedagogical literature to describe our efforts of enabling the user to fluently switch among different types of spatiotemporal visualizations and to more efficiently solve analysis tasks. The extensions in this paper include:

- A detailed explanation of the mechanism that selects portions from the time series curves and links to spatial visualizations. This mechanism was revised and extended.

- A new mechanism that reversely selects and links spatial visualizations and data views to time series curves.

- New use cases to demonstrate these two mechanisms.

- Redesigned interaction operations that allow the user to access information more smoothly.

To achieve smooth transitions across interactive visualizations, techniques such as brush and linking have been widely used in VA systems. This paper provides a practical technical mechanisms to link multiple visualizations and aims to help users gain better experience and improve performance when analyzing the big network security data visually.

2. Related Work

To enhance the user experience of cybersecurity visual analytics, we suggest adopting representational fluency in designing the structure of spatiotemporal visualizations because "users of this information will need fluency in the tools of digital access, exploration, visualization, analysis, and collaboration[10]". The literature review inspects two main components: representational fluency and spatiotemporal data visualization methods.

2.1. Representational Fluency of Visualizations

The concept of fluency is originally associated with the ability to express oneself in both spoken and written language and to move effortlessly between the two representations. Although fluency is often associated with language, researchers have extended fluency to other fields such as physics, chemistry, engineering, and mathematics. In these fields, fluency is the ability to understand and translate among commonly used modes of representation, such as verbal, mathematical, graphical, and manipulatable. In the context of information systems, fluency is the ability to access, make sense of, and use information to build new understandings[11]. Defined by Irving Sigel[9], representational fluency is the ability to (1) comprehend equivalence in different modes of expression; (2) comprehend information presented in different representations; (3) transform information from one representation to another: and (4) learn in one representation and apply that learning to another.

Representational fluency is an important aspect of deep conceptual understanding. It was mainly discussed in pedagogical literature about promoting the transfer between learning and the development of "expertise". In our context of visual analytics, we borrowed this concept to describe how to let the analyst better comprehending the multiple visualizations of "when, where, and what" for cybersecurity situational awareness. Representational fluency is more skillfulness than skill[12]. Skillfulness connotes continuous adaptation and dynamism along with the ability to perform with facility, adeptness, and expertise. Skillfulness of representational fluency in visual analytics includes several capabilities, such as abstractly visualizing and conceptualizing transformation processes, qualifying quantitative data, working with patterns, and working with continuously changing qualities and trends. To achieve these goals, analysts should be supported with proper tools to interpret visualizations more efficiently.

2.2. Visualization Methods of W3 Attributes

Much previous research has been devoted to exploring different methods to visualize the large-scale high-dimensional datasets. Keim et al.[13] reviewed and summarized recent visualization techniques to deal with large multivariate datasets. One of their own techniques is a hybrid approach that is scalable with "big-data" visualization[14]. Guo et al.[15] proposed to use multiple-linked views to visualize the multivariate data. Andrienko et al.[4] created a structured inventory of existing exploratory spatiotemporal visualization techniques related to the types of data and tasks they are appropriate for. Based on the W3 attributes, Foresti et al.[3][16] developed a novel visualization paradigm, VizAlert, to visualize network intrusion from all three "when", "where", and "what" perspectives. Con-centric rings were used to represent different time periods, from inside to outside. Because of the limited screen space, the VizAlert system may be unable to display the history for a long period. The user needs to rely on interaction to pan and zoom for shifting between different periods.

Some significant approaches were to analyze spatiotemporal patterns by making separate use of multiple maps and statistical graphs. Alan M. MacEachren's GeoVISTA Center[9] uses highlighting, brushing, and linking, and filtered and linked selections to help users analyze geo- referenced time-varying multivariate data. IEEE VAST 2012 Mini-Challenge 1 (MC1) asked researchers to analyze a high-dimensional spatiotemporal dataset[17]. Most of the challenge entries used maps and statistical graphs. For example, Chen et al.[18] and Choudury et al.[19] used one 2-D map to visualize the overall computer statuses in a given time and a slider to adjust the time. Dudas et al.[20] used time series curves to show the aggregate trend of certain qualities.

2.3. Analysis Process for Spatiotemporal Cybersecurity Data Sets

The analysis process on a spatiotemporal dataset often happens conditionally and in sequence[21]. At first the temporal aspect is analyzed, and then the spatial aspect, or vice versa. It is difficult to have a joint integral modeling approach. We have observed such sequential analysis processes in our own practice[22] while solving the VAST 2012 challenge MC2[17], and in other winning entries[23][24] when they tried to solve the VAST 2013 challenge MC3[25]. Many times when looking for issues, the user first examined the temporal aspect by looking at the time series curves to find out the anomalies (e.g., huge peak in the curve), then checked out other detailed visualizations to allocate the affecting hosts (IP addresses). Sometimes the analysis starts with a detailed visualization, e.g., an IP address showing abnormal behavior. To understand the overall picture of the affected computers, the user will then need to examine the time series curves. Sometimes this process happens iteratively. The user starts from one visualization, then goes to others, returns to the first visualization with a different parameter (e.g. time or place), and goes on to gain comprehensive cybersecurity awareness. To investigate the detail, the analyst usually need to narrow down and even to read the raw data such as the log file.

3. Context - Data and System

Our implementation of representational fluency was developed on a visual analytics system SemanticPrism[18]. It won the award of "outstanding integrated analysis and visualization" in the VAST 2012 MC1. From 2011 to 2013, the IEEE VAST challenges committee created three cyber-network visual analytics tasks[25] to simulate the complex nature of cyber security. VAST 2011 MC2 data contain 3-day logs of a small computer network. VAST 2012 MC1 data record 2-day logs of a huge global network. VAST 2013 MC3 data include 2-week logs of a 1200-computer network. All the datasets provided are spatiotemporal.

The high-dimensional spatiotemporal dataset we used in this paper was from the VAST 2012 MC1. It simulates a large enterprise network named the BankWorld, which contains approximately a million computers in about 4000 offices. Offices have latitude and longitude information that can be marked on the map. Computers are divided into three classes, server, workstation, and ATM (Automated teller machine). By their functions, Servers are further divided as web, email, file server, compute, or multiple, and Workstations are further divided into teller, loan, or office. Every 15 minutes each computer generates a status log. Within the 48 hour period, the network accumulated approximately 160 million logs. Each log contains a time stamp, IP address, activity flag, policy status, and number of connections (NOC). Policy status has a value range from1 to5 to represent healthy status from normal to severe condition. Value 1 means the machine is healthy. 2 means the machine is suffering from mild policy deviation. 3 means the machine has non-critical patches failing and is suffering from serious policy deviations. 4 means critical policy deviations and many patches are failing. 5 means the machine may be infected by virus or unknown files are found. Activity flags also have 5 possible values range from 1 to 5. Value 1 means normal activities have been detected on the machine. 2 means the machine is going down for maintenance. It may appear offline for the next couple time slots. 3 means there were more than 5 invalid login attempts. 4 means the machine's CPU is running at 100% capacity. 5 means a device (e.g. an USB drive or a DVD) has been added to the machine.

The spatial part of this VAST 2012 MC1 dataset contains two layers, the physical geographic location and virtual IP addresses. Its IP space ranges from 172.1.1.2 to 172.56.39.254. The information of both has a hierarchical structure enabling the top-level larger range to be divided into several lower-level smaller ranges. In a real computer network, the geographic locations may range from a continent, a country, a state, or a province to a specific office in a building. For IP addresses, the network can be divided into multiple levels of subnetworks that are connected through gateways. Each subnetwork occupies a partial IP address space.

With the system SemanticPrism, the analyst is able to see and compare data of different dimensions at multiple granularities. We chose visualization methods and designed interactions based on the nature of the data and the problems faced. SemanticPrism uses a multilinkedview approach to explore the data from different perspectives. Using different transformation methods, data are visualized by using the geospatial map, time series curves, and pixel-oriented visualizations. The technique of semantic zooming[26] was used as the basic interaction technique to navigate through these visualizations. Each visualization has multiple zoom levels to present different levels of details. The analyst can scan to quickly understand the overall situation of the enterprise network and navigate further to read more details of regions, offices, and even the level of individual computers.

3.1. Geospatial Visualization with a Time Slider

The default view in SemanticPrism is a geospatial visualization with a time slider that helps to aware the network status at a given time (**Figure 1**). Offices of the BankWorld are marked as square dots on the analogous world map. Their

Figure 1. Default view of the visual analytics system SemanticPrism.

different color shades indicate the maximum policy violation statuses of the computers within the offices at that time. The analyst can slide the pointer on the time slider to update the geospatial visualization to a different time frame. Different dimensions of the information (e.g. policy status and activity flags) were stacked on the map as different layers. To let the analyst see the global status, SemanticPrism provides a time-zone layer to indicate the local times of different regions in this global organization.

Besides zooming in on and out of the map, the analyst can focus and investigate the data at different levels of details through semantic zooming. Depending on the size of available space, an office can be dynamically visualized at four levels: (1) an individual dot when using the default full-map view or when the space is still quite dense after zooming; (2) a horizontal color bar to show the percentage of computers with different policy statuses in the office; (3) a series of growth curves of all policies in the office where the X axis presents the temporal direction and the Y axis the number of computers; (4) history diagrams of each computer within the office.

3.2. Time Series Curves

We adopted Ben Shneiderman's visual information- seeking mantra to guide the design of the SemanticPrism's information query process, "overview first, then zoom and filter, and lastly details on demand"[27]. The time series curves (the curve graph in **Figure 1**) can be configured to provide an overview of the growth trends of policy statuses, activities, server populations, and NOC (number of connections) over the given period. **Figure 2** and **Figure 3** curves show the total number of workstations in different policies (2 5) and activities (2 5) over the 2-day period. With the support of time series curves, the analyst can easily identify the overall trend of policy violation growth and patterns of activities. By relying solely on the curves, however, the analyst cannot see the cause, details, and effects of an event. Usually he/she must manually switch to other views to investigate, such as what causes the curve to change, and where the change takes place. This significant user-experience problem motivates our new development of extending user interactions from semantic zooming to marking interesting segments on the curve.

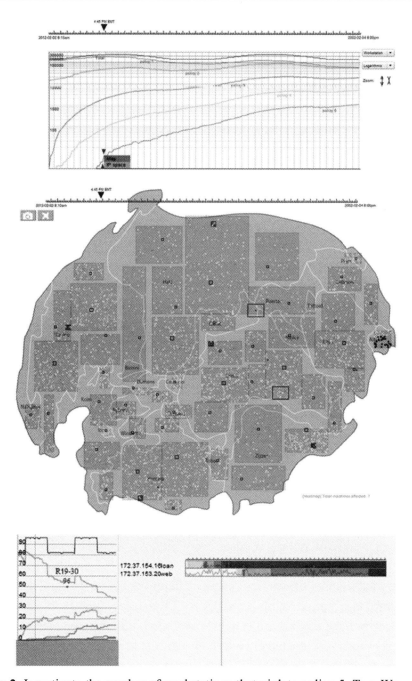

Figure 2. Investigate the number of workstations that violate policy 5. Top: We want to check out which workstations are new to violation of policy 5 at 2012-02002 4:45 p.m. by clicking on the segment in the curve. Middle: The map marks by red squares the two new offices with policy 5 violations. Bottom: Clicking on the top marked square to see the details.

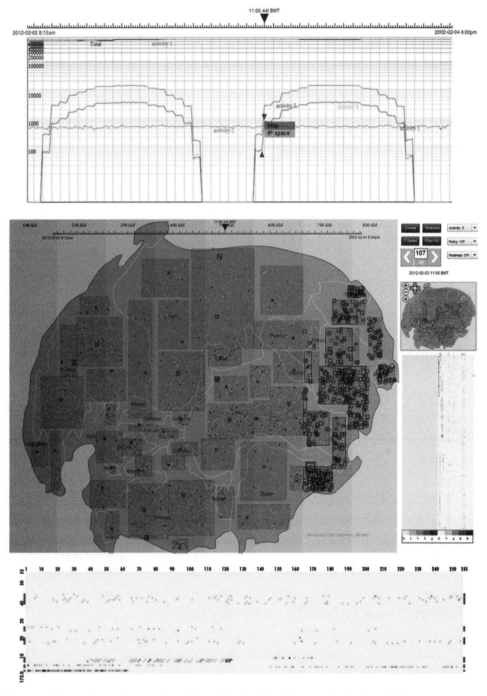

Figure 3. Investigate the number of workstations that have activity flag 5. Top: we want to check out what happened on activity 5 at that time. Middle: The map marks offices by regions. Bottom: These computers are also marked by their IP addresses.

3.3. Pixel-Based Visualizations

IP addresses indicate the virtual locations of network computers. For cybersecurity issues, they provide a different perspective of spatial information than physical locations. The classification of IP addresses also partially reflect the organization's network structure. SemanticPrism incorporated a pixel-based visualization to show many IP blocks. In the default zoom level, five rectangular panels show the number of computers within an IP block that are affected by each activity and policy. In the panels, each pixel represents a group of computers in a particular class-C block. The X axis consists of the IP's class-B block, and the Y axis consists of the values of class- C blocks. The colors of the pixels encode the number of computers that carry the selected policy status or activity flags in the C-level blocks. Through semantic zooming, the analyst is able to overview time series curves of all C-level blocks within one B-level block and all individual computers within a C-level bock.

4. Mechanism and Implementation

SemanticPrism's comprehensive visualizations and interactions show multiple visualizations of where, when, and what data components (**Figure 4**). With it, we were able to discover all anomalies hidden within the large dataset in the competition. In this paper, we implement the representational fluency concept by extending the interaction design in this system. We consider three important representations for the user to be truly aware of the situation – raw data, spatial visualization, and temporal visualization. We seek to allow the user to shift fluently back and forth among these three representations of the cybersecurity information without losing the analysis context.

4.1. Dimension Hierarchy in SemanticPrism

To enhance the efficiency while analyzing a large multidimensional dataset, we adopt the OLAP (online analytical processing)[21] approach to execute analytical queries. OLAP's slicing operation enables the user to take out one specific part of data. SemanticPrism[23] pre-computed the aggregation values along necessary dimensions and storing them into several database tables. The dimension

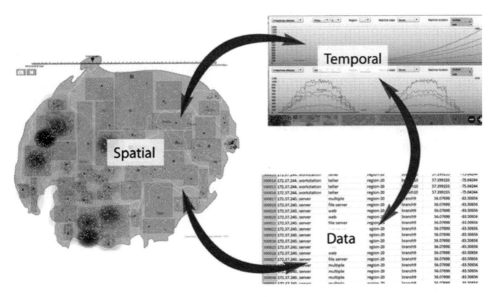

Figure 4. Three components in spatiotemporal data. SemanticPrism's visualizations show: where (spatial data in geospatial visualization), when (temporal data in time curves), and what (data objects in data tables).

hierarchy is essential for these computations. Pre-computing all possible aggregations on all different granularities, however, will use too many resources. We selected several dimensions to compute in certain granularities.

SemanticPrism maintains a set of dimension hierarchies so that the analyst can have multiple navigation paths to narrow down and examine computers with a certain status (e.g., policy or activity) at a certain time slot and in a certain region. Spatially on a map, a computer is located at the following hierarchy:

$$Company \Rightarrow Region \Rightarrow Office \Rightarrow Computer\ class$$

As virtual IP space, an IP address is located at

$$Whole\ IP\ space \Rightarrow B\text{-}level\ IP\ blocks \Rightarrow C\text{-}level\ IP\ Blocks$$

In this dataset, computers within one C-level block belong to the same office and are in the same class of server or workstation. But one office may contain many C-level blocks. Therefore the basic aggregation of computers we choose is the number of computers in one C-level IP block with a given policy/activity status at a given time. From such basic units, we can compute the number of computers of on-policy status at one office at one time, then the policy status at the region level, then to the whole company level. Thus we can have different levels of time series curves of different activity/policy statuses, from the basic level of computer classes, to regions, and lastly the entire company. The number of connections (NOC) are more related to IP-related attacks (e.g., port scan); thus we can simply use the IP address hierarchy to divide it.

With the spatial hierarchy and policy/activity status, the system has multiple paths to aggregate the basic units based on the user's analysis needs.

4.2. Link Data Query to Visualizations

For visual analysis systems, the raw data are the resource of everything. The more details the dataset contains, the more insights and discoveries can be found. For solving cybersecurity issues, the datasets are usually very large and comprehensive. It is impossible for human beings to read through, compare, and identify issues in the large-scale datasets. Visualization becomes the only feasible way to allow the analyst to make sense of the large amount of data. However, visualizations cannot show all the information in the dataset. Through categorization, aggregation, and visualization, only part of the information has been presented in graphs. An analyst still needs to frequently examine the original raw data (e.g., a recorded log or an event report) to determine the exact issue. Thus we should provide a direct-query interface for the user to search the raw data. Based on the searched criteria, the user can pop up visualizations, for example, to display the locations of the computers under investigation in the geographic visualization. Also from the visualizations, the analyst should be able to open, allocate, and read the piece of the raw data of an interesting point.

4.3. Link from Time Series Curves to Spatial Visualization

The time series curve is the visualization to show the plot of the data narration. The data are measured at successive points in the temporal direction at uniform time intervals. In computer networks, it is a common strategy to aggregate (or count) certain network incidents at a given time interval (e.g., 15 minutes in the

VAST 2012 data). Thus a series of data points along the time will be generated and can be visualized as time series curves. In our implementation, the curves are plotted on a 2-D Cartesian system with line segments connecting a series of points. X axis is the time direction and Y axis is the value of data. Thus such data have a natural temporal ordering. The user should be able to see the overall trend of the network status through the temporal curve. For a running system, its temporal curve can present certain kinds of patterns (e.g., fixed frequency and amplitude, or various grow rates). For a complex system, such patterns are sometimes hard to define by mathematic equations, and therefore hard to be detected solely by machines. Temporal visualizations rely on a human's visual perception and pattern recognition to help the analyst to detect such potential attacks through recognizing abnormal patterns in time series curves.

Figure 5 lists six popular abnormal situations, including a sudden jump, dive, peak, valley, slop gradient change, and frequency or amplitude change in oscillating curves. In the figure, blue squares mark the data points, and red line segments label the abnormal sections. Such abnormal segments on a curve imply that there are some computers behaved abnormally during that time period. This

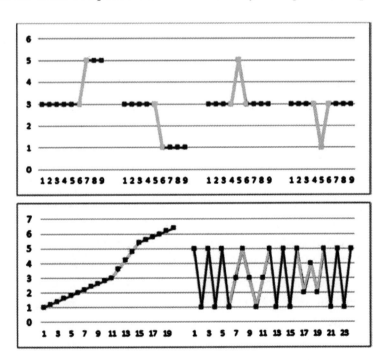

Figure 5. Examples of abnormal sections (marked in red) on curves.

paper only focuses on the abnormal segments as the four scenarios on the top image of **Figure 5**. After detecting an abnormal segment on the curve, the analyst needs to investigate what caused the change. He/she should switch from the overview curve to more detailed curves or other visualizations to investigate its when, where, and what details. Although sometimes the abnormal behavior may happen globally, in our observation such behaviors most of the time happen on computers within a small region. Again, such a region could be a physical location or in the virtual space of IP addresses. It is essential for the analyst to find out which region(s) causes these problems in detail.

While working on a large-scale complex dataset, an analyst will find it tedious and exhausting to examine each individual curve section to learn the related spatial information. This VAST 2012 Challenge dataset includes 4,000 offices and 13,000 C-level blocks. Manually examining each office or C-level block is simply impossible.

A time series curve of higher level granularity (e.g., a region) can be divided into several curves of its subgranularities (e.g., all sub-regions). Spatial data have hierarchy and can be divided into many levels of sub-regions. The aggregated value of the upper-level region is the total of all its sub-regions. For example, the total number of computers in one company must be equal to adding up all computers in its regional offices. Thus an anomaly (e.g., a jump) on a higher level curve must appear on some of its sub-curves. According to our observation, usually only a few sub-curves contribute most of the change in the higher-level curve. Thus it is essential for us to find these sub-curves and allocate the spatial information from them. In reality, curves will fluctuate slightly even in normal conditions. While finding the cause for the anomalies in the curve, we must filter out these small fluctuations.

We store the time series data according to the dimension hierarchies we discussed in the previous section. Using the OLAP slicing operation, we are able to divide an aggregated value into different granularity levels. The overall process of detecting anomaly from aggregated time series curves to geospatial details can be described as follows:

- The system maintains the hierarchies and relations of different levels of subdimensions in different directions. This information enables the system to iteratively check all subdimensions until it reaches the grounded basic granularity.

- The analyst anchors a suspicious segment on the curve. In this operation, the user defines the following parameters: start-and-end times, start-and-end data values, and value difference at this dimension.

- The detailed spatiotemporal data of the suspicious segment can be shown in two ways: locations on the geospatial map at the current time, or many curves "sliced" from the original curve. Based on the nature of the sliced curve, the system may automatically select one direction to show the details, or prompt to ask the user to select a direction to show the captured segment in detail.

- The system checks all of its subdimensions to learn which contribute the most to the overall value difference. This is done by sorting them by their percentages of value changes in the given period. If the percentages are within the same range, the system will rank the subdimensions by their absolute values. The higher the percentage of value change, the more contribution we must consider it will be giving to the overall value change. In some scenarios, some subdimensions may have much smaller values than others. The absolute value change in one smaller region might be too small to contribute to the overall, although it is still significant enough locally. Thus we consider the relative change of numbers instead of the absolute change.

- The value changes might be caused by one or several subdimensions, or by most (or all) subdimensions. In the evenly distributed scenario, the relative percentage value of each subdimension should be very close.

- The system sums up the absolute number of changes from the highest-ranked subdimension to the lower-ranked subdimensions and tracks all subdimensions until the summary value reaches a certain user-defined threshold (e.g., 95%) of the original value. If we consider threshold as 100%, all small fluctuations will be counted and may blur the focus of the problem.

- If these subdimensions can be divided further to the next level of subdimensions, the system will iterate back to step 4 until the subdimensions are the most-grounded basic granularity.

- Group these basic granularities into a cluster if necessary. Based on the nature of these granularities, some spatial-clustering algorithms, such as DBSCAN (Density-Based Spatial Clustering of Application with Noise)[28], can be used.

- Mark these clusters on the map if these basic granularities are geographic based, or display them as a collection of time series curves if they are still time based.

4.4. Link from Spatial Visualization to Time Series Curves

From a spatial visualization back to a time series curve is relatively straightforward. The spatial visualization normally presents the geographic distribution of different types of data. In the SemanticPrism (and many other spatial visualization systems), which type of data to be visualized can be controlled by menus that turn the data layers on and off. Also with zooming technology, the analyst can zoom ina smaller area on the visualization to open the view of a region or a subnetwork. Therefore from one spatial visualization we can capture a list of parameters, including the type of data items being visualized, current time, and current display area/region/subnetwork. Based on these parameters, popping up the related time series curve is simple.

4.5. Interaction Design to Support the Fluency

An analyst may start the investigation by analyzing the curves. The abnormal segments in a curve, like sudden jumps, dives, peaks, or valleys, reflect the value change and therefore present us with a hint that something worthwhile is waiting to be investigated further. As discussed earlier, the time series curve can be seen as a series of vertices with connecting line segments. Thus the analyst can interact with two types of objects on the curve, the vertices and line segments, and can mark the suspicious segments in two ways. The first way is to mark a suspicious one-time-unit segment by simply clicking on the segment. To mark a segment across several time-unit periods, the user clicks on the starting point and end point on the curve and leaves two red triangle marks (top screen shot in **Figure 2**).

After selecting one segment or two vertices, the system will present a pop-up menu for the analyst to select from if there are several possible sub- dimensions. In the example shown by **Figure 3**, further details can be shown in either the map or the IP pixel-based visualization. If there is only one subdimension, the system will automatically jump to the detailed view and display the marked area. Because the data are discrete with time intervals, selecting a partial segment is unnecessary. The minimal selectable range should be one segment (or the two neighboring data vertices).

The area of interest on the map is indicated by a red rectangle. Although the offices are spatially spread across the map, they are hierarchically grouped by regions. Therefore we did not use particular spatial clustering algorithms, but rather cluster offices by regions. If two or more offices are in one region, they will be marked together within one block. The boundary of the rectangle is defined by the spatial elements (offices in the middle screen shot, **Figure 2** and **Figure 3**). Sometimes the affected area will be tiny, for example, containing only one office. Marking the tiny area may not be visually significant enough to be noticed. Thus we define the minimum size of a marking as a rectangle measuring 45 × 35 pixels (**Figure 2**). The analyst can click on the rectangle to zoom in. The semantic zoom mechanism will automatically display details of the affected offices (middle and bottom screen shots).

5. Use Cases

We use some examples below to show how we implement visual fluency in SemanticPrism, which tries to provide the user a smooth and efficient method to link information from different visualizations.

5.1. From Time Series Curve to Spatial Visualization

In **Figure 2**, from the time series curve, the analyst saw the increasing number of computers are falling into high policy statuses. To accommodate multiple curves in one graph, we used thin lines in SemanticPrism to draw the curves. To identify how the policy violence spread spatially, the analyst needs to examine the locations of the computers. The user can inspect each segment on the policy-5 curve to check new computers that violate the policy. After clicking on the segment between 4:30 to 4:45 p.m., the user chooses the map from the pop-up menu to see which offices have new computers are new in policy status 5 starting at 4:45 p.m. The spatial view highlights the two new offices as the middle image in **Figure 2**. It is possible for the user to highlight all computers by selecting one time point. The user can simply click on the vertex in the curve to highlight all computers having the problem at that time.

In our current implementation, we simply use regions to cluster offices. Thus the two offices are marked separately. Clicking on the red marked boundary will lead the spatial view to zoom to that area. But because only one office is in that region, the system automatically zooms to the maximum level, which shows the detailed information of the office, including time series curves about policies in this office, and shows all computers with that policy 5 violation. We can see the IP 172.37.154.15 just started in policy 5 status at the given time (marked by the gray vertical line to indicate the current time).

Figure 3 demonstrated how the same curve jumps can be marked on either maps or IP addresses. The top image has 6 curves about the number of computers at different activities status (including total number of online computers) along the two days. At each hour there is a step (up or down) on the curves of activity 3 and 4. To find out what causes these steps, the analyst selects and examines one of the jumping segments (top image). By checking out the affected area on the map, he/she can see that they are actually caused by time zones – Offices open at 7 a.m. and close at 5 p.m. As time passes, offices open to turn on computers and close to shut off computers, which causes the sudden steps on the curve. The red squares mark the offices with computers that are newly emerging in activity 5. However, in here the marks are not 100% accurately aligned with the time zone because of the threshold we used (defined in step 6 the previous mechanism section). Small fluctuations happen all the time everywhere, especially for these computer activities, such as log-in errors. We assume that within a large area (e.g., a region), these small fluctuations that happened in small sub regions (e.g., in an office) will be counter-acted with each other and make the regional number relatively stable in a normal situation. Therefore smaller areas might sometimes be neglected, or mis-marked, as shown in the middle image of **Figure 3**. But the areas that contributed much to the change will be clearly marked out.

The bottom image of **Figure 3** shows the distribution of new computers in the IP space. Each small square in the image represents a C-level IP block. Rows from bottom to top are the 2nd byte of the IP address (from 172.0.xx.xx to 172.55.xx.xx). Columns from left to right are the 3rd byte of the IP address (0 to 255). Besides marking each C-level block with blue squares, we also mark the B-blocks on both the left and right sides with red indicators (**Figure 3** bottom).

5.2. From Spatial Visualization to Time Series Curves

SemanticPrism provides a semantic zooming mechanism to change the details of display while the user is zooming in[22]. Offices on the map can change into 4 levels of details, depending on the available on-screen space. When zoomed in enough, the user is able to see the time series curves for individual offices (**Figure 6**).

Besides using semantic zooming to check out time series curves of different offices, the user can also click on a region or an office to see the temporal summary. Region 25 (the right-most region Alta at the top image of **Figure 6**) has many blacked-out offices, which means that these offices are disconnected from the Internet, possibly because of a power outage in the area. We can see that the distribution of blacked-out offices changes as time passes. To get to the affected computers over time, we can click on the region to bring out a regional time series curve (top image of **Figure 7**). The black curve shows that the overall computers sent out status reports during the period. A big valley on the curve shows that more and more computers lost connections in the middle of the first day. The worst time was at 11 p.m. BMT (Bankworld Mean Time). The situation recovered in the next 4 hours. The analyst can also choose to turn on the layers to highlight one activity status or one policy status. The green squares surrounding offices on the top image show offices with computers at activity 2 (going down for maintenance). Since the activity 2 layer is currently turned on, the time series curve of activity 2 is also included in the curves. However, the number of computers with activity 2 is so small, at the pixel level it is at the baseline and hides behind the policy 5 curve. The middle image in **Figure 7** uses a logarithmic scale to boost these curves with extremely small values on the screen. Zooming in on this curve will break down the time series data of the region into individual offices. These curves for all offices are displayed in a grid as the bottom image of **Figure 7**.

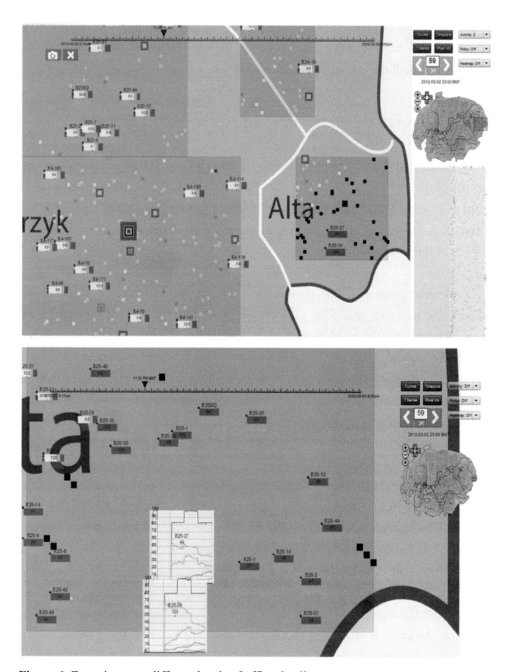

Figure 6. Zoom in to see different levels of office details.

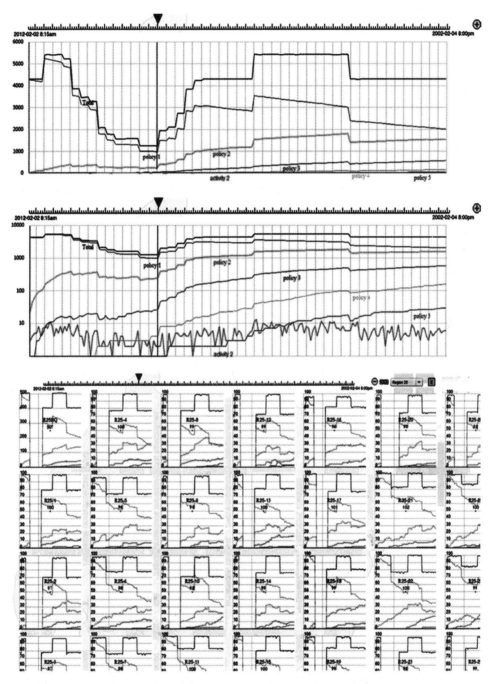

Figure 7. Time series curves corresponding to the region 25 in **Figure 6**. Top: Overall number of computers in different policy flags in region 25. Middle: The logarithmic scale version of the top visualization. Bottom: Zoom in on the curve to break down the regional time series curves into offices.

6. Discussions

Visual analytics is the process for an analyst to learn the facts from the large volume of raw data through different forms of visualization. Representational fluency is the ability to comprehend equivalence in different modes of expression[9]. We borrow this term from psychology and pedagogical literature to describe our efforts to enable the analyst to fluently switch among different types of visualizations and data views to build up the understanding of facts. Cybersecurity issues can be visualized in temporal, in geospatial, in structural, or in raw data as logs. Visual analytics fluency allows the ability (1) to transform information from one representation to another; (2) to comprehend the equivalence in different modes of representations, including data and visualizations; and (3) to comprehend information presented in different representations.

In this paper, we propose an auto linking mechanism that can smoothly transfer the analyst from one view to the other and thus effectively improve the speed of visual data analysis. Cognitively, a person can pay attention to only 3 or 4 things at one time. Our fluency metaphor may also reduce the cognitive load, helping the analyst to focus on some important incidents. At the stage of submitting SemanticPrism to the VAST 2012 challenge (July 2012), the four team members needed several days to identify all the anomalies by manually going over suspicious areas on all the curves and jumping across different views to examine and filter information. Most of the energy and time was exhausted during the back-and-forth navigation. With this newly developed linking mechanism, on one hand an analyst can mark suspicious segments on the time series curves and go directly to its related spatial visualization and data view. On the other hand, the analyst can simply right-click on the map, opening the menu to show one or several related time series curves.

We plan to improve this mechanism and its direct interaction design in the following directions.

First, we should extend our approach to other types of data and visualizations. The VAST 2012 MC1 dataset contains no data about computer network connectivity. In some cybersecurity analysis scenarios, visualizing such connections as the network intrusions from external IPs to internal hosts is crucial. Most often, connection data of these kinds can be visualized as a tree, or a network graph, with different layout variations (e.g., layout nodes in radial fashion). How to anchor parts of such spatial visualizations and link them to their related time series curves, geographic visualizations, or data views comprise the new domain we want to explore.

Second, we should find a method to automatically detect anomalies on the curves. A curve must be displayed at a certain resolution to allow the analyst to identify problematic areas. However, because the curves are mostly based on aggregation, the user sometimes cannot visually detect the problem when the number is too small to cause a significant visual change on the curve. Some literature on data mining and statistics[29][30] shows that allowing the system to detect anomalies on the curves by itself is possible. We will consider integrating this effective approach.

This approach can also be easily extended to handle streaming data such as real time analysis. In such case, the time series curve will become dynamic by updating itself in regular time intervals. Visually the curve will grow, extend, and slide from right to left (if the new data starts from the right end) just like the electrocardiography. Old part of curve will disappear on the left end. The user still be able to notice the anomaly happened during the recent past time intervals. For the just past time interval, the aggregations should be computed across the hierarchy of the spatial structure from top to bottom. The computing resource needed for pre-compute the aggregation depends on the length of the time interval and the complexity of the spatial structure. For this VAST 2012 MC1 data, since the time interval is pretty long as 15 minutes and there are only several thousands of spatial units, computing aggregations for one time interval is very fast. For existing computed aggregations of each time interval, there is no need to re-compute them. The only aggregations need to be updated are the aggregations about recent past history (e.g. recent two days). But normally there is no urgent need to get the aggregation for the past history in real-time.

The inspiration and implementation of this fluency mechanism were based on the visual analytics system SemanticPrism and the VAST 2012 challenge dataset. To understand its generalizability and limits, we will use other datasets to test the possibility of linking the W3 structure visualizations. Furthermore, we aim to study the possibility of representational fluency being a suitable and valid design goal in the context of visual analytics and how to promote it to different platforms and systems.

Competing Interests

The authors declare that they have no competing interests.

Source: Qian Z C, Chen Y V. Fluency of visualizations: linking spatiotemporal visualizations to improve cybersecurity visual analytics[J]. Security Informatics, 2014, 3(6): 1–15.

References

[1] G Jiang, G Cybenko, Temporal and spatial distributed event correlation for network security, in *American Control Conference, 2004, Proceedings of the 2004*, vol. 2 (IEEE Boston, MA, USA, 2004), pp. 996–1001.

[2] DJ Peuquet, It's about time: A conceptual framework for the representation of temporal dynamics in geographic information systems. Ann. Assoc. Am. Geographers. **84**(3), 441–461 (1994).

[3] S Foresti, J Agutter, Y Livnat, Moon S, R Erbacher, Visual correlation of network alerts. IEEE Comput. Graphics Appl. **26**(2), 48–59 (2006).

[4] N Andrienko, G Andrienko, P Gatalsky, Exploratory spatio-temporal visualization: an analytical review. J. Visual Languages & Comput. **14**(6), 503–541 (2003).

[5] J Booker, T Buennemeyer, A Sabri, C North, High-resolution displays enhancing geo-temporal data visualizations, in *Proceedings of the 45th Annual Southeast Regional Conference* (ACM New York, NY, USA, 2007), pp. 443–448.

[6] H Shiravi, A Shiravi, AA Ghorbani, A survey of visualization systems for network security. IEEE Trans. Visualization Comput. Graphics. **18**(8), 1313–1329 (2012).

[7] J Heer, B Shneiderman, Interactive dynamics for visual analysis. Mag. Queue - Microprocessors. **55**(4), 45–54 (2012).

[8] YV Chen, ZC Qian, From when and what to where: Linking spatio-temporal visualizations in visual analytics, in *IEEE International Conference on Intelligence and Security Informatics* (IEEE Seattle, WA, USA, 2013), pp. 39–45.

[9] IE Sigel, Approaches to representation as a psychological construct: a treatise in

diversity, in *Development of Mental Representation: Theories and Applications* (Psychology Press East Sussex, UK, 1999), pp. 3–12.

[10] M Stone, Challenge for the humanities, in *Working Together or Apart: Promoting the Next Generation of Digital Scholarship* (Washington, DC, USA, 2009). The Council on Library and Information Resources and The National Endowment for the Humanities.

[11] B Stripling, Assessing information fluency: gathering evidence of student learning. School Library Media Activities Monthly. **23**(8), 25–29 (2007).

[12] (R Lesh, H Doerr, eds.), *Beyond constructivism: a models and modeling perspective on mathematics teaching, learning, and problem solving.* (Lawrence Erlbaum Associates, Hillsdale, NJ). ISBN 0-8058-3822-8.

[13] D Keim, C Panse, M Sips, ed. by Dykes J, Maceachren A, and Kraak M, Information visualization: Scope, techniques and opportunities for geovisualization, in *Exploring Geovisualization* (Elsevier Ltd Oxford, UK, 2005), pp. 23–52.

[14] DA Keim, C Panse, M Sips, SC North, Pixelmaps: a new visual data mining approach for analyzing large spatial data sets, in *Proceedings of the Third IEEE International Conference on Data Mining* (IEEE Los Alamitos, CA, USA, 2003), pp. 565–568.

[15] D Guo, J Chen, Eachren MacAM, K Liao, A visualization system for space-time and multivariate patterns (vis-stamp). IEEE Trans. Visualization Comput. Graphics. **12**(6), 1461–1474 (2006).

[16] Y Livnat, J Agutter, S Moon, S Foresti, Visual correlation for situational awareness, in *Information Visualization, 2005. INFOVIS 2005. IEEE Symposium on* (IEEE Minneapolis, MN, USA, 2005), pp. 95–102.

[17] KA Cook, G Grinstein, M Whiting, M Cooper, M Havig, K Liggett, B Nebesh, CL Paul, VAST challenge 2012, visual analytics for big data, in *Visual Analytics Science and Technology, 2012 IEEE Conference on* (IEEE Seattle, WA, USA, 2012), pp. 151–155.

[18] VY Chen, AM Razip, S Ko, CZ Qian, DS Ebert, SemanticPrism: a multi-aspect view of large high-dimensional data: VAST 2012 mini challenge 1 award: Outstanding integrated analysis and visualization, in *Visual Analytics Science and Technology, 2013 IEEE Conference on* (IEEE Seattle, WA, USA, 2012), pp. 259–260.

[19] S Choudury, N Kodagoda, P Nguyen, C Rooney, S Attfield, K Xu, Y Zheng, BLW Wong, R Chen, G Mapp, L Slabbert, M Aiash, A Lasebae, M-sieve: a visualisation tool for supporting network security analysts, in *VisWeek 2012*, 165–166 (2012).

[20] L Dudas, Z Fekete, J Gobolos-Szabo, A Radnai, A Salanki, A Szabo, G Szucs, OWLAP - using OLAP approach in anomaly detection, in *Visual Analytics Science and Technology, 2012 IEEE Conference on* (IEEE Seattle, WA, USA, 2012), pp. 167–168.

[21] O Schabenberger, CA Gotway, *Statistical Methods for Spatial Data Analysis.* (CRC Press, Boca Raton, FL, USA, 2004).

[22] VY Chen, AM Razip, S Ko, ZC Qian, DS Ebert, Multi-aspect visual analytics on large-scale high-dimensional cyber security data, in *Information Visualization 2013* (Sage Publications Thousand Oaks, CA, 2013).

[23] Y Zhao, X Liang, Y Wang, M Yang, F Zhou, X Fan, MVSec: a novel multi-view visualization system for network security, in *VisWeek* (2013).

[24] S Chen, F Merkle, H Schaefer, C Guo, H Ai, X Yuan, T Ertl, Annette - collaboration oriented visualization of network data, in *VisWeek* (2013).

[25] M Whiting, KA Cook, CL Paul, K Whitley, G Grinstein, B Nebesh, K Liggett, M Cooper, J Fallon, VAST challeng 2013: Situation awareness and prospective analysis, in *Visual Analytics Science and Technology, 2013 IEEE Conference on* (IEEE Atlanta, GA, USA, 2013).

[26] K Perlin, D Fox, Pad: an alternative approach to the computer interface, in *Proceedings of the 20th Annual Conference on Computer Graphics and Interactive Techniques* (ACM New York, NY, USA, 1993), pp. 57–64.

[27] B Shneiderman, The eyes have it: A task by data type taxonomy for information visualizations, in *Proceedings of 1996 IEEE Symposium on Visual Languages* (IEEE Boulder, CO, USA, 1996), pp. 336–343.

[28] M Ester, H-P Kriegel, J Sander, X Xu, A density-based algorithm for discovering clusters in large spatial databases with noise, in *KDD, vol. 96 1996* (AAAI Portland, OR, USA), pp. 226–231.

[29] RS Tsay, Outliers, level shifts, and variance changes in time series. J. Forecasting. 7(1), 1–20 (1988).

[30] JD Hamilton, *Time series analysis*, vol. 2. (Cambridge University Press, Cambridge, UK, 1994).

Chapter 26

Waterwall: A Cooperative, Distributed Firewall for Wireless Mesh Networks

Leonardo Maccari, Renato Lo Cigno

Department of Information Engineering and Computer Science (DISI), University of Trento, Via Sommarive 5, Povo, Trento 38123, Italy

Abstract: Firewalls are network devices dedicated to analyzing and filtering the traffic in order to separate network segments with different levels of trust. Generally, they are placed on the network perimeter and are used to separate the intranet from the Internet. Firewalls are used to forbid some protocols, to shape the bandwidth resources, and to perform deep packet inspection in order to spot malicious or unauthorized contents passing through the network. In a wireless multihop network, the concept of perimeter is hard to identify and the firewall function must be implemented on every node together with routing. But when the network size grows, the rule-set used to configure the firewall may grow accordingly and introduce latencies and instabilities for the low-power mesh nodes. We propose a novel concept of firewall in which every node filters the traffic only with a portion of the whole rule-set in order to reduce its computational burden. Even if at each hop we commit some errors, we show that the filtering efficiency measured for the whole network can achieve the desired precision, with a positive effect on the available network resources. This approach is different from the protection of a space behind a wall: we use the term *waterwall* to indicate a distributed and homogeneous filtering function spread among all the nodes in the network.

1. Introduction

Protecting a network from unsolicited, often malicious traffic is one of the constant concerns of any network administrator. Apart from standard networking devices as switches and routers, middleboxes as NATs and firewalls are normally installed on the network boundary to separate trusted portions of the network from the global Internet and in general from less trusted ones.

In some cases, however, even the separation between the internal and the external network is not straightforward, and identifying boundaries and points of interconnection is even more difficult. A typical example is a wireless mesh network, in which a collection of subnets are interconnected through a backbone of mesh nodes, but each subnet is only loosely coupled with the others. Moreover, many points of access to the global Internet may exist (see **Figure 1** for a pictorial representation). Mesh networks are often used with this configuration in order to bring connectivity in a cost-effective way to areas where other technologies would be too expensive[1]. As a concrete example, community networks[2][3] use this approach to share network resources between hundreds or even thousands of users and represent one of the most successful application of mesh networking. Projects like Guifi or Awmn (see http://guifi.net and http://awmn.gr) are examples of how this technology integrates with standard networks and how successful this approach can be. Future advances will open new possibilities for this technology[4].

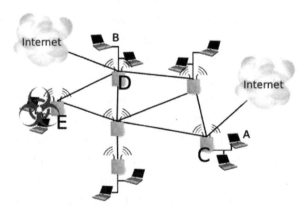

Figure 1. A commonmesh network structure.

In this paper, we tackle the problem of firewalling in large mesh networks. In such networks, each mesh node applies a specific firewall rule-set to the traffic directed to itself (or to subnets attached to it). The firewall is used to defend the local network from attacks, to shape the access to the Internet across a connection, or to forbid the access to certain logical resources. If all nodes share their rule-sets and enforce them also on the outgoing traffic, the traffic is not filtered at the destination but directly at the source. This reduces the waste of network resources but forces each node to filter with a global rule-set made of thousands of rules, which is not practical for most of low-power Linux-based mesh routers. We propose to split the global rule-set in pieces and enforce only a portion of it at every hop with the goal of filtering the packets as close as possible to their source node in order to save network resources.

A correctly configured traditional firewall does not introduce false-positives (packets that should be dropped but are instead forwarded). Instead, with our approach, each node singularly introduces some false-positives, but as a global network function, the firewall will work with an arbitrary accuracy that can be tuned to the needs. To stress the difference from a typical firewall we chose the term *waterwall* to indicate a distributed and homogeneous filtering function spread among all of the nodes in the network. Note that we use filtering as target application for the sake of simple explanation and by way of example, but the same logic can be applied to other traffic analysis functions such as intrusion detection.

2. Related Works and Motivation

Distributed firewalling has not received much attention in the literature, but an initial model has been proposed by Bellovin *et al.* in[5], where the firewall was moved from a bastion host to endpoints in a traditional architecture network. Recently, the subject has been investigated with more attention. Bellovin again, followed by other authors, proposed a distributed policy enforcement platform[6]–[10]. These works are not focused on the complexity introduced by large rule-sets. Other works focus on the application of hash functions to speed up rule matching[11]–[13] or on limiting the nodes that enforce the firewall[14]. None of these works focus on techniques to reduce the rule-sets on single nodes.

The work whose idea comes closer to the contribution of this paper is[15], where the most recently matching rules are stored in a cache that is used to enforce filtering, thus using only a subset of the entire rule-set. The cache is split in two halves, with each one regulated with a different policy in order to ensure efficiency and fairness. This approach requires a feedback from the nodes that generate the rule-sets in order to organize the cache; moreover, as with every caching strategy, its performance depends on the characteristics of the underlying traffic.

Our work assumes that the default filtering policy is to forward a packet. Packets are dropped only when there is a rule that matches them. This approach is more viable in a mesh network than a deny-by-default one: for the last one to be usable, the rule-sets must be perfectly synchronized and updated; otherwise, there is the risk of dropping legitimate traffic. In networks where a high security level is required, also deny-by-default rule-set can be used as in[16][17], but it does not match our network scenario.

Additional similarities can be found in the field of intrusion detection since filtering with large rule-sets presents the same difficulties of traffic inspection with a large database of fingerprints. An approach to distributed intrusion detection systems (IDS) like[18] could benefit from the solution we propose. More affinities can be found in[19] that, as our work, exploits the distributed nature of an *ad hoc* network to spread the IDS function over the entire network.

2.1. Motivation

Figure 1 shows a widely used configuration for a wireless mesh network where a set of mesh routers interconnects separated local area networks (LANs). Each LAN has its own internet protocol (IP) addressing, and the routing protocol running on the mesh routers allows the clients of distinct LANs to communicate. In some cases, nodes may physically roam from one LAN to another, depending on the kind of routing protocol they may or may not maintain their initial IP addresses to keep their sessions alive. Finally, some of the LANs have a direct access to the Internet and share it with other users that are not equipped with it.

In this scenario, the owner of a mesh node is generally also the manager of the corresponding LAN, and he is interested in protecting it. We take into consideration three use cases applicable to the simple network in **Figure 1**:

1) The manager of network C wants to protect its network from unwanted traffic coming from the outside. For instance, he does want to block connections to remote shell protocols coming from the mesh network to host A in LAN C.

2) With a finer granularity, he may want to limit access only to some logical resources; for instance, host A may have some folders that are shared only on the LAN while some others are shared with the whole mesh network. The access to these resources can be denied or simply limited to a maximum bit rate.

3) The manager of network C wants to forbid some traffic types that come from the mesh network and are directed to the Internet using its connection. This is normally due to the commercial agreements that the manager has with his network service provider. Again, traffic can be forbidden or it can be limited to a certain maximum bit rate.

Now imagine that a node in the network labeled E starts an attack against, let us say, host A. This may be due to a malicious user or to a virus that took control of a host in the network and starts a denial of service (DoS) or a brute force attack. We can add a fourth use case:

4) The manager of network C detects an attack and reactively enforces network filters to protect its resources.

The issues described in the use cases can be partially resolved by configuring a firewall on each mesh node in order to filter the traffic directed to its LAN. The first three use cases can be tackled by setting up a mixture of layer-4 and application layer firewall rules on the mesh router in C, which will drop or shape some traffic. The fourth one can be approached with dynamic rules that are activated when the firewall detects an anomaly in the usage of the resources, for instance, an abnormal number of internet control message protocol packets. Modern Linux-based firewalls support all these features. What remains unsolved are the consequences for the rest of the mesh network and for the other LANs. Clearly, the malicious traffic coming from network E will still traverse the mesh network and subtract useful resources to the other allowed communications. Considering that in a mesh network the available bandwidth is shared between upload and download, this can severely impact not only the victim LAN but also the other networks on the way from the attacker to the victim. This example shows how the concept of border firewall does not correctly apply to the mesh network scenario, where the border itself of the network is very hard to define.

To solve this problem the mesh routers can share their rule-sets in order to apply them directly on the other mesh routers. The rule-set of mesh router C applies only to the traffic directed to LAN C, to some logical resources it controls or to the internet traffic flowing across its connection. Each mesh router will publish its rule-set and collect all the other rule-sets in a global rule-set. Then it will enforce it directly on the packets that it is forwarding so that the traffic is filtered as close as possible to the source. This approach indeed protects not only the resources of each LAN but also the shared resources of the mesh network. It's not the goal of this paper to investigate how the rule-sets are securely distributed, in the simplest case rule-sets can be known in advance and every node just sponsors the identifier (ID) of one or more predefined rule-set in routing messages.

Now imagine that this model is applied to a large mesh network. As an extreme but realistic use case, imagine that this model is applied to a community wireless mesh network like the Guifi network. Guifi is made up of thousands of nodes[a] and used by tens of thousands of users that daily access the network from various places (see[20] for a characterization of its topology features). What happens if even only 10% of the mesh routers start distributing a rule-set made of, let us say, 30 rules each? The result will be a global rule-set with tens of thousands of rules. Corporate firewalls can handle large rule-sets up to tens of thousands of rules, but this is not the case for wireless routers that are generally low-cost devices designed for minimal energy consumption. The most used products are commercial devices that embed a low-power processor (e.g., a 133-MHz Intel or AMD low-end device), one or more IEEE 802.11b/g/a/n wireless cards, and run a customized Linux kernel. The whole hardware is enclosed in an outdoor shell powered over LAN and costs no more than 100 e. A 133-MHz processor cannot easily handle a rule-set made up of thousands of rules organized in a linear list; it will introduce processing delays and packet dropping.

To improve filtering performance, rule-sets can be preprocessed with various approaches, none of which is easy to port in this context. For instance, once the whole rule-set has been created, wildcards and numeric ranges can be used to group rules and reduce their total number. This involves a complex and costly preprocessing of the rule-set but speeds up the lookup time during the routing decision. It is convenient when the rule-set is mostly static and when the hardware is powerful enough for the preprocessing. In the case we consider, rules can be dynamically generated, nodes can be added to or removed from the network, and links may be temporarily unavailable. Each of these events will change the rule-sets or the network topology (and consequently add/remove rule-sets associated with nodes). Assuming that the nodes are powerful enough to perform the preprocessing, they would spend most of their CPU time repeating this task.

Techniques based on complex data structures, such as trees or graphs, can be used instead of using a linear list. The more complex the data structure, the more memory and preprocessing are needed. The less complex the data structure, the less the technique will be flexible and high performing. For instance, rules can be grouped using their target netmask, but this is meaningless for application layer rules, for multicast rules, or when a node that has a certain resource to be filtered roams to a new network. Moreover, with a mesh network made of thousands of nodes, there are thousands of netmasks, so filtering is still cumbersome. This gets even worse with networks based on IPv6 addresses. Both these approaches are hardly applicable when the rules do not match IP addresses and TCP/UDP ports but layer-7 data inside a packet.

In this work, we take a different direction. We keep the simplicity of linear lists, but we exploit the cooperative nature of mesh networks to reduce the overhead for each single node.

2.2. Firewalls with Large Rule-Sets

Before we detail the proposed approach, we further investigate the consequences of large rule-sets on the performance of the network. **Figure 2** reports the increment in the processing time of a single packet when the rule-set size grows. The data have been measured using an embedded system equipped with a 400-MHz processor and 128 Mbytes of RAM over a wired network. Fifty percent of the rules matches the network and transport layer fields; the rest matches the packet contents at layer 7. Contrary to the results we obtained in a previous work[14], where the tests were carried without traffic, the measures have been taken when the node is under a load of 1 Mbit/s.

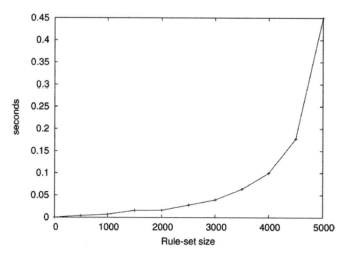

Figure 2. Delay introduced by a growing rule-set size when the host forwards approximately 1 Mbit/s of traffic.

With up to 3,000 rules, the delay grows almost linearly, meaning that the system is able to handle the load as expected. After that threshold, the delay grows at a faster pace and arrives close to 0.5 s with 5,000 rules. Since this delay is introduced by every node for every hop, the total round-trip time in a mesh network using large rule-sets makes the network unusable. Filtering is simply not a function that can be introduced 'for free' when the rule-sets get large.

3. Filtering Based on Route Length

Consider a network N like the one in **Figure 1** where a proactive routing protocol is running (from now on, we refer to mesh nodes simply as 'nodes'). Each node j is connected to a subnet, and for each node j, there exists a rule-set rj that is used to filter the traffic directed to its own subnet, to itself, or to the Internet across the connection attached to its subnet. Node j will sponsor its own rule-set to the rest of the nodes so that every node is aware of a global rule-set $R = U_j r_j$; $\forall j \in N$. The routing table of any node i contains the next hop and the distance in terms of hops to reach j (and all the nodes in the subnet of j). This is the usual configuration of a mesh network configured, for instance, with optimized link state routing (OLSR) protocol[21].

Now consider a packet p coming from the subnet of node k that is forwarded by node i and is destined to the subnet of node j. Assume that for this packet, there exists a rule in R that will drop it when it arrives to j. The aim of the waterwall is to drop the packet as close as possible to the source node k. The simplest solution is to enforce the whole R directly in k. This solution has two drawbacks: it is impossible if R is made up of thousands of rules for the considerations introduced in Section 2.2, and it would be extremely easy to circumvent since, when the packet leaves its own subnet, it is not filtered anymore. A node k that behaves in a malicious way can start an attack against a node j, and all the traffic will arrive at the destination[b]. To tackle the second issue, more nodes on the path from k to j will have to apply the filter, thus aggravating the first issue. The strategy we propose is to filter at each hop with only a subset of the global rule-set that is dynamically chosen for each packet and for each hop. We aim to use larger rule-sets for nodes close to the source and smaller rule-sets for nodes far from the source. The definition of the strategy behind this intuition, however, requires some more discussion and formalization. **Table 1** contains a set of definitions that are used (and detailed) in the rest of the paper.

First of all, how should a node i estimate its distance from the source node k of a packet p with destination to j? The simplest way is to look at the time-to-live (TTL) field in the IP header, but it is also the easiest to circumvent. The attacker could simply forge packets with a low TTL and avoid the waterwall to be effective.

Table 1. Definitions and formal notation.

Notation	Definition
N	The set of nodes in the network
$sp(i,j)$	A set of nodes that form the shortest path between node i and node j
$sp_l(i,j)$	The length of the shortest path $sp(i,j)$
δ	A parameter that determines the maximum size of a rule-set enforced on a single node
$m(i)$	The average distance of node i from all the other nodes in N
r	The average size of a rule-set used by a node in N
R	The global rule-set, i.e., the union of all rule-sets
$t(i)$	The average $sp_l(i,j)$ computed on node i for every packet with source k and destination j and for every (k,j) for which $i \in sp(k,j)$ (see Equation 2)

Another way is to use the distance from the source node k to i, but the attacker can set the source IP to the address of another node w and, contrary to what happens in the Internet, it would still be able to intercept the replies provided it is in the shortest path between w and i. Summing up, node i cannot trust the contents of a packet coming from a node that is possibly an attacker, so the distance from the source must be estimated with other means.

What we propose is that each node uses a subset of rules R_i whose size depends on the ratio between the distance from the destination and $m(i)$, the average distance of node i to any node in the network. In practice, node i compares the length of the remaining path to the destination with the average length of the path of packets generated by i itself. We define P^f as the probability that node i filters a packet going from node k to node j:

$$P^f(k,i,j) \triangleq \begin{cases} \dfrac{sp_l(i,j)}{m(i)} \delta & \text{if } sp_l(i,j) \leq m(i) \\ 1 \cdot \delta & \text{if } sp_l(i,j) > m(i) \end{cases} \quad (1)$$

δ is a parameter that can be used to limit the maximum number of rules enforced in a single node. Node i will use a random subset R_i of R of size $P^f(k,i,j) \times \|R\|$ ensuring that $P^f(k,i,j)$ is the probability that i filters p. If R is organized as a linear list, this can be implemented as starting to scan the list from a random point for a portion of the list of size $\|R_i\|$. When i is close to j, the fraction $\dfrac{sp_l(i,j)}{m(i)}$ decreases; in contrary, when i is close to k, the value of $P^f(k, i, j)$ is close to $1 \cdot \delta$. We also define $t(i)$ as the value of $spl(i, j)$ averaged on all routes passing through i between any couple (k, j) If we call $C(i)$ the set of all the couples (k, j) for which $i \in sp(k, j)$, then

$$t(i) \triangleq \sum_{(k,j \in C(i))} \dfrac{sp_l(i,j)}{\|C(i)\|} \quad (2)$$

To understand how our approach scales with the size and shape of the network graph, we have to understand the behavior of $t(i)$. Let us define m and t as the average $m(i)$ and $t(i)$ computed on every node, respectively; m is the average number of hops in the network, and t is the average number of hops remaining after a packet is forwarded by any node, averaged on all the nodes. Intuitively, t must be smaller than m, but how do $t(i)$ and $m(i)$ change depending on the position of i in the network? In the next sections, we will first present the results based on an example of linear topology, then we will analyze a more complex two-dimensional (2D) topology.

3.1. 1D Linear Topology

As a clarifying example, we take a linear topology with 10 nodes and report the average values of $t(i)$, $m(i)$, and $t(i)/m(i)$ in **Figure 3**.

It can be noticed that the values of $m(i)$ are influenced by the position of i in the topology. In particular, nodes that are close to the periphery will have larger values compared to nodes that are in the center of the topology. This can be explained noting that when i is in the periphery of the network, its average distance from the other nodes is larger than when i is in the center, so $m(i)$ is higher on the periphery. It is also easy to see that in this simple topology, if we compute $t(i)$ excluding the packets that are generated by i itself, $t(i)$ is constant. This would make the ratio $t(i)/m(i)$ decrease for nodes close to the extreme ends of the network.

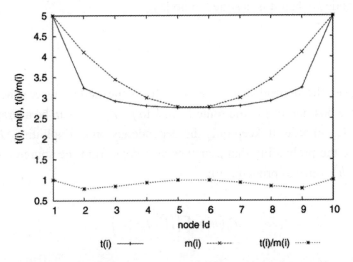

Figure 3. Values of $t(i)$, $t(m)$, and their ratio on the sample linear topology with 10 nodes. Each node is identified by an ID corresponding to its position in the line.

In the figure, instead, we plot $t(i)$ including also the packets generated by node i, which increases the values of $t(i)$ on the periphery. This takes into account that in our scenario, each node is a gateway for its own subnet, so the first hop is counted in its own subnet. Even in this case, $t(i)/m(i)$ is still larger for nodes that are central in the topology.

We expect the central nodes of the network to be more congested than the nodes in the borders since the number of shortest paths that pass across them is higher. Considering this, the shape of $t(i)/m(i)$ introduces a positive effect: the more p gets close to the center of the network, the higher is the chance of being filtered. The practical consequence is that when p is moving from the periphery to the center of the network, that is more congested, its chances of being filtered are increased. When p has already passed the central region of the network, the chances of being filtered decrease. If we look it from a different perspective, we impose a larger filtering effort for packets that are going towards the most loaded area of the network because we want to save resources in that area where they are more precious. When the packets have passed the central area, we spend less effort to filter them since they are directed to the periphery of the network, which is less congested; in any case, packets will be filtered at the destination.

We now define the probability that a packet p is filtered after h hops from the source node k when it is destined to node j:

$$P_h^f(k,h,j) = P^f(k,i,j) \text{ where } sp_l(k,i) = h \qquad (3)$$

$P_h^f(\ )$ moves the dependency of $P^f(\ )$ from the node i where the packet is filtered to the position of i in the route from k to j. $P_h^f(\ )$ can be averaged for all the couples k,j in order to keep only the dependency on h. Exploiting $P_h^f(\ )$, we can compute the probability that p arrives at h hops from the source node, which we call $P_a(h)$ (arrival probability):

$$P_a(h) = \prod_{i=0}^{h-1}\left(1 - P_h^f(i)\right) \qquad (4)$$

In **Figure 4**, we report P_a for the same network considered in **Figure 3** when $\delta = 0.5$.

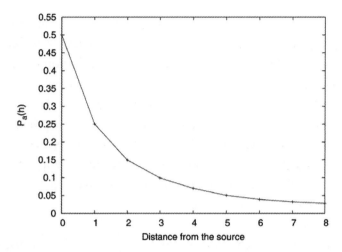

Figure 4. The values of $P_a(h)$ on the linear topology with 10 nodes.

The diameter of the network is equal to nine hops. When the packet arrives at the destination, it is filtered with a destination-specific rule-set, so we do not include the last hop in the curve. We have numbered them from 0 to 8, indicating that the first chance of being filtered is on node k itself. **Figure 4** shows that we obtain indeed the desired effect: the chances of a packet to be filtered are higher close to the source and decrease when it gets close to destination.

3.2. 2D Topologies

In the linear topology described so far, the distance between two nodes is given by the modulus of the difference between their node IDs, so the results are obtained by means of simple algebra. When the network topology is defined on a 2D plane, more complex instruments must be used. The most suitable instrument to study the behavior of a mesh network with a 2D topology is computer simulations; nevertheless, we want to test our technique against networks that may grow up to hundreds of nodes. Network simulators cannot handle scenarios of such size; thus, we use Python NetworkX library to evaluate the characteristics associated with large topologies. For some applications, approximating a wireless mesh network with an abstract graph may be a simplification that is too far from reality. In our case, we rely on the existence of a proactive routing protocol running in the mesh network. We are not interested in physical layer and MAC layer performances (that are more sensitive to the simplifications introduced by graph analysis); we operate directly on the graph that the routing protocol generates, assuming that it is able to find neighbor nodes, to identify and use only symmetric links, and to build the routing table from any source k to any destination j. This is perfectly compatible with, for instance, the widely used OLSR protocol. Note also that we assume the routing protocol uses a shortest-path metric, and we use NetworkX functions in order to compute the values of $sp_l(i,j)$ directly on the graph. It is out of the scope of this paper to show it, but we believe that the same approach can be applied even when the routing is not a simple minimum hop. In this case, the graph will be a weighted one where it is still possible to compute $m(i)$ and $t(i)$ taking into account the weights of each graph edge.

To test the performance of the waterwall, we will use two metrics introduced in previous works[14][22] and defined as follows:

- $M_1(k, j)$. It counts each false-positives on the route from k to j, that is, it is incremented each time an unwanted packet is forwarded on the path from the sender to the destination. It is normalized on the route length from k to j, so it expresses the fraction of the path that p is able to reach before being filtered.

- $M_2(k, j)$. It counts each false-positives end-to-end, that is, it is incremented each time an unwanted packet arrives to j. It is normalized to 1, so it represents the probability of unwanted traffic to arrive to destination j.

When averaged on every couple (k, j), M_1 gives an estimation of the impact of false-positives on the whole network traffic. For instance, when a node that has been infected by a worm starts a DoS attack against another host, M_1 tells how much the waterwall fails to mitigate this attack in terms of wasted network resources.

M_2 instead measures the inefficiency in filtering traffic directed against a specific host. In our scenario, the destination node j applies its own rule-set so that M_2 always goes to zero when p arrives to its destination. But we consider it since it is useful in other scenarios (for instance, for intrusion detection or when some traffic is forbidden by a network administrator but not all nodes support filtering).

$M_2(k, j) = P_a(k, j)$ as it is the probability of not being filtered on the whole path from k to j. M_1 is defined as the average number of hops that p makes before being discarded:

$$M_1(k,j) = \left(\sum_{i=0}^{sp_l(k,j)-1} sp_l(k,i) \times M_2(k,i) \times P^f(k,i,j) + sp_l(k,j) \times M_2(k,j) \right) \frac{1}{sp_l(k,j)} \quad (5)$$

The first term of the equation takes into account packets that are filtered before they arrive to the destination (including node k). It is the sum of the path length from k to i, multiplied by the probability of reaching i and multiplied again by the probability of being filtered on node i. The second terms takes into consideration the packets that arrive to the destination j.

One more evaluation parameter we consider is the average end-to-end delay for every route in the network. For a network in which every node j has a rule-set of size $r_j = 30$, for each route, we compute the average end-to-end delay introducing at every node a processing delay d that depends on $P^f(k, i, j)$. The value of d is taken directly from the data measured on a real platform and reported in **Figure 2**. The delay thus depends on the total number of nodes and on the value of the δ parameter.

Figure 5, **Figure 6**, and **Figure 7** report the value of the metrics M_1, M_2, and delay for a 2D topology with random placement of nodes, increasing the network size and varying δ. The nodes are placed in an area of growing size with constant spatial density of nodes, and each node is connected to the neighbors that fall inside a radius of 70 m. Using NetworkX primitives, we are able to compute the shortest paths on the considered graphs and compute the equations we have defined so far.

We can see that, as expected, M_1 and M_2 decrease when δ is increased (recall that M_1 and M_2 measure false-positives, so they are measures of badness). This is intuitive since a larger δ corresponds to less false-positives. Less intuitive is the fact that given a certain δ, a larger network has smaller values of M_1 and M_2. In the previous section, we have shown that the values of $t(i)$ are smaller if i is close to the periphery of the network; this is true also in 2D topologies. As a consequence, the ratio $t(i)/m(i)$ is smaller in the periphery of the network as can be seen in **Figure 3**. In a 2D topology, the periphery of the network is represented by nodes that are placed on the perimeter of the covered area and that have fewer neighbors compared to the ones that are at the center of the area. If we keep the density constant and increase the number of nodes, we increase the covered area and, consequently, its perimeter. But the perimeter of the network grows more slowly compared to the area, so in larger networks the fraction of the nodes on the perimeter becomes less relevant. As a consequence, a larger network will have a larger average value of the t/m ratio and will filter more packets per hop. **Figure 7** shows the average end-to-end delay for the networks under consideration. A larger δ corresponds to higher processing delays introduced at every hop.

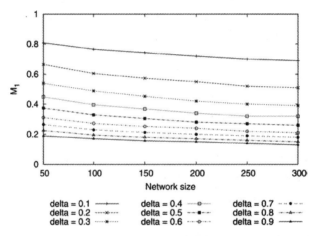

Figure 5. Metric M_1 for increasing network size and δ ranging from 0.1 to 0.9.

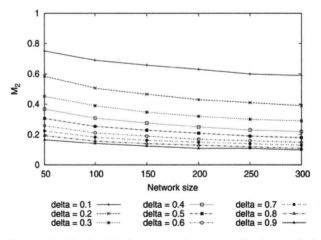

Figure 6. Metric M_2 for increasing network size and δ ranging from 0.1 to 0.9.

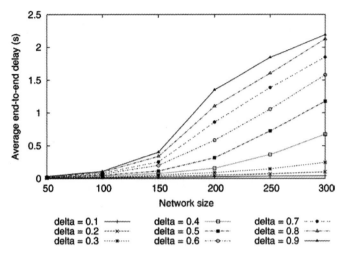

Figure 7. Estimated delay for increasing network size and δ ranging from 0.1 to 0.9.

To interpret these results, consider a network with 200 nodes and 30 rules per node, thus $\|R\| = 6{,}000$. With such a large rule-set, each hop would introduce a delay larger than 0.45 s, as can be seen in **Figure 2**. If we consider that m in such a network has an average larger than 8, this would produce an average delay larger than 3.6 s, which would make the network unusable. Instead, with the waterwall approach, we can configure the δ parameter in order to find the right equilibrium between latency and filtering efficiency; for instance, if $\delta = 0.4$ we obtain an average M_1 lower than 50% and keep the delay around s. That is, we decrease the filtering efficiency to one half, but we reduce the delay by a factor of 24.

Still, if a higher performance of the firewall is needed with a large network size, the delay introduced by the waterwall must be further reduced. In the next section, we introduce an optimization that, at the cost of a simple ordering function applied to the rule-set, can further reduce the false-positive rate.

3.3. Smart Rule-Set Partition

When node i processes a packet p from k to j, it randomly chooses a position λ in R and uses a portion R_i of the rule-set of size $P^f(k, i, j) \times \|R\|$ starting from position λ. Packet p is tested against all the rules in Ri. The probability of evaluating the same rule twice in the path from source to destination given that each choice of λ is independent at each hop is high due to the so-called *birthday paradox*. If we are able to use minimum overlapping R_i sets, then we can expect that M_1 and M_2 decrease faster with the distance from the source. The results obtained with disjoint rule-sets are reported in **Figure 8** and **Figure 9** and represent an upper bound of the gain reachable with this improvement. Comparing **Figure 5** and **Figure 8**, we can see that to obtain similar results, a lower value of δ is sufficient; for instance, to have M_1 below 50%, $\delta = 0.3$ is sufficient (even if $\delta = 0.2$ is below 50% for a network larger than 100 nodes) which corresponds in **Figure 7** to a tolerable delay even for a network with 300 nodes.

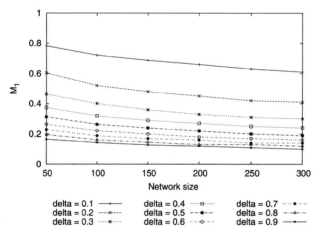

Figure 8. Metric M_1 for increasing network size and δ ranging from 0.1 to 0.9 with ordered rule-set.

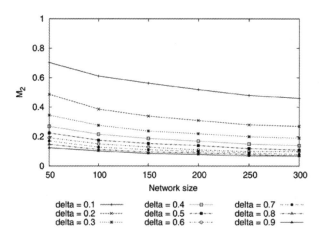

Figure 9. Metric M_2 for increasing network size and δ ranging from 0.1 to 0.9 with ordered rule-set.

We can thus try to find a smarter way to choose λ in order to minimize the intersection between different R_i along the path. In this paper, we introduce two proposals to be further evaluated in future works.

The first is to choose λ as a function of specific network parameters of p and node i:

$$x = \left(\text{IP}_{\text{dest}} \oplus \text{IP}_{\text{src}} \oplus \left(\text{IP}_{\text{chk}} | \text{IP}_{\text{id}} \right) \oplus \widehat{\text{IP}_i} \right) \tag{6}$$

$$\lambda = x \left(\text{mod} \| R \| \right) \tag{7}$$

where

- $\oplus$ is the XOR operator, | is the concatenation operator, and (mod) is the modulo operation

- IP_{dest} and IP_{src} are the destination and source IP addresses of p, respectively

- IP_{chk} and IP_{id} are checksum and identification field of the IP header, respectively. Those fields are immutable from source to destination, and their combination is unique for each packet. They are concatenated since their size is just half of the size of an IP address.

- $\widehat{\text{IP}_i}$ is the IP address of node i with inverted byte order. Bytes are swapped since we want the host identifier of the IP that has a larger variability to be the most significant byte. Otherwise, the modulo operation may just return the same value for each host.

The rationale of this choice is to produce a λ that changes from hop to hop depending on a unique parameter of node i. In this way, we spread the choices of λ with a deterministic algorithm and try to get a better coverage of R. Nevertheless, we have to avoid that a node i deterministically selects the same λ for all the packets belonging to the same flow (identified by IPs and ports). If this condition does not hold and a node always chooses the same rule-set to filter the packets, then there is a chance that portions of R are never covered.

If this condition does not hold, then an attacker may try to precompute the behavior of the nodes in between the attacker and the destination and choose the route with the highest probability of not being filtered. For this reason, we introduce in Equation (6) the unique identifier IP_{id} and the checksum IP_{chk} in order to make λ hard to predict along the evolution of the traffic flow.

As an alternative approach, we could use sp_l (i,j) to determine not only the size of R_i, but also its position in R. This way, λ would not depend on some identifier of the node that is performing the filtering (IP_i) but on the estimation of its position in the path from source to destination. This proposal, as the previous one, is an initial design that needs further analysis.

For both these approaches to be applicable, every node must keep the rules in its rule-set in an ordered list. Nevertheless, we do not lose the generality of the approach since the ordering is independent on the semantics of the rule, so it can be applied to rules of any kind. For instance, given the data structure that is used to store the rule in the operating system, an ordering based on a fingerprint on this data structure is sufficient.

Note that in all the results we have shown so far, $M1$ and $M2$ hardly reach values lower than 0.1. This is due to the fact that $Pf\,(k, i, j)$ in Equation (1) may not be equal to 1 even if $\delta = 1$ at the first hop for the case $sp_l(i, j) > m(i)$. To have values of $P^f\,(k, i, j)$ closer to 1, we can use $\delta > 1$. In this case, Equation (1) must be modified in order to make the value of $P^f\,()$ bounded by 1, as follows:

$$P^f(k,i,j) \triangleq \begin{cases} \dfrac{sp_l(i,j)}{m(i)}\delta & \text{if } sp_l(i,j) \leq m(i) \text{ and } \dfrac{sp_l(i,j)}{m(i)} < 1/\delta \\ 1 & \text{if } sp_l(i,j) \leq m(i) \text{ and } \dfrac{sp_l(i,j)}{m(i)} > 1/\delta \\ 1 & \text{if } sp_l(i,j) > m(i) \end{cases} \quad (8)$$

The rest of the equations do not change. In **Figure 10**, we report M_1 and M_2 when δ is larger than 1; for the sake of clarity, we report only the smallest scenario (50 nodes). It can be noticed that the metrics follow the same trend observed for values lower than 1 and reach values lower than 0.1.

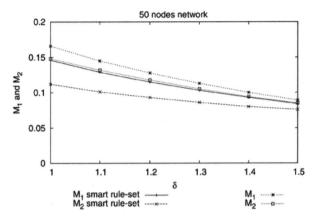

Figure 10. Metric M_1 and M_2 with both filtering strategies when the value of δ is larger than 1.

4. Conclusions

In this paper, we have introduced a new model to perform distributed firewalling in mesh networks that take advantage of the multihop nature of those networks to share the load needed for the filtering function. To stress the difference with a traditional firewall, we chose the term waterwall indicating a fluid and distributed network function, instead of a single filtering host. We have shown that the waterwall can be used to greatly reduce the unwanted traffic in a mesh network. To quantify the cost of the filtering function, we used the delay measured on an embedded processor by large rule-sets and have shown that our approach scales well up to mesh networks of hundreds of nodes. The source code used to realize the test is available on the website of the main project financing this work (www.pervacy.eu).

As future work, we intend to implement the filtering strategy on a network simulator in order to test and optimize the enhancement described in Section 3.3. Afterwards, we plan to embed this technique in some widely used routing protocol implementation, such as OLSR, in order to test on real networks.

Endnotes

[a]At the time of writing, the Guifi network is made up of about 22,000 nodes

and growing at a pace of a hundred nodes per week. The network is divided in zones, each one can be formed by hundreds of nodes.

[b]The attacker we take into consideration is able to mangle the contents of packets, but we imagine that the routing protocol implements some security measures to avoid, or at least identify, attacks on network routing.

Competing Interests

Both authors declare that they have no competing interests.

Acknowledgements

This work has been financed by Provincia di Trento under *The Trentino programme of 360 research, training and mobility of post-doctoral researchers, incoming Post-docs 361 2010* CALL 1, PCOFUND-GA-2008-226070. Renato Lo Cigno has been partially funded by the European Commission under grant agreement no. FP7-288535 'CONFINE': Open Call 1, *Open Source P2P Streaming for Community Networks –OSPS–*.

Source: Maccari L, Cigno R L. Waterwall: a cooperative, distributed firewall for wireless mesh networks [J]. Eurasip Journal on Wireless Communications & Networking, 2013, 2013(1): 1–12.

References

[1] IF Akyildiz, X Wang, W Wang, Wireless mesh networks: a survey. Elsevier Comput. Netw. **47**(4), 445–487 (2005).

[2] P Frangoudis, G Polyzos, V Kemerlis, Wireless community networks: an alternative approach for nomadic broadband network access. IEEE Commun. Mag. **49**(5), 206–213 (2011).

[3] D Vega, L Cerda-Alabern, L Navarro, R Meseguer, in *IEEE International Conference on Wireless and Mobile Computing, Networking and Communications (WiMob)*. Topology patterns of a community network: Guifi.net (Barcelona, 8–10 Oct 2012).

[4] G Min, Y Wu, AY Al-Dubai, Performance modelling and analysis of cognitive mesh networks. IEEE Transactions on Communications. **60**(6), 1474–1478 (2012).

[5] S Ioannidis, AD Keromytis, SM Bellovin, JM Smith, in *7th ACM Conference on Computer and Communications Security (CCS)*. Implementing a distributed firewall (Athens, 01–04 Nov 2000).

[6] H Zhao, CK Chau, SM Bellovin, in *Workshop on New security paradigms (NSPW)*. ROFL: routing as the firewall layer (Lake Tahoe, 22–25 Sept 2008).

[7] H Zhao, SM Bellovin, Source prefix filtering in ROFL. Technical report CUCS-033-09, Columbia University, 2009.

[8] H Zhao, SM Bellovin, in *IEEE International Conference on Mobile Ad-hoc and Sensor Networks*. High performance firewalls in MANETs (Hangzhou, 20–22 Dec 2010).

[9] H Zhao, J Lobo, A Roy, SM Bellovin, in *IFIP/IEEE International Symposium on Integrated Network Management (IM)*. Policy refinement of network services for MANETs (Dublin, 23–27 May 2011).

[10] M Alicherry, A Keromytis, A Stavrou, Distributed firewall for MANETs. Technical report. Columbia University (Computer Science Technical Report Series), 2008.

[11] R Fantacci, L Maccari, P Ayuso, R Gasca, Efficient packet filtering in wireless ad hoc networks. IEEE Commun. Mag. **46**(2), 104–110 (2008).

[12] L Maccari, R Fantacci, P Neira, R Gasca, in *IEEE International Conference on Communications (ICC)*. Mesh network firewalling with bloom filters (Glasgow, 24–28 June 2007).

[13] P Neira, R Gasca, L Maccari, L Lefevre, in *International Conference on New Technologies, Mobility and Security, 2008 (NTMS)*. Stateful firewalling for wireless mesh networks (Tangier, 5–7 Nov 2008).

[14] L Maccari, in *International Conference on Security and Cryptography (SECRYPT)*. A collaborative firewall for wireless ad-hoc social networks (Rome, 24–27 July 2012).

[15] M Taghizadeh, A Khakpour, A Liu, S Biswas, in *IEEE International Conference on Computer Communications (INFOCOM)*. Collaborative firewalling in wireless networks (Shanghai, 10–15 Apr 2011).

[16] H Zhang, B DeCleene, J Kurose, D Towsley, in *IEEE Military Communications Conference (MILCOM)*. Bootstrapping deny-by-default access control for mobile ad-hoc networks (San Diego, 17–19 Nov 2008).

[17] M Alicherry, AD Keromytis, A Stavrou, in *IEEE International Conference on Internet Multimedia Services Architecture and Applications (IMSAA)* Evaluating a collaborative defense architecture for MANETs (Bangalore, 9–11 Dec 2009).

[18] M Esposito, C Mazzariello, F Oliviero, L Peluso, SP Romano, C Sansone, in *Intrusion Detection Systems,* ed. by R di Pietro, LV Mancini. Advances in Information Security, vol. 38 172–210, 2008. Intrusion detection and reaction: an integrated ap-

proach to network security (Springer, New York).

[19] C Panos, C Xenakis, I Stavrakakis, in *International Conference on Security and Cryptography (SECRYPT)*. A novel intrusion detection system for MANETs (Athens, 26–28 July 2010).

[20] L Cerda-Alabern, in *IEEE International Conference on Wireless and Mobile Computing, Networking and Communications (WiMob)*. On the topology characterization of Guifi.net (Barcelona, 8–10 Oct 2012).

[21] T Clausen, P Jacquet (eds.), *3626-Optimized Link State Routing Protocol (OLSR)* (The Internet Society, Geneva, 2003).

[22] L Maccari, R Lo Cigno, in *IEEE/IFIP Conference on Wireless On demand Network Systems and Services (WONS), Poster Session*. Privacy in the pervasive era: a distributed firewall approach (Courmayeur, 23 Dec 2012).

Chapter 27
WRSR: Wormhole-Resistant Secure Routing for Wireless Mesh Networks

Rakesh Matam, Somanath Tripathy

Dept. of Computer Science and Engineering, Indian Institute of Technology Patna, Patna, Bihar 800013, India

Abstract: Wormhole attack is one of the most severe security threats in wireless mesh network that can disrupt majority of routing communications, when strategically placed. At the same time, most of the existing wormhole defence mechanisms are not secure against wormhole attacks that are launched in participation mode. In this paper, we propose WRSR, a wormhole-resistant secure routing algorithm that detects the presence of wormhole during route discovery process and quarantines it. Unlike other existing schemes that initiate wormhole detection process after observing packet loss, WRSR identifies route requests traversing a wormhole and prevents such routes from being established. WRSR uses unit disk graph model to determine the necessary and sufficient condition for identifying a wormhole-free path. The most attractive features of the WRSR include its ability to defend against all forms of wormhole (hidden and Byzantine) attacks without relying on any extra hardware like global positioning system, synchronized clocks or timing information, and computational intensive traditional cryptographic mechanisms.

Keywords: Wormhole Attack, Secure Routing, Unit Disk Graph, Wireless Mesh

Network

1. Introduction

Wireless mesh networks (WMNs) have emerged as a promising technology to provide low-cost, high-bandwidth, wireless access services in a variety of application scenarios[1]. A typical WMN as shown in **Figure 1** is comprised of a set of stationary mesh routers (MRs) that form the mesh backbone and a set of mesh clients that communicate via mesh routers. Security is a critical component that contributes to the performance of WMN. The major challenges that need to be dealt with in addressing security issues mainly arise due to open nature of the wireless medium and multi-hop cooperative communication environment. These factors make network services more vulnerable, specifically due to attacks coming from within the network.

Routing protocols in WMN are susceptible to various security attacks. A detailed survey of such attacks can be found in[2]. In this paper, we focus on a particularly devastating form of attack called wormhole attack[3], on hybrid wireless

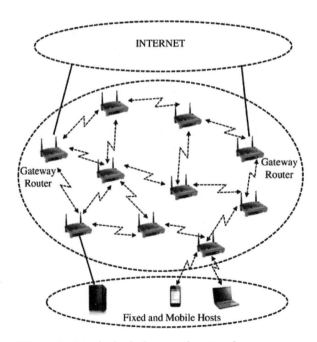

Figure 1. A typical wireless mesh network.

mesh protocol (HWMP), the default path-selection protocol for IEEE 802.11-based WMN[4]. Wormhole attacks can be broadly categorized into two types depending on the type of adversary involved. Wormhole attack launched by colluding external adversaries is called as hidden wormhole attack. Similarly, a wormhole attack launched by malicious colluding internal nodes is called as an exposed/Byzantine wormhole attack. Wormhole attacks (both hidden and exposed) in general are challenging to defend against[3]. However, Byzantine wormhole attack is relatively much difficult to detect than a hidden wormhole, as the nodes involved in the former form legitimate part of the network, and can bypass existing security mechanisms[5]. To launch a wormhole attack, the colluded malicious nodes establish a direct communication channel between themselves and thereby bypass several intermediate nodes. The established channel can be an out-of-band high-speed communication link or an in-band logical tunnel. The wormhole link is usually established between nodes that are located far away from each other. Once established, the wormhole link attracts most of the traffic as the control packets traversing through a wormhole link advertise much better link metric. Selection of such links results in denial of service (DoS), affecting the performance of the network severely.

It has been shown that a strategic placement of the wormhole can disrupt on average 32% of all communication across the network[6]. In this work, we consider both hidden and exposed wormhole attacks. Hereinafter whenever we refer to wormhole attack, it means both hidden and exposed wormhole attack, unless specified explicitly.

There are several potential ways of defending against a wormhole attack, each of which exploits a different unique feature exhibited by a wormhole node/link. For example, schemes like[3][7]–[13] exploit the abnormal length of a wormhole. As previously stated, a wormhole link is usually established between nodes that are physically separated by large distance, thereby bypass several intermediate nodes. Therefore, the simplest way to defend a wormhole attack is by preventing nodes from being tricked into forming a wormhole link by equipping nodes with location systems (GPS) and verifying the relative position of a transmitter during peer-link establishment. Location-based schemes can successfully defend hidden wormhole attacks but cannot prevent Byzantine wormholes from being established as the colluded nodes involved in the attack are legitimate part of the network.

Clock-based mechanisms can restrict the distance travelled by packets but are constrained by clock synchronization issues. Even though alternate mechanisms exist that overcome the synchronization issue, they cannot prevent malicious nodes from forming a Byzantine wormhole for the aforementioned reason.

The other unique characteristic of a wormhole link is that it abnormally increases the node's neighbourhood, and this feature is being exploited in [8][14]–[16] to detect hidden wormhole. Let W_1 be a wormhole node that shares an out-of-band channel with another wormhole node W_2. Now, W_1 can relay its neighbourhood information to W_2 and trick W_2's neighbours into believing that they share direct neighbourhood with W_1's neighbours. This abnormally increases the neighbour count of a node-sharing neighbourhood with a wormhole node. Unfortunately, such schemes fail to detect Byzantine wormholes as Byzantine wormhole link (established between colluded internal nodes) does not alter the neighbourhood information of their respective neighbours. On similar lines, protocols exist that exploit abnormal path attractions of wormhole nodes[17].

In this paper, we present a novel routing protocol (*WRSR*) that addresses both hidden and exposed wormhole attacks in WMN. It depends on neighbourhood connectivity information and relies on existence of shorter alternate subpaths. To the best of our knowledge, this is the first of such kind that prevents Byzantine wormhole attacks using neighbourhood connectivity information. A part of this work is published in[18].

The rest of the paper is organized as follows. The related work is presented in Section 2. Network assumptions and adversarial model are presented in Section 3. Section 4 presents the proposed wormhole resistant secure routing (WRSR) protocol. The proof of concept supporting WRSR is presented in Section 5. The performance of WRSR is analysed in Section 6. Section 7 presents a brief discussion on WRSR and other existing approaches addressing wormhole attacks. Finally, Section 8 concludes the paper.

2. Related Work

Most of the existing approaches that address wormhole attacks rely on specialized hardware like GPS, synchronized clocks or directional antennas. These protocols have been specifically designed to address hidden wormhole attacks. The very first countermeasure developed by Hu et al. in[3] requires GPS and tightly synchronized clocks. To overcome the clock synchronization issues, alternate schemes[9]-[12] have been proposed based on the message round trip time (RTT). Wormhole attack prevention algorithm (WAP) presented in[7] is based on timing information that requires each node to maintain a Wormhole Prevention Timer and overhear its neighbour's retransmission. WAP assumes that wormhole nodes only use wormhole link (in-band-tunnel/out-of-band channel) and do not re-broadcast control messages in their local neighbourhood. Recently, Zhou et al. proposed a wormhole detection mechanism called neighbour-probe—acknowledge (NPA) that is based on standard deviation of RTT (stdev(RTT))[19]. NPA is triggered when node detects change in network topology. To obtain RTTs, each node sends probe messages locally for T times to all its neighbours and gets T acknowledge messages from each neighbour. A wormhole is detected by identifying large deviations in stdev(RTT).

The end-to-end wormhole detection algorithm presented in[8] is based on euclidean distance estimation technique that requires GPS. The source estimates the minimum hop count of the shortest path between source to destination and compares it with shortest route received. If it is much smaller than the estimated value, the source node raises an alert of wormhole attack and initiates a wormhole *TRACING* procedure to identify the end points of a wormhole.

The protocol presented in[16] uses local neighbourhood information to detect wormholes. It is based on the observation that formation of a wormhole link changes the network topology. It assumes that, in a sufficiently dense network, for every given pair of neighbours, there exists at least one common neighbour. Nodes sharing neighbourhood with wormhole node (w_1) can detect a wormhole if it cannot reach the subsequent wormhole node (w_2) through any other node except (w_1). Thaier et al. proposed DeWorm[20] that uses routing discrepancies between neighbours along a path from the source to destination to detect a wormhole. It is based on the observation that, to have a successful impact on the network, the wormhole must attract significant amount of traffic and the length of the wormhole must be significantly large.

Few protocols exist that specifically address Byzantine wormhole attacks.

On-demand secure Byzantine resilient routing protocol (ODSBR)[5] is one such protocol based on DSR[21] that addresses Byzantine attacks. ODSBR relies on explicit network layer acknowledgement on the received data and on a binary search-based probing mechanism to detect malicious dropping of packets. The detection mechanism is instantiated by a source node after observing *log n* number of faults, where *n* is the length of the path. The source node probes intermediate nodes in a binary search fashion to determine the faulty link.

SPROUT[22] is another source-routed, link-state, multipath probabilistic routing protocol that operates in two stages: route generation and route selection. In the route generation stage, a large number of routes are probabilistically generated without taking any routing metric to account. The reliability and round-trip time of each active route are then analysed to choose an optimal route. In this scheme, the route performance feedback is used to select an optimal route which leads to high route establishment latency.

WARP[17] is a wormhole-avoidance routing protocol based on *ad hoc* on-demand routing protocol (AODV)[23], which avoids wormhole attacks by anomaly detection. It is based on the fact that wormhole nodes have abnormal path attractions. WARP considers link-disjoint multiple paths during path discovery but eventually selects only one path to transmit data. Each node in WARP maintains the anomaly values of its neighbours in its routing table. It computes the percentage of routing decisions in which a particular neighbour is involved. That is, it determines the anomaly value by computing the ratio of number of actual routes established through that neighbour to the number of route replies transmitted by that neighbour. If its above a certain threshold, routes replies transmitted by such a node are ignored and thus wormhole nodes are isolated.

A key point to note is that in most of the above existing work, the discussed approaches are restricted to hidden wormhole attacks. Works specifically addressing Byzantine wormhole attacks like ODSBR[5], SPROUT[22] and WARP[17] depend upon the existence of multiple disjoint paths between source S and destination D. In this paper, we propose a wormhole-resistant secure routing protocol (WRSR) that detects and prevents the selection of wormhole paths based on neighbourhood connectivity information and alternate shorter paths.

3. Network Assumptions and Adversary Model

3.1. Network Assumptions

We consider a typical WMN architecture as shown in **Figure 1**, where a set of MRs forms the backbone of WMN. Few of the MRs are equipped with access point functionality to provide access services to its clients. In addition to that, a few of the MRs are designated as gateways and are connected to the Internet. MRs are more or less static and communicate in a multi-hop fashion to provide access services to its clients. MCs are typical wireless clients connected to specific MRs with access point functionality.

3.2. Adversary Model

We assume that an adversary is capable of launching various kinds of wormhole attacks. To begin with, an adversary is assumed to be capable of establishing a high-speed low-latency communication link, required to launch a hidden wormhole attack. Further, an adversary can compromise a few MRs in the mesh backbone to launch a wormhole attack in participation mode. These compromised MRs exhibit Byzantine behaviour and can manipulate routing metric to influence route selection decisions. We mainly focus on wormhole attacks launched by MRs, since route discovery is carried out by MRs on behalf of mesh clients.

An attacker launches a hidden wormhole attack by recording packets at one location, relays them to another location through a wormhole link and retransmits them there into the network[3]. In a WMN, where nodes establish secure peer links and process packets only from peer stations, an adversary needs to target the Authenticated Mesh Peer-link Exchange (AMPE) protocol to successfully launch a wormhole attack[24]. In such a case, the attacker tries to convince two far-away nodes as peers by relaying AMPE protocol messages. On successfully establishing falsified non-existent peer links, an adversary can launch various kinds of active DoS attacks and passive attacks like traffic analysis. Various kinds of defence mechanisms have been proposed, out of which, some of them depend upon extra hardware such as GPS, synchronized clocks or directional antennas, while a few others exploit the features exhibited by wormhole (like neighbourhood information, large discrepancies in path metric, etc.).

A Byzantine wormhole attack is launched by colluding malicious nodes that are legitimate part of the network and can participate in normal network operations. Therefore, securing AMPE protocol cannot prevent colluding nodes from launching a Byzantine wormhole attack. Moreover, the colluded nodes do not alter neighbourhood information of their respective neighbours, therefore detection schemes based on neighbourhood information fail to detect such a wormhole. The major challenge lies in dealing with nodes that are part of the network that can bypass the existing security mechanisms.

4. WRSR: The Proposed Secure Routing

WRSR, the proposed wormhole-resistant secure routing protocol prevents the selection of route requests traversing the wormhole link. WRSR is based on HWMP and therefore inherits majority of its characteristics. The operation of HWMP can be found in the 'Appendix' section.

The operating principle of WRSR is to allow nodes to monitor the two-hop sub-path on a received route request (RREQ) and identify a RREQ that traverses a wormhole. A route request that traverses via a wormhole link would not satisfy the necessary wormhole-free path criterion, which can be detected at the neighbours of a wormhole node and can easily be quarantined.

A path is said to be free from wormhole links if and only if for each sub-path of length $2R$ there exists an alternate sub-path of maximum length $4R$, where R is the transmission range of a node. WRSR thrives on the fact that the probability of finding alternate routes between nodes separated by a distance $d(R < d \leq 2R)$ is high. This proof of concept is presented in Section 5.

Since, nodes in WRSR need to monitor two-hop subpaths on a received RREQ, they need to maintain neighbourhood relations with all the nodes in their two-hop range. To facilitate this, the IEEE 802.11s beacon frame can be extended, as shown in **Figure 2**, to obtain necessary neighbourhood information. The extended beacon frame includes two additional fields, a flag bit and a variable length neighbour address field. The neighbour address field accommodates addresses of varying number of neighbours, and the flag bit is used to indicate presence/absence of neighbourhood information. WRSR employs an extended RREQ element and a modified routing entry as shown in **Figure 3** to accommodate additional addresses for discovering wormhole-free routes. The notations used in this paper are depicted in **Table 1**. WRSR operates in following three processes discussed subsequently.

Frame Control	Duration	Addr 1	Addr 2	Addr 3	Sequence Control	HT Control	←Beacon Frame Body→		FCS
							Flag	Neighbor Address	

Figure 2. Extended beacon frame format.

(a)
Element-ID
Length
Flags
Hop Count
Element TTL
RREQ ID
Originator Mesh STA Address
...
Two Hop Address
Third Hop Address
Fourth Hop Address
Per Target Flags #1
Target Address #1
Target Sequence Number #1

(b)
Destination Mesh STA Address
Sequence Number
Next Hop Address
Two Hop Address
Third Hop Address
Fourth Hop Address
Path Metric
Hop Count
Precursor List
Life Time
State

Figure 3. Route request (a) and routing entry (b) in WRSR.

Table 1. Notations and their meaning in WRSR.

Notation	Meaning
RREQ	Route request
$RREQ_{ID}$	Route request identity
SEQ_No	Route request sequence number
$RREQ_N$	Newly received route request
$RREQ_O$	Existing routing table entry
2Hop	Two-hop address in route request
$RREQ_{N\_iHop}$	ith hop address in $RREQ_N$
RENTRY	Entry in routing table
$RENTRY_{iHop}$	ith hop address of routing entry
$RREQ_{N\_Metric}$	Routing metric in $RREQ_N$
$RENTRY_{Metric}$	Routing metric of a particular routing entry

4.2. Route Discovery Process

Route discovery process employed by WRSR is almost similar to that of HWMP. The source node S initiates route discovery process for establishing a path to the destination D by broadcasting an RREQ. The RREQ processing rules of WRSR are similar to that of HWMP apart from an additional verification required to validate an RREQ. Any intermediate node that receives a broadcasted RREQ verifies the validity of two-hop address present in the RREQ (*i.e.* node I checks whether the two-hop address received in RREQ is present in its neighbourhood information) along with the usual validation process of HWMP. On validating the RREQ, node I creates a routing entry for the corresponding RREQ-ID, sets its state as transient and rebroadcasts it. Otherwise, it drops the RREQ.

4.3. Route Selection Process

The primary goal of WRSR route selection process is to select wormhole-free paths. Nodes monitor the received RREQ's for a necessary and sufficient condition to classify a path to be free from wormholes. Once an RREQ is verified to be wormhole free, the corresponding routing entry is elevated to stable state from transient state. The route selection process is shown in **Algorithm 1**.

Consequent upon receiving a new RREQ (*RREQN*), the intermediate node I processes it to take an appropriate action. Initially, the intermediate node I verifies if a transient routing entry corresponding to the RREQ-ID and sequence number received in *RREQN* exists. Multiple transient routing entries may exist for the same RREQ-ID that are received through a unique two-hop node. Node I then compares the two-hop address present in *RREQN* with the two-hop addresses of a set of existing routing entries represented by {*RREQO*}. If it matches with any of the existing routing entries *RREQO*, it is updated with *RREQN* provided that it offers better metric. In case of no matching address, node I further compares the three and four-hop addresses present in *RREQN* with the two-hop addresses of routing entries represented by *RREQO* or vice versa. If any one of the two addresses match (two-hop address in *RREQN* with three-/four-hop addresses of *RREQO* or vice versa), provided the 2HA of *RREQN* does not match with transmitter address of *RREQO* or vice versa, an optimal of the two RREQ's (*RREQO* or

Algorithm 1. WRSR: route selection process. On receiving $RREQ_N$ by an intermediate node I.

1: **if** no routing entry exists for **S then**
2:   **if** (2Hop is valid) **then**
      create corresponding RENTRY
      state ← transient
      broadcast $RREQ_N$
3:   **else**
    drop $RREQ_N$
4:   **end if**
5: **else**
6:   **if** ($RREQ_{ID}$, SEQ_No, $2Hop$ are valid) **then**
7:     **for** (all routing table entries)
8:       **if** ($RREQ_{N\_2Hop}$==$RENTRY_{2Hop}$) **then**
9:         **if** ($RREQ_{N\_Metric}$ < $RENTRY_{Metric}$) **then**
          update(RENTRY ← $RREQ_N$)
10:         **else**
          drop $RREQ_N$
11:       **end if**
12:     **else**
13:       **if** ($RREQ_{N\_2Hop}$ ==($RENTRY_{3Hop}$ | $RENTRY_{4Hop}$)) **then**
        update(RENTRY ← $RREQ_N$)
        state ← stable
14:       **else**
        update(RENTRY ← better($RREQ_N$, RENTRY))
        state ← stable
15:     **end if**
16:     **else**
      create routing entry for $RREQ_N$
      state ← transient
      broadcast $RREQ_N$
17:   **end if**
18:   **else**
    drop $RREQ_N$
19:   **end if**
20: **end if**

$RREQ_N$) is selected and state of the routing entry is set to stable. If none of the comparisons match, a new transient routing entry is created for the corresponding RREQ-ID.

This matching of addresses is carried out to select an optimal wormhole-free path. The necessary and sufficient condition for detecting a wormhole-free path is presented in Section 5. Finally, if an intermediate node *I* receives an *RREQN* when it already has a stable routing entry to a destination *D*, *I* processes the *RREQN* only if the new route request offers a better metric than the existing route. WRSR creates a separate routing entry for *RREQN* and updates the existing stable entry with *RREQN*, only after it has been verified to be free from wormholes. This process assures the selection of a wormhole free path.

4.4. Route Reply Process

Like any intermediate node *I*, the destination node *D* processes multiple RREQs before selecting an optimal wormhole-free path, satisfying the route selection criteria. It unicasts an RREP through which a stable the RREQ has been received. Subsequently, intermediate nodes propagate the RREP through wormhole-free routes.

4.5. Route Maintenance

Route maintenance in WRSR is similar to that of HWMP. Whenever a node *I* discovers a link failure, it initiates the route maintenance process by transmitting a RERR message addressed to the source. Node *I* can optionally initiate route discovery process on behalf of the source to reduce the route selection latency. Intermediary nodes on receiving a RERR message mark the corresponding routing entry and propagate the RERR message towards the source *S*. The source *S* on receiving the RERR message can re-initiate the route discovery process by broadcasting an RREQ.

5. Proof of Concept

In this section, we show that the route selection process employed by WRSR avoids the wormhole path by verifying the necessary and sufficient condition for wormhole-free path. For simplicity reason, let us assume that each node is equipped with an omnidirectional antenna with unit transmission range. This can be easily fitted into the unit disk graph (UDG) model.

5.1. Unit Disk Graph

UDGs have been extensively employed to create an idealized communication model for a multi-hop wireless network[14][20][25]. In UDGs, each node can be modelled as a disk of unit radius in a plane. Each node is a neighbour of all nodes located within its disk. We assume that the network consists of a large number of nodes distributed uniformly with density ρ (number of nodes in a circle) inside a disk of radius R (considered to be unity in our model). Two nodes can directly communicate with each other if the distance between them is less than or equal to R.

5.2. Problem Formulation

Hop count is an important field in the routing process. Therefore, many popular routing protocols including HWMP, DSDV-ETX[26], MR-LQSR[27], etc., use hop-count as an important field in the RREQ element, even though the metric employed to select an optimal route is different. Essentially, a wormhole bypasses the multiple intermediary hops to cover a distance of W_d between two wormhole nodes W_1 and W_2 that are usually separated by a large distance $d_{W1,2}$ ($d_{W1,2} > 2R$) in a single hop. A typical wormhole path in a network is shown in **Figure 4**. The length of a wormhole sub-path which connects two distant nodes u and v that are neighbours of wormhole nodes w_1 and w_2, respectively, is three hops apart from each other; that is, a node v can be effectively reached from $w1$ in two hops through wormhole link. It has been observed that, in an uniformly distributed network, alternate paths exist between nodes separated by a distance d ($> R$). Therefore, in a genuine case (absence of a wormhole), it is possible to reach v that is two-hop away from w_1 in at most four hops with a high probability. This characteristic can be exploited by node v to differentiate a wormhole link from a genuine link. The following lemma tries to prove the existence of an alternate shortest path between nodes separated by a maximum distance of $2R$.

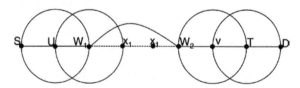

Figure 4. A typical path passing through a wormhole.

Lemma 1. *Lemma 1 A path is said to be free from wormhole links if the following condition is satisfied: "for each sub-path of length 2R there exists an alternate sub-path of maximum length 4R with a probability of* $\left(1-e^{-\rho\left(\pi\left(\frac{R}{4}\right)^2\right)}\right)^t$ *where 't' is the number of disks on a selected path."*

Proof. Consider a network where large number of nodes are uniformly and independently distributed with density ρ, inside a disk of radius R. In such a network, the number of nodes in a region $\Re$ with area $\Delta_\Re$ follows Poisson distribution that can be realised as follows.

$$\Pr(\Re \text{ contains } n \text{ nodes}) = e^{-\rho\Delta_\Re} \frac{(\rho\Delta_\Re)^n}{n!} \tag{1}$$

Let $N_{u,v}$ be the number of hops on the shortest path between u and v. Then, clearly we have

$$N_{u,v} \geq \frac{d_{u,v}}{R} \tag{2}$$

In a network where density of nodes ρ is high, with high probability we should obtain

$$N_{u,v} \leq 2\frac{d_{u,v}}{R} \tag{3}$$

If $d_{u,v} \geq \frac{R}{2}$, then there are $t = \left[2\frac{d_{u,v}}{R}\right] - 1$ disk with radius $\frac{R}{4}$ and origins at distances $d_i = \frac{R}{2}i + \frac{R}{4}, 1 \leq i \leq t$, from u on a line going through u to v, as shown in **Figure 5**. Clearly, the distance between two nodes in adjacent disks is at most R. Using Equation (1) we obtain,

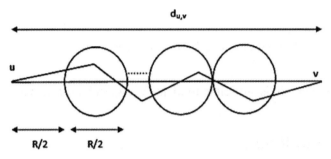

Figure 5. Finding a path between u and v.

$$\Pr(\text{at least one node in each disk})$$

$$= \left(1 - P(\text{no node in a disk})\right)^t = \left(1 - e^{-\rho\pi\left(\left(\frac{R}{4}\right)^2\right)}\right)^t \quad (4)$$

Therefore, there is a path of length $t+1 = \left\lceil 2\frac{d_{u,v}}{R}\right\rceil$ with probability at least $\left(1 - e^{-\rho\pi\left(\left(\frac{R}{4}\right)^2\right)}\right)^t$. Thus, obtaining such a path is possible with high probability if $\left(e^{-\rho\pi\left(\left(\frac{R}{4}\right)^2\right)}\right)^t \ll 1$ that implies

$$\rho \gg \frac{\ln\left(\left\lceil 2\frac{d_{u,v}}{R} - 1\right\rceil\right)}{\pi\left(\frac{R^2}{4}\right)} \quad (5)$$

In **Figure 4**, $d_{w_1,v} \leq 2R$ i.e. the minimum number of hops required to cover a distance of $2R$ between w_1 and v is 2 (if traversed through a wormhole). Therefore, the nodes w_1 and v, separated by a distance $2R$, can reach each other with the help of a common neighbour w_2 traversing through a shortest path $w_1 \to w_2 \to v$. Based on the above observations, if one can obtain a necessary condition that there exists a sub-path of maximum length 4R with probability $\left(1-e^{-\rho\pi\left(\left(\frac{R}{4}\right)^2\right)}\right)^t$ which is computed to be high, is sufficient to identify a wormhole-free path.

6. Simulation Results

In this section, we present the simulation results to showcase the effectiveness of proposed wormhole-resistant secure routing protocol. The experiments were carried out on OMNeT++4.2.1, a discrete event network simulator[28]. The performance of WRSR is compared with WARP[17] and ODSBR[5]. To carry out the following experiments, we set up a network of 400 MRs over an area of 2,000 m^2, forming the mesh backbone. The transmission range of each MR is set to 100 m. The IEEE 802.11 MAC protocol is employed with a channel data rate of 54 Mbps.

To begin with, we analyse the effect of wormhole attack on the performance of WRSR, WARP and ODSBR on a network, where nodes are independently and uniformly distributed based on Poisson distribution. To maximize the impact of wormhole attack, the wormhole link is centrally placed, as a centrally placed wormhole link can attract higher number of route selection decisions[6]. The wormhole link is simulated as a high-speed low-latency communication link between two malicious nodes. The comparison results are based on percentage of packets (PDR) delivered in presence of multiple wormholes. Source and destination nodes are chosen randomly, and the total simulation time was set to 3,000 s. The experiment was designed in such a way that each source transmits 0.5 MB of video traffic to a corresponding destination in presence of wormhole links. Initially, the length of wormhole was set to 4R, where R is the transmission range of a node. The packet length was set to 1,024 bytes. The same experiment was repeated by increasing the number of wormhole links. For any given combination of simulation parameters, we ran 150 different simulations and finally averaged over all 150 different topologies.

Figure 6 shows the performance comparison of different protocols. We mainly focus on the result of WARP and ODSBR, as HWMP is devoid of any

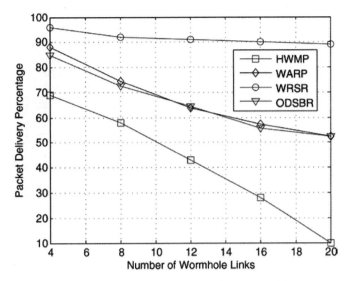

Figure 6. Performance comparison of WRSR with WARP, ODSBR and HWMP.

defence mechanism. The lower percentage of packets delivered by WARP and ODSBR is due to the latency in detecting the wormhole link. Since, WARP is an anomaly based wormhole detection scheme, it initially suffers from packet loss due to the possible selection of wormhole nodes in the initial route discovery process. Its performance is enhanced once a wormhole node is detected and isolated. Similarly, ODSBR enters into a probing state only when there is a violation in packet loss threshold. Therefore, both WARP and ODSBR suffer from initial packet losses as conformed in **Figure 6**. The results depict cumulative packet loss registered in the network at the end of simulation time. The lower packet loss percentage of WRSR is attributed to the zero latency in detecting a wormhole link and therefore registers consistent performance over rest of the protocols.

Our second experiment, carried out on the same network topology, analyses the performance of different protocols by varying the length of the wormhole link. Similar to the first experiment, the wormhole link is centrally placed to increase its effectiveness. The results shown in **Figure 7** clearly indicate that the length of the wormhole has no impact on the performance of WRSR, whereas performance of WARP and ODSBR falls consistently with increase in wormhole length. This characteristic is attributed to the fact that large wormholes can influence more route selection decisions.

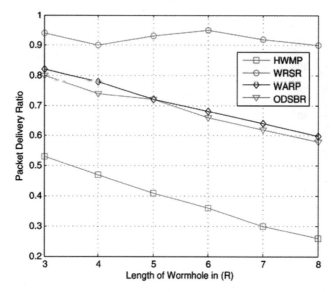

Figure 7. Performance comparison of WRSR with WARP, ODSBR and HWMP.

In both the cases (varying percentage of wormhole links and varying length of wormhole link), WRSR clearly performs better in comparison to WARP and ODSBR. WRSR avoids route requests traversing a wormhole link at all times whereas WARP and ODSBR selects route requests traversing a wormhole node, until the route-building rate is higher than the allowed threshold or the packet loss threshold falls below the allowed limit, respectively. Therefore, choosing a threshold value plays a vital role in detection of wormhole node in both the protocols. The threshold values of ODSBR and WARP are obtained from the experiments conducted in[5] and[17], respectively. The loss threshold of ODSBR is considered to be *log n* faults, where *n* is the hop-count of the selected path, whereas a threshold value of 0.51 is considered for WARP.

The above experiments were carried out for network topologies formed using Poisson distribution. On the other hand, to analyse the performance of WRSR on randomly distributed network topologies, we carry out similar kind of experiments as above with an only difference that the nodes are randomly distributed. The other network parameters are unaltered. **Figure 7** and **Figure 8** show the performance comparison of different protocols in a network under wormhole attack. It can be observed that the random distribution of nodes does not have any impact on the percentage of packets delivered by WRSR. The rationale behind this is the fact

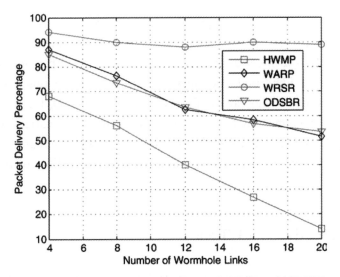

Figure 8. Performance comparison of WRSR with WARP, ODSBR and HWMP.

that WRSR selects routes only if the RREQs meet the necessary wormhole-free path criterion. Failing to meet the same results in dropping of an RREQ that may be genuine. However, existence of alternate paths due to higher node density allows WRSR to select alternate routes thereby not effecting the packet delivery ratio. At the same time, it can be observed that there is no variation in performance of WARP and ODSBR when compared to their performance in uniformly distributed network, as the protocols are not dependent on underlying network topology.

Finally, **Figure 9** shows the impact of wormhole length on the packet delivery ratio of different protocols. The results clearly show that the length of wormhole link has no impact on PDR of WRSR. However, PDR of WARP and ODSBR falls with increase in length of wormhole link.

6.1. Detection Rate of WRSR

The success achieved by a wormhole detection algorithm is measured in terms of percentage of wormholes detected. Every successful detection of a wormhole link contributes to the success rate of WRSR. To compute the detection rate, we set up a network of 400 MRs that are uniformly distributed over an area of 2,000 m^2 using Poisson distribution function. Wormhole nodes are randomly selected, and the density of nodes ρ (number of nodes in a disk) is varied between 4 to 7. A successful wormhole detection event comprises of an RREQ traversing a wormhole that fails to meet the required wormhole-free path criterion, which is then appropriately identified and quarantined. To achieve this, each node in WMN monitors the two-hop sub-path traversed by an RREQ, for existence of an alternate sub-path connecting the two-hop node. **Table 2** summarizes the detection rate of WRSR. WRSR reports higher detection rate with increasing value of ρ. This is due to the fact that, for higher values of ρ, the probability of finding an alternate path is high.

6.2. False Positives in WRSR

The amount of false positives reported by WRSR is computed in a similar fashion as the computation of the detection rate. A false positive in WRSR is a

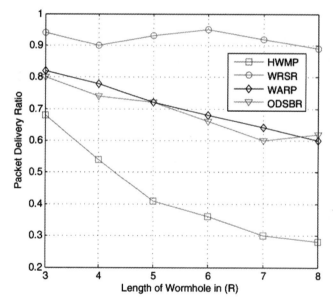

Figure 9. Performance comparison of WRSR with WARP, ODSBR and HWMP.

Table 2. Wormhole detection rate of WRSR.

Density (ρ)	4	5	6	7
Detection rate (%)	94.67	99.33	100	100

situation where an RREQ traversing a genuine link is dropped for failing to meet the necessary wormhole-free path criterion. This situation arises when no alternate path exists within a maximum of four hops connecting a two-hop node traversed by the RREQ.

However, the existence of an alternate path only depends on the density (ρ) of nodes in the network. Therefore, the impact of false positives reported by WRSR is studied by varying the density of nodes in the network. The simulation set up is such that for varying configurations of the network, the performance of WRSR is evaluated for various node densities (number of nodes in a disk). We specifically monitor for scenarios where a genuine link is falsely considered as a wormhole and such an RREQ is dropped. **Table 3** summarizes the percentage of false positives reported by WRSR. The amount of false positives reported is relatively high for a lower density values. But, as shown in the **Table 3**, negligible false positives occur for higher values of ρ. The values represented are rounded off to the nearest ceiling value.

6.3. Impact of Wormhole Length

The length of a wormhole link has negligible impact on the performance of WRSR. Wormholes greater than $2R$ have almost 100% detection rate. We evaluated the performance of WRSR by varying the length of wormhole link for a constant node density $\rho(=4)$. **Table 4** summarizes the impact of wormhole length on WRSR. Results clarify that WRSR reports almost 100% detection rate for wormhole links greater than $2R$.

Table 3. False positives reported by WRSR.

Density (ρ)	4	5	6	7
False positive (%)	6	3	0	0

Table 4. Impact of wormhole length on WRSR.

Length of wormhole	3R	4R	5R	6R
Detection rate (%)	98.33	100	100	100

6.4. Impact of Node Degree

Node degree plays an important role in the detection rate achieved by WRSR. It requires a minimum average node degree (number of neighbours) of 3. However, the percentage of false positives is high for a node degree of 3, as shown in **Table 3**. This is due to non-availability of alternate links between nodes of interest. WRSR reports 100% detection rate for an average node degree of 4 and above. A node degree of 4 is justified in a network like WMN, where nodes are strategically placed to provide access services to its clients.

7. Discussion

Routes traversing through a wormhole link are relatively much shorter and offer better metric when compared to genuine routes. The wormhole link essentially bypasses intermediary nodes to create non-existent routes in the network. WRSR successfully identifies a wormhole link during route discovery due to the existence of alternate paths between nodes separated by a distance d ($R < d \leq 2R$). In a uniformly distributed network with density ρ, we have analytically and experimentally shown that the probability of finding at least one alternate path is very high (98%). The static nature of mesh topology in mesh backbone contributes to the higher detection probability as links are more stable.

Existing protocols like SPROUT[22] and WARP[17] that address Byzantine wormhole attacks rely on existence of link-disjoint multiple paths between source and destination. SPROUT probabilistically generates multiple routes to destination and monitors the performance of each active route by means of signed end-to-end acknowledgements. It computes the reliability and round-trip time of an active route using the fraction of packets sent over it. SPROUT allows nodes to establish routes through wormhole links but based on their performance they are ignored. On the other hand, WARP considers link-disjoint multiple paths during path discovery and provides greater path selections to avoid malicious nodes but eventually uses only one path to transmit data. WARP allows nodes to monitor the number of routes created through each neighbour and isolates a particular neighbour whose anomaly value (route-building rate) is greater than the threshold. WARP suffers from rapid fluctuation of anomaly values which results in frequent isolation and recovery of nodes. Isolation of a node simply involves ignoring the route replies transmitted by a particular neighbour. Even though the route replies are not processed from such a neighbour whose anomaly value is greater than threshold, the route reply count maintained in the routing table for that neighbour is still incremented. This allows a node to recover from the isolation phase. But since the anomaly value changes rapidly for each received RREP, a malicious node can also recover very quickly from isolation phase. Therefore, determining the threshold value for a network is one of the major limitations of WARP.

ODSBR[5] addresses all kinds of Byzantine attacks including Byzantine wormhole. However, ODSBR has several limitations. One of the major limitation of ODSBR is that it can only work with a source routing protocol, such as DSR, where the source knows all the intermediate nodes on a selected path. Second, the diagnosis packets have to be encrypted with a shared key between the source and the intermediate nodes. Third, the isolation is done per link rather than per node, *i.e.* blacklisting a malicious node results in isolating a honest node. Finally, the blacklisting of malicious nodes is done at the source of packet and not locally at the neighbours of the malicious node. Therefore, even if a malicious node is already blacklisted by some nodes in the network, it continues to be active and cause harm to traffic from other sources.

WRSR performs better as compared to other wormhole detection mechanisms. This is because of its ability to prevent a wormhole link from being selected during the path establishment. Thus, by avoiding malicious wormhole nodes, WRSR prevents malicious packet loss. The only scenario in which a path traversing through the wormhole link may be selected by WRSR is if it satisfies the necessary wormhole-free path criterion. This case can arise only when the wormhole nodes are neighbours to each other, *i.e.* wormhole link of length ($< 2R$), in which, an alternate path within a maximum of four hops may exist. However, as a wormhole link between two neighbouring nodes has very less impact on path selection decisions, it can be safely ignored.

Apart from that, WRSR overcomes the limitations of WARP and SPROUT, without relying on multiple link-disjoint paths. It does not require multiple routes to be considered before selecting an optimal route, thus reducing the route discovery latency. Lastly, the static nature of wireless mesh routers makes WRSR well suited to WMN. The comparison of security of the existing protocols is summarised in **Table 5**.

Table 5. Security comparison of various existing protocols.

Protocol	Employed mechanism	Extra hardware	Hidden mode	Participation mode
WRSR	Connectivity information	No	Yes	Yes
Packet leashes [3]	GPS & clock	Yes	Yes	No
Wang et al. [14]	Neighbourhood information	No	Yes	No
DeWorm [20]	Neighbourhood information	No	Yes	No
ODSBR [5]	Binary search	No	Yes	Yes
SPROUT [22]	Multipath routing	No	Yes	Yes
EDWA [7]	Neighbourhood information	No	Yes	No
Znaidi et al. [16]	Neighbourhood information	No	Yes	No
WARP [17]	Multiple link-disjoint paths	No	Yes	Yes

8. Conclusions

Addressing wormhole attacks is a crucial issue to ensure security in a wireless mesh network. In this paper, we proposed a novel wormhole-resistant secure routing (WRSR) protocol that relies on shorter alternate paths to detect a wormhole link. During route discovery, WRSR monitors for alternate paths for a cached RREQ and quarantines such RREQ that fails to meet the necessary and sufficient condition. The necessary and sufficient condition to differentiate a normal path from wormhole link is derived using unit disk graphs. The probability of finding such alternate paths has been analytically computed and shown to be high in a uniformly distributed network.

Appendix

Hybrid Wireless Mesh Protocol

Hybrid wireless mesh protocol (HWMP) is the default path-selection (routing) protocol for IEEE 802.11-based WMN. As the name implies, HWMP is a combination (hybrid) of on-demand route selection mode and proactive tree-based approach. The on-demand path-selection mode of HWMP is based on AODV. The set of protocol elements (like route request, route reply and route error), their generation and processing rules of HWMP are similar to AODV. HWMP supports two modes of operation depending upon the network configuration. The on-demand route selection mode does not require root MR support and can be employed by any node that needs to establish a route. Whereas, the proactive tree building mode that compliments the existing on-demand mode can be employed only when a root MR is configured. The proactive tree routes can be established either using proactive route request (PREQ) or route announcement (RANN) messages. The proactive and on-demand modes are not exclusive and can be used concurrently, because the proactive modes are extensions to the on-demand mode.

Whenever a source node needs to find a route to a destination using the on-demand path-selection mode, it broadcasts a route request with the target address set to the address of destination and the metric initialized to initial value of the active route selection metric. The default route selection metric employed by HWMP is airtime. Airtime reflects the amount of channel resources consumed to transmit a frame over a particular link. The essence of airtime metric is to capture the status of a wireless link in terms of required time units to transmit a frame. On the other hand, there are two ways to proactively establish routes to the root node. The first method uses a proactive route request element and is intended to create paths between all mesh routers and the root node in the network. The second method uses a root announcement message and is intended to distribute path information for reaching the root node. The mesh router configured as root node sends either PREQ or RANN messages periodically.

The proactive tree building process begins when a root MR sends out a PREQ message with the target address set to all ones, implying all the MRs in WMN. Similarly, the root MR can also periodically propagate a RANN into the network. On receiving a RANN message, each MR that needs to create or refresh a path to the root MR, sends an individually addressed route request to the root MR via the MR from which it received the RANN. The root MR sends a route reply in response to each received route request. Thus, the proactive and reactive path-selection elements collectively allow HWMP to meet the path-selection requirements of WMN.

Competing Interests

Both authors declare that they have no competing interests.

Acknowledgements

We would like to sincerely thank Dr. Sukhamay Kundu, Associate Professor, Dept. of Computer Science, Louisiana State University, for his valuable comments and suggestions. We would also like to thank the anonymous reviewers for their suggestions which helped us to improve this paper.

Source: Matam R, Tripathy S. WRSR: wormhole-resistant secure routing for wireless mesh networks [J]. Eurasip Journal on Wireless Communications & Networking, 2013, 2013(1): 1–12.

References

[1] IF Akyildiz, X Wang, W Wang, Wireless mesh networks: a survey. J. Comput. Netw. ISDN Syst. **47**(4), 445-487 (2005).

[2] W Zhang, Z Wang, SK Das, M Hassan, in *Book, Wireless Mesh Networks: Architectures and Protocols*. Security Issues in Wireless Mesh Networks (Springer, New York, 2008).

[3] Y Hu, A Perrig, D Johnson, in *Proceedings of the 22nd IEEE International Conference on Computer Communications*. Packet leashes: a defence against wormhole attacks in wireless networks (IEEE, San Francisco, 30 March–3 April 2003).

[4] IEEE P802.11s Part 11: Wireless LAN Medium Access Control (MAC) and Physical Layer (PHY) specifications, Amendment 10: Mesh Networking. (IEEE, New York, 2011).

[5] B Awerbuch, R Curtmola, D Holmer, C Nita-Rotaru, H Rubens, ODSBR: an on-demand secure Byzantine resilient routing protocol for wireless *adhoc* networks. ACM Trans. Inf. Syst. Security. **10**(4), (2008).

[6] M Khabbazian, H Mercier, VK Bhargava, Severity analysis and countermeasure for the wormhole attack in wireless *ad hoc* networks. IEEE Trans. Wireless Commun. **8**(2), 736-745 (2009).

[7] S Choi, DY Kim, DH Lee, JI Jung, in *Proceedings of IEEE International Conference on Sensor Networks, Ubiquitous and Trustworthy Computing*. WAP: Wormhole attack prevention algorithm in mobile *ad hoc* networks (IEEE, Taichung, 11–13 June 2008), pp. 343–348.

[8] X Wang, J Wong, in *Proceedings of the Thirty-First Annual International Computer Software and Applications Conference*. An end-to-end detection of wormhole attack in wireless *ad hoc* networks (IEEE, Beijing, 24–27 July 2007), pp. 39–48.

[9] Z Tun, AH Maw, Wormhole attack detection in wireless sensor networks. J. World Acad. Sci. Eng. Technol. **46**(2), 545–550 (2008).

[10] P Tran, LX Hung, YK Lee, S Lee, H Lee, in *Proceedings of Fourth IEEE Consumer Communication and Networking Conference*. TTM: an efficient mechanism to detect wormhole attacks in wireless *ad-hoc* networks (IEEE, Las Vegas, January 2007), pp. 593–598.

[11] T Korkmaz, in *Proceedings of International Conference on Information Technology: Coding and Computing*. Verifying physical presence of neighbors against replay-based attacks in wireless *ad hoc* networks (IEEE, Las Vegas, 4–6 April 2005), pp. 704–709.

[12] S Capkun, L Buttyan, JP Hubaux, in *Proceedings of the First ACM Workshop on Security of ad hoc and Sensor Networks*. SECTOR: Secure tracking of node encounters in multi-hop wireless networks (ACM, Virginia, 31 October 2003), pp. 21–32.

[13] R Poovendran, L Lazos, A graph theoretic framework for preventing the wormhole attack in wireless *ad hoc* networks. ACM J. Wireless Netw. **13**(1), 27–59 (2007).

[14] Y Wang, Z Zhang, J Wu, in *Proceedings of IEEE Fifth International Conference on Networking, Architecture and Storage*. A Distributed Approach for Hidden Wormhole Detection with Neighborhood Information (IEEE, Macau, 15–17 July 2010), pp. 63–72.

[15] X Ban, R Sarkar, J Gao, in *Proceedings of the 12th ACM International Symposium on Mobile Ad Hoc Networking and Computing*. Local connectivity tests to identify wormholes in wireless networks (ACM, Paris, 16–20 May 2011).

[16] W Znaidi, M Minier, JP Babau, in *Proceedings of IEEE international Symposium on Personal, Indoor and Mobile Radio Communications*. Detecting wormhole attacks in wireless networks using local neighborhood information (IEEE, Cannes, 15–18 September 2008), pp. 1–5.

[17] MY Su, WARP: a wormhole-avoidance routing protocol by anomaly detection in mobile ad hoc networks. Comput. Security. **29**(2), 208–224 (2010).

[18] R Matam, S Tripathy, in *Proceedings of the 8th International Conference on Information Systems and Security*. Defence against wormhole attacks in wireless mesh networks (Springer LNCS, Guwahati, 15–19 December 2012), pp. 181–193.

[19] J Zhou, J Cao, J Zhang, C Zhang, Y Yu, in *Proceedings of IEEE 26th International Conference on Advanced Information Networking and Applications*. Analysis and Countermeasure for Wormhole Attacks in Wireless Mesh Networks on a Real Testbed (IEEE, Fukuoka, 26–29 March 2012), pp. 59–66.

[20] T Hayajneh, P Krishnamurthy, D Tipper, in *Proceedings of Third International Conference on Network and System Security*. DeWorm: A simple protocol to detect wormhole attacks in wireless *ad hoc* networks (IEEE, Gold Coast, 19–21 October 2009), pp. 73–80.

[21] DB Johnson, DA Maltz, YC Hu, The dynamic source routing protocol for mobile ad-hoc network (DSR), IETF internet draft. (IETF MANET Working Group, Fremont, 2004).

[22] J Eriksson, M Faloutsos, S Krishnamurthy, in *Proceedings of IEEE International Conference on Network Protocols*. Routing amid colluding attackers (IEEE, Beijing, 16–19 October 2007).

[23] CE Perkins, EM Royer, SR Das, *Ad hoc* on-demand distance vector (AODV) routing, IETF internet draft. (IETF MANET Working Group, Fremont, California USA, 2004).

[24] B Swathi, S Tripathy, R Matam, Secure peer-link establishment in wireless mesh networks. Adv. Intell. Syst. Comput. **176**, 189–198 (2012).

[25] BN Clark, CJ Colbourn, DS Johnson, Unit disk graphs. Discrete Math. J. **86**(1-3), 165–177 (1990).

[26] D De Couto, D Aguayo, J Bicket, R Morris, in *Proceedings of MobiCom*. A high-throughput path metric for multi-hop wireless routing (ACM, San Diego, 14–19 September 2003), pp. 134–146.

[27] R Draves, J Padhye, B Zill, in *Proceedings of ACM MobiCom*. Routing in multi-radio multi-hop wireless mesh networks (ACM, Philadelphia, 26 September–1 October 2004), pp. 114–128.

[28] The OMNeT++ Network Simulator. www.omnetpp.org. Accessed 23 September 2011.

Chapter 28

An Effective Implementation of Security Based Algorithmic Approach in Mobile Adhoc Networks

Rajinder Singh[1], Parvinder Singh[2], Manoj Duhan[3]

[1]Deenbandhu Chhotu Ram University of Science & Technology, Murthal, Haryana, India
[2]Department of Computer Science and Engineering, Deenbandhu Chhotu Ram University of Science & Technology, Murthal, Haryana, India
[3]Department of Electronics and Communications, Deenbandhu Chhotu Ram University of Science & Technology, Murthal, Haryana, India

Abstract: Mobile Ad-hoc Network one of the prominent area for the researchers and practitioners in assorted domains including security, routing, addressing and many others. A Mobile Ad-hoc Network (MANET) refers to an autonomous group or cluster of mobile users that communicate over relatively bandwidth constrained wireless links. Mobile ad hoc network refers to the moving node rather than any fixed infrastructure, act as a mobile router. These mobile routers are responsible for the network mobility. The history of mobile network begin after the invention of 802.11 or WiFi they are mostly used for connecting among themselves and for connecting to the internet via any fixed infrastructure. Vehicles like car, buses and trains equipped with router acts as nested Mobile Ad-hoc Network. Vehicles today consists many embedded devices like build in routers, electronic devices like Sensors PDAs build in GPS, providing internet connection to it gives, information and infotainment to the users. These advances in MANET helps the vehicle to communicate with each other, at the time of emergency like accident, or during climatic changes like snow fall, and at the time of road block, this information will be informed to the nearby vehicles. Now days technologies rising to provide efficiency to MANET users like providing enough storage space, as we all know the cloud computing is the next generation computing paradigm many researches are conducting experiments on Mobile Ad-hoc Network to provide the cloud service securely. This paper attempts to propose and implement the security based algorithmic approach in the mobile ad hoc networks.

Keywords: MANET, Network Security, Wormhole Attack, Secured Algorithm

1. Introduction

Now days, lots of research is going on in the domain of mobile ad hoc networks. One of the major issues in the mobile ad hoc networks is the performance - in a dynamically varying topology; the nodes are expected to be power-aware because of the bandwidth constrained network. Another matter in such networks is security—as each node participates in the operation of the network equally, malicious nodes are intricate to identify. There are several applications of mobile ad hoc networks such as disaster management, ware field communications, etc. To analyze and detailed investigation of these issues, the scenario based simulation of secure protocol is done and compared with classical approaches. The scenarios used for the simulation and predictions depict critical real-world applications including battlefield and rescue operations but these can be used in many other applications also.

In ad hoc networks all nodes are responsible of running the network services meaning that every node also works as a router to forward the networks packets to their destination. It is very challenging for researchers to provide comprehensive security for ad hoc networks with the desired quality of service from all possible threats. Providing security becomes even more challenging when the participating nodes are mostly less powerful mobile devices.

Wireless Ad Hoc networks have been an interesting area of research for more than a decade now. What makes ad hoc networks interesting and challenging is its potential use in situations where the infrastructure support to run a normal network does not exist. Some applications include a war zone, an isolated remote area, a disaster zone like earthquake affected area and virtual class room etc.

In ad hoc networks all nodes are responsible of running the network services meaning that every node also works as a router to forward the networks packets to their destination. It is very challenging for researchers to provide comprehensive security for ad hoc networks with the desired quality of service from all possible threats. Providing security becomes even more challenging when the participating nodes are mostly less powerful mobile devices. In this paper an effort has been made to evaluate various security designs proposed.

2. Security Aspects in Mobile ad hoc Networks

In any classical fixed or wireless network, the security is implemented at three stages: prevention, detection and cure. The key parts of prevention stage include authentication and authorization. The authentication is concerned with authenticating the participating node, message and any other meta-data like topology state, hop counts etc. Authorization is associated with recognition. The point where detection is the ability to notice misbehavior carried out by a node in the network, the ability to take a corrective action after noticing misbehavior by a node is termed as cure.

Assorted possible attacks that are implemented on ad hoc networks are eavesdropping, compromising node, distorting message, replaying message, failing to forward message, jamming signals etc. The central issues behind many of the possible attacks at any level of security stage are authentication, confidentiality, integrity, non repudiation, trustworthiness and availability.

3. Assumption and Dependencies

- Basically Ad-hoc Networks depends upon any fixed infrastructure or any other mobile node to communicate, through forwarding and receiving packets.

- Comparing the security issues of wireless ad-hoc network with wired ad-hoc network, wired network has the proper infrastructure for forward and receiving packets, whereas in wireless network there is no proper infrastructure and it is accessible by both authorized users and hackers.

- In this wireless ad-hoc network there is no particular design to monitor the traffic and accessibility, these leads to third party intervention like malicious users.

In this manuscript, various issues are focused that affect the ad-hoc networks security mechanism and also to concentrate on pros and cons of Mobile networks protocols. The focus on enhancing security and reliability to Mobile Ad-hoc Network (MANET)[1] is also addressed.

Many researches were done before to provide security to MANET[1] but none of the protocol shines in providing security and performance. There are many defects in the Mobile framework; this may cause unknown nodes to connect frequently without any proper routing. In order to prevent other nodes from trespassing we are going to concentrate on providing more security to Mobile Ad-hoc network.

There were so many research areas in MANET[1] in that security is the major concern among others.

The scope of securing MANET[1] is mentioned here.

- Securing MANETs[1] is great challenge for many years due to the absence of proper infrastructure and its open type of network.

- Previous security measures in MANETs[1] are not effective in the challenging world with advancement in technology.

- Many layers often prone to attacks man in middle attack or multilayer attack, so proposal should concentrate on this layers.

- The proper intelligent approach[2] of securing MANETs[1] has not yet discovered.

- In this project we are going to concentrate on applying bio inspired intelligence[2] techniques for securing MANETs.

4. Problem Identification

- The main objective of the manuscript is providing security to the existing systems mainly on the network layer to prevent the attacks like wormhole attacks[3] etc.

- To analyze the scope of multi layer attacks[4].

- To evaluate the techniques like Genetic Algorithms[5], Swarm Intelligence[6], Memetic Algorithms[7] etc.

- To analyze the needs of above mentioned techniques in different network layers especially in the multi link layer.

- To propose a unique technique for above mentioned attacks.

- Intelligent MANET[6] proposal to deal with all kinds of attacks.

- To validate the above techniques by implementing and analyzing its results with the existing systems.

5. Applications

- It provides a relative study of the systems under the parameters packet loss, packet delivery rate and network connectivity.

- A better understanding of the Quality of Service (QoS) parameters can be obtained and they can be used for solving various networking complexities.

6. Hardware Requirements

The minimum requirements needed to perform operations are

- Intel Pentium Processor at 2 GHz or Higher
- RAM 256 MB or more
- Hard disk capacity 10 GB or more

7. Software Requirements

The software required to perform the implementation are

- Linux Operating System (Ubuntu, Fedora)
- NS2, NAM tools
- GNU Plot

8. Manet Security Attacks

Malicious node[8] is one which causes attacks on various layers on MANET like application layer, data link layer, physical and network layer.

There were two types of attacks on MANET, they are

- Active attacks
- Passive attacks

9. Active Attacks

In this attack, some harmful information is injected into the network, which causes malfunctioning of the other nodes or network operation. For performing this harmful information it consumes some sort of energy from other nodes, those nodes are called as malicious node.

9.1. Passive Attacks

In this passive attack, the malicious nodes disobey to perform its task for some sort reasons like saving energy for its own use of moving randomly, by diminishing the performance of the network.

9.2. Network layer attack

Let us concentrate on various attacks on the network layer.

9.3. Wormhole attack

Wormhole attack[3] is also known as tunnelling attack, in this tunnelling attack the colluding attackers build tunnel between the two nodes for forwarding packets claiming that providing shortest path between the nodes and taking the full control of the nodes, which is invisible at the higher layers.

Figure 1 represents the wormhole attack, where S and D nodes are the source and destination, A B and C are the connecting nodes providing path between source and destination. M and N are the malicious nodes, tunnelled by colluding attackers.

10. Existing Technique for Preventing Wormhole Attack

In the previous techniques wormhole attack is prevented using the Location based Geo and Forwarding (LGF) Routing Protocol.

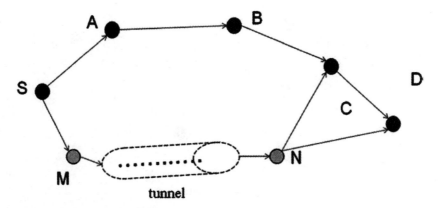

Figure 1. Wormhole attack.

11. Implementation of LGF Routing Protocol

There are several steps in implementing LGF routing protocol, consider source node S wants to communicate with destination node D (**Figure 2**).

- The Source node multicast the RREQ message to all the intermediate which contains the IP address of the destination node based on distance of the destination node.

- This protocol is tested with source node 100 M away from the destination node and the intermediate nodes as
  DIST (S, 1) = 40 M
  DIST (S, 2) = 53 M
  DIST (S, 5) = 48 M
  DIST (1, 3) = 60 M
  DIST (2, 3) = 130 M
  DIST (3, D) = 180 M
  DIST (4, 6) = 45 M
  DIST (S, 4) = 62 M
  DIST (5, 6) = 85 M
  DIST (6, D) = 78 M

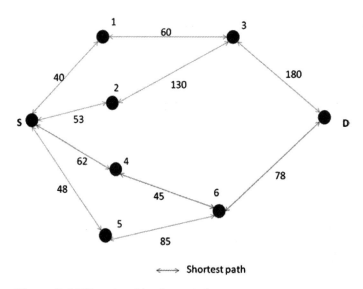

Figure 2. LGF protocol implementation.

- Compare distance between source and destination using the following code If (intermediate nodes < source node S to destination node D distance)
{
These are the nodes in between S to D, can conditionally transfer the RREQ packet to D.
}
Else
{
The intermediate node is out of transmission area, so send RREQ error message to S node
}

- RREQ has been received in destination node, start D node sending RREP packet towards the intermediate node to reach the source node.

- S node received RREP packet from different intermediate nodes, compare the distance from different intermediate nodes.

- Select the shortest path between the source and destination node with respect to the received RREP packet and then send the original packets between S and D node this was the technique used in LGF protocol.

However the preventive measures of wormhole attack with this LGF protocol was not solved clearly.

12. Black Hole Attack

Black hole attack[8] is the serious problem for the MANETs, in this problem a routing protocol has been used by malicious node reports itself stating that it will provides shortest path.

In flooding based protocol, a fake route is created by the malicious node rather than the actual node, which results in loss of packets as well as denial of service (DoS).

In the **Figure 3**, S and D nodes are the source and destination nodes, A B C

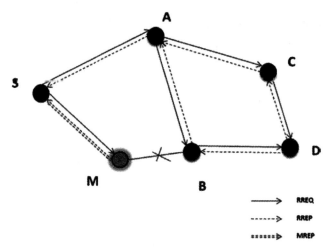

Figure 3. Black hole attack.

are the intermediate nodes and M is the malicious node. RREQ and RREP are the key terms for route request and route reply respectively. MREP is abbreviation for malicious reply.

13. Existing Technique

Two Tier Secure AODV (TTSAODV)

TTSAODV protocol is proposed earlier to prevent the black hole attack. In these protocol two levels of security is provided

1) During route discovery mechanism and
2) During data transfer mechanism

In this technique, black hole attack is easily identified either of these two techniques, even it fails in any of the mechanism. The major drawback in this technique causes enormous packet loss and delay in transferring packet.

14. Resource Consumption Attack

In the resource consumption attack, a malicious node can try to consume more battery life demanding too much of route discovery, or by passing unwanted packets to the source node.

15. Location Disclosure Attack

In the location disclosure based attack, the malicious node collects the information of routes map and then focus on further attacks. This is one of the unsolved security attacks against MANETs.

16. Multi Layer Attacks in Manet

There are different types of multilayer attacks in MANET, they are as follows

- Denial of Service (DoS)
- Jamming
- SYN flooding
- Man In Middle attacks
- Impersonation attacks

17. Alpha Numeric Based Secure Reflex Routing

In this, proposed algorithm prevents the worm-hole attacks by routing the data through the authorized nodes like LN, and AN nodes through this way the communication takes place.

In the proposed algorithm the worm-hole tunnel is prevented through the following steps (**Figure 4**).

17.1. Step 1

Since every connection through nodes is possible only through Leader Node and Access node so there is impossible for a malicious node to make tunnel from the source node.

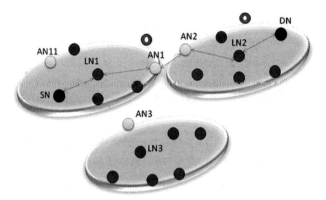

Figure 4. Proposed Worm-hole prevention technique.

17.2. Step 2

The Leader Node manages the routing table and also the details of all the nodes in its group, it also contains the details of whether the particular node is Access Node or normal node. The Leader node also maintains details about other groups Leader Node and its address with the help of its Access Nodes.

17.3. Step 3

The normal node in a group maintains a table that contains information of its Leader Node address and the common identifier generated by the Leader Node. The Access nodes have a table that maintains the other Leader Nodes common identifiers.

17.4. Step 4

The address of the Leader Node that has already involved in routing has stored in every packet, it is used for verification by other Leader Nodes.

17.5. Step 5

When a source node in a need of route to deliver packets to the destination node, it sends Route Request message to the Leader node, the Leader Node uses its common identifier to verify the packet with alpha numeric values.

17.6. Step 6

The leader Node checks whether the destination node is in house, if the destination node is present under the leader node, then it sends the packet directly. If the destination node is not in house then it sends Route Request message to all its Access nodes, The Access nodes using their common identifier verifies the alpha numeric values from Leader node then transfers that packet to the neighbours Access Node.

17.7. Step 7

The neighbour Access node checks whether the packet came from its neighbour Leaders node or from any malicious node by common identifier that has previously exchanged, then it sends the Route Request message to its Leader Node, this Leader Node verifies the Leader node details and include its details in that packet and forwards the original packet until it reaches the destination.

17.8. Step 8

Finally the destination node checks whether the packet came from its Leader node or from any malicious node using the identifier, after verification process is over it accepts the packet.

17.9. Step 9

Destination node sends the Reply Request message (RREP) to source node through the same route already followed for transferring packet.

17.10. Step 10

In case the any node involved in the routing moves away from one group into the another group, the previous process is not needed as it is already registered in that network, some other node in that group replace the previous node.

17.11. Step 11

Suppose if the source node or destination node moves away from its group, the foreign Access Node acts as a relay node for forwarding packets this process minimizes the time for authenticating in newer group.

18. Proposed Architecture

18.1. Worm-Hole Attack Prevention Using Alpha Numeric Reflex Routing Algorithm

In this technique, there won't be any possibilities for a malicious node to make tunnelling between the source and the destination nodes, as it is not included in the either of any groups. The packets are safe to reach the destination node efficiently.

18.2. Pseudocode for Alpha Numeric Reflex Routing Algorithm

```
BEGIH
    Initialize nodes
    Initialize source and destination nodes
    FOR i = 0 to n DO
        LNᵢ ← Node with higher battery power, ability to manage
        other nodes
        IF (nodes in range of LN) THEN
            Transmit common identifier
        ELSE
            The node is under other LN
        END IF
    END FOR
```

```
        FOR i = 0 to n DO
        FOR j = j + 1 to n DO
            ANᵢⱼ ←Nodes receive common identifier from other LN
            IF (node accepts the common identifier and replies its details
            to LN) THEN
                Node = trusted
            ELSE
                Node=malicious
            END IF
            Source node → Forward RREQ
            IF (source node and destination node is under same LN)
            THEN
                Forward RREQ → destination node
            ELSE
                Forward RREQ → ANᵢⱼ
                ANᵢⱼ → LNᵢ
                LNᵢ → destination node
            END IF
            END FOR
        END FOR
END
```

18.3. Proposed Algorithm to Prevent Black Hole Attack

In this proposed algorithm, the Expected broadcast count algorithm is introduced. With the help of this algorithm highest throughput is possible between the nodes but however the actual algorithm does not prevent the black hole attack.

Throughput refers to the average number of message transmitted in a given time, it is usually measured in bps or bits per second, and it is also mentioned as packet delivery ratio. Malicious node plays a major role in affecting throughput in black hole attacks.

Secure mesh network measurement technique is proposed in this project to prevent the black hole attacks during route discovery process between the source and destination node with the help of the throughput measurement values, this makes the routing process more consistent and efficient communication between the nodes.

18.4. Expected Broadcast Count Algorithm

This EBX algorithm is used to increase throughput in MANETs, it is referred as the expected number of packets transmission and retransmission required to successfully deliver a packet in the network.

It is calculated using the delivery ratio of packets in destination node d_d and delivery ratio of packets in the source node d_s, d_d is the prospect of forward packet transmission and d_s is the reverse packet transmission.

These d_s and d_d values are calculated from the acknowledgement packets known as query, nodes commonly exchanges their query message with their neighbours after delivering each packet.

Suppose consider a link from $A \rightarrow B$ where A and B are the nodes, these two nodes determined themselves to send query message for particular time gap period g/τ, where as τ = jitter (packet delay variations).

A and B counts the number of query they received from each other during gap period count $(t - g, t)$ then A calculates the d_d from the equation.

$$d_d = count(t-g,t) \bigg/ \left(\frac{g}{t}\right) \qquad (1)$$

where count $(t - g, t)$ is the number of query commenced by node B and received by node A.

The node B calculates the d_s in similar way to d_d.

$$d_s = count(t-g,t) \bigg/ \left(\frac{g}{t}\right) \qquad (2)$$

A and B swaps the d_s and d_d values to calculate the *EBX*.

$$EBX_{A \to B} = \frac{1}{d_s * d_d} \tag{3}$$

This equation s used to find *EBX* value for more routes, *EBX* value has more hops, and the routes with more number of hops may have lesser throughput due to the intrusion among hops in the same path.

Source and Destination nodes *EBX* value can be calculated through the following formula.

$$EBX_{S \to D} = EBX_{A \to B} \tag{4}$$

Less EBX value in the routes have fewer possibility of packet loss, and that route is more preferable than others routes (**Figure 5, Table 1**).

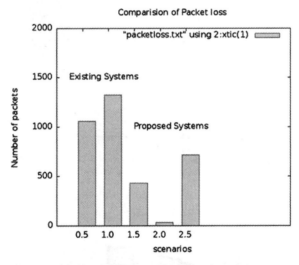

Figure 5. Packet loss comparison graph.

Table 1. Packet loss comparisons.

Scenarios	Time (in seconds)	Packet drop (in bits)
Existing system 1	6.5	10581
Existing system 2	6.5	13221
Proposed system 1	6.5	4372
Proposed system 2	6.5	322
Proposed system 3	6.5	715

19. Intelligent Manet Algorithm

In this intelligent approach, nodes connected to this network is monitored by server agent, the server agent manages the details of the mobile nodes in a network like

- Behaviour of the node
- Speed of the node
- Direction of the node
- Position of the node

This technique prevents the malicious node from attacking other nodes (**Figure 6**).

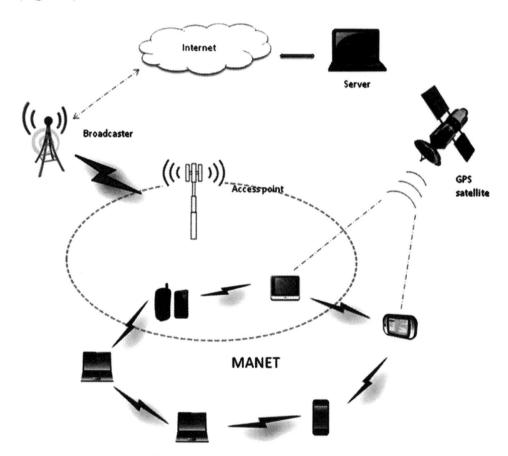

Figure 6. Intelligent MANET architecture.

19.1. Step 1

The nodes participating in the networks to access service like internet registers its identity with the server agent, the server agent replies with unique ID to the requesting node.

19.2. Step 2

The source node request route with the current access point to the destination node the current access point forwards the route request to the server agent.

19.3. Step 3

The server agent verifies the source ID, then it accepts the route request from sender then it gathers the information of receiver using destination ID from the list.

19.4. Step 4

The server agent then broadcasts the route request message using destination ID, the registered adjacent nodes that are nearer to the destination node which are ready to provide the service replies with the acknowledgement message to the server agent.

19.5. Step 5

The server agent chooses the adjacent node with the longest life time (the ability of the nodes to stay connected with the destination node) using the details collected from the ID, Such as nodes position, direction of motion and speed of the node.

19.6. Step 6

Then the server agent provides route reply message for the source node, after this authentication process, source node starts sending data packets in a secure way.

19.7. Step 7

In case any node moves away from the network, immediately the server agent replaces it with some other nodes to maintain the continuity of connection.

19.8. Step 8

In this technique, the malicious node or selfish nodes are completely eliminated from the network, as the server agent takes full control of the ad-hoc network.

20. Conclusion

Mobile adhoc networks are facing vulnerability and security issues from a long time. Assorted protocols and algorithmic approaches has been developed and implemented so far to avoid and remove the issues associated. In this manuscript, we have implemented an empirical and effective approach to optimize the packet loss frequency. The algorithmic approach is implemented in the network simulator ns2 to execute the scenarios and results.

Competing Interests

The authors declare that they have no competing interests.

Authors' Contributions

RS carried out the development of algorithmic approach, actual logic and implementation. PS and MD finally analyzed the results. All authors read and approved the final manuscript.

Source: Singh R, Singh P, Duhan M. An effective implementation of security based algorithmic approach in mobile adhoc networks [J]. Human-centric Computing and Information Sciences, 2014, 4(1): 1–14.

References

[1] Clausen TH (2007) Introduction to mobile ad-hoc networks, Internet Draft.

[2] Yu C-F (1989) Security safeguards for intelligent networks. In: IEEE International Conference on World Prosperity Through Communications. ICC '89, BOSTO-NICC/89. Conference record, vol 3. GTE Lab. Inc, Waltham, MA, USA, pp 1154–1159.

[3] Choi S, Kim DY, Lee DH, Jung J-i (2008) WAP: wormhole attack prevention algorithm in mobile ad hoc networks, SUTC '08. IEEE International Conference on Sensor Networks, Ubiquitous and Trustworthy Computing pp 343–348.

[4] Li JH, Das S, McAuley A, Lee J, Stuhrmann T, Gerla M (2010) A multi-layer approach for seamless soft handoff in mobile ad hoc networks. Hui Zeng Intell. Autom., Inc. (IAI), Rockville, MD, USA, pp 21–26, GLOBECOM Workshops (GC Wkshps), IEEE.

[5] Leonard J (1997) Interactive Game Scheduling with Genetic Algorithms, Minor Thesis, RMIT (Royal Melbourne Institute of Technology University). Department of Computer Science.

[6] Prasad S, Singh YP, Rai CS (2009) Swarm based intelligent routing for MANETs. Int J Recent Trends Eng 1(1).

[7] Garg P (2009) "A comparison between memetic algorithm and genetic algorithm for the cryptanalysis of simplified data encryption standard algorithm". Int J Netw Secur Appl (IJNSA) 1(1).

[8] Sanjay R, Huirong F, Manohar S, John D, Kendall N (2003) Prevention of Cooperative Black Hole Attack in Wireless Ad Hoc Networks". International Conference on Wireless Networks (ICWN'03), Las Vegas, Nevada, USA.

Chapter 29
How to Make a Linear Network Code (Strongly) Secure

Kaoru Kurosawa, Hiroyuki Ohta, Kenji Kakuta

Ibaraki University, 4-12-1 Nakanarusawa, Hitachi, Ibaraki 316-8511, Japan

Abstract: A linear network code is called k-secure if it is secure even if an adversary eaves-drops at most k edges. In this paper, we show an efficient deterministic construction algorithm of a linear transformation T that transforms an (insecure) linear network code to a k-secure one for any k, and extend this algorithm to strong k-security for any k. Our algorithms run in polynomial time if k is a constant, and these time complexities are explicitly presented. We also present a concrete size of |F| for strong k-security, where F is the underling finite field.

Keywords: Information Theory, Network Code, Security

1. Introduction

The notion of network code was introduced by Ahlswede et al.[1]. Li et al.[11] proved that the source node s can multicast n field elements $(m_1, ..., m_n)$ to a set of sink nodes Sink = $\{t_1, ..., t_q\}$ by using a linear network code if $|F| \geq |\text{Sink}|$, where

$$n = \min_i \text{max-flow}(s, t_i)$$

and F is a finite field such that $m_i \in$ F. (**Figure 1** shows an example of a linear network code.)

Jaggi et al.[8] proposed a polynomial time algorithm which can construct a linear network code from any network instance $(G(V, \varepsilon), s, \text{Sink}, n)$, where $G(V, \varepsilon)$ is the underlying network.

Consider a model such that the source node s multicasts $(m_1, ..., m_{n-k}, r_1, ..., r_k)$ instead of $(m_1, ..., m_n)$, where ri is chosen uniformly at random from the field F. We say that a linear network code is k-secure if an adversary learns no information on $(m_1, ..., m_{n-k})$ even by eavesdropping at most k edges.

Figure 2 shows a 1-secure linear network code. For example, $d_1 = m_1 + r$ is transmitted on the edge (s, v_1). It leaks no information on m_1 because the random element r works as one-time pad.

Cai and Yeung[4] proved that there exists a linear transformation T that makes any linear network code k-secure if $|F| > \binom{|\varepsilon|}{k}$, where ε is the set of edges. In fact, T is an $n \times n$ nonsingular matrix.

The advantage of this method is that it does not require changing the underlying linear network code[6]. The source node s only has to multicast $(\tilde{m}_1, ..., \tilde{m}_n) = (m_1, ..., m_{n-k}, r_1, ..., r_k) \times T$.

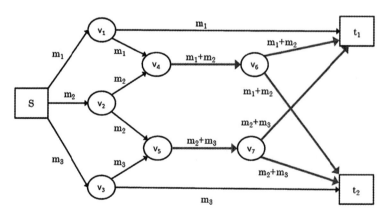

Figure 1. Linear network coding scheme with $n = 3$.

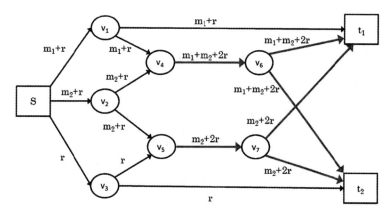

Figure 2. 1-Secure linear network coding scheme (mod 3).

Cai and Yeung, however, only showed the *existence* of T based on a counting argument (see[4] [Sect. V]). They did not show how to construct T efficiently.

Harada and Yamamoto[7] extended the notion of k-security to strong k-security. Consider any $A \subset \varepsilon$ and any $B \subset \{m_1, ..., m_n\}$ such that $|A| = |B| \leq k$. Then a linear network code is called *strongly k-secure* if A leaks no information on $\{m_1, ..., m_n\} \setminus B$.

Figure 3 shows a strongly 1-secure linear network code. For example, $d_1 = 2m_1 + m_2 + m_3$ is transmitted on the edge (s, v_1) and

- d_1 leaks no information on (m_1, m_2) since m_3 works as one-time pad.
- d_1 leaks no information on (m_2, m_3) since $2m_1$ works as one-time pad.
- d_1 leaks no information on (m_1, m_3) since m_2 works as one-time pad.

In this model, it is assumed that each m_i is independently random.

Harada and Yamamoto proved that for sufficiently large $|F|$, there exists a strongly k-secure linear network code for any network instance $(G(V, \varepsilon), s, \text{Sink}, n)$ if $k < n$. However, they did not explicitly state the time complexity of their algorithm explicitly. In addition, they did not suggest a concrete size for $|F|$, leaving the derivation of a sufficient condition on $|F|$ as an open problem. (See "open problem" in Table 1 of[7]. They considered strong $k!$-security with $k! \leq k$, where . is used instead of k in[7].)

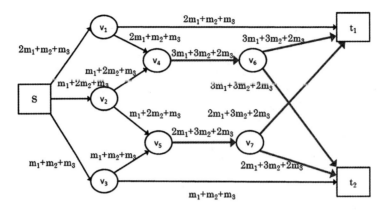

Figure 3. Strongly 1-secure linear network coding scheme (mod 5).

In this paper, we first show an efficient deterministic construction algorithm of a linear transformation T that transforms an (insecure) linear network code to a k-secure one for any $1 \leq k < n$. We then extend this algorithm for strong k-security for any $1 \leq k < n$. Both of our algorithms run in polynomial time if k is a constant.

We explicitly present the time complexities of our algorithms. We also present a concrete size of |F| for strong k-security, thereby solving the open problem of Harada and Yamamoto[7].

By applying our methods to **Figure 1**, we can obtain the 1-secure linear network code shown in **Figure 2**, the strongly 1-secure linear network code shown in **Figure 3** and the strongly 2-secure linear network code shown in **Figure 4**.

1.1. Related Works

Rouayheb et al.[5] proposed a direct method to construct a k-secure linear network code from a network instance $(G(V, \varepsilon), s, \text{Sink}, n)$. This method does not use a linear transformation T, and therefore requires changes to the underlying linear network code.

Bhattad and Narayanan[2] showed how to construct weakly secure linear network codes. Silva and Kschischang[14][15] introduced the notion of universal k-secure codes and universal strongly k-secure codes. Kurihara et al.[9] improved the universal strongly k-secure codes of[14][15]. In these schemes[9][14][15],

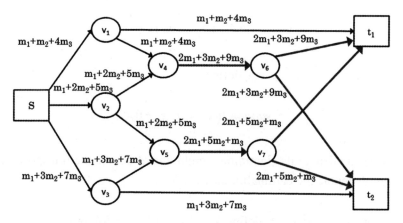

Figure 4. Strongly 2-secure linear network coding scheme (mod 11).

a vector (instead of a field element) is transmitted over an edge. Because each element of the vector is transmitted over multiple time slots, it is assumed that the k tapped edges are fixed during the transmission period. Shioji et al.[13] considered a stronger eavesdropping model where the adversaries possess the ability to re-select the tapping edges during the transmission. They then showed that the scheme in[15] is not secure under this eavesdropping model.

Matsumoto and Hayashi[12] considered a random linear precoder at the source node and proved that it is strongly secure and universal secure if we allow arbitrary small but nonzero mutual information on the transmission symbols to the eavesdropper. In their scheme, they showed that this mutual information is upper bounded by some small quantity.

Tang et al.[16] showed a probabilistic method to construct a linear transformation T that transforms a linear network code to a k-secure one. (See "Time Complexity" of[16] [p. 313].) However, they did not show the success probability and claimed (without proof) that the time complexity is $\left(\binom{|\varepsilon|}{k} \right)$.

2. Preliminaries

Let $H(\cdot)$ denote the Shannon entropy. For a tuple of random variables $\Lambda = (\tilde{a}_1, ..., \tilde{a}_\ell)$ and a subset $A = \{i_1, ..., i_j\} \subset \{1 ... \ell\}$, define

$$\Lambda A = \left(\tilde{a}_{i_1}, \ldots, \tilde{a}_{i_\ell}\right).$$

For a vector $\boldsymbol{x} = (x_1, \ldots, x_N)$, define

$$support(\boldsymbol{x}) = \{i \mid x_i \neq 0\}.$$

Let $w_H(\boldsymbol{x})$ denote its Hamming weight and $\boldsymbol{x}^t$ denote the transpose of $\boldsymbol{x}$. For a set A, let $|A|$ denote the cardinality of A.

For an $n \times \ell$ matrix X, define X_A, $X_{A,k}$ and $X_{A,B}$ as follows.

- For $A \subset \{1, \ldots, \ell\}$, let X_A denote an $n \times |A|$ submatrix of X such that the columns are restricted to A.

- For $k < n$, let $X_{A,k}$ denote a $k \times |A|$ submatrix of X_A such that the rows are restricted to the last k rows. Namely

$$X_A = \begin{pmatrix} Y \\ X_{A,k} \end{pmatrix}$$

for some Y.

- For $B \subset \{1, \ldots, n\}$, $X_{A,B}$ denotes a $|B| \times |A|$ submatrix of X_A such that the rows are restricted to B.

Definition 1 For an $n \times \ell$ matrix X, $\text{Rank}_p(X)$ denotes the set of all $A \subset \{1, \ldots, \ell\}$ such that

$$|A| = rank(X_A) = p.$$

It is easy to see that the following lemma holds.

Lemma 1 *Suppose that T is a $n \times n$ nonsingular matrix T. Then* $\text{Rank}_p(T \cdot X) = \text{Rank}_p(X)$ *for $p = 1, \ldots, n-1$.*

$\mathcal{I}_\ell$ denotes the $\ell \times \ell$ identity matrix. F denotes a finite field and, in particular, F_p denotes a finite field of order p.

3. *k*-Secure Linear Network Code

3.1. Linear Network Code

We define a network instance by $(G(\mathcal{V}, \mathcal{E}), s, \text{Sink}, n)$:

- $G(\mathcal{V}, \mathcal{E})$ is a directed acyclic network such that each edge $e \in \mathcal{E}$ has a unit capacity, *i.e.*, each edge can transmit one field element per time unit. G may include multiple parallel edges.

- $s \in \mathcal{V}$ is a source node.

- $\text{Sink} = \{t_1, ..., t_q\} \subset \mathcal{V}$ is a set of sink nodes.

- n is defined as $n = \min_i \text{max-flow}(s, t_i)$, where $\text{max-flow}(s, t_i)$ denotes the maximum flow from s to $t_i \in \text{Sink}$.

A linear network code for a network instance $(G(\mathcal{V}, \mathcal{E}), s, \text{Sink}, n)$ is defined by an $n \times |\mathcal{E}|$ linear network coding matrix U such that

$$(m_1, ..., m_n) \times U = (d_1, ..., d_{|\mathcal{E}|}), \tag{1}$$

where $(m_1, ..., m_n)$ is the message that s multicasts to Sink, and d_i is the field element that is transmitted on an edge $e_i \in \mathcal{E}$. For example,

$$U = \begin{pmatrix} 1001100001011100 \\ 0100011001111111 \\ 0010000110100011 \end{pmatrix} \tag{2}$$

is the linear network coding matrix used in **Figure 1**.

Formally we say that an $n \times |\mathcal{E}|$ matrix $U = (\boldsymbol{u}_1, ..., \boldsymbol{u}_{|\mathcal{E}|})$ over F is a linear network coding matrix for a network instance $(G(\mathcal{V}, \mathcal{E}), s, \text{Sink}, n)$ if the following conditions are satisfied, where each u_i is indexed by an edge $e_i \in \mathcal{E}$.

1) If $e_i \in \mathcal{E}$ is an outgoing edge of a node v ($\neq s$) and v has incoming edges $e_{i_1},...,e_{i_j}$, then u_i is a linear combination of $u_{i_1},...,u_{i_j}$.

2) Let $\{e_{i_1},...,e_{i_j}\}$ be the set of incoming edges of a sink node $t_i \in$ Sink. Then $Rank(u_{i_1},...,u_{i_j}) = n$ for each $t_i \in$ Sink. This condition guarantees that t_i can reconstruct $(m_1, ..., m_n)$.

Proposition 1[8] *There exists a polynomial time algorithm which can construct a linear network coding matrix U from a network instance (G, s, Sink, n) if $|F| \geq |\text{Sink}|$.*

Proposition 2[6] [Sect. 3] *Suppose that U is a linear network coding matrix for a network instance (G, s, Sink, n). Then for any $n \times n$ nonsingular matrix T, $T \times U$ is also a linear network coding matrix for (G, s, Sink, n).*

Proposition 2 results from

$$(m_1,...,m_n) \times T \cdot U = (\tilde{m}_1,...,\tilde{m}_n) \times U,$$

where $(\tilde{m}_1,...,\tilde{m}_n) = (m_1,...,m_n) \times T$. Namely using $T \cdot U$ as a linear network coding matrix is equivalent to using U as a linear network coding matrix such that s multicasts $(\tilde{m}_1,...,\tilde{m}_n)$. Each t_i can reconstruct $(m_1, ..., m_n)$ because T is nonsingular.

3.2. k-Secure Linear Network Code

Consider the use of a linear network coding matrix U such that

$$(m_1,...,m_{n-k},r_1,...,r_k) \times U = (d_1,...,d_{|\mathcal{E}|}), \tag{3}$$

where $(m_1, ..., m_{n-k})$ is chosen according to some probability distribution, and each r_i is independently and uniformly chosen from F. Then, we say that U is k-secure if any k edges leak no information on $(m_1, ..., m_{n-k})$.

More formally, let $\tilde{M} = (\tilde{m}_1, ..., \tilde{m}_{n-k})$, where $\tilde{m}_i$ is the random variable induced by m_i for $i = 1, ..., n - k$, and let $\tilde{D} = (\tilde{d}_1, ..., \tilde{d}_{|\mathcal{E}|})$, where $\tilde{d}_j$ is the random variable induced by d_j for $j = 1, ..., |\mathcal{E}|$. Then

Definition 2 A linear network coding matrix U is k-secure if for any probability distribution on $(m_1, ..., m_{n-k})$, it holds that

$$H(\tilde{M}|\tilde{D}_A) = H(\tilde{M})$$

for any $A \subset \{1 ... |\mathcal{E}|\}$ such that $|A| = k$.

Proposition 3[3] [Lemma 3.1] *A linear network coding matrix* $U = (u_{i,j})$ *is k-secure if and only if*

$$rank(U_A) = rank(U_{A,k})$$

for any $A \subseteq \{1, ..., |\mathcal{E}|\}$ such that $|A| \leq k$.

In particular, U is 1-secure if and only if $u_{n,i} \neq 0$ for all i.

Cai and Yeung proved the following proposition.

Proposition 4[4] [Theorem 2] *Suppose that* $|F| > \binom{|\mathcal{E}|}{k}$. *Then for any* $n \times |\mathcal{E}|$ *linear network coding matrix U, there exists an $n \times n$ nonsingular matrix T such that $V = T \times U$ is k-secure.*

The advantage of this proposition is that no changes to U are required. The source node s only has to multicast $(\tilde{m}_1, ..., \tilde{m}_n) = (m_1, ..., m_{n-k}, r_1, ..., r_k) \times T$. (See the paragraph at the end of Sect. 3.1.)

Cai and Yeung, however, only showed the *existence* of T based on a counting argument (see[4] [Sect. V]). They did not show how to efficiently construct T.

4. Tools

4.1. Reduced Coding Matrix

We say that a matrix $A = (a_1, ..., a_h)$ is pairwise column independent if each pair of columns (a_i, a_j) are linearly independent.

For an $n \times |\mathcal{E}|$ linear coding matrix U, we say that an $n \times L$ matrix $\tilde{U}$ is a reduced coding matrix of U if $\tilde{U}$ is a maximal submatrix of U such that $\tilde{U}$ is pairwise column independent.

This is formally define as follows.

Definition 3 $\tilde{U} = (\tilde{u}_1, ..., \tilde{u}_L)$ is a reduced coding matrix of $U = (u_1, ..., u_{|\mathcal{E}|})$ if

- $(\tilde{u}_1, ..., \tilde{u}_L) \subset (u_1, ..., u_{|\mathcal{E}|})$,
- $\tilde{U}$ is pairwise column independent, and
- For any u_i, there exists some $\tilde{u}_j$ such that $u_i = \beta \tilde{u}_j$ for some $\beta \neq 0$.

We say that L is the reduced size of U.

For example, the following matrix is a reduced coding matrix of U given by Equation (2)

$$\tilde{U} = \begin{pmatrix} 10010 \\ 01011 \\ 00101 \end{pmatrix} \tag{4}$$

and $L = 5$ is the reduced size of U.

Lemma 2 *We can compute* $(\tilde{U}, L)$ *from* $U = (u_1, ..., u_{|\mathcal{E}|})$ *in time* $O(n|\mathcal{E}|^2 \cdot poly(\log|F|))$. *Proof* We can check if u_i and u_j are linearly independent in $O(n \cdot poly(\log|F|))$ time. (Addition, subtraction, multiplication and devision in F takes $O(poly(\log|F|))$ times.) Therefore we can compute $(\tilde{U}, L)$ in $O(n|\mathcal{E}|^2 \cdot poly(\log|F|))$ time.

It is easy to see that the following lemma holds.

Lemma 3 *Suppose that T is an n × n nonsingular matrix. Then, $T \cdot \tilde{U}$ is a reduced coding matrix of $T \cdot U$ if and only if $\tilde{U}$ is a reduced coding matrix of U.*

It is clear that the following corollaries hold from Proposition 3.

Corollary 1 *A linear network coding matrix U is k-secure if and only if*

$$rank(\tilde{U}_A) = rank(\tilde{U}_{A,k})$$

for any $A \subset \{1, ..., L\}$ such that $|A| \leq k$.

Corollary 2 *A linear network coding matrix U is k-secure if and only if for $i = 1, ..., k$, $rank(\tilde{U}_{A,i}) = i$ for any $A_i = \text{Rank}_i(\tilde{U})$.*

Corollary 3 *A linear network coding matrix U is 1-secure if and only if the last element of $\tilde{u}_i$ is nonzero for each column vector $\tilde{u}_i$ of $\tilde{U}$.*

4.2. How to Increase Hamming Weight

For two vectors $x = (x_1, ..., x_N)$ and $y = (y_1, ..., y_N)$, we present an algorithm MaxWeight(x, y) that finds α such that

$$support(\alpha x + y) = support(x) \cup support(y)$$

in $O(N)$ time. To do so, we first show an algorithm that finds $\alpha \notin S$ in $O(|S|)$ time, where $S \subset F$.

Procedure: Outside(S).

Let the elements of F be $a_0, a_1, \ldots$

1) Set $c(0) = \cdots = c(|S|) = 0$.
2) For each $a \in S$, do:
3)   If $a = a_i$ for some $i \leq |S|$, then set $c(i) = 1$.
4) Let i_0 be the least i such that $c(i) = 0$. Return a_{i_0} as α.

For example,

- If $S = \{a_0, a_1, a_2\}$, then $c(0) = c(1) = c(2) = 1$ and $c(3) = 0$. Hence the above procedure returns $\alpha = a_3$.

- If $S = \{a_0, a_1, a_4\}$, then $c(0) = c(1) = 1$ and $c(2) = c(3) = 0$. Hence the above procedure returns $\alpha = a_2$.

It is easy to prove the following lemma.

Lemma 4 *If $|F| > |S|$, then* Outside(S) *returns* $\alpha \in F$ *such that* $\alpha \notin S$ *in* $O(|S|)$ *time.*

We present the procedure for MaxWeight(x, y) as follows.

Procedure: MaxWeight (x, y).

Let $x = (x_1, \ldots, x_N)$ and $y = (y_1, \ldots, y_N)$.

1) Let $S_0 = \{-y_i/x_i \mid x_i \neq 0\}$.
2) $\alpha \leftarrow$ Outside(S_0).
3) Return α.

Lemma 5 *For two vectors $x = (x_1, \ldots, x_N)$ and $y = (y_1, \ldots, y_N)$,* MaxWeight(x, y) *returns α such that*

$$\text{support}(\alpha x + y) = \text{support}(x) \cup \text{support}(y) \tag{5}$$

in $O(N \cdot poly(\log|F|))$ if $|F| > N$ time.

Proof Suppose that $|F| > N$. Then because $|F| > N \geq |S0|$, we have $\alpha \notin S_0$ at line 2 of the above procedure according to Lemma 4. This means that $\alpha x_i + y_i \neq 0$ if $x_i \neq 0$.

Hence if $\alpha x_i + y_i = 0$, then $x_i = 0$. This means $y_i = 0$. Conversely if $x_i = y_i = 0$, then $\alpha x_i + y_i = 0$. Therefore $\alpha x_i + y_i = 0$ if and only if $x_i = y_i = 0$. In other words, $\alpha x_i + y_i \neq 0$ if and only if $x_i \neq 0$ or $y_i \neq 0$. Consequently we obtain Equation (5).

Line 1 of the above procedure takes $O(N \cdot poly(\log|F|))$ time because computing y_i/x_i needs $O(poly(\log|F|))$ time. At line 2, Outside(S_0) runs in $O(N)$ time from Lemma 4. Therefore the algorithm runs in $O(N \cdot poly(\log|F|))$ time.

For example, consider $\boldsymbol{x} = (1, 1, 1, 0)$ and $\boldsymbol{y} = (0, 4, 2, 3)$ over F5. Then $S_0 = \{0, -4, -2\} = \{0, 1, 3\}$ at line 1. At line 2, we obtain $\alpha = 2 \notin S_0$. Then

$$\alpha \boldsymbol{x} + \boldsymbol{y} = 2 \times (1, 1, 1, 0) + (0, 4, 2, 3) = (2, 1, 4, 3).$$

Thus Equation (5) is satisfied.

4.3. Making a Nonzero Row

Let X and Y be two $n \times \ell$. matrices. Let $\boldsymbol{xi}$ denote the i th row of X and $\boldsymbol{y}_i$ denote the i th row of Y for $i = 1, ..., n$. For $c \in \{1,...,n\}$, we write

$$Y \cong_{except(c)} X$$

if $\boldsymbol{y}_i = \boldsymbol{x}_i$ for all $i \neq c$. We also write

$$Y \cong_{nonzero(c)} X$$

if $w_H(\boldsymbol{y}_c) = \ell$. and $\boldsymbol{y}_i = \boldsymbol{x}_i$ for all $i \neq c$. We present a deterministic polynomial time algorithm that outputs an $n \times n$ nonsingular matrix T such that

$$T \cdot X \cong_{\text{nonzero}(c)} X$$

for X that does not contain a column vector $(0, ..., 0)^t$ and

$$T \cdot Y \cong_{\text{except}(c)} Y$$

for any Y.

Procedure: NonZeroRow(X,c)

Let x_i denote the ith row of X for $i = 1, ..., n$.

1) $y := x_c$.
2) $\alpha_c \leftarrow 0$.
3) For $i = 1, ..., n$, do:
4)   If $i \neq c$, do:
5) $\alpha_i \leftarrow \text{MaxWeight}(x_i, y)$.
6) $y \leftarrow \alpha_i x_i + y$.
7) Let Q be an $n \times n$ matrix such that the ith row is
$$q_i = \begin{cases} (\alpha_1, ..., \alpha_n) & \text{if } i = c \\ (0, ..., 0) & \text{if } i \neq c \end{cases}$$
8) $T \leftarrow I_n + Q$.
9) Return T.

Theorem 1 *Let X be an $n \times \ell$. matrix that does not contain a column vector $(0, ..., 0)^t$. Then the above algorithm outputs nonsingular matrix T such that*

$$T \cdot X \cong_{\text{nonzero}(c)} X$$

and

$$T \cdot Y \cong_{\text{except}(c)} Y$$

for any matrix Y in $O(n\ell \cdot \text{poly}(\log|F|))$ time if $|F| > \ell$.

Proof Let y_i denote the i th row of $T \cdot X$ for $i = 1, ..., n$. Note that

$$T \cdot X = (\mathcal{I} + Q) \cdot X = X + Q \cdot X.$$

Therefore $y_i = x_i$ for $i \neq c$ and y_c is given as follows.

$$y_c = x_c + \sum_{i=1}^{n} \alpha_i x_i$$
$$= \left(...\left(\left(x_c + \alpha_1 x_1\right) + \alpha_2 x_2\right) + ...\right) + \alpha_n x_n$$

Suppose that $|F| > \ell$. Then, MaxWeight outputs α_i correctly at line 5 according to Lemma 5. Therefore

$$w_H(y_c) = |support(y_c)| = \left|\bigcup_{i=1}^{n} support(x_i)\right| = \ell$$

because X does not include $(0, ..., 0)^t$. Therefore

$$T \cdot X \cong_{\text{nonzero}(c)} X.$$

By the same argument, it is easy to see that

$$T \cdot Y \cong_{\text{except}(c)} Y$$

for any matrix Y.

Furthermore, it is clear that T is nonsingular. Finally, line 5 takes $O(\ell \cdot poly(\log|F|))$ time according to Lemma 5. Line 6 also takes $O(\ell \cdot poly(\log|F|))$ time. Hence the algorithm runs in $O(n\ell \cdot poly(\log|F|))$ time.

5. Making a Linear Network Coding Matrix k-Secure

In this section, we propose the first efficient deterministic algorithm to compute a nonsingular matrix T such that $T \times U$ is k-secure from a given linear network coding matrix U. Our algorithm runs in polynomial time if k is a constant.

Furthermore, our algorithm succeeds if $|F| > \binom{L}{k}$, where L is the reduced size of U. Note that in Proposition 4 and $|\mathcal{E}| \geq L$. Therefore our sufficient condition on |F| is usually much smaller than that of Proposition 4.

Let $\tilde{U}$ be a reduced coding matrix of U.

5.1. Making a Linear Network Coding Matrix 1-Secure

We begin by showing a polynomial time algorithm to compute an $n \times n$ nonsingular matrix T such that $T \cdot U$ is 1-secure. Our algorithm outputs T such that the last row of $T \cdot \tilde{U}$ consists of nonzero elements. Then, $T \cdot U$ is 1-secure according to Corollary 3.

Algorithm: 1-Secure($\tilde{U}$).[1]

1) $T \leftarrow$ NonZeroRow($\tilde{U}$, n).
2) Return T.

Theorem 2 *The above algorithm outputs a nonsingular matrix T such that the last row of $T \cdot \tilde{U}$ consists of nonzero elements in $O\left(nL \cdot poly(\log|F|)\right)$ time if $|F| > L$, where $\tilde{U}$ is an $n \times L$ matrix.*

Proof Follows from Theorem 1. Note that no column vector of $\tilde{U}$ is $(0, ..., 0)^t$.

Corollary 4 *We can construct a nonsingular matrix T such that $T \cdot U$ is 1-secure from any $n \times |\mathcal{E}|$ linear coding matrix U in $O\left(n|\mathcal{E}|^2 \cdot poly(\log|F|)\right)$ time if $|F| > L$, where L is the reduced size of U.*

Proof Suppose that $|F| > L$ and let T be the output of 1-Secure ($\tilde{U}$). Note

---

[1] A similar idea was used in[16] [p. 309]. However, they did not show how to find α that appears in MaxWeight(x, y) in polynomial time. They did not extend it to $k \geq 2$ like this paper, either.

that $T \cdot \tilde{U}$ is a reduced coding matrix of $T \cdot U$ according to Lemma 3. Therefore $T \cdot U$ is 1-secure according to Theorem 2 and Corollary 3.

We can compute $\tilde{U}$ from $U = (u_1,...,u_{|\mathcal{E}|})$ in $O(n|\mathcal{E}|^2 \cdot poly(\log|F|))$ time according to Lemma 2. Note that $\max(n|\mathcal{E}|^2 \cdot nL) = n|\mathcal{E}|^2$. Therefore we can compute T from U in $O(n|\mathcal{E}|^2 \cdot poly(\log|F|))$ time.

We now transform the insecure linear network code of **Figure 1** into a 1-secure one. By applying the above algorithm to $\tilde{U}$ of Equation (4), we obtain.[2]

$$T = \begin{pmatrix} 1 & 0 & 0 \\ 0 & 1 & 0 \\ 1 & 1 & 1 \end{pmatrix} \text{ over } F_3$$

The linear network coding matrix of **Figure 2** is $T \cdot U$, where U is given by Equation (2). Namely **Figure 2** is 1-secure.[3]

5.2. Making a Linear Network Coding Matrix 2-Secure

We now show a polynomial time algorithm to compute an $n \times n$ nonsingular matrix T such that $V = T \cdot U$ is 2-secure.

Lemma 6 *For an $n \times L$ matrix X, suppose that*

$$rank(X_{B,h-1}) = h - 1$$

for any $B \in Rank_{h-1}(X)$. Then for any $A \in Rank_h(X)$, we can find a nonzero vector a of length h such that

---
[2] We can often compute T with smaller $|F|$ than the sufficient condition stated in Corollary 4 (see Sect. 8.)
[3] We cannot use F_2 instead of F_3 because $m_1 + m_2 + 2r = m_1 + m_2 \mod 2$. Hence **Figure 2** is not 1-secure if we use F_2.

$$X_A \cdot a = (c, 0^{h-1})^t$$

for some c in $O(h^3 \cdot poly(\log|F|))$ *time. Furthermore,*

$$rank(X_{A,h}) = h$$

if and only if the last element of c is nonzero.

Proof Suppose that $rank(X_{B,h-1}) = h - 1$ for any $B \in Rank_{h-1}(X)$. Fix $A \subset \{1, ..., L\}$ such that $A \in Rank_h(X)$ arbitrarily. Then an $h \times h$ matrix $X_{A,h}$ includes an $(h - 1) \times (h - 1)$ submatrix $X_{B,h-1}$ such that $rank(X_{B,h-1}) = h - 1$. Therefore $rank(V_{A,h}) = h$ or $h - 1$.

Let

$$X_A = \begin{pmatrix} Y_1 \\ X_{A,h-1} \end{pmatrix} = \begin{pmatrix} Y_2 \\ X_{A,h} \end{pmatrix}$$

where $X_{A,h-1}$ is an $(h - 1) \times h$ submatrix and $X_{A,h}$ is an $h \times h$ submatrix. For a vector a of length h, we have

$$X_A \cdot a = \begin{pmatrix} Y_1 \cdot a \\ X_{A,h-1} \cdot a \end{pmatrix} = \begin{pmatrix} Y_2 \cdot a \\ X_{A,h} \cdot a \end{pmatrix}$$

It is easy to see that $rank(X_{A,h-1}) = h - 1$ according to our assumption. Therefore we can find a nonzero vector a of length h such that

$$X_{A,h-1} \cdot a = (0,...,0)^t$$

in $O(h^3 \cdot poly(\log|F|))$ time. We then have

$$X_A \cdot a = (c, 0^{h-1})^t,$$

where $c = Y_1 \cdot a$. Let the last element of c be $c0$. Hence

$$X_{A,h} \cdot a = (c_0, 0, ..., 0)^t.$$

1) If $c_0 = 0$, then it is clear that $rank(X_{A,h}) = h - 1$.

2) Suppose that $rank(X_{A,h}) = h - 1$. Then for some $B \subset A$ such that $|B| = h - 1$, there exists a vector b of length $h - 1$ such that

$$X_{B,h} \cdot b = X_{A,h} \cdot a = (c_0, 0, ..., 0)^t. \qquad (6)$$

This means that

$$X_{B,h-1} \cdot b = (0, ..., 0)^t.$$

However, $rank(X_{B,h-1}) = h - 1$ according to our assumption. Hence we have $b = (0, ..., 0)^t$. This means that $c_0 = 0$ according to Equation (6).

Therefore $c_0 = 0$ if and only if $rank(X_{A,h}) = h - 1$. Hence $c_0 \neq 0$ if and only if $rank(X_{A,h}) = h$.

Algorithm: 2-Secure($\tilde{U}$).

1) $T_1 \leftarrow$ NonZeroRow($\tilde{U}$, n).
2) $V_1 \leftarrow T_1 \cdot \tilde{U}$.
3) For each $A_i \in$ Rank2($\tilde{U}$), find a nonzero vector a_i of length 2 such that

$$y_i = (V_1)_{A_i} \cdot a_i = (c_i, 0)^t \qquad (7)$$

for some c_i.
4) Let $X = (y_1, y_2, ...)$.
5) $T_2 \leftarrow$ NonZeroRow(X, $n - 1$).
6) $T \leftarrow T_2 \cdot T_1$.
7) Return T.

As an example, suppose that $\tilde{U}$ is a 3×2 matrix such that $rank(\tilde{U}) = 2$.

(1) At step 2, we obtain V_1 such that

$$V_1 = T_1 \cdot \tilde{U} = \begin{pmatrix} * & * \\ * & * \\ nonezero & nonezero \end{pmatrix}$$

because T_1 makes the last row of $\tilde{U}$ nonzero according to Theorem 1.

(2) At step 3, we find $(a_1, a_2) \neq (0, 0)$ such that

$$y = V_1 \cdot \begin{pmatrix} a_1 \\ a_2 \end{pmatrix} = \begin{pmatrix} c_1 \\ c_2 \\ 0 \end{pmatrix}$$

for some c_1, c_2.

(3) At step 5, we compute T_2 such that

$$T_2 \cdot y = \begin{pmatrix} c_1 \\ nonzero \\ 0 \end{pmatrix}$$

where T_2 makes the second element of y nonzero and does not change the other elements according to Theorem 1.

Now let

$$V_2 = \begin{pmatrix} v_{11} & v_{12} \\ v_{21} & v_{22} \\ v_{31} & v_{32} \end{pmatrix} = T \cdot \tilde{U} = T_2 \cdot T_1 \cdot \tilde{U}$$

Then

(a) First we have

$$V_2 = T_2 \cdot (T_1 \cdot \tilde{U}) = T_2 \cdot \begin{pmatrix} * & * \\ * & * \\ nonzero & nonzero \end{pmatrix} = \begin{pmatrix} * & * \\ *' & *' \\ nonzero & nonzero \end{pmatrix}$$

because T_2 does not changes the third row of $T_1 \cdot \tilde{U}$ according to Theorem 1. Hence,

$$v_{31} \neq 0 \text{ and } v_{32} \neq 0 \tag{8}$$

(b) Second we have

$$V_2 \cdot \begin{pmatrix} a_1 \\ a_2 \end{pmatrix} = T_2 \cdot V_1 \cdot \begin{pmatrix} a_1 \\ a_2 \end{pmatrix} = T_2 \cdot y = \begin{pmatrix} c_1 \\ nonzero \\ 0 \end{pmatrix}$$

Therefore

$$rank\left(\begin{pmatrix} v_{21} & v_{22} \\ v_{31} & v_{32} \end{pmatrix}\right) = 2 \tag{9}$$

according to Lemma 6.

- Thus V_2 satisfies the condition of Corollary 2 from Equations (8) and (9).

- Furthermore, $V_2 = T \cdot \tilde{U}$ is a reduced coding matrix of $T \cdot U$ from Lemma 3, where U is the underlying linear network coding matrix.

- Consequently $T \cdot U$ is 2-secure from Corollary 2.

Theorem 3 *Let $\tilde{U}$ be an $n \times L$ matrix. If $|F| > \binom{L}{2}$, then the above algorithm outputs a nonsingular matrix T in $O(nL^2 \cdot poly(\log|F|))$ time as follows. Let $V_2 = T \cdot \tilde{U}$. Then*

1) *$rank((V_2)_{A,1}) = 1$ for any $A \in Rank_1(V_2)$, and*
2) *$rank((V_2)_{A,2}) = 2$ for any $A \in Rank_2(V_2)$*

Proof Suppose that $|F| > \binom{L}{2}$. At line 4 of 2-Secure($\tilde{U}$), the number of columns of X is equal to $\binom{L}{2}$. Therefore, according to Theorem 1, NonZeroRow outputs T_2 correctly at line 5. Namely, T_2 makes the $(n-1)$th row of $X = (y_1, y_2, ...)$ nonzero and does not change the other rows. Also, NonZeroRow outputs T_1 correctly at line 1.

Note that $V_1 = T_1 \cdot \tilde{U}$ and

$$V_2 = T \cdot \tilde{U} = T_2 \cdot T_1 \cdot \tilde{U} = T_2 \cdot V_1.$$

Therefore

$$\text{Rank}_p(V_2) = \text{Rank}_p(V_1) = \text{Rank}_p(\tilde{U}) \tag{10}$$

for any p according to Lemma 1. We can thus write $A \in \text{Rank}_p$ instead of $A \in \text{Rank}_p(\tilde{U})$.

(a) The last row of $V_1 = T_1 \cdot \tilde{U}$ consists of nonzero elements according to Theorem 1. This means that $rank((V_1)_{A,1}) = 1$ for any $A \in \text{Rank}_1$.

(b) The last row of $V_2 = T_2 \cdot V_1$ also consists of nonzero elements because T_2 does not change the last row of V_1 according to Theorem 1. Therefore $rank((V_2)_{A,1}) = 1$ for any $A \in \text{Rank}_1$.

(c) Consider $A_i \in \text{Rank}_2$ at step 3. From Lemma 6 and (a), we can find a nonzero vector $\boldsymbol{a}_i$ that satisfies Equation (7). Then we have

$$\begin{aligned}
(V_2)_{A_i} \cdot \boldsymbol{a}_i &= (T_2 \cdot V_1)_{A_i} \cdot \boldsymbol{a}i \\
&= T_2 \cdot \left((V_1)_{A_i} \cdot \boldsymbol{a}i\right) \\
&= T_2 \cdot \boldsymbol{y}_i \\
&= T_2 \cdot (\boldsymbol{c}_i, 0)^t \\
&= (\boldsymbol{c}'_i, 0)^t
\end{aligned}$$

for some c'_i. The last equality holds because T_2 does not change the last element of $(c_i, 0)^t$.

(d) The last element of is nonzero because T_2 makes the $(n-1)$th element of $y_i = (c'_i, 0)^t$ nonzero. Thus we have $\text{rank}((V_2)_{A_i,2}) = 2$ for any $A_i \in \text{Rank}_2$ from Lemma 6.

From (b) and (e), we see that (1) and (2) of this theorem hold.

It is clear that T is nonsingular. Finally the most time consuming part is line 5. The number of columns of X is $O(L^2)$. Hence line 5 runs in $O(nL^2 \cdot poly(\log|F|))$ time according to Theorem 1. Therefore the algorithm runs in $O(nL^2 \cdot poly(\log|F|))$ time.

Corollary 5 *We can construct a nonsingular matrix T such that $T \cdot U$ is 2-secure from any $n \times |\mathcal{E}|$ linear coding matrix U in $O(n|\mathcal{E}|^2 \cdot poly(\log|F|))$ time if $|F| > \binom{L}{2}$, where L is the reduced size of U.*

Proof Suppose that $|F| > \binom{L}{2}$ and let T be the output of 2-Secure ($\tilde{U}$). Then $T \cdot \tilde{U}$ is a reduced coding matrix of $T \cdot U$ according to Lemma 3. Therefore, $T \cdot U$ is 2-secure according to Theorem 2 and Corollary 2.

According to Lemma 2, we can compute $\tilde{U}$ from $U = (u_1, ..., u_{|\mathcal{E}|})$ in $O(n|\mathcal{E}|^2 \cdot poly(\log|F|))$ time. Note that $\max(n|\mathcal{E}|^2, nL^2) = n|\mathcal{E}|^2$. Therefore we can compute T in $O(n|\mathcal{E}|^2 \cdot poly(\log|F|))$ time from U.

5.3. Making a Linear Network Coding Matrix k-Secure

Finally we show an efficient deterministic algorithm to compute an $n \times n$ nonsingular matrix T such that $V = T \cdot U$ is k-secure for $3 \leq k < n$.

Algorithm: k-Secure($\tilde{U}$).

1) $T_1 \leftarrow$ NonZeroRow($\tilde{U}$, n).
2) $V_1 \leftarrow T_1 \cdot \tilde{U}$.
3) For $h = 2, ..., k$, do:
4)   For each $A_i \in \text{Rank}_h(\tilde{U})$, find a nonzero vector $\boldsymbol{a}_i$ of length h such that
$$\boldsymbol{y}_i = (V_{h-1})_{A_i} \cdot \boldsymbol{a}_i = (c_i, 0^{h-1})^t$$
   for some c_i.
5)   Let $X = (\boldsymbol{y}_1, \boldsymbol{y}_2, ...)$.
6) $T_h \leftarrow$ NonZeroRow(X, $n - h + 1$).
7) $V_h \leftarrow T_h \cdot V_{h-1}$.
8) $T \leftarrow T_k ... T_1$.
9) Return T.

Theorem 4 *Let $\tilde{U}$ be an $n \times L$ matrix. If $|F| > \binom{L}{k}$, then the above algorithm outputs a nonsingular matrix T in $O(knL^k \cdot \text{poly}(\log|F|))$ time as follows. Let $V_k = T \cdot \tilde{U}$. Then*

$$\text{rank}\left((V_k)_{A,h}\right) = h \tag{11}$$

for any $A \in \text{Rank}_h(V_k)$ for $h = 1, ..., k$.

Proof Suppose that $|F| > \binom{L}{k}$. At line 5 of k-Secure($\tilde{U}$), the number of columns of X is at most $\binom{L}{k}$. Therefore from Theorem 1, NonZeroRow outputs T_h correctly at line 6. Furthermore NonZeroRow also outputs T_1 correctly at line 1.

We say that an $n \times L$ matrix V is bottom h full independent if $\text{rank}(V_{A,i}) = i$ for any $A \in \text{Rank}_i(V)$ for $i = 1, ..., h$. Suppose that T is an $n \times n$ nonsingular matrix. Then $T \cdot V$ is bottom h full independent if V is bottom h full independent.

$V_1 = T_1 \cdot \tilde{U}$ is bottom 1 full independent from the property of T_1. Suppose that V_h is bottom h full independent. By applying the same argument as in the proof of Theorem 3, we can see that V_{h+1} is bottom $h + 1$ full independent. Therefore V_k is bottom k full independent by induction. Hence Equation (11) holds for any $A \in \text{Rank}_h(V_k)$ for $h = 1, ..., k$.

It is clear that T is nonsingular. Finally the most time consuming part is line 6. The number of columns of X is $O(L_k)$. Hence line 6 runs in time $O(nL^k \cdot poly(\log|F|))$ from Theorem 1. Therefore the algorithm runs in in time $O(knL^k \cdot poly(\log|F|))$.

Corollary 6 *We can construct a nonsingular matrix T such that $T \cdot U$ is k-secure from any $n \times |\mathcal{E}|$ linear coding matrix U in time*
$$O\left((n|\mathcal{E}|^2 + knL^k) \cdot poly(\log|F|)\right) \text{ if } |F| > \binom{L}{k},$$ *where L is the reduced size of U.*

Proof Similar to the proof of Corollary 5.

6. Strongly k-Secure Network Coding Scheme

In Equation (1), suppose that each m_i is independently and uniformly distributed over F. Let $\tilde{m}_i$ denote this random variable induced by m_i for $i = 1, ..., n$, and let $\tilde{d}_j$ denote the random variable induced by d_j for $j = 1, ..., |\mathcal{E}|$. For $B \subset \{1, ..., n\}$, let $\bar{B} = \{1, ..., n\} \backslash B$. Then we define a strongly k-secure network coding matrix as follows.

Definition 4 A linear network coding matrix U is strongly k-secure if

$$H\left((\tilde{m}_1,...,\tilde{m}_n)_{\bar{B}} \middle| (\tilde{d}_1,...,\tilde{d}_{|\mathcal{E}|})_A\right) = H\left((\tilde{m}_1,...,\tilde{m}_n)_{\bar{B}}\right) \tag{12}$$

for any $A \subset \{1,...,|\mathcal{E}|\}$ and any $B \subset \{1 ... n\}$ such that $|A| = |B| \leq k$.

Harada and Yamamoto[7] proved the following proposition.

Proposition 5[7] [Theorem 3] *For sufficiently large |F|, there exists a strongly k-secure linear network coding matrix for any network instance (G, s, Sink, n) if k < n.*

However, they did not show the time complexity of their algorithm explicitly. They did not present a concrete size of |F| either. Indeed, they left it as an open problem to derive a sufficient condition on |F|. (See "open problem" in Table 1 of[7]. They considered strong k'-security with $k' \leq k$, where ℓ. is used instead of k in[7].)

6.1. Necessary and Sufficient Condition

We can generalize Proposition 3 to strong k-security as follows.

Lemma 7 *Consider $A \subset \{1,...,|\mathcal{E}|\}$ and $B \subset \{1 \ldots n\}$ such that $|A| = |B|$. Then Equation (12) holds if and only if*

$$rank(U_A) = rank(U_{A,B}). \tag{13}$$

The proof is similar to that of Proposition 3[3] Lemma 3.1].

Proof Without loss of generality, let $A = B = \{1, ..., k\}$. Then we have

$$(d_1,...,d_k) = (m_1,...,m_n) \cdot UA$$
$$= (m_1,...,m_k) \cdot U_{A,B} + (m_{k+1},...,m_n) \cdot U_{A,\bar{B}}$$
$$= (r_1,...,r_k) + (s_{k+1},...,s_n)$$

where

$$(r_1,...,r_k) = (m_1,...,m_k) \cdot U_{A,B}$$
$$(s_{k+1},...,s_n) = (m_{k+1},...,m_n) \cdot U_{A,\bar{B}}$$

Let L_0 be the image space of U_A and L_1 be the image space of $U_{A,B}$.

Case 1 Suppose that $rank(U_{A,B}) = rank(U_A)$. Then $L_1 = L_0$. Since $(m_1, ..., m_k)$

is a random vector, $(r_1, ..., r_k)$ is uniformly distributed over $L_1 = L_0$. Therefore $(r_1, ..., r_k)$ works as one-time pad to mask $(s_{k+1}, ..., s_n)$. Hence Equation (12) holds because one-time pad implies perfect secrecy.

Case 2 Suppose that $rank(U_{A,B}) < rank(U_A)$. Then $L_1 \subset L_0$ and there exists some $y \in L \backslash L_1$. Let $\tilde{r}_i$ denote the random variable induced by r_i for $i = 1, ..., k$, and $\tilde{s}_i$ denote the random variable induced by s_i for $i = k + 1, ..., n$. Then we have

$$\Pr\left(\left(\tilde{d}_1, ..., \tilde{d}_k\right) = y \middle| \left(\tilde{s}_1, ..., \tilde{s}_k\right) = (0, ..., 0)\right)$$
$$= \Pr\left(\left(\tilde{r}_1, ..., \tilde{r}_k\right) = y\right)$$
$$= 0$$

On the other hand, $= \Pr\left(\left(\tilde{d}_1, ..., \tilde{d}_k\right) = y\right) > 0$ because each m_i is independently random. Therefore

$$\Pr\left(\left(\tilde{d}_1, ..., \tilde{d}_k\right) = y \middle| \left(\tilde{s}_1, ..., \tilde{s}_k\right) = (0, ..., 0)\right) \neq \Pr\left(\left(\tilde{d}_1, ..., \tilde{d}_k\right) = y\right)$$

Hence $\left(\tilde{d}_1, ..., \tilde{d}_k\right)$ and $\left(\tilde{s}_1, ..., \tilde{s}_k\right)$ are not independent. This means that Equation (12).

Theorem 5 *A linear network coding matrix $U = (u_{i,j})$ is strongly k-secure if and only if*

$$rank(U_A) = rank(U_{A,B})$$

for any $A \subset \{1, ..., |\mathcal{E}|\}$ *and any* $B \subset \{1, ..., n\}$ *such that* $|A| = |B| \leq k$.

Proof From Lemma 7.

6.2. Relation to Reduced Coding Matrix

Corollary 7 *A linear network coding matrix U is strongly 1-secure if and only if each element of $\tilde{U}$ is nonzero, where $\tilde{U}$ be a reduced coding matrix of U.*

Corollary 8 *A linear network coding matrix U is strongly k-secure if and only if*

$$rank(\tilde{U}_A) = rank(\tilde{U}_{A,B})$$

for any $A \in \text{Rank}_p(\tilde{U})$ *and any* $B \subset \{1, ..., n\}$ *such that* $|B| = p$ *for* $p = 1, ..., k$, *where* $\tilde{U}$ *be a reduced coding matrix of U.*

7. How to Make a Linear Network Coding Matrix Strongly 1-Secure

In this section, we show the first efficient deterministic algorithm which computes a non-singular matrix T such that $T \times U$ is strongly k-secure from a given linear network coding matrix U. In particular, it runs in polynomial time if k is a constant.

Let $\tilde{U}$ be a reduced coding matrix of U and let L be the reduced size of U. Then our algorithm succeeds if

$$|F| > L + \sum_{i=1}^{k-1} \binom{n-1}{i}\binom{L}{i+1}$$

7.1. How to Make a Linear Network Coding Matrix Strongly 1-Secure

We first show a polynomial time algorithm which computes an $n \times n$ non-singular matrix T such that $T \cdot U$ is strongly 1-secure. In fact, our algorithm outputs T such that each element of $T \cdot \tilde{U}$ is nonzero. Then $T \cdot U$ is strongly 1-secure from Corollary 7.

Algorithm: Strongly 1-Secure($\tilde{U}$).

1) $T_0 \leftarrow \text{NonZeroRow}(\tilde{U}, n)$.
2) $V \leftarrow T_0 \cdot \tilde{U}$.

Let x_i be the i th row of V.
3) For $i = 1, ..., n - 1$, do:
4) $\beta_i \leftarrow \text{MaxWeight}(x_n, x_i)$.
5) Let

$$T_1 \leftarrow \begin{pmatrix} & & & \beta_1 \\ & T_{n-1} & & \vdots \\ & & & \beta_{n-1} \\ 0 & \cdots & 0 & 1 \end{pmatrix}$$

6) Return $T = T_1 \times T_0$.

Theorem 6 *Let $\tilde{U}$ be an $n \times L$ matrix. Then the above algorithm outputs a nonsingular matrix T such that each element of $T \cdot \tilde{U}$ is nonzero in time $O(nL \cdot poly(\log|F|))$ if $|F| > L$.*

Proof Note that

$$T \cdot \tilde{U} = T_1 \cdot T_0 \cdot \tilde{U} = T_1 \cdot V = \begin{pmatrix} \beta_1 x_n + x_1 \\ \vdots \\ \beta_{n-1} x_n + x_{n-1} \\ x_n \end{pmatrix}$$

Suppose that $|F| > L$. Then from Theorem 1, we have $w_H(x_n) = L$ at line 2. Then from Lemma 5, we have $w_H(\beta_i x_n + x_i) = L$ for $i = 1, ..., n - 1$. Therefore each element of $T \cdot \tilde{U}$ is nonzero.

Further it is clear that T is nonsingular. Finally, line 1 takes time $O(nL \cdot poly(\log|F|))$ from Theorem 2, line 2 takes time $O(nL \cdot poly(\log|F|))$, line 4 takes time $O(L \cdot poly(\log|F|))$ from Lemma 5 and line 6 takes time $O(n^2 \cdot poly(\log|F|))$. Hence the algorithm runs in time $O(nL \cdot poly(\log|F|))$.

Corollary 9 *We can construct a nonsingular matrix T such that $T \cdot U$ is strongly 1-secure from any $n \times |\mathcal{E}|$ linear coding matrix U in time $O(n|\mathcal{E}|^2 \cdot poly(\log|F|))$ if $|F| > L$, where L is the reduced size of U.*

Proof Similar to the proof of Corollary 5 where we use Corollary 7.

We transform the insecure linear network code of **Figure 1** into a strongly 1-secure one. By applying the above algorithm to $\tilde{U}$ of Equation (4), we obtain[4]

$$T = \begin{pmatrix} 2 & 1 & 1 \\ 1 & 2 & 1 \\ 1 & 1 & 1 \end{pmatrix} \text{ over } F_5$$

The linear network coding matrix of **Figure 3** is $T \cdot U$, where U is given by Equation (2). Namely **Figure 3** is strongly 1-secure.[5]

7.2. How to Make a Linear Network Coding Matrix Strongly 2-Secure

We next show a polynomial time algorithm which computes an $n \times n$ non-singular matrix T such that $V = T \cdot U$ is strongly 2-secure.

Definition 5 For a subset $D \subset \{1, ..., n\}$ and an $n \times 1$ matrix E, we say that a vector $\boldsymbol{y} = (y_1, ..., y_n)^t$ is a D-zero image of E if (1) $\boldsymbol{y} = E \cdot \boldsymbol{a}$ for some nonzero vector $\boldsymbol{a}$ and (2) $y_i = 0$ for each $i \in D$.

For example,

$$\boldsymbol{y} = E \cdot \boldsymbol{a} = (0, y_2, ..., y_n)^t$$

is a $\{1\}$-zero image of E.

Lemma 8 *For a $p \times p$ matrix X, suppose that*

$$\mathrm{rank}\left(X_{B, \{1,...,p-1\}}\right) = p - 1$$

---

[4] We can often compute T with smaller $|F|$ than the sufficient condition stated in Corollary 9. See Sect. 8.
[5] We cannot use F_3 instead of F_5 because $3m_1 + 3m_2 + 2m_3 = m_3 \mod 3$. Hence **Figure 3** is not strongly 1-secure if we use F_3.

for any $B \in \text{Rank}_{p-1}(X)$. Then we can find a nonzero vector **a** of length p such that

$$X \cdot \boldsymbol{a} = \left(0^{p-1}, c\right)^t$$

for some c in time $O\left(p^3 \cdot poly(\log|F|)\right)$. Further

$$rank(X) = p$$

if and only if c is nonzero.

Proof Similar to the proof of Lemma 6.

Algorithm: Strongly 2-Secure($\tilde{U}$).

1) $T_0 \leftarrow$ NonZeroRow($\tilde{U}$, 1).
2) $V_0 \leftarrow T_0 \cdot \tilde{U}$.
3) For $h = 1, ..., n - 1$, do:
4)   For each $A_i \in \text{Rank}_2(\tilde{U})$, find a $\{h\}$-zero image y_i of $(V_{h-1})A_i$.
5) $W_h \leftarrow (y_1, y_2, ...)$.
6) $T_h \leftarrow$ NonZeroRow$((V_{h-1}, W_1, ..., W_h), h + 1)$.
7) $V_h \leftarrow T_h \cdot V_{h-1}$.
8) $W_1 \leftarrow T_h \cdot W_1$.
9) $\vdots$
10) $W_h \leftarrow T_h \cdot W_h$.
11) $T \leftarrow T_{n-1} \cdot ... \cdot T_0$.
12) Return T.

We illustrate how the algorithm proceeds for $n = 3$.

1) At line 2:

$V_0 = T_0 \tilde{U}$
non-zero
***
***

2) In the first loop, at line 5,

$V_0 = T_0 \tilde{U}$	W_1
non-zero	0, ..., 0
***	***
***	***

3) In the first loop, after line 8,

$V_1 = T_1 T_0 \tilde{U}$	$W1$
non-zero	0,..., 0
non-zero	non-zero
***	***

4) In the second loop, at line 5,

$V_1 = T_1 T_0 \tilde{U}$	W_1	W_2
non-zero	0, ..., 0	***
non-zero	non-zero	0, ..., 0
***	***	***

5) In the second loop, after line 10,

$V_2 = T_2 T_1 T_0 \tilde{U}$	W_1	W_2
non-zero	0,..., 0	***
non-zero	non-zero	0,..., 0
non-zero	non-zero	non-zero

We can see the following from the last table.

- Each element of V_2 is nonzero.

- From Lemma 8, for any $A_i \in \text{Rank}_2(V_2)$,

- $rank\left((V_2)_{A_i}, B\right) = 2$ for $B = \{1, 2\}$ and $B = \{1, 3\}$.
- $rank\left((V_2)_{A_i}, B\right) = 2$ for $B = \{2, 3\}$.

Hence $V = (T_2 T_1 T_0)U$ is strongly 2-secure from Corollarys 7 and 8.

Theorem 7 Let $\tilde{U}$ be an $n \times L$ matrix. Define

$$\lambda = L + (n-1)\binom{L}{2}$$

If $|F| > \lambda$. then the above algorithm outputs an $n \times n$ nonsingular matrix T in time $O(n^2 \lambda \cdot poly(\log|F|))$ such as fllows. Let $V_{n-1} = T \cdot \tilde{U}$. Then

1) each element of V_{n-1} is nonzero and

2) $$rank\left((V_{n-1})_{A,B}\right) = 2 \tag{14}$$

for any $A \in \text{Rank}_2(V_{n-1})$ and any $B \subset \{1, ..., n\}$ such that $|B| = 2$.

Proof Suppose that $|F| > \lambda$. At line 6, the number of columns of $(V_{h-1}, W_1, ..., W_h)$ is at most λ. Hence NonZeroRow outputs T_h correctly from Theorem 1. Namely T_h makes the $(h + 1)$th row of $(V_{h-1}, W_1, ..., W_h)$ nonzero and does not change the other rows. Also NonZeroRow outputs T_0 correctly at line 1, too.

For $1 \leq j \leq n - 1$, it holds that

$$\text{Rank}_2(V_j) = \text{Rank}_2(\tilde{U}) \tag{15}$$

from Lemma 1 because $V_j = T_j...T_0 \cdot \tilde{U}$. In this sense, we write $A \in \text{Rank}_2$ instead of $A \in \text{Rank}_2(\tilde{U})$.

(a) The first row of $V_0 = T_0 \cdot \tilde{U}$ consists of nonzero elements from Theorem 1.

(b) Suppose that

- the first h rows of V_{h-1} consists of nonzero elements, and
- $rank((V_{h-1})_{A,B}) = 2$ for any $A \in \text{Rank}_2$ and any $B \subset \{1,..., h\}$ such that $|B| = 2$.

Then we will show that

- the first $h + 1$ rows of V_h consists of nonzero elements, and
- $rank((V_h)_{A,B}) = 2$ for any $A \in \text{Rank}_2$ and any $B \subset \{1, ..., h + 1\}$ such that $|B| = 2$.

(c) T_h makes the $(h + 1)$th row of $(V_{h-1}, W_1, ..., W_h)$ nonzero and does not change the other rows. Therefore the first $h + 1$ rows of $V_h = T_h \cdot V_{h-1}$ consists of nonzero elements from our assumption.

Since T_h does not change the first h rows, we have $rank((V_h)_{A,B}) = 2$ for any $A \in \text{Rank}_2$ and any $B \subset \{1, ..., h\}$ such that $|B| = 2$ from our assumption.

Now consider $B = \{j, h + 1\}$ such that $j \in \{1, ..., h\}$.

- Each column vector y_i of W_h is a $\{h\}$-image of $(V_{h-1})A_i$, where $A_i \in \text{Rank}_2$. Therefore

$$y_i = (y_{i,1}, 0, y_{h+1,i}, y_{i,2})^t = (V_{h-1})_{A_i} \cdot a_i$$

for some nonzero vector a_i, where

$$y_{i,1} = (y_{1,i}, ..., y_{h-1,i})$$

$$y_{i,2} = (y_{h+2,i}, ..., y_{n,i})$$

Multiply T_h to the both hand sides. Then we have

$$T_h \cdot y_i = (y_{i,1}, 0, \text{nonzero}, y_{i,2})^t$$
$$= (T_h \cdot V_{h-1})_{A_i} \cdot a_i$$
$$= (V_h)_{A_i} \cdot a_i$$

because T_h only makes the $(h + 1)$th element of y_i nonzero. Therefore we have

$$(V_h)_{A_i} \cdot a_i = (y_{i,1}, 0, \text{nonzero}, y_{i,2})^t.$$

Then from Lemma 8, we have $rank((V_h)_{Ai}, B) = 2$ for $B = \{h, h + 1\}$ and $A_i \in \text{Rank}_2$.

- By applying the same argument to $W_1, ..., W_{h-1}$, we can see that $rank((V_h)_{A,B}) = 2$ for any $A \in \text{Rank}_2$ and $B = \{j, h + 1\}$ such that $j \in \{1, ..., h - 1\}$.

Therefore $rank((V_h)_{A,B}) = 2$ for any $A \in \text{Rank}_2$ and any $B \subset \{1, ..., h + 1\}$ such that $|B| = 2$.

Consequently (1) and (2) of this theorem hold by induction.

It is clear that T is nonsingular. Finally, the most time consuming part is line 6. It takes time $O(n\lambda \cdot poly(\log|F|))$ from Theorem 1. Hence the algorithm runs in time $O(n^2\lambda \cdot poly(\log|F|))$.

Corollary 10 *We can construct a nonsingular matrix T such that $T \cdot U$ is strongly 2-secure from any $n \times |\mathcal{E}|$ linear coding matrix U in time* $O((n|\mathcal{E}|^2 + n^2\lambda) \cdot poly(\log|F|))$ *if $|F| > \lambda$, where*

$$\lambda = L + (n-1)\binom{L}{2}$$

and L is the reduced size of U.

Proof Similar to the proof of Corollary 5, where we use Corollary 8.

We transform the insecure linear network code of **Figure 1** into a strongly 2-secure one. By applying the above algorithm to $\tilde{U}$ of Equation (4), we obtain[6]

---

[6] We can often compute T with smaller $|F|$ than the sufficient condition stated in Corollary 10. See Sect. 8.

$$T = \begin{pmatrix} 1 & 1 & 1 \\ 1 & 2 & 3 \\ 4 & 5 & 7 \end{pmatrix} \text{ over } F_{11}$$

The linear network coding matrix of **Figure 4** is $T \cdot U$, where U is given by Equation (2). Namely **Figure 4** is strongly 2-secure[7].

7.3. How to Make a Linear Network Coding Matrix Strongly k-secure

We finally show an efficient algorithm which computes an $n \times n$ nonsingular matrix T such that $V = T \cdot U$ is strongly k-secure for $3 \leq k < n$.

Algorithm: Strongly k-Secure $\tilde{U}$).

1) $T_0 \leftarrow \text{NonZeroRow}(\tilde{U}, 1)$.
2) $V_0 \leftarrow T_0 \cdot \tilde{U}$ and $X_0 \leftarrow V_0$.
3) For $h = 1, ..., n-1$, do:
4)   For each $D_j \subset \{1, ..., h\}$ such that $h \in D_j$ and $|D_j| < k$, do:
5)   For each $A_i \in \text{Rank}|D_j| + 1(\tilde{U})$, find a D_j-zero image y_i of $(V_{h-1})A_i$.
6) $W_{D_j} \leftarrow (y_1, y_2, ...)$.
7) $T_h \leftarrow \text{NonZeroRow}((V_{h-1}, X_{h-1}, W_{D1}, W_{D2}, ...), h+1)$.
8) $V_h \leftarrow T_h \cdot V_{h-1}$.
9. $X_h \leftarrow T_h \cdot (X_{h-1}, W_{D1}, W_{D2}, ...)$.
10. $T \leftarrow T_{n-1} \cdot ... \cdot T_0$.
11. Return T.

We illustrate how the algorithm proceeds for $k = 3$ and $n = 4$.

1) At line 2,

---

[7]We cannot use F_7 instead of F_{11} because $m_1 + 3m_2 + 7m_3 = m1 + 3m_2 \mod 7$. Hence **Figure 4** is not strongly 2-secure if we use F_7.

$V_0 = T_0 \tilde{U}$
non-zero
***
***

2) At the end of the 1st loop,

$V_1 = T_0 T_1 \tilde{U}$	$W_{\{1\}}$
non-zero	0, ..., 0
non-zero	non-zero
***	***
***	***

3) At the end of the 1st loop,

$V_2 = V_2 V_1 V_0 \tilde{U}$	$W_{\{1\}}$	$W_{\{2\}}$	$W_{\{1,2\}}$
non-zero	0, ..., 0	***	0, ..., 0
non-zero	non-zero	0, ..., 0	0, ..., 0
non-zero	non-zero	non-zero	non-zero
***	***	***	***

4) At the end of the 3rd loop,

$V_3 = T_3 \ldots T_0 \tilde{U}$	$W_{\{1\}}$	$W_{\{2\}}$	$W_{\{1,2\}}$	$W_{\{3\}}$	$W_{\{1,3\}}$	$W_{\{2,3\}}$
non-zero	0,..., 0	***	0,..., 0	***	0 ... 0	***
non-zero	non-zero	0,..., 0	0,..., 0	***	***	0 ... 0
non-zero	non-zero	non-zero	non-zero	0 ... 0	0 ... 0	0 ... 0
non-zero	non-zero	non-zero	non-zero	non-zero	non-zero	non-zero

We can see the following from the last table.

- Each element of V_3 is nonzero.

- By applying Lemma 8 to the columns indexed by $W_{\{1\}}$, $W_{\{2\}}$ and $W_{\{3\}}$, we can see that $rank((V_3)_{A_i,B}) = 2$ for any $A_i \in \text{Rank}_2(V_3)$ and for any $B \subset \{1, 2, 3, 4\}$ with $|B| = 2$.

- By applying Lemma 8 to the columns indexed by $W_{\{1,2\}}$, $W_{\{1,3\}}$ and $W_{\{2,3\}}$, we can see that $rank((V_3)_{A_i,B}) = 3$ for any $A_i \in \text{Rank}_3(V_3)$ and for any $B \subset \{1, 2, 3, 4\}$ with $|B| = 3$.

Therefore $V = (T_3 T_2 T_1 T_0) U$ is strongly 3-secure from Corollarys 7 and 8.

Theorem 8 *Let $\tilde{U}$ be an $n \times L$ matrix. Define*

$$\lambda = L + \sum_{i=1}^{k-1} \binom{n-1}{i} \binom{L}{i+1}$$

If $|F| > \lambda$, Then the above algorithm outputs an $n \times n$ nonsingular matrix T in time $O\left(n^2 \lambda \cdot \text{poly}\left(\log|F|\right)\right)$ such that

$$rank\left((T \cdot U)_{A,B}\right) = p \tag{16}$$

for any $A \in \text{Rank}_p(\tilde{U})$ and any $B \subset \{1, ..., n\}$ with $|B| = p$ for $p = 1, ..., k$.

Proof Suppose that $|F| > \lambda$. At line 7, the number of columns of the matrix which is the input to NonZeroRow is at most λ. Hence NonZeroRow outputs T_h correctly from Theorem 1. Further NonZeroRow outputs T_0 correctly at line 1, too.

We say that an $n \times L$ matrix V is top (h, k) full independent if $rank(V_{A,B}) = |B|$ for any $B \subset \{1, ..., h\}$ such that $|B| \leq k$ and any $A \in \text{Rank}_{|B|}(V)$. Suppose that T is an $n \times n$ nonsingular matrix. Then $T \cdot V$ is top (h, k) full independent if V is top (h, k) full independent.

$V_0 = T_0 \cdot \tilde{U}$ is top $(1, 1)$ full independent because the first row of V_0 consists of nonzero elements from the property of T_0. Suppose that V_{h-1} is top (h, k) full independent. By applying the same argument as in the proof of Theorem 7, we can see that V_h is top $(h + 1, k)$ full independent. Therefore V_{n-1} is top (n, k) full independent by induction. Hence Equation (16) holds for any $A \in \text{Rank}_p(\tilde{U})$ and any $B \subset \{1, ..., n\}$ with $|B| = p$ for $p = 1, ..., k$.

Furthermore, it is clear that T is nonsingular. Finally, the most time consuming part is line 7. It takes $O(n\lambda \cdot poly(\log|F|))$ time according to Theorem 1. Hence, the algorithm runs in $O(n^2\lambda \cdot poly(\log|F|))$ time.

Corollary 11 *We can construct a nonsingular matrix T from any $n \times |\mathcal{E}|$ linear coding matrix U such that $T \cdot U$ is strongly k-secure in $O\left(\left(n|\mathcal{E}|^2 + n2\lambda\right) \cdot poly(\log|F|)\right)$ time if $|F| > \lambda$, where*

$$\lambda = L + \sum_{i=1}^{k-1} \binom{n-1}{i}\binom{L}{i+1}$$

Proof Similar to the proof of Corollary 5, where we use Corollary 8.

8. Summary

We have proposed an efficient deterministic construction algorithm of a linear transformation T that transforms a linear network code to a k-secure one for any $1 \leq k < n$. We have also extended this algorithm to strong k-security for any $1 \leq k < n$. Our algorithms run in polynomial time if k is a constant, and these time complexities are explicitly presented. We also have presented a concrete size of $|F|$ for strong k-security.

The condition on $|F|$ in Lemma 5 is a sufficient condition. Therefore Max-Weight(x, y) succeeds with smaller $|F|$ as long as $|F| > |S_0|$. For example, if the Hamming weight of y is small, then $|S_0|$ is small. Hence $|F|$ can be small. For the same reason, our construction algorithms for the transformation matrix T may succeed with smaller $|F|$ than the sufficient condition on $|F|$ that is stated in each theorem.

Further work should explore if there exists a network instance such that our sufficient condition on $|F|$ is tight.

Source: Kurosawa K, Ohta H, Kakuta K. How to make a linear network code (strongly) secure [J]. Designs Codes & Cryptography, 2016: 1–24.

References

[1] Ahlswede R., Cai N., Li S.-Y.R., Yeung R.W.: Network information flow. IEEE Trans. Inf. Theory **46**(4), 1204–1216 (2000).

[2] Bhattad K., Narayanan K.R.: Weakly secure network coding. In: Proceedings of NetCod 2005, April (2005).

[3] Cai N., Yeung R.W.: A security condition for multi-source linear network coding. In: Proceedings of IEEE ISIT, 24–29 June, pp. 561–565 (2007).

[4] Cai N., Yeung R.W.: Secure network coding on a wiretap network. IEEE Trans. Inf. Theory **57**(1), 424–435 (2011).

[5] El Rouayheb S., Soljanin E., Sprintson A.: Secure network coding for wiretap networks of type II. IEEE Trans. Inf. Theory **58**(3), 1361–1371 (2012).

[6] Feldman J., Malkin T., Servedio R., Stein C.: On the capacity of secure network coding. In: 42nd Annual Allerton Conference on Communication, Control, and Computing (2004).

[7] Harada K., Yamamoto H.: Strongly secure linear network coding. IEICE Trans. **91–A**(10), 2720–2728 (2008).

[8] Jaggi S., Sanders P., Chou P.A., Effros M., Egner S., Jain K., Tolhuizen L.M.G.M.: Polynomial time algorithms for multicast network code construction. IEEE Trans. Inf. Theory **51**(6), 1973–1982 (2005).

[9] Kurihara J., Uematsu T., Matsumoto R.: Explicit construction of universal strongly secure network coding via MRD codes. In: Proceedings of IEEE ISIT 2012, Cambridge, MA, pp. 1488–1492 (2012).

[10] Kurosawa K., Ohta H., Kakuta K.: How to construct strongly secure network coding scheme. In: ICITS, pp. 1–17 (2013).

[11] Li S.Y.R., Yeung R.W., Cai N.: Linear network coding. IEEE Trans. Inf. Theory **49**(2), 371–381 (2003).

[12] Matsumoto R., Hayashi M.: Universal Strongly Secure Network Coding with Dependent and Non-uniform Messages. (2012). arXiv:1111.4174.

[13] Shioji E., Matsumoto R., Uyematsu T.: Vulnerability of MRD-code-based universal secure network coding against stronger eavesdroppers. IEICE Trans. **93-A**(11), 2026–2033 (2010).

[14] Silva D., Kschischang F.R.: Universal weakly secure network coding. In: Proceedings

of IEEE ITW 2009, pp. 281–285 (2009).

[15] Silva D., Kschischang F.R.: Universal secure network coding via rank-metric codes. IEEE Trans. Inf. Theory **57**(2), 1124–1135 (2011).

[16] Tang Z., Lim H.W., Wang H.: Revisiting a secret sharing approach to network codes. In: ProvSec, pp. 300–317 (2012).

Chapter 30

Privacy and Information Security Risks in a Technology Platform for Home-Based Chronic Disease Rehabilitation and Education

Eva Henriksen[1], Tatjana M. Burkow[1], Elin Johnsen[1], Lars K. Vognild[2]

[1]Norwegian Centre for Integrated Care and Telemedicine, University Hospital of North Norway, Tromsø, Norway
[2]Norut, Northern Research Institute, Tromsø, Norway

Abstract: Background: Privacy and information security are important for all healthcare services, including home-based services. We have designed and implemented a prototype technology platform for providing home-based healthcare services. It supports a personal electronic health diary and enables secure and reliable communication and interaction with peers and healthcare personnel. The platform runs on a small computer with a dedicated remote control. It is connected to the patient's TV and to a broadband Internet. The platform has been tested with home-based rehabilitation and education programs for chronic obstructive pulmonary disease and diabetes. As part of our work, a risk assessment of privacy and security aspects has been performed, to reveal actual risks and to ensure adequate information security in this technical platform. Methods: Risk assessment was performed in an iterative manner during the development process. Thus, security solutions have been incorporated into the design from an early stage instead of being included as an add-on to a nearly completed system. We have adapted existing risk management methods to our own environment, thus creating our own method. Our method conforms to ISO's standard for information security risk management. Results: A total of approximately 50 threats and possible unwanted incidents were identified and analysed. Among the threats to the four information security aspects: confidentiality, integrity, availability, and quality; confidentiality threats were identified as most serious, with one threat given an unacceptable level of High risk. This is because health-related personal information is regarded as sensitive. Availability threats were analysed as low risk, as the aim of the home programmes is to provide education and rehabilitation services; not for use in acute situations or for continuous health monitoring. Conclusions: Most of the identified threats are applicable for healthcare services intended for patients or citizens in their own homes. Confidentiality risks in home are different from in a more controlled environment such as a hospital; and electronic equipment located in private homes and communicating via Internet, is more exposed to unauthorised access. By implementing the proposed measures, it has been possible to design a home-based service which ensures the necessary level of information security and privacy.

Keywords: Privacy, Confidentiality, Information Security, Risk Assessment, Pulmonary Rehabilitation, Diabetes Self-Management Education, Video Conference, Tele-Homecare

1. Background

The increasing prevalence of chronic conditions has become a major concern to healthcare systems throughout the world. Healthcare provision faces a shift in emphasis from the management of acute illnesses to the provision of long-term services for people with chronic conditions[1][2]. In this shift, the home environment may have an important role in chronic disease management; for secondary prevention and follow-up, improved self-management skills and quality of life, stabilisation of the chronic disease, prevention of exacerbations, and less hospitalisation.

We have designed and implemented a prototype technology platform for providing home-based healthcare services. It supports a personal electronic health diary and enables secure and reliable communication and interaction with peers and healthcare personnel. The platform runs on a small computer (the Residential Patient Device, RPD) with a dedicated remote control. It is connected to the patient's TV and to a broadband Internet connection. A web camera and a microphone are used during videoconferencing. By making the technical platform appear as a "smart TV" solution, we also aimed at reaching people less familiar with computers. Our technical platform, including its security measures, has been used in real settings in several trials[3][4].

Privacy and security are important for all healthcare services, including home-based services. As part of our work, a risk assessment of privacy and information security aspects was performed. The motivation for the risk assessment was to reveal actual information security risks and to ensure that the security of the technical platform was adequate. The risk assessment was first performed early in the design process; it was revisited and updated later in the development process. In this way, information security was embedded into the design, in line with the concept of "Privacy by design"[5].

This paper describes the risk assessment and its findings, and the security measures which were implemented. We also discuss some general privacy concerns relating to home-based systems in healthcare.

1.1. The Services and the Technical Platform

The two services subject to risk assessment were two home programmes: a comprehensive pulmonary rehabilitation programme and an education programme for diabetes self-management, both for virtual online groups. The aim of the home programmes is improved access to education and rehabilitation. The home programmes are based on conventional programmes, respectively: 1) a group-based pulmonary rehabilitation programme which encompasses education, exercising, and psychosocial support over several weeks, and 2) a two-day group-based course in diabetes self-management, with a telephone service for individual diabetes consultation.

Both home programmes had weekly group education sessions and individual consultation sessions. In addition, the pulmonary rehabilitation programme had weekly group exercise sessions. An individual consultation with a physiotherapist, pulmonary nurse, or diabetes nurse was introduced as an opportunity for addressing potential personal health matters which they did not want to discuss in the presence of peers. An exercise video (rehabilitation only) and educational videos were also available for the participants.

The patients had a personal electronic health diary for daily use, with a health-related multiple-choice questionnaire, manual input of pulse oximetry values from a self- administered pulse oximeter (Nonin Medical, Inc), and input of blood glucose values from a self-administered OneTouch Ultra Blood Glucose Meter (LifeScan, Inc.) connected through a wireless Bluetooth adapter.

The electronic health diary stores the entered information. The health diary is intended both to support patient self-management and to provide supplementary information for the weekly individual consultation. Each patient explicitly made one week of the health diary data available to the healthcare personnel ahead of each consultation, and the encrypted data were transferred over the Internet from home to a server at the hospital. The healthcare personnel could then access the patient's health diary before the individual weekly consultation.

The RPD has components for the following functionality: multiparty videoconferencing, storage and visualisation of health diary information, storage and playback of videos, storage and presentation of textual information, wireless incorporation of data from the LifeScan blood glucose meter, and data encryption and transfer of information from home to the hospital information system. The multiparty videoconference component was used for both educational and exercise group sessions and for the individual consultations. The RPD performed no actions or manipulation on the data, except for storage, archival, and communication. Therefore, the RPD is not considered to be a medical device, according to directives 93/42/EEC[6] and 2007/47/EC[7], and based on the EC guidance document MEDDEV 2.1/6[8].

More information on the platform and services can be found in[3][4][9].

1.2. Legal Baseline and Security Requirements

Requirements concerning the electronic communication of personal information are imposed by national legislation. In European countries these are based on the EU directive on processing of personal data[10]. In the US, similar requirements are based on the Health Insurance Portability and Accountability Act of 1996 (HIPAA)[11]. Under both international and Norwegian legislation, all health-related information concerning an identifiable person is considered sensitive (Personal Data Act §2)[12]. No one else than the health professionals who have a treatment relationship to the person should be able to access this person's health information, unless the person has given his or her consent. In Norway, strong emphasis is put on privacy considerations, also in the healthcare sector.

Most chronically ill patients need, at least to some extent, to have their next-of-kin involved in their care. These individuals are often considered a valuable resource in the treatment and follow-up of the patient. In terms of Norwegian legislation, it is up to the patient to decide whether their next-of-kin should be informed about the details of their condition[13]. If that happens, there is a corresponding exemption from the commitment to professional confidentiality for the healthcare workers[14].

Information security is usually defined as including the three main aspects confidentiality, integrity, and availability[12][15][16]. In the Norwegian healthcare legislation[17], quality is included as a fourth aspect of information security. Quality can be seen as overlapping with integrity, and these aspects are often merged in analyses of security threats and risks. However, for applications involving images, video, and audio, it is often useful to manage these two aspects separately. Quality may also be a relevant security aspect in the analysis of applications where usability, ease of use, and user interface are important. The following is a brief definition of the four security aspects:

- **Confidentiality** is the property that information is not made available for or disclosed to unauthorised persons, entities, or processes.

- **Integrity** relates to the trustworthiness of the information. It is the property that data have not been deliberately tampered with, nor accidentally changed.

- **Availability** is the property that information is accessible and usable upon demand by an authorised entity.

- **Quality** refers to the information being correct and not misleading.

Because of the sensitivity of personal health information, confidentiality is particularly important within the healthcare sector. Confidentiality requirements originate from the professional secrecy and non-disclosure undertaking required of all healthcare workers—all the way back to Hippocrates 2400 years ago. The Hippocratic Oath states: "What I may see or hear in the course of the treatment or even outside of the treatment in regard to the life of men, which on no account one must spread abroad, I will keep to myself, holding such things shameful to be spoken about[18]."

Privacy is another side of confidentiality. Article 1 of the EU Directive on processing of personal data[10] specifies the "right to privacy with respect to processing of personal data". A rule of thumb could be to define privacy as the right of the client and confidentiality the duty of the service provider[19].

2. Methods

Security risk analysis is a basic requirement of ISO 27002[15], internationally recognised as the generic information security standard. In our case, risk assessment is performed with respect to the information security aspects confidentiality, integrity, availability, and quality.

There are two basic types of risk analysis methods: qualitative and quantitative. Quantitative methods use mathematical and statistical tools to represent risk, e.g. by including a calculation of the expected loss of value to an asset. In qualitative risk analysis, risk is analysed by using linguistic variables (adjectives) rather than using mathematics[20]. Qualitative assessment techniques base the risk assessment on anecdotal or knowledge-driven factors[21]. Our risk assessment approach is qualitative. We regard this to be the best approach when assessing new technology being developed for a new model of service delivery. In our case, the final design of technology and service was still to be decided, and there was not much relevant quantitative information available. A qualitative approach opened for discussions of risk aspects in groups involving relevant stakeholders. In these circumstances it is more appropriate to express likelihood and consequence in terms of "adjectives", instead of unfounded numbers for percentage and money. In addition, methods based on qualitative measures may be more suitable for today's complex environment of information systems[20].

There are several methods and guidelines for how to conduct risk analysis; they all include the central tasks of

- identifying threats and possible unwanted incidents.
- analysing impacts and likelihood of the identified threats.
- evaluating risks with respect to acceptance criteria.

There seem to be a lack of relevant literature which present risk assessment of information security in healthcare settings, especially within home-based e-health services. Standards for risk assessment in general, as well as textbooks describing ways to perform risk assessment, are available. Organisations often adapt existing risk management methods to their own environment and culture, thus creating their own method[22]. Our method has evolved through several risk assessments[23]-[25], and it conforms to ISO's standards for risk management, ISO 31000[26] and ISO 27005[27].

The standards set out the risk analysis process in five main steps:

1) **Context establishment:** describe the subject of the analysis, *i.e.* the system to be analysed and its environment.

2) **Risk identification:** identify which unwanted incidents might possibly happen, *i.e.* identify threats and vulnerabilities.

3) **Risk analysis:** consider the consequences of the threats and the likelihood that these consequences may occur, *i.e.* impact and probability analysis.

4) **Risk evaluation:** relate the resulting risk level to the risk acceptance criteria, *i.e.* decide which risks might be acceptable or not.

5) **Risk treatment:** identify and assess risk treatment options.

Steps 2, 3, and 4 are collectively referred to as "Risk assessment". In our method, these steps are often performed in a series of risk assessment meetings. Meeting participants are persons with the necessary knowledge about the analysed system. The process is led by a risk assessment expert.

By doing risk assessment in an iterative manner, it becomes an important part of the design and development process. We performed the first risk assessment at an early stage in the design phase, and repeated it as a review and update later on in the development process. In this way, security requirements have been identified, and security solutions were incorporated into the design from the early development phase instead of being included as an add-on to a nearly completed system.

The following sections describe our use of the risk assessment method, step by step.

2.1. Step 1: Context Establishment

A high-level description of the system and service and the environment for its use, is necessary in order to perform a risk assessment at an architectural level[21]. In addition to a written description, which is included in the risk assessment report, this is often performed as a walk-through of the system and service, presented at the first risk assessment meeting.

It is also necessary to identify the assets to be protected. In most health-related projects, the patients' life and health is the main asset. However, in this case the services in question have hardly any potential to harm the patient's health. But there are other assets to protect, such as the patients' right to privacy; the hardware and software of the system itself; sensitive information in the system; other healthcare systems and their information, which could be affected from this system; and trust in, and the reputation of, the services and the service provider.

As a part of establishing the context, the risk acceptance criteria are stated during this first phase of the risk assessment. The acceptance criteria are related to the security requirements. It is not possible to completely avoid any risk—to have a risk level of zero. But it is difficult to define that small level of risk which might be acceptable, especially at an early stage of the service and technology development.

It is, however, important to have a common understanding of the acceptance criteria. The following criteria were discussed for our risk assessment:

- A patient should not die or have a reduction in health as a result of using these services.

No such risk would be accepted in these types of services, as for conventional rehabilitation and education programmes.

- Unauthorised persons should not be able to acquire a patient's health information (confidentiality breach).

An "unauthorised person" is anyone without a treatment role towards the patient, so this includes family members and relatives. However, for patients with a chronic disease, family members are often considered resource persons in the treatment and follow-up care. In our case, it may be difficult for patients to conceal that they have a chronic disease, especially for COPD patients who are receiving long-term oxygen therapy.

To understand the type and amount of information that might be disclosed is vital for a decision on the acceptability of the risk. In addition to being a legal question, it may also be an individual and subjective question, as to which information should be kept secure. Users may want information to remain private even if it does not concern a matter which is legally defined as "sensitive" information.

- A patient's health information should not be modified or deleted by unauthorised persons or as a result of software or hardware errors (integrity breach).

By "modify" we mean to insert wrong information, or to alter or remove existing information. As a worst case, this might cause that a patient receive a wrong medical advice or a wrong treatment, but in most cases the only result is destroyed data.

- Stored information should not be permanently lost (availability breach).

In the context of our study, there would be no serious consequences for the patient if data registered in this system were deleted, but it would indicate that the system is not trustworthy. The confidence in and reputation of this system and the services would diminish with an increasing number of such incidents.

- Access to the system and/or its information should not be rendered impossible for those who are entitled to such access (availability breach).

There would be no serious consequence for the patient or the healthcare worker if the service or information were not accessible, but trust in, and the reputation of, the system and the services would diminish on the frequency of unavailability.

At this first step of the risk assessment process, definitions of levels for consequence, likelihood and risk were presented and discussed. In the absence of statistical information or other quantitative information from an operational system, we describe the different levels using qualitative values, sometimes referred to as "linguistic variables" or adjectives[22]. It can be difficult to define the values before threats are identified, and the definitions usually need to be reviewed and updated when they are in use during step 3 of the process. Definitions used in our risk assessment are presented in **Table 1**.

2.2. Step 2: Threat Identification

The second step of the risk assessment process, the identification of possible threats and unwanted incidents, was performed as a structured brainstorming between the project members. The brainstorming process was "structured" in the way that a walkthrough of the intended services was performed, using predefined keywords and attributes. Keywords related to the security aspects confidentiality, integrity, quality, and availability, and to attributes such as "internal" and "external" (threats), and "deliberate" and "accidental" (actions). The identified threats and the discussion were summarised in a threat table.

The production of a threat table is a documentation technique which we use throughout the risk assessment process. Brainstorming, with the use of a table to document the threats, corresponds to the technique used in the risk analysis

Table 1. Definitions of values for consequence, likelihood, and risk level used in the risk assessment.

Consequence:	
Small	For the hospital or the service: No violation of law; offence that does not lead to reaction; or negligible financial loss which can be recovered; or small reduction of reputation in the short run. For the patient: No impact on health; or negligible financial loss which can be recovered; or small reduction of reputation in the short run.
Moderate	For the hospital or the service: Offence, less serious violation of law which results in a warning or a reprimand; or financial loss which can be recovered; or reduction of reputation that may influence trust and respect. For the patient: No direct impact on health or a minor temporary impact; or financial loss which can be recovered; or some loss of reputation caused by revelation of less sensitive or offensive health information.
Severe	For the hospital or the service: Violation of law which results in minor penalty or fine; or a large financial loss which cannot be recovered; or serious loss of reputation that will affect trust and respect for a long time. For the patient: Reduced health; or some financial loss which cannot be recovered; or serious loss of reputation caused by revealing of sensitive and offending information.
Catastrophic	For the hospital or the service: Serious violation of law which results in a penalty or fine; or considerable financial loss which cannot be recovered; or serious loss of reputation which is devastating for trust and respect. For the patient: Death or permanent damage of health; or considerable financial loss which cannot be recovered; or serious loss of reputation which permanently affects life, health, and finances.
Likelihood:	
Low	Rare, occurs less frequently than every 10th year, or less than 10 % of the times the system/service is used. Detailed knowledge about the system is needed; or special equipment is needed; or it can only be performed deliberately and with the help of internal personnel.
Medium	May happen, occurs not more than once a year, or between 10 % and 30 % of the times the system/service is used. Normal knowledge about the system is sufficient; or normally available equipment can be used; or it can be performed deliberately.
High	Fairly often, occurs several times a year, or between 30 % and 50 % of the times the system/service is used. Can be done with minor knowledge about the system; or without any additional equipment being used; or it can occur because of wrong or careless usage.
Very high	Very often, occurs several times a month or more frequent than 50 % of the times the system/service is used. Can be done without any knowledge about the system; or without any additional equipment being used; or it can occur because of wrong or careless usage.
Risk level:	
Low	Acceptable risk. The service can be used with the identified threats, but the threats must be observed to detect changes that could increase the risk level.
Medium	Possibly an acceptable risk for this particular service, but each threat must be considered separately and the development of the risk must be monitored on a regular basis, with an assessment of whether remedial measures should be implemented.
High	Unacceptable risk. Cannot start using the service before risk reducing measures have been implemented.

method HazOP[28][29]. We have, however, found it convenient to use some other columns in our threat table. These are:

1) unique identifier of threat (threat number),

2) textual description of threat or unwanted incidence,

3) consequence value,

4) likelihood value,

5) risk value (as a product of consequence and likelihood),

6) any other comments from the brainstorming (including ideas for risk treatment).

During the brainstorming, all possible threats were written into the threat table (column 2), together with any relevant comments (column 6), including those related to consequence and likelihood. Subsequently, the threat table was refined, by grouping related threats, and putting threats into a relevant sequence. At this stage, each threat was given a unique identifier (column 1). Values for consequence, likelihood, and risk were added in later steps of the process.

The threats identified in our risk assessment are listed in **Table 2** which corresponds to the first two columns of our threat table.

2.3. Step 3: Impact and Probability Analysis

In step 3 of the process, the identified threats and unwanted incidents were analysed. For each possible threat we asked the project participants to evaluate the impact or consequence and the likelihood that the threat or incident would occur. The qualitative values for consequence and likelihood (as defined in **Table 1**) were inserted in the threat table, columns 3 and 4, for each identified threat.

It is particularly difficult to set a value for likelihood when the system is at an early stage of development; it is difficult to identify vulnerabilities before the system is built and tested. When deciding on a value for likelihood one could ask relevant questions, for instance related to the ease and motivation for the threats. In our case, the following questions were useful:

- Knowledge, acquaintance: Must an attacker know the configuration of the system? Does he or she need to know a password or PIN? Or is it sufficient to turn on the TV or the computer? Is the system "always on"?

- Remedies: Does an attacker need special equipment, e.g. a smartcard or a code-calculator? Does he or she need access to special software, or would access to normally available software and hardware be sufficient?

- Mistakes: How easy is it to make mistakes? (Mistakes can be caused by insufficient user education or training, or by a poor user interface.)

- Motivation: Would any third party be interested in viewing the information? Would anyone have an interest in deliberately modifying the information?

Table 2. List of threats identified in the risk assessment.

ID	Threats/unwanted incidents
Locally – at the patient's home	
c1	Unauthorised persons can view/read personal (sensitive) health information because the user has forgotten to switch off (or "log out" from) the RPD.
c3	Unauthorised persons can view/read personal (sensitive) health information because the PIN code (or password or another authentication mechanism) is available/known – e.g. too weak/simple (a general problem)
c4	Unauthorised persons can view/read personal (sensitive) health information because the RPD with stored information is stolen, then restarted and accessed without authorisation.
c5	Video conference (VC) to participant at home (individual sessions): Unauthorised persons present in the patient's home, outside camera view, may happen to hear personal information given to **this patient** by health personnel (e.g. instructions/education regarding his/her own disease) Remember: Unauthorised persons are persons (including family members and visitors) with whom the patient does not want to share that information.
c6	Group education via VC (all patients in their own home): Unauthorised persons in a patient's home, outside camera view, can see and hear **other** patients/participants without their knowledge.
c7	The RPD is compromised because of software weaknesses, making it possible for unauthorised persons to see/log ongoing activity.
c8	Wireless data transfer from sensor to RPD can be intercepted by others.
i1	Unauthorised persons (e.g. grandchildren who play with the sensor) can *by accident* (i.e. unintentionally) insert false values if the system is not fail safe. That is, measures taken from other persons than the registered user are entered.
i3	Unauthorised persons (e.g. other family members or visitors) can *deliberately* insert fake values.
i4	The patient him-/herself can *by mistake* modify inserted values or insert erroneous values (e.g. it is easy to type in wrong O_2 values).
i7	The patient him/herself can *deliberately* insert fake values or modify inserted values.
i8	Data in the RPD is corrupted – e.g. wrong clock time from a sensor may follow the sensor value and cause existing data to be overwritten.
i9	SW/HW-weaknesses in the RPD that can be exploited (e.g. by malware) in such a way that the information is being damaged or modified.
i10	The RPD is stolen and software, keys or configuration are being exploited for unauthorised communication.
i12	The RPD is being compromised because of SW weaknesses and becomes a relay for attacking healthcare systems, e.g. by sending messages containing executable payload.
i14	Unauthorised persons can remotely configure the RPD, install/update software, etc., thus making the system behave differently than specified.
a1	The service is unavailable for both the patient and the health personnel because the RPD has been stolen.
a2	Data from the RPD cannot be retrieved locally by the patient (SW or HW errors, e.g. disk crash).
a3	Data from the RPD cannot be sent to the health personnel (SW or HW errors).
a4	The RPD is damaged (crushed, fire, dropped to the floor etc.) so that data cannot be retrieved or inserted.
a5	Shutdown because of electricity power failure in the patient's home.
a6	The patient forgets his PIN-code (or other authentication method) so that data cannot be retrieved from the RPD at home. (Information sent is available at the central server.)
a7	PKI certificates expire. If this happens, it is not possible to send data with valid signatures or to encrypt correctly for the specified recipient.
a8	SW/HW weaknesses in the RPD that can be exploited (e.g. by malware) in such a way that stored information is destroyed/deleted or access is blocked (e.g. Denial of Service attack, DoS)
a9	Patients will not use the system: "Too high-tech". Fear of surveillance. Feeling of lack of control. Afraid of damaging the system. Think it is difficult to use.
a10	Patients will not use the service because too many errors occur, too often. E.g. in the case of an alert function, error which leads to triggering of the alert.
During data transfer	
c9	Unauthorised persons obtain access to personal (sensitive) information during transfer: measurement values from sensors, textual information from patient at home
c10	Unauthorised persons obtain access to personal (sensitive) information being transferred in the two-way video conference, both audio (what is said) and video (see patients in their homes).
i15	Unauthorised persons can modify or delete personal health information during transfer.
i18	Errors during transfer lead to duplication of messages.
a11	Unauthorised persons can delete personal health information during transfer so that it does not reach the intended recipient.
a13	Low network quality (QoS): the quality of the connection is so low that the remote education and exercising is useless.
a14	DoS attack (on the network or a network component) so that the information does not reach the intended recipient.
a15	Low network quality (QoS): data is not transferred, is lost during transfer, or is delayed.
a16	Information corrupted or lost during transfer (caused by errors), i.e. cannot be used by the intended recipient.

Continued

Data in the central server/database, in the health institution

c11	Unauthorised persons obtain access to personal health information (in server/database) in the health institution. The server contains information about all patients/participants. If unauthorised persons obtain access, information about several patients can be seen at a time, not just that concerning a single patient.
i21	Information stored on the central server is deliberately manipulated (modified, deleted) by unauthorised persons.
i22	Information stored on the central server is manipulated (modified, deleted) by mistake (e.g. wrong usage)
a17	Permanent loss of data from central server (because of SW errors or HW failures), data are lost or destroyed
a18	Data on the central server are unavailable for a short or a longer time period (e.g. electricity power failure)

Quality of video communication

q1	The video quality from the patient's home is inadequate (e.g. because of limited bandwidth, camera type, use of camera, placement of camera, lighting, etc.) for the healthcare workers to be able to instruct the patients. They do not see clearly enough what the patient is doing (exercise, use of medical equipment)
q2	Unacceptable audio quality, e.g. echo, jitter, drop-out. The healthcare workers can hear their own echo in the sound from the participants. The patients at home can hear an echo if the healthcare workers do not use an extra microphone

2.4. Step 4: Risk Evaluation

The risk of a threat is defined as the product of consequence and likelihood for that threat. This is illustratively represented by a two-dimensional matrix (like **Figure 1**).

In step 4 of the process, the unique ID of each identified threat was written into the corresponding cell of the matrix. The result of our risk assessment is shown in the risk matrix in **Figure 1**. The shading of the cells in the matrix indicates the three risk levels defined in **Table 1**. The risk value of each threat, indicated by the position in the matrix, was also inserted into column 5 of the threat table.

2.5. Step 5: Risk Treatment

For all threats with a non-acceptable risk level, ris-reducing treatment was proposed and discussed (step 5 of the risk assessment process). The responsibility for follow-up of the proposed measures is outside the scope of the risk assessment. That responsibility belongs to the project management.

There are basically four different approaches to managing a risk[27].

- Risk retention—accept the risk in accordance with the organisation's security policy. This applies to those risks that are deemed low enough to be acceptable. It is worth remembering that accepting the risk does not mean that the unwanted incident indicated by the threat is acceptable.

Consequence → / ↓ Likelihood	Small	Moderate	Severe	Catastrophic
Low	i1 a8, a9, a11, a14	c5, c8 i9, i17 a17	c4, c7, c9, c11 i3, i7, i10, i12, i14, i15, i21	
Medium	i4, i18 a1, a2, a3, a4, a5, a6, a7, a10, a13, a15, a16, a18	c6 i8, i22 q1	c3, c10	
High		q2	c1	
Very high				

Figure 1. Risk matrix showing risk level for identified threats.

- Risk reduction—reduce the risk to an acceptable level. Since risk is a product of likelihood and consequence, this implies reduction of the likelihood, the consequence, or both. It is most difficult to reduce the consequence of a threat; the focus should therefore first of all be on reduction of likelihood.

- Risk avoidance—not be exposed to the risk, do not do the things that could lead to the risk.

- Risk transfer—transfer the risk to a third party (e.g. to an insurance company).

Risk reduction measures should be considered and evaluated with reference to the cost-benefit for the service. Some measures might reduce the risk level for several threats at the same time. Cheap and easy-to-implement measures that are likely to reduce the risk of even an acceptable threat should be implemented as a matter of course.

3. Results

3.1. Identified Threats

A total of approximately 50 threats and potential unwanted incidents were identified and analysed. Each threat was given a unique identifier whereby the initial letter shows the security category the threat belongs to: c = confidentiality; i = integrity; q = quality; a = availability.—The relevant threats identified are listed in **Table 2** (a few threats which were considered irrelevant have been excluded from this summary). Threats to health information can occur in the patient's home, at the remote server, or during transfer between the system in the patient's home and the server. The intermediate headings in the table indicate a grouping according to subsystems of the analysed system in question and its context.

3.2. Analysis and Treatment of Risk

The identified threats were given values for likelihood and consequence. The unique ID of each threat was written into the corresponding cell of the risk matrix, as shown in **Figure 1**. The risk level indicated by the matrix is analysed, and appropriate treatment of the risk is proposed.

The risk matrix shows that none of the threats was regarded as having a catastrophic consequence. For these particular services, this would seem to be a reasonable conclusion, as the aim of the home programmes is to provide education and rehabilitation services; it is not for use in acute situations or for continuous health monitoring. The matrix also shows that none of the threats was considered to have a very high likelihood, according to our definitions in **Table 1**.

The risk matrix indicates that confidentiality threats have been identified as the most serious threats. This is because all health-related personal information is defined to be sensitive information in the legal regulations[10][12]. It is also worth noting that all the availability threats were identified as low risk. This reflects the fact that the services are not intended to be used for real-time monitoring or for acute situations. The health diary acts as supplementary information for the weekly individual consultations. The worst consequence of these threats is probably that the system will lose credibility and the users will fail to trust it, and will stop using the service if there are too many such problems.

The only threat that scored an unacceptable High risk level, c1, is related to confidentiality. It concerns the case that a third party, e.g. a visitor in the patient's home, by chance could access personal health information on the TV screen if the system is left switched on and logged-in. The High risk is a result of the severe consequence of confidentiality breaches, and the high likelihood of this particular threat.

A technical solution to threat c1 is to implement a timeout mechanism to be activated after a specified period without interaction between the user and the system. The user has to be authenticated again when reactivating the service. The length of the timeout must be considered carefully; there is a trade-off between security and usability: Too short a timeout will be an annoyance to the user, who has to log in repeatedly, while a too long timeout has limited value.

The rest of the threats discussed here are threats with medium risk. These threats may be acceptable according to the risk level definitions in **Table 1**. It is, however, important to monitor the development of these threats, as their risk may rise to an unacceptable high level in case of increased likelihood. For many of them, it would be appropriate to implement risk-reducing measures.

In addition to c1, there are other confidentiality threats to health information in the patient's home:

Unauthorised persons, e.g. visitors, in a patient's home can see and hear other persons participating in the video conference (c6). This can happen without the knowledge of the other participants; they will only see persons within the camera's coverage area. On the other hand, it can be argued that participating in centre-based training courses also increases the likelihood that other persons will know about your chronic disease, since it is possible that family or next-of-kin also participate in some of these courses.

The user has to be authenticated when logging in to the RPD. If a PIN is introduced for logging on to the system, it is quite likely that the patient will write the code on a paper, and thus make it possible for others to access the system (c3).

For most of the threats involving the patients in their homes there are also non-technical measures for risk reduction. It is important to give information to the users, make them aware of the risks and explain the reason for restrictions imposed by technical solutions and routines for use. Such education to the users is relevant for threats c1, c3 and c6.

Threat c4 concerns the security of the information stored in the RPD in case the dedicated computer is stolen. The information stored consists of both the values from sensor measurements and manual input from the patient, as well as the education videos made by the healthcare workers. In the design of the system it was decided that medical data stored in the RPD should be encrypted, which gives a very low likelihood for this threat to occur. The education videos were not encrypted; they do not contain information that is sensitive in terms of any legislation.

Unauthorised persons might exploit potential SW weaknesses to hack into the RPD and reveal health information (c7). The likelihood of this is greater than zero because this is a prototype system; on the other hand, the opportunities to access the system have been greatly reduced by setting it up as a dedicated PC, with access only via dedicated ports and protocols.

Confidentiality threats also arise during information transfer and when the information is stored in the hospital's server.

Sensitive information is transferred in both directions over the Internet. A security requirement which originates from the legislation is that sensitive data transmitted over an insecure network, *i.e.* a network over which the organisation does not have full control, should be encrypted during transfer. So textual data and sensor values are encrypted during transfer, and it is therefore less likely that these data will be revealed (c9). We cannot disregard the risk that someone is able to wiretap audio and video during video conferencing sessions (c10). This risk was reduced by using Virtual Private Network (VPN) to achieve private encrypted video conferencing over the Internet.

Data residing in the hospital's server are covered by the same security as the rest of the hospital's systems, and we consider the likelihood of unauthorised access to these data to be low (c11).

Most of the threats related to integrity and availability are considered to have less consequence for the patient, as long as similar incidents are not repeated over a long period. This is because decisions for the patients concerned are not based on single measurements. On the other hand, if there were systematic incidents over time (such as threats i3 and i7 which could concern deliberate registration of wrong values for a long time), this could perhaps lead to more severe consequences for the patient's health. This is also the case for the integrity of the health information during transfer or while it resides in the hospital server (threats i15, i21). Taking into account the security measures at the hospital, the likelihood is low that the information stored on the server will be deliberately manipulated. However, the data on the server could also be modified by mistake (i22), for instance if the user interface is not good enough.

Threats i10, i12 and i14 all relate to the possibility of obtaining control over the RPD and its content and using it for illegitimate purposes, either by stealing it (i10) or by exploiting SW weaknesses to attack the system over the Internet (i12). If the dedicated computer can be reached via the network, it might be reconfigured and thus made available for processes other than those initially intended (i14). All these are threats with severe consequence, but we regard their likelihood to be very low.

Threat i8 concerns the possibility that data in the RPD are corrupted because of SW weaknesses or HW problems. If the information still appears to be credible, this might, in the worst case, cause incorrect advice given to the patient; if not, it is just unavailable information as in threat a16. The consequence is considered to be moderate. The information from the electronic health diary acts as a supplement to the individual consultation where the patient and the health professional discuss health issues. The information is readable and can therefore be verified by the patient. No automated decision is made based on the information from the diary.

Two threats to quality were also identified in the risk assessment, q1 concerning the video quality and q2 concerning the audio quality. The quality may be diminished if the bandwidth is too low or the equipment is too poor. It might result in blurred pictures, or delayed video. If, for instance, there is a delay in the video signals, this might be misinterpreted by the physiotherapist as the patient not keeping up with the speed. The audio quality is usually considered more important than the video quality. The most annoying problem would be echoing of the sound, a participant receiving his/her own voice back through the TV set. Training health workers and patients in the use of the system will help to reduce the risk of threats such as q1 and q2.

4. Discussion

A risk assessment process is continuous and iterative, and it applies to different phases of the system development[21]. Our risk assessment was performed following this approach. Thus, security aspects were incorporated into the design from an early stage instead of being included as an add-on to a nearly completed system. This is also referred to as "Privacy by design", information security is embedded into the design; it is proactive instead of reactive, preventive instead of remedial[5].

4.1. Principle Results

Of the approximately 50 threats and potential unwanted incidents that were identified and analysed, none were regarded as having catastrophic consequence, and only one threat was identified as having an unacceptable High risk level. Threats to confidentiality were deemed most serious. This is because all health-related personal information is regarded as sensitive. Most of the threats to integrity and quality were analysed to have Medium risk, while threats to availability were regarded Low risk.

Some security measures were planned already in the design of the system, and these were taken into account as prerequisites in the risk assessment. Additional measures were added as a result of the risk assessment. In summary, the following security measures were included in the design and implementation of the technical platform[3][4][9], reducing the risk accordingly:

- Patient authentication at log-on to the service (PIN, 4 digits, input using the remote control).
- Timeout: Automatic logoff of user after a given period of idleness.
- Encrypted storage of sensitive user data in the Residential Patient Device (RPD).
- Encrypted transfer of messages.
- Encrypted VPN for video conferencing.

- The RPD was configured as a dedicated computer permitting network access only via selected ports and protocols. The device could only be reached from outside through a VPN or SSH port.
- Health diary information is stored both locally and remotely in order to secure availability.
- Education of users, including awareness raising, information about privacy risks, and education in secure behavior.

4.2. Limitations

There are always questions about completeness when conducting a risk assessment: Did we identify all risks? Did we find the most important risks? The answer to the first question is probably "no", but we consider the answer to the second question to be "yes". The systematic, structured process for threat identification, together with thorough documentation, helps to ensure consistency and improve completeness. The most important effect of the risk assessment is perhaps the awareness it imposes on the designers and developers—and on the users, of possible risks to their system.

The concern about completeness may also be related to performing a qualitative risk assessment. Our approach is qualitative, which we find more suitable for the types of services and information systems we assess. However, one drawback of qualitative risk analysis methods is that their nature yields inconsistent results. The results may be subjective when using qualitative risk analysis methods[20], and there is uncertainty associated with this subjective judgment[22].

A question that may be asked regarding the subjective judgment in our risk assessment is: Were we too strict in the evaluation of risks? For instance, it may be debated whether threats with severe consequence but low likelihood (see risk matrix in **Figure 1**) should be defined as moderate or low risk. In our case we concluded that this is moderate risk, because of the severe consequence.

4.3. Generalisation

Another relevant question concerning the risk assessment is related to generality: Can the result be reused for other technical solutions or similar services?

The validation of impact of confidentiality threats relative to threats to other information security aspects may vary between countries. The privacy legislation in most of the European countries is based on the European privacy directive (95/46/EC,[10]), but it is a matter of the national authorities to implement it in their own legislation and manage the compliance of the regulations. In Norway, strong emphasis is put on privacy and confidentiality, also in the healthcare sector.

Within healthcare policy, there is an ongoing discussion about how to achieve a shift from centre-based rehabilitation to rehabilitation at home with the involvement of healthcare workers; and further to self-managed rehabilitation, training, and exercise; reducing the involvement of healthcare staff[1][2]. —This risk assessment had a focus on risks in the home environment, and many of the identified threats will apply to home-based solutions in general. The risks to confidentiality are different at home than in a more controlled (closed) environment: on the one hand, one is more private indoors at home; on the other hand it is the persons in the immediate entourage who can most easily obtain sensitive information about the patient. The patient may want to hide sensitive information from close relatives, yet these may be the very people who are best able to support patients with chronic conditions.

On the technical side, equipment located in private homes is more exposed to access by unauthorised persons. For instance, other persons in the same room, whether they are family members or casual visitors, may obtain sensitive information from this service. The equipment is physically available and thus more likely to be used (or tampered with) by other people in the family. Enabling electronic communication between equipment in the patients' homes, connected to the open Internet, and equipment placed in the secure zone of a network (e.g. at a hospital), is another security challenge.

Until now, the capabilities of commercial TV set-top-boxes (STB) have been somewhat limited and usually tailored to TV functionality, such as program guides, simple web browsing, and recording of broadcast TV to hard disk. However, the next generation of TVs or STBs will support Hybrid Broadcast Broadband TV or "HbbTV". The latest innovations include smart TV technologies where the TV has its own broadband Internet connection and supports applications (apps) of the type we know from the mobile world (Apple TV, Google TV, Samsung smart TV). These TV-apps provide rich functionality on the TV, with content and functions from the back-end Internet. These smart TVs do not have touch-sensitive screens, so user interactions are controlled by means of the TV's remote control. Most of the threats identified in this risk assessment will apply to STBs and smart TVs, as well.

Many of the threats also apply to solutions and services intended for mobile devices. There are, however, other typical threats when dealing with mobile equipment; these devices are easily mislaid, lost or stolen, thus creating extra risks, especially to confidentiality and availability.

The use of home technology and mobile solutions will increasingly supplement and to some extent replace the physical encounter between patient and healthcare professionals. With proper use, this can enhance the privacy of individuals, but it also provides new challenges. Some of the responsibility must remain with the users. The patient must be made aware about potential risks, and the way in which new technology used in home healthcare might affect the individual's privacy[30].

The increased use of personal and mobile electronic devices inevitably leads to changing attitudes among the population, regarding privacy – ranging from higher awareness on one extreme, to greater carelessness on the other extreme. Similarly, attitudes concerning what are regarded as private or sensitive information may also be changing. As well as being a question of legislation, this is an individual/subjective question. Some users may want information to remain private even if it is not legally defined as sensitive information. For example, a woman who has been badly injured by a violent husband will regard her new address as more "sensitive" than information about her broken arm. On the other hand, many users will say "I have nothing to hide", and some patients do already present their whole case history on the open Internet (Facebook, YouTube, Patients-like-me, etc.)[31]-[33].

5. Conclusion

A total of approximately 50 threats and potential unwanted incidents were identified and analysed in the risk assessment. The confidentiality threats were evaluated as being the most serious threats. This reflects the fact that health-related personal information is regarded as sensitive. All the availability threats were classified as having a low risk. That is because the aim of the home programmes is to provide education and rehabilitation services; it is not intended for use in acute situations or for continuous health monitoring.

Only one threat was ascribed an unacceptable high risk level. It concerns the case that a third party, e.g. a visitor in the patient's home, might by chance access personal health information on the TV screen if the system was left switched on and logged in. A solution to this threat is to implement a timeout mechanism to be activated after a specified period without interaction between user and system. The rest of the threats were identified as medium or low risk.

Most of the threats identified are representative of healthcare services intended for patients or citizens in their homes. The risks to confidentiality are different at home than in a more controlled environment such as a hospital, and electronic equipment situated in private homes and communicating via the open Internet is more exposed to access by unauthorised persons.

By implementing the proposed measures, it has been possible to design a home-based service which ensures the necessary level of information security and privacy.

Abbreviations

COPD: Chronic obstructive pulmonary disease; DoS: Denial of service; HIPAA: Health insurance portability and accountability act; HW: Hardware; ID: Identifier; ISO: International Standards Organisation; PIN: Personal identification number; PKI: Public key infrastructure; QoS: Quality of service; RPD: Residential patient device; STB: Set-top-box; SW: Software; VC: Video conference; VPN: Virtual private network.

Competing Interests

There are no financial or personal relationships with people or organisations that inappropriately influence this work as far as we know.

Authors' Contribution

TMB, LVK, and EJ all participated in the risk assessment. EH was the risk assessment leader. EH, TMB, LKV, and EJ all took part in preparing the manuscript, and all authors have approved the final version.

Acknowledgements

We would like to thank Per Bruvold, privacy ombudsman from UNN (University Hospital of North Norway), and the other participants who also took part in the risk assessment: Njål Borch from Norut, Geir Østengen and Trine Krogstad from NST (Norwegian Centre for Integrated Care and Telemedicine), and Thomas Strandenæs from Well Diagnostics. This work was financed partly by the Norwegian Research Council (the HØYKOM programme, and TTL, Tromsø Telemedicine Laboratory) and partly by EU (the eTEN programme).

Source: Henriksen E, Burkow T M, Johnsen E, *et al.* Privacy and information security risks in a technology platform for home-based chronic disease rehabilitation and education [J]. Bmc Medical Informatics & Decision Making, 2013, 13(1): 3234–3234.

References

[1] Epping-Jordan JE, Pruitt SD, Bengoa R, Wagner EH: Improving the quality of health care for chronic conditions. Qual Saf Health Care 2004, 13:299–305. doi:10.1136/qshc.2004.010744.
http://qualitysafety.bmj.com/content/13/4/299.full. (Accessed 2012-11-06).

[2] Pruitt SD, Epping-Jordan JE: Preparing the 21st century global healthcare workforce. BMJ 2005, 330:637. doi:10.1136/bmj.330.7492.637.
http://www.bmj.com/content/330/7492/637.full. (Accessed 2012-11-06).

[3] Vognild LK, Burkow TM, Luque LF: The MyHealthService approach for chronic disease management based on free open source software and low cost components. Engineering in Medicine and Biology Society (EMBC), Minneapolis, US, 2009.

IEEE, pp. 1234–1237. doi:10.1109/IEMBS.2009.5333475.
http://ieeexplore.ieee.org/xpl/articleDetails.jsp?arnumber=5333475.
(Accessed 2013-05-16).

[4] Burkow TM, Vognild LK, Krogstad T, Borch N, Østengen G, Bratvold A, Risberg MJ: An easy to use and affordable home-based personal eHealth system for chronic disease management based on free open source software. Stud Health Technol Inform 2008, 136:83–88.

[5] Cavoukian A, Hoffman DA, Killen S: Remote Home Health Care Technologies: How to Ensure Privacy? Build It In: Privacy by Design. Information and Privacy Commissioner of Ontario, Canada; 2009.
http://www.ipc.on.ca/images/Resources/pbd-remotehomehealthcarew_Intel_GE.pdf. (Accessed 2013-05-15).

[6] Council Directive 93/42/EEC of 14 June 1993 concerning medical devices. The European Parliament and the Council of the European Union; 1993.
http://eur-lex.europa.eu/LexUriServ/LexUriServ.do?uri=CONSLEG:1993L0042:20071011:en:PDF. (Accessed 2013-05-15).

[7] DIRECTIVE 2007/47/EC OF THE EUROPEAN PARLIAMENT AND OF THE COUNCIL of 5 September 2007 amending Council Directive 90/385/EEC on the approximation of the laws of the Member States relating to active implantable medical devices, Council Directive 93/42/EEC concerning medical devices and Directive 98/8/EC concerning the placing of biocidal products on the market. The European Parliament and the Council of the European Union; 2007.
http://eur-lex.europa.eu/LexUriServ/LexUriServ.do?uri=OJ:L:2007:247:0021:0055:en:PDF. (Accessed 2013-05-15).

[8] MEDDEV 2.1/6 Guidelines on the Qualification and Classification of Stand Alone Software used in Healthcare within the Regulatory Framework of Medical Devices. European Commission DG Health and Consumer, Directorate B, Unit B2 "Health Technology and Cosmetics"; 2012.
http://ec.europa.eu/ health/medical-devices/files/meddev/2_1_6_ol_en.pdf. (Accessed 2013-05-15).

[9] Burkow TM, Vognild LK, Ostengen G, Johnsen E, Risberg MJ, Bratvold A, Hagen T, Brattvoll M, Krogstad T, Hjalmarsen A: Internet-enabled pulmonary rehabilitation and diabetes education in group settings at home: a preliminary study of patient acceptability. BMC Medical Informatics and Decision Making 2013, 13(33). doi:10.1186/1472-6947-13-33.
http://www.biomedcentral.com/1472-6947/13/33. (Accessed 2013-05-02).

[10] Directive 95/46/EC of the European Parliament and of the Council of 24 October 1995 on the protection of individuals with regard to the processing of personal data and on the free movement of such data. The European Parliament and the Council of the European Union; 1995.
http://eur-lex.europa.eu/LexUriServ/LexUriServ.do?uri=CELEX:31995L0046:EN:HTML. (Accessed 2013-05-15).

[11] Health Insurance Portability and Accountability Act of 1996 (HIPAA). US Depart-

ment of Health & Human Services; 1996.
http://aspe.hhs.gov/admnsimp/pl104191.htm. (Accessed 2012-11-08).

[12] Lov om behandling av personopplysninger [personopplysningsloven]. (Norwegian Act of 14 April 2000 no. 31 relating to the processing of personal data [Personal Data Act]). Det norske justisog beredskapsdepartement (Norway's Ministry of Justice and Public Security), 2000. http://www.lovdata.no/all/hl-20000414-031.html. (English version: http://www.ub.uio.no/ujur/ulovdata/lov-20000414-031-eng.pdf). (Accessed 2012-11-08).

[13] Lov om pasientog brukerrettigheter [pasientrettighetsloven]. (The Act of 2 July 1999 No. 63 relating to Patients' Rights [the Patients' Rights Act]). Det norske helseog omsorgs departement (Norway's Ministry of Health and Care Services); 1999. http://www.lovdata.no/all/hl-19990702-063.html. (English version: http://www.ub.uio.no/ujur/ulovdata/lov-19990702-063-eng.pdf). (Accessed 2012-11-08).

[14] Lov om helsepersonell m.v. [helsepersonelloven]. (Act of 2 July 1999 No. 64 relating to Health Personnel etc. [The Health Personnel Act]). Det norske helseog omsorgsdepartement (Norway's Ministry of Health and Care Services); 1999. http://www.lovdata.no/all/hl-19990702-064.html. (English version: http://odin.dep.no/hd/engelsk/regelverk/p20042245/042051-200005/index-dok000-b-n-a.html). (Accessed 2012-11-08).

[15] ISO/IEC 27002:2005, Information technology - Security techniques—Code of practice for information security controls. International Organization for Standardization (ISO) and International Electrotechnical Commission (IEC); 2005. http://www.iso.org/iso/catalogue_detail?csnumber=50297. (Accessed 2012-11-08).

[16] ISO/IEC 27000:2009, Information technology—Security techniques—Information security management systems—Overview and vocabulary. International Organization for Standardization (ISO) and International Electrotechnical Commission (IEC); 2009. http://www.iso.org/iso/catalogue_detail?csnumber=41933. (Accessed 2012-11-08).

[17] Lov om helseregistre og behandling av helseopplysninger [helseregisterloven]. Norwegian Act of 18 May 2001 no 24 on personal health data filing systems and the processing of personal health data [Personal Health Data Filing System Act]). Det norske helse- og omsorgsdepartement (Norway's Ministry of Health and Care Services); 2001. http://www.lovdata.no/all/hl-20010518-024.html (English version: http://www.ub.uio.no/ujur/ulovdata/lov-20010518- 024-eng.pdf). (Accessed 2012-11-08).

[18] Wikipedia. http://en.wikipedia.org/wiki/Hippocratic_Oath. (Accessed 2012-11-06).

[19] Human Factors (HF); User experience guidelines; Telecare services (eHealth). ETSI EG 202 487, European Telecommunications Standards Institute (ETSI); 2008. http://pda.etsi.org/pda/home.asp?wki_id=Ro1rPxqLlNCEDCHECsIwG. (Accessed 2012-11-08).

[20] Karabacak B, Sogukpinar I: ISRAM: information security risk analysis method.

Computers & Security 2005, 24(2):147–159. doi:10.1016/j.cose.2004.07.004.
http://www.sciencedirect.com/science/article/pii/S0167404804001890.
(Accessed 2012-11-08).

[21] Verdon D, McGraw G: Risk Analysis in Software Design. IEEE Security and Privacy 2004, 2(4):79–84. doi:10.1109/MSP.2004.55.

[22] Papadaki K, Polemi N: Towards a systematic approach for improving information security risk management methods. PIMRC 2007. IEEE 18th International Symposium on Personal, Indoor and Mobile Radio Communications, Athens, Greece 2007. doi:10.1109/PIMRC.2007.4394150.
http://ieeexplore.ieee.org/xpl/freeabs_all.jsp?arnumber=4394150.
(Accessed 2012-11-08).

[23] Bolle SR, Hasvold P, Henriksen E: Video calls from lay bystanders to dispatch centers-risk assessment of information security. BMC Health Services Research 2011, 11:244. doi:10.1186/1472-6963-11-244.
http://www.biomedcentral.com/1472-6963/11/244. (Accessed 2013-05-16).

[24] Henriksen E, Johansen MA, Baardsgaard A, Bellika JG: Threats to Information Security of Realtime Disease Surveillance Systems. 22nd International Conference on Medical Informatics Europe MIE 2009. Sarajevo, 2009. IOS Press, volume 150, p. 710–714. doi:10.3233/978-1-60750-044-5-710.
http://ebooks.iospress.nl/publication/12754. (Accessed 2013-05-16).

[25] Bønes E, Hasvold P, Henriksen E, Strandenæs T: Risk analysis of information security in a mobile instant messaging and presence system for healthcare. International Journal of Medical Informatics 2007, 76(9):677–687. doi:10.1016/j.ijmedinf.2006.06.002.
http://www.ijmijournal.com/article/S1386-5056%2806%2900162-6/abstract.
(Accessed 2013-05-16).

[26] ISO 31000:2009, Risk management—Principles and guidelines. International Organization for Standardization (ISO); 2009.
http://www.iso.org/iso/catalogue_detail?csnumber=43170. (Accessed 2013-05-02).

[27] ISO/IEC 27005:2011, Information technology—Security techniques—Information security risk management. International Organization for Standardization (ISO) and International Electrotechnical Commission (IEC); 2011.
http://www.iso.org/iso/home/store/catalogue_ics/catalogue_detail_ics.htm?csnumber=56742. (Accessed 2012-11-08).

[28] IEC 61882, Hazard and operability studies (HAZOP studies)—Application guide. International Electrotechnical Commission (IEC); 2001.

[29] Redmill F, Chudleigh M, Catmur J: System Safety: HAZOP and Software HAZOP. Wiley, Chichester; 1999.

[30] Strategi for godt personvern i helsesektoren (Strategy for good privacy in the healtcare sector). Datatilsynet (The Norwegian Data Protection Authority); 2011.
http://www.datatilsynet.no/Global/04_veiledere/Strategi_for_godt_personvern_i_helsesektoren_-_17_NOV_2011.pdf. (Accessed 2012-11-08) (in Norwegian only).

[31] Fernandez-Luque L, Elahi N, Grajales FJ: An Analysis of Personal Medical Information Disclosed in YouTube Videos Created by Patients with Multiple Sclerosis. 22nd International Conference on Medical Informatics Europe MIE 2009, Sarajevo, 2009. IOS Press, volume 150:292–296. doi:10.3233/978-1-60750-044-5-292. http://ebooks.iospress.nl/publication/12659. (Accessed 2013-05-16).

[32] Gómez-Zúñiga B, Fernandez-Luque L, Pousada M, Hernández-Encuentra E, Armayones M: ePatients on YouTube: Analysis of Four Experiences From the Patients' Perspective. Medicine 2.0, Boston; 2012. doi:10.2196/med2.2039. http://www.medicine20.com/2012/1/e1/. (Accessed 2012-11-06).

[33] Johnsen E, Fernandez-Luque L, Hagen R: Det sosiale nettet og helse. In Digitale pasienter. Edited by Tjora A, Sandaunet AG. Gyldendal Akademisk; 2010:26–65 (in Norwegian only).